Robert Bosch GmbH

Diesel-Engine Management

3rd
Edition, completely revised and extended

Published by:
© Robert Bosch GmbH, 2004
Postfach 1129
D-73201 Plochingen
Automotive Aftermarket Division,
Product Marketing – Diagnostics &
Test Equipment (AA/PDT5)

Editor in Chief:
Dipl.-Ing. (FH) Horst Bauer.

Editorial team:
Dipl.-Ing. Karl-Heinz Dietsche,
Dipl.-Ing. (FH) Thomas Jäger.

Translated by:
STAR Deutschland GmbH
Member of Star Group

2nd edition updated and expanded, 1998
3rd edition, completely revised and extended,
January 2004
All rights reserved
Printed in Germany. Imprimé en Allemagne.

Bentley Publishers
1734 Massachusetts Avenue
Cambridge, MA 02138
USA
Tel: (800) 423-4595
Fax: (617) 876-9235
Internet: www.bentleypublishers.com
E-Mail: sales@bentleypublishers.com

ISBN 0-8376-1051-6

Diesel-Engine Management

Robert Bosch GmbH

"My engine is still making enormous progress ..."
This quotation from Rudolf Diesel in 1895 still holds true today. In fact, there is a lot of movement – also in a figurative sense – when it comes to the diesel engine and diesel-fuel injection,
in particular. These developments are now described in this completely revised and updated
3rd Edition of the "Diesel-Engine Management" reference book. This compilation of the Bosch
Yellow Jackets of the "Expert Know-How on Automotive Technology" series on diesel fuel-injection technology provides a detailed description of all Bosch diesel fuel-injection systems.

75 years after the first production fuel-injection pump for fast-running diesel engines left the
assembly lines of Robert Bosch, this technical reference work explains the capabilities of diesel
engine management today. The major part of this book concentrates on detailed descriptions
of fuel-injection systems and their components. The systems treated include:
- The PE in-line fuel-injection pumps and governors
- All variants of the VE..F, VE..EDC, and VE..MV (VP29, VP30)
  distributor injection pumps and VR (VP44)
- The PF single-plunger fuel-injection pump
- The Unit Injector System (UIS)
- The Unit Pump System (UPS) and
- The common-rail fuel-injection system (CR)

The description of each fuel-injection system is preceded by a system overview of its area of application and its special features. This will help you find your way in this technical reference work.

The second focal point of this work is Electronic Diesel Control (EDC) and its associated sensors.
This technical field is explained in detail in separate chapters.

The chapter on workshop technology provides an insight into test methods and test equipment
for diesel fuel-injection systems.

The book ends with a chapter on emissions-control engineering.

An improved Table of Contents and Index of Technical Terms, in conjunction with an extensive
Index of Abbreviations, make this book a comprehensive reference work on the subject of
"Diesel Engine Management".

The Editorial Team

Contents

Authors

Areas of Application of Diesel Engines
Dipl.-Ing. (FH) Hermann Grieshaber;
Dipl.-Ing. Joachim Lackner;
Dr.-Ing. Herbert Schumacher

Basic Principles of the Diesel Engine,
Fuel Injection
Dipl.-Ing. (FH) Hermann Grieshaber

Cylinder-Charge Control Systems
Dr.-Ing. Michael Durst, Filterwerk Mann+Hummel;
Dr.-Ing. Thomas Wintrich

Fuel Supply
Dipl.-Ing. (FH) Rolf Ebert;
Dr.-Ing. Ulrich Projahn;
Ing. Dipl.-Betr.-Wirt (FH) Matthias Schmidl;
Dr. tech. Theodor Stipek

In-Line Fuel-Injection Pumps and Governors
Henri Bruognolo;
Dipl.-Ing. Ernst Ritter

Port-Controlled Distributor Injection Pumps
and Add-On Modules
Dipl.-Ing. (FH) Helmut Simon;
Prof. Dr.-Ing. Helmut Tschöke

Solenoid-Valve-Controlled
Distributor Injection Pumps
Ing. (grad.) Heinz Notdurft;
Dr.-Ing. Uwe Reuter;
Dipl.-Ing. Nestor Rodriguez-Amaya;
Dipl.-Ing. (FH) Karl-Friedrich Rüsseler;
Dipl.-Ing. Werner Vallon;
Dipl.-Ing. Burkhard Veldten

Common Rail
Dipl.-Ing. Ralf Isenburg;
Dipl.-Ing. (FH) Michael Münzenmay, STZ System
and Simulation Technology, Prof. H. Kull, Esslingen

Single-Plunger Fuel-Injection Pumps,
Unit Injector System and Unit Pump System
Dr.-Ing. Ulrich Projahn;
Dipl.-Ing. Nestor Rodriguez-Amaya;
Dr. tech. Theodor Stipek

Nozzles and Nozzle-Holder Assemblies
Dipl.-Ing. Thomas Kügler

High-Pressure Connections
Kurt Sprenger

Electronic Diesel Control and
Electronic Control Unit
Dr.-Ing. Stefan Becher;
Dipl.-Ing. Johannes Feger;
Dipl.-Ing. Lutz-Martin Fink;
Dipl.-Ing. Wolfram Gerwing;
Dipl.-Ing. (BA) Klaus Grabmaier;
Dipl.-Ing. Martin Grosser;
Dipl.-Inform. Michael Heinzelmann;
Dipl.-Math. techn. Bernd Illg;
Dipl.-Ing. (FH) Joachim Kurz;
Dipl.-Ing. Felix Landhäußer;
Dipl.-Ing. Rainer Mayer;
Dipl.-Ing. Thomas Nickels,
MAN Nutzfahrzeuge AG
Dr. rer. nat. Dietmar Ottenbacher;
Dipl.-Ing. (FH) Andreas Werner;
Dipl.-Ing. Jens Wiesner

Sensors
Dipl.-Ing. Joachim Berger

Data Transmission
Dr.-Ing. Michael Walther

Actuators
Dipl.-Ing. Werner Pape

Workshop Technology
Hans Binder;
Dipl.-Ing. Rainer Rehage;
Rolf Wörner

Exhaust-Gas Emissions
Dipl.-Ing. Eberhard Schnaibel;
Dipl.-Ing. (FH) Hermann Grieshaber

Exhaust-Gas Treatment
Priv.-Doz. Dr.-Ing. Johannes K. Schaller

Emissions-Control Legislation
Dr.-Ing. Stefan Becher;
Dr.-Ing. Michael Eggers

Customer Service Academy
Günter Haupt;
Albert Lienbacher

and the Editorial Team in collaboration with the
relevant specialist departments at Bosch.

Unless otherwise stated, the above are employees
of Robert Bosch GmbH, Stuttgart.

Areas of use for diesel engines

No other internal-combustion engine is as widely used as the diesel engine [1]. This is due primarily to its high degree of efficiency and resulting fuel economy.

The chief areas of use for diesel engines are
- Fixed-installation engines
- Cars and light commercial vehicles
- Heavy goods vehicles
- Construction and agricultural machinery
- Railway locomotives and
- Ships

Diesel engines are produced as inline or V-configuration units. They are ideally suited to turbocharger or supercharger aspiration as – unlike the gasoline engine – they are not susceptible to knocking (refer to the chapter "Cylinder-charge control systems").

Suitability criteria

The following features and characteristics are significant for diesel-engine applications (examples):
- Engine power
- Specific power output
- Operational safety
- Production costs
- Economy of operation
- Reliability
- Environmental compatibility
- User-friendliness
- Convenience (e.g. engine-compartment design)

The relative importance of those characteristics affect engine design and vary according to the type of application.

Applications

Fixed-installation engines
Fixed-installation engines (e.g. for driving power generators) are often run at a fixed speed. Consequently, the engine and fuel-injection system can be optimized specifically

---

[1] Named after Rudolf Diesel (1858 to 1913) who first applied for a patent for his "New rational thermal engines" in 1892. A lot more development work was required, however, before the first functional diesel engine was produced at MAN in Augsburg in 1897.

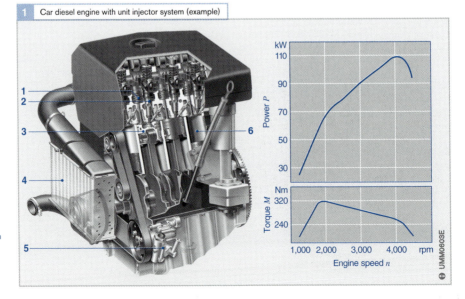

1   Car diesel engine with unit injector system (example)

Fig. 1
1 Valve gear
2 Injector
3 Piston with gudgeon pin and con rod
4 Intercooler
5 Coolant pump
6 Cylinder

UMM0603E

for operation at that speed. An engine governor adjusts the quantity of fuel injected dependent on engine load. For this type of application, mechanically governed fuel-injection systems are still used.

Car and commercial-vehicle engines can also be used as fixed-installation engines. However, the engine-control system may have to be modified to suit the different conditions.

Cars and light commercial vehicles
Car engines (Figure 1) in particular are expected to produce high torque and run smoothly. Great progress has been made in these areas by refinements in engine design and the development of new fuel-injection with **E**lectronic **D**iesel **C**ontrol (EDC). Those advances have paved the way for substantial improvements in the power output and torque characteristics of diesel engines since the early 1990s. And as a result, the diesel engine has forced its way into the executive and luxury-car markets.

Cars use fast-running diesel engines capable of speeds up to 5,500 rpm. The range of sizes extends from 10-cylinder 5-liter units used in large saloons to 3-cylinder 800-cc models for small subcompacts.

In Europe, all new diesel engines are now direct-injection (DI) designs as they offer fuel consumption reductions of 15 to 20 % in comparison with indirect-injection engines. Such engines, now almost exclusively fitted with turbochargers, offer considerably better torque characteristics than comparable gasoline engines. The maximum torque available to a vehicle is generally determined not by the engine but by the power-transmission system.

The ever more stringent emission limits imposed and continually increasing power demands require fuel-injection systems with extremely high injection pressures. Improving emission characteristics will continue to be a major challenge for diesel-engine developers in the future. Consequently, further innovations can be expected in the area of exhaust-gas treatment in years to come.

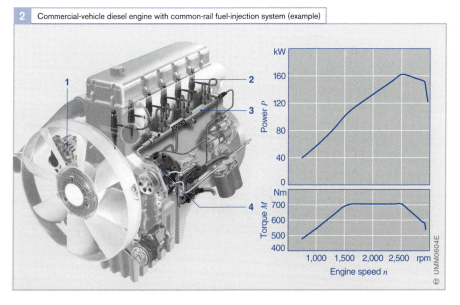

2   Commercial-vehicle diesel engine with common-rail fuel-injection system (example)

Fig. 2
1   Alternator
2   Injector
3   Fuel rail
4   High-pressure pump

Heavy goods vehicles

The prime requirement for engines for heavy goods vehicles (Figure 2) is economy. That is why diesel engines for this type of application are exclusively direct-injection (DI) designs. They are generally medium-fast engines that run at speeds of up to 3,500 rpm.

For large commercial vehicles too, the emission limits are continually being lowered. That means exacting demands on the fuel-injection system used and a need to develop new emission-control systems.

Construction and agricultural machinery

Construction and agricultural machinery is the traditional domain of the diesel engine. The design of engines for such applications places particular emphasis not only on economy but also on durability, reliability and ease of maintenance. Maximizing power utilization and minimizing noise output are less important considerations than they would be for car engines, for example. For this type of use, power outputs can range from around 3 kW to the equivalent of HGV engines.

Many engines used in construction-industry and agricultural machines still have mechanically governed fuel-injection systems. In contrast with all other areas of application, where water-cooled engines are the norm, the ruggedness and simplicity of the air-cooled engine remain important factors in the building and farming industries.

Railway locomotives

Locomotive engines, like heavy-duty marine diesel engines, are designed primarily with continuous-duty considerations in mind. In addition, they often have to cope with poorer quality diesel fuel. In terms of size, they range from the equivalent of a large truck engine to that of a medium-sized marine engine.

Ships

The demands placed on marine engines vary considerably according to the particular type of application. There are out-and-out high-performance engines for fast naval vessels or speedboats, for example. These tend to be 4-stroke medium-fast engines that run at speeds of 400...1,500 rpm and have up to

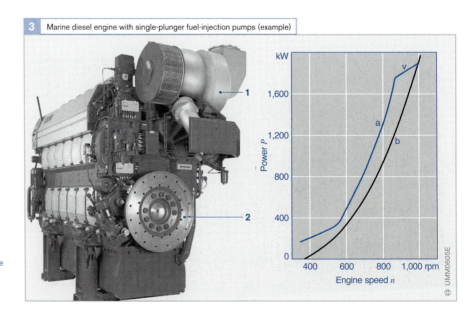

3 Marine diesel engine with single-plunger fuel-injection pumps (example)

Fig. 3

1   Turbocharger
2   Flywheel

a   Engine power output
b   Running-resistance curve
v   Full-load limitation zone

UMM0605E

History of the diesel engine

In 1892 Rudolf Diesel (1858 to 1913) embarked on research work at MAN in Augsburg based on his idea of a totally new engine in which the fuel would be ignited by compression. After many years of hard work, the world's first diesel engine was produced in 1897. It developed 20 horsepower at 175 rpm.

Compared with the conventional power units of the time (steam engines and gasoline engines), this engine had a number of advantages: It used substantially less fuel (which itself was cheaper to begin with) and could be dimensioned for much higher power outputs.

Diesel's invention rapidly established itself in the marine and fixed-installation sectors. However, there were problems in achieving higher engine speeds. The more widespread the diesel engine became, and the more widely known its advantages were, the more insistent were the demands for a smaller, faster-running version.

The biggest obstacle to the development of a fast-revving diesel engine was the fuel supply system. The air-blast method used at that time, where fuel was sprayed into the combustion chamber by compressed air, was not capable of adaptation to higher engine speeds. In addition, the compressor required was very large so that size and weight could not be significantly reduced.

In the latter part of 1922, Robert Bosch decided to direct its attention to the development of a fuel-injection system for diesel engines. By the beginning of 1923, around a dozen different designs for fuel-injection pumps had been produced. The first tests of the system fitted to the engine started in the middle of that year. By the summer of 1925, the design of the injection pump was complete. The first series-production units left the Stuttgart factory in 1927.

Those Bosch fuel-injection pumps were the breakthrough in achieving higher running speeds in diesel engines. Since that time it has conquered ever wider areas of application. The first volume-production car to be fitted with a diesel engine was the Mercedes-Benz 260D in 1936 (2,580 cc, 50 hp). Rudolf Diesel's vision had at last become reality.

One of the first series-production Bosch Type PE..A fuel-injection pumps

⊕ SMK1752Y

Fig. 1
This fuel-injection pump was tested out in a Stoewer motor car in 1927. The engine had a capacity of 2,580 cc and produced 27 horsepower (about 20 kW).

24 cylinders (Figure 3). At the other end of the scale there are 2-stroke heavy-duty engines designed for maximum economy in continuous duty. Such slow-running engines (< 300 rpm) achieve effective levels of efficiency of up to 55 %, which represent the highest attainable with piston engines.

Large-scale engines are generally run on cheap heavy oil. This requires pretreatment of the fuel on board. Depending on quality, it has to be heated to temperatures as high as 160 °C. Only then is its viscosity reduced to a level at which it can be filtered and pumped.

Smaller vessels often use engines originally intended for large commercial vehicles. In that way, an economical propulsion unit with low development costs can be produced. Once again, however, the engine management system has to be adapted to the different service profile.

Multi-fuel engines

For specialized applications (such as operation in regions with undeveloped infrastructures or for military use), diesel engines capable of running on a variety of different fuels including diesel, gasoline and others have been developed. At present they are of virtually no significance whatsoever within the overall picture, as they are incapable of meeting the current demands in respect of emissions and performance characteristics.

Engine characteristic data

Table 1 shows the most important comparison data for various types of diesel and gasoline engine.

The average pressure in petrol engines with direct fuel injection is around 10 % higher than for the engines listed in the table with inlet-manifold injection. At the same time, the specific fuel consumption is up to 25 % lower. The compression ratio of such engines can be as much as 13:1.

1   Comparison of diesel and gasoline engines

Fuel-injection system	Rated speed n_{rated} [rpm]	Compression ratio ε	Mean pressure [1] p_e [bar]	Specific power output $P_{e,spec}$ [kW/l]	Power-to-weight ratio m_{spec} [kW/kg]	Specific fuel consumption [3] b_e [g/kWh]
Diesel engines						
IDI[3] conventionally aspirated car engines	3,500...5,000	20...24:1	7...9	20...35	1:5...3	320...240
IDI[3] turbocharged car engines	3,500...4,500	20...24:1	9...12	30...45	1:4...2	290...240
DI[4] conventionally aspirated car engines	3,500...4,200	19...21:1	7...9	20...35	1:5...3	240...220
DI[4] turbocharged car engines with i/clr[5]	3,600...4,400	16...20:1	8...22	30...60	1:4...2	210...195
DI[4] convent. aspirated comm. veh. engines	2,000...3,500	16...18:1	7...10	10...18	1:9...4	260...210
DI[4] turbocharged comm. veh. engines	2,000...3,200	15...18:1	15...20	15...25	1:8...3	230...205
DI[4] turboch. comm. veh. engines with i/clr[5]	1,800...2,600	16...18:1	15...25	25...35	1:5...2	225...190
Construct. and agricultural machine engines	1,000...3,600	16...20:1	7...23	6...28	1:10...1	280...190
Locomotive engines	750...1,000	12...15:1	17...23	20...23	1:10...5	210...200
Marine engines (4-stroke)	400...1,500	13...17:1	18...26	10...26	1:16...13	210...190
Marine engines (2-stroke)	50...250	6...8:1	14...18	3...8	1:32...16	180...160
Gasoline engines						
Conventionally aspirated car engines	4,500...7,500	10...11:1	12...15	50...75	1:2...1	350...250
Turbocharged car engines	5,000...7,000	7...9:1	11...15	85...105	1:2...1	380...250
Comm. veh. engines	2,500...5,000	7...9:1	8...10	20...30	1:6...3	380...270

Table 1
1) The average pressure, p_e, can be used to calculate the specific torque, M_{spec} [Nm], by means of the following equation:

$$M_{spec} = \frac{25}{\pi \cdot p_e}$$

2) Best consumption
3) Indirect Injection
4) Direct Injection
5) Intercooler

▶ Diesel aircraft engines of the 1920s and 30s

In the 1920s and 1930s numerous two and four-stroke diesel engines were developed for use as aircraft engines. Apart from their economical consumption and the lower price of diesel fuel, diesels had a number of other features in their favor such as a lower fire risk and simpler maintenance due to the absence of carburetor, spark plugs and magneto. Engineers also hoped that the compression-ignition engine would provide good performance at high altitudes. In those days, spark-ignition engines were liable to misfire because the ignition system was subject to atmospheric pressure. The main problems associated with the development of a diesel aircraft engine involved controlling the fuel/air mixture effectively and handling the higher mechanical and thermal stresses.

The most successful production aircraft diesel engine was the Jumo 205 6-cylinder two-stroke opposed-piston heavy-oil engine (see illustration). Following its introduction in 1933 it was fitted in numerous planes. It had a take-off power output of up to 645 kW (880 hp). Its strengths primarily lay in its suitability for long-distance flights at constant speeds, e.g. for transatlantic postal services. Around 900 units of this reliable engine were built.

The fuel injection system for the Jumo 205 consisted of two pumps and two injectors for each cylinder. The injection pressure was in excess of 500 bar. It was that fuel-injection system which was a major factor in the breakthrough of the Jumo 205. Based on the experience gained from that engine, development work was also started on direct fuel-injection for spark-ignition aircraft engines in the 1930s.

The Jumo 205 was followed in 1939 by the Jumo 207 high-altitude engine which also had a take-off power output of 645 kW (880 hp). Thanks to its turbocharger aspiration, aircraft with the new engine could reach altitudes of up to 14,000 metres.

The technical high point in the development of diesel aircraft engines was the experimental 24-cylinder opposed-piston Jumo 224 produced in the early 1940s which developed as much as 3,330 kW (4,400 hp) take-off power. This "square configuration" engine had its cylinders arranged in a cross formation driving four separate crankshafts.

A whole series of diesel aircraft engines were developed by other manufacturers as well. However, none of them progressed beyond the experimental stage. In later years interest in diesel aircraft engines waned because of progress made with high-performance spark-ignition engines with fuel injection.

▼ Junkers Jumo 205 two-stroke opposed-piston diesel aircraft engine

(Source: Deutsches Museum, Munich)

SMM0606Y

Basic principles of the diesel engine

The diesel engine is a compression-ignition engine in which the fuel and air are mixed inside the engine. The air required for combustion is highly compressed inside the combustion chamber. This generates high temperatures which are sufficient for the diesel fuel to spontaneously ignite when it is injected into the cylinder. The diesel engine thus uses heat to release the chemical energy contained within the diesel fuel and convert it into mechanical force.

The diesel engine is the internal-combustion engine that offers the greatest overall efficiency (more than 50 % in the case of large, slow-running types). The associated low fuel consumption, its low-emission exhaust and quieter running characteristics assisted, for example, by pre-injection have combined to give the diesel engine its present significance.

Diesel engines are particularly suited to aspiration by means of a turbocharger or supercharger. This not only improves the engine's power yield and efficiency, it also reduces pollutant emissions and combustion noise.

In order to reduce NO_x emissions on cars and commercial vehicles, a proportion of the exhaust gas is fed back into the engine's intake manifold (exhaust-gas recirculation). An even greater reduction of NO_x emissions can be achieved by cooling the recirculated exhaust gas.

Diesel engines may operate either as two-stroke or four-stroke engines. The types used in motor vehicles are generally four-stroke designs.

Method of operation

A diesel engine contains one or more cylinders. Driven by the combustion of the air/fuel mixture, the piston (Figure 1, Item 3) in each cylinder (5) performs up-and-down movements. This method of operation is why it was named the "reciprocating-piston engine".

The connecting rod, or conrod (11), converts the linear reciprocating action of the piston into rotational movement on the part of the crankshaft (14). A flywheel (15) connected to the end of the crankshaft helps to maintain continuous crankshaft rotation and reduce unevenness of rotation caused by the periodic nature of fuel combustion in the individual cylinders. The speed of rotation of the crankshaft is also referred to as engine speed.

1   Four-cylinder diesel engine without auxiliary units (schematic)

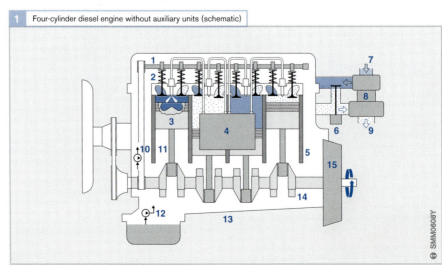

Fig. 1
1 Camshaft
2 Valves
3 Piston
4 Fuel-injection
   system
5 Cylinder
6 Exhaust-gas
   recirculation
7 Intake manifold
8 Turbocharger
9 Exhaust pipe
10 Cooling system
11 Connecting rod
12 Lubrication system
13 Cylinder block
14 Crankshaft
15 Flywheel

2 Operating cycle of a four-stroke diesel engine

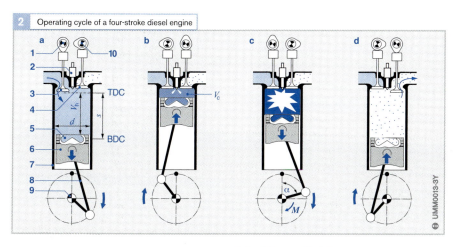

Fig. 2
a Induction stroke
b Compression stroke
c Ignition stroke
d Exhaust stroke

1   Inlet-valve camshaft
2   Fuel injector
3   Inlet valve
4   Exhaust valve
5   Combustion
    chamber
6   Piston
7   Cylinder wall
8   Connecting rod
9   Crankshaft
10 Exhaust-valve
    camshaft

α Crankshaft angle
     of rotation
d Bore
M Turning force
s Piston stroke
V_c Compression
     volume
V_h Swept volume

TDC Top dead center
BDC Bottom dead
    center

Four-stroke cycle

On a four-stroke diesel engine (Figure 2), inlet and exhaust valves control the intake of air and expulsion of burned gases after combustion. They open and close the cylinder's inlet and exhaust ports. Each inlet and exhaust port may have one or two valves.

1. Induction stroke (a)

Starting from top dead center (TDC), the piston (6) moves downwards increasing the capacity of the cylinder. At the same time the inlet valve (3) is opened and air is drawn into the cylinder without restriction by a throttle valve. When the piston reaches bottom dead center (BDC), the cylinder capacity is at its greatest (V_h+V_c).

2. Compression stroke (b)

The inlet and exhaust valves are now closed. The piston moves upwards and compresses the air trapped inside the cylinder to the degree determined by the engine's compression ratio (this can vary from 6:1 in large-scale engines to 24:1 in car engines). In the process, the air heats up to temperatures as high as 900 °C. When the compression stroke is almost complete, the fuel-injection system injects fuel at high pressure (as much as 2,000 bar in modern engines) into the hot, compressed air. When the piston reaches top dead center, the cylinder capacity is at its smallest (compression volume, V_c).

3. Ignition stroke (c)

After the ignition lag (a few degrees of crankshaft rotation) has elapsed, the ignition stroke (working cycle) begins. The finely atomized and easily combustible diesel fuel spontaneously ignites and burns due to the heat of the compressed air in the combustion chamber (5). As a result, the cylinder charge heats up even more and the pressure in the cylinder rises further as well. The amount of energy released by combustion is essentially determined by the mass of fuel injected (quality-based control). The pressure forces the piston downwards. The chemical energy released by combustion is thus converted into kinetic energy. The crankshaft drive translates the piston's kinetic energy into a turning force (torque) available at the crankshaft.

4. Exhaust stroke (d)

Fractionally before the piston reaches bottom dead center, the exhaust valve (4) opens. The hot, pressurized gases flow out of the cylinder. As the piston moves upwards again, it forces the remaining exhaust gases out.

On completion of the exhaust stroke, the crankshaft has completed two revolutions and the four-stroke operating cycle starts again with the induction stroke.

Valve timing

The cams on the inlet and exhaust camshafts open and close the inlet and exhaust valves respectively. On engines with a single camshaft, a rocker-arm mechanism transmits the action of the cams to the valves.

Valve timing involves synchronizing the opening and closing of the valves with the rotation of the crankshaft (Figure 4). For that reason, valve timing is specified in degrees of crankshaft rotation.

The crankshaft drives the camshaft by means of a toothed belt or a chain (the timing belt or timing chain) or sometimes by a series of gears. On a four-stroke engine, a complete operating cycle takes two revolutions of the crankshaft. Therefore, the speed of rotation of the camshaft is only half that of the crankshaft. The transmission ratio between the crankshaft and the camshaft is thus 2:1.

At the changeover from exhaust to induction stroke, the inlet and exhaust valves are open simultaneously for a certain period of time. This "valve overlap" helps to "flush out" the remaining exhaust and cool the cylinders.

Fig. 3
TDC Top dead center
BDC Bottom dead
        center

Fig. 4
EO Exhaust opens
EC Exhaust closes
SOC Start of
        combustion
IO Inlet opens
IC Inlet closes
IP Injection point
TDC Top dead center
BDC Bottom dead
        center

⬛ Valve overlap

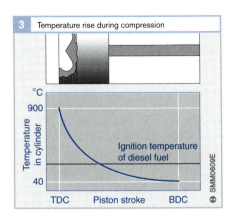

3 Temperature rise during compression

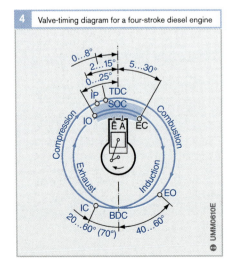

4 Valve-timing diagram for a four-stroke diesel engine

Compression

The compression ratio, ε, of a cylinder results from its swept volume, V_h, and its compression volume, V_c, thus:

$$\varepsilon = \frac{V_h + V_c}{V_c}$$

The compression ratio of an engine has a decisive effect on the following:
● The engine's cold-starting characteristics
● The torque generated
● Its fuel consumption
● How noisy it is and
● The pollutant emissions

The compression ratio, ε, is generally between 16:1 and 24:1 in engines for cars and commercial vehicles, depending on the engine design and the fuel-injection method. It is therefore higher than in gasoline engines ($\varepsilon = 7:1...13:1$). Due to the susceptibility of gasoline to knocking, higher compression ratios and the resulting higher combustion-chamber temperatures would cause the air/fuel mixture to spontaneously combust in an uncontrolled manner.

The air inside a diesel engine is compressed to a pressure of 30...50 bar (conventionally aspirated engine) or 70...150 bar (turbocharged/supercharged engine). This generates temperatures ranging from 700 to 900 °C (Figure 3). The ignition temperature of the most easily combustible components of diesel fuel is around 250 °C.

Torque and power output

Torque

The conrod converts the linear movement of the piston into a rotational movement on the part of the crankshaft because its point of action is offset from the crankshaft's center of rotation. The force with which the expanding air/fuel mixture forces the piston downwards is thus translated into a turning force or torque by the leverage of the crankshaft.

The output torque, M, of the engine is therefore dependent on the mean pressure, p_e (mean piston or operating pressure). It is expressed by the equation:

$$M = \frac{p_e \cdot V_H}{4\pi}$$

where
V_H is the cubic capacity of the engine and $\pi \approx 3.14$.

The mean pressure can reach levels of 8...22 bar in small turbocharged diesel engines for cars. By comparison, gasoline engines achieve levels of 7...11 bar.

The maximum achievable torque, M_{max}, that the engine can deliver is determined by its design (cubic capacity, method of aspiration, etc.). The torque output is adjusted to the requirements of the driving situation essentially by altering the fuel and air mass and the mixing ratio.

Torque increases in relation to engine speed, n, until maximum torque, M_{max}, is reached (Figure 1). As the engine speed increases beyond that point, the torque begins to fall again (maximum permissible engine load, desired performance, gearbox design). Engine design efforts are aimed at generating maximum torque at low engine speeds (under 2,000 rpm) because at those speeds fuel consumption is at its most economical and the engine's response characteristics are perceived as positive (good "pulling power").

Power output

The power, P (work per unit of time), generated by the engine increases in relation to torque, M, and engine speed, n. The relationship is expressed by the equation:

$$P = 2 \cdot \pi \cdot n \cdot M$$

Figure 1a shows a comparison between the power curves of diesel engines made in 1968 and in 1998. Engine power output increases with engine speed until it reaches its maximum level, or rated power P_{rated} at the engine's rated speed, n_{rated}.

The power-output and torque characteristics of the internal-combustion engine require the use of a gearbox that can adapt engine output to the varying requirements of different driving situations.

Because of their low maximum engine speeds, diesel engines have a lower specific power output than gasoline engines. Modern diesel engines for cars have rated speeds of between 3,500 and 5,000 rpm.

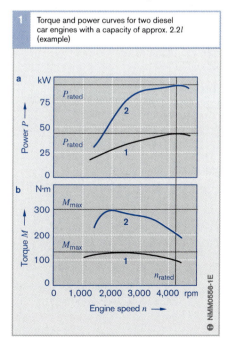

1 Torque and power curves for two diesel car engines with a capacity of approx. 2.2l (example)

Fig. 1
a   Power curve
b   Torque curve

1   1968 engine
2   1998 engine

M_{max} Maximum torque
P_{rated} Rated power
n_{rated} Rated speed

Engine efficiency

Pressure-volume diagram (p-V diagram)

The changes in gas pressure and consequent variations in volume that take place inside a diesel engine are responsible for the theoretical work, W, that it performs. They are represented by a pressure-volume work diagram, or p-V diagram.

Seiliger process

The Seiliger process (Figure 1) describes the thermodynamic comparison process and therefore the work theoretically achievable by the diesel engine. The aim of engine design is to achieve a real process that approximates the Seiliger process as closely as possible. The ideal process is based on the following simplifications:

• Ideal gas quality
• Constant specific heat
• Infinite speed of heat input and dissipation
• No flow-related losses as the theoretical process does not take account of the processes involved in charge cycles

The enclosed area in the p-V diagram describes the work, W, theoretically achievable in the course of an operating cycle. The following individual stages make up the process:

Isentropic compression (1–2)

During isentropic compression (compression at constant entropy, i.e. without transfer of heat) pressure in the cylinder increases while the volume of the gas decreases.

Isochoric heat propagation (2–3)

The air/fuel mixture starts to burn. The heat propagation (q_{BV}) that takes place as a result does so at a constant gas volume (isochoric). Gas pressure also increases.

Isobaric heat propagation (3–3')

Further heat propagation (q_{Bp}) takes place when the piston moves downwards (the gas volume increases); the pressure remains constant (isobaric).

Isentropic expansion (3'–4)

The piston continues to move downwards to bottom dead center. No further heat transfer takes place. The gas volume increases.

Isochoric heat dissipation (4–1)

During the gas-exchange phase, the remaining heat is removed (q_A). This takes place at a constant gas volume (completely and at infinite speed). The initial situation is thus restored and a new operating cycle begins.

Real process

The real process can also be represented by a p-V diagram (indicator diagram, Figure 2). The indicated (generated) work is the upper enclosed area on the diagram (W_M). For assisted-aspiration engines, the gas-exchange area (W_G) has to be added to that since the compressed air delivered by the turbocharger/supercharger also helps to press the piston downwards on the induction stroke. The process is also frequently represented by a graph of cylinder pressure versus crankshaft rotation (Figure 3).

Fig. 1
1–2 Isentropic compression
2–3 Isochoric heat propagation
3–3' Isobaric heat propagation
3'–4 Isentropic expansion
4–1 Isochoric heat dissipation

TDC Top dead center
BDC Bottom dead center

q_A Quantity of heat dissipated during gas exchange
q_{Bp} Combustion heat at constant pressure
q_{BV} Combustion heat at constant volume
W Theoretical work

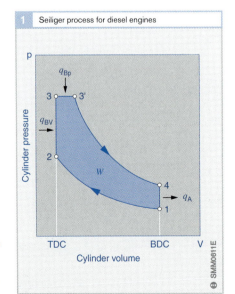

1   Seiliger process for diesel engines

p

q_{Bp}

3 — 3'

q_{BV}

2

W

4

q_A

1

Cylinder pressure

TDC   BDC V

Cylinder volume

SMM0611E

2 Real process in a turbocharged/supercharged diesel engine represented by *p-V* indicator diagram (recorded using a pressure sensor)

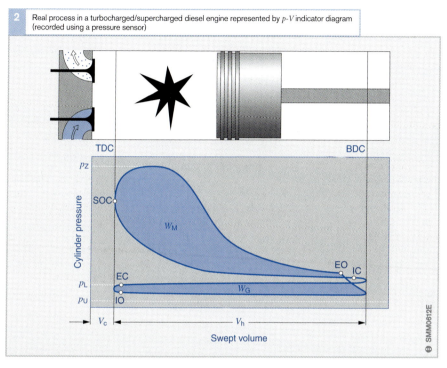

Fig. 2
EO   Exhaust opens
EC   Exhaust closes
SOC Start of
     combustion
IO   Inlet opens
IC   Inlet closes
TDC Top dead center
BDC Bottom dead
     center

p_U Ambient pressure
p_L Charge-air
     pressure
p_Z Maximum cylinder
     pressure
V_c Compression
     volume
V_h Swept volume
W_M Useful work
W_G Work during
     gas exchange
     (turbocharger/
     supercharger)

3 Pressure vs. crankshaft rotation curve (*p-α* diagram) for a turbocharged/supercharged diesel engine

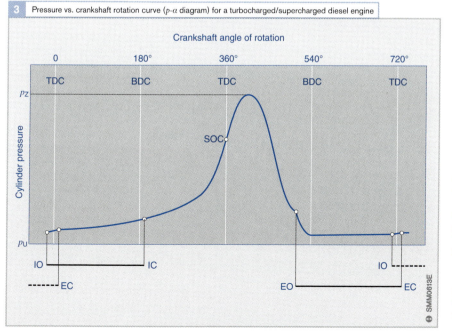

Fig. 3
EO   Exhaust opens
EC   Exhaust closes
SOC Start of
     combustion
IO   Inlet opens
IC   Inlet closes
TDC Top dead center
BDC Bottom dead
     center

p_U Ambient pressure
p_L Charging pressure
p_Z Maximum cylinder
     pressure

Efficiencies

The overall efficiency, η_e, of the diesel engine is represented by the equation

$$\eta_e = \frac{W_e}{W_B}$$

where W_e is the work effectively available at the flywheel and W_B is the energy content of the fuel consumed.

That overall efficiency figure is the combined result of a series of individual efficiency ratings (Figure 4) which all constitute energy losses:

$$\eta_e = \eta_{th} \cdot \eta_g \cdot \eta_m$$

Theoretical efficiency, η_{th}

η_{th} is the theoretical efficiency of the Seiliger process. It represents the theoretical work in relation to the energy content of the fuel consumed and is around 42.5 MJ/kg for diesel engines.

As previously outlined, the parameters of this "ideal process" are:
- Ideal gas quality
- Constant specific heat
- Infinite speed of heat propagation and dissipation
- No flow-related losses

Efficiency of high-pressure work process, η_g

η_g describes the real high-pressure work process in relation to the theoretical process (Figure 2). This efficiency figure takes account of the heat and flow-related losses of the real gas-exchange phase. Its parameters are:
- Real gas quality
- Heat losses
- Finite rate of heat propagation and dissipation and
- Variable specific heat

All air/fuel mixture parameters have an effect on combustion and therefore a decisive influence on thermal efficiency.

Mechanical efficiency, η_m

η_m defines the mechanical losses due to friction including ancillary systems with reference to the indicated process. It therefore describes the real engine. Frictional and power-transmission losses increase with engine speed. At rated speed, the frictional losses are made up as follows:
- Pistons and piston rings approx. 50 %
- Bearings approx. 20 %
- Oil pump approx. 10 %
- Coolant pump approx. 5 %
- Valve-gear approx. 10 %
- Fuel-injection pump approx. 5 %

If the engine has a supercharger, this must also be included.

Comparison of diesel engine and gasoline engine

The higher overall efficiency of the diesel engine compared with the conventional gasoline engine is essentially due to three factors:
- Higher compression ratio (giving a larger area on the p-V indicator diagram)
- Greater excess air (made possible by heterogeneous internal air/fuel mixing) and
- Absence of throttle flap – and consequently no throttle-related losses in the part-load range

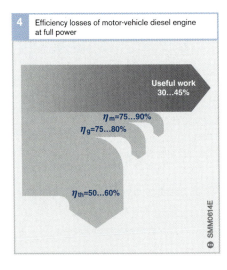

4   Efficiency losses of motor-vehicle diesel engine at full power

Useful work 30...45%

η_m=75...90%
η_g=75...80%

η_{th}=50...60%

SMM0614E

Operating statuses

Starting
Starting an engine involves the following stages: cranking, ignition and running up to self-sustained operation. The hot, compressed air produced by the compression stroke has to ignite the injected fuel (combustion start). The minimum ignition temperature required for diesel fuel is approx. 250 °C.

That temperature must be achievable with a sufficient degree of certainty at low engine speeds and in cold weather conditions with a cold engine. There are a number of physical parameters which tend to oppose that aim:
- The lower the engine speed, the lower is the ultimate pressure at the end of the compression stroke and accordingly, the ultimate temperature (Figure 1). The reasons for this phenomenon are the leakage losses through the piston ring gaps between the piston and the cylinder wall and the fact that when the engine is first started, an oil film is not present. Because of the heat loss during compression, the

maximum compression temperature is reached a few degrees before TDC (thermodynamic loss angle, Figure 2).
- When the engine is cold, heat loss occurs during the compression stroke. On indirect-injection (IDI) engines, that heat loss is particularly high due to the larger surface area of the combustion chamber.
- In addition, the internal friction of the engine is higher at low temperatures than at normal operating temperature because of the higher viscosity of the engine oil.
- Furthermore, the speed of the starter motor is slower when it is cold because the battery voltage drops at low temperatures.

There are a number of measures that can be employed in order to counteract those physical factors as outlined below.

Fuel modification
A filter heater or direct fuel heater (Figure 3 overleaf) can prevent fuel problems which generally occur at low temperatures due to the precipitation of paraffin crystals. The oil industry also supplies fuels suitable for use

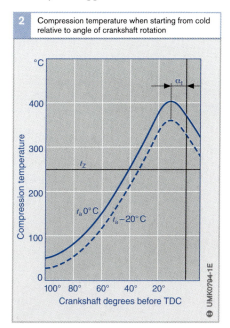

1 Compression pressure and ultimate temperature relative to engine speed

2 Compression temperature when starting from cold relative to angle of crankshaft rotation

Fig. 2
t_a Outside temperature
t_z Ignition temperature of diesel fuel
α_T Thermodynamic loss angle

$n \approx 200$ rpm

in cold temperatures. The addition of paraffin or gasoline is no longer necessary with these "winter-grade fuels" (refer to the section "Diesel fuels").

Start-assist systems

On direct-injection (DI) engines, assisted starting is achieved partially by pre-heating the intake air (commercial vehicles) or by the use of sheathed-element glow plugs (cars) (refer to the section "Actuators"). On indirect-injection (IDI) engines, assisted starting is achieved exclusively by means of glow plugs in the prechamber or swirl chamber. Both the above methods assist fuel vaporization and air/fuel mixing and therefore facilitate reliable combustion of the air/fuel mixture.

The most technically advanced glow plugs require only a few seconds to preheat to the required temperature and thus enable quick starting (Figure 4). The lower post-glow temperature of the latest generation of glow plugs also enables even longer post-glow periods. This reduces not only harmful pollutant emissions but also noise levels during the engine's warm-up period.

Injection adaptation

Another means of assisted starting is the injection of an *excess amount of fuel for starting* to compensate for condensation and leakage losses and to increase the engine torque in the running-up phase.

A further method involves *advancing* the start of injection to offset ignition lag and to ensure reliable ignition at top dead center, i.e. at the maximum final compression temperature. The optimum start of injection must be achieved as precisely as possible within tight tolerance limits.

If the fuel is injected too soon, it condenses on the cold cylinder walls. Only a small proportion of it vaporizes, since at that point the temperature of the air charge is too low.

If the fuel is injected too late, ignition occurs during the downward stroke (expansion phase) and the piston is not fully accelerated.

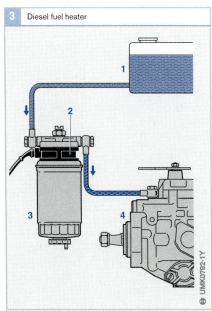
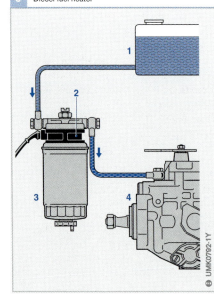

3 Diesel fuel heater

Fig. 3
1   Fuel tank
2   Fuel heater
3   Fuel filter
4   Fuel-injection pump

Fig. 4
Filament material:
1   Nickel (conventional glow plug type S-RSK)
2   CoFe alloy (2nd-generation glow plug type GSK2)

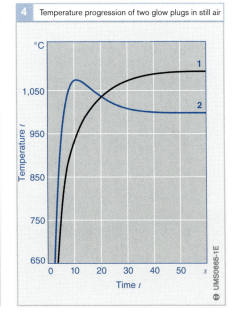

4 Temperature progression of two glow plugs in still air

°C

Temperature *t*

1,050

950

850

750

650

0   10   20   30   40   50   *s*

Time *t*

The injection system has to ensure that the correct fuel-droplet size for optimum speed and efficiency of air/fuel mixing inside the combustion chamber is achieved by optimum fuel atomization and delivery (see the chapter "Basic principles of diesel fuel injection").

No load

No load refers to all engine operating statuses in which the engine overcomes only its own internal friction. It does not produce any torque output. The accelerator pedal may be in any position. All engine speeds up to and including breakaway speed are possible.

Idle

The engine is said to be idling when it is running at the lowest no-load speed. The accelerator pedal is not depressed. The engine is not generating any output torque. It overcomes only internal friction. Some sources refer to the entire no-load range as idle. The upper no-load range (breakaway speed) is then called the upper idle speed.

Full load

At full load, the accelerator pedal is fully depressed or the governor acts independently within the range up to fuel shutoff. The maximum possible fuel volume is injected and the engine generates its maximum possible torque output under steady-state conditions. Under non steady-state conditions (limited by turbocharger/supercharger pressure) the engine develops the maximum possible (lower) full-load torque with the quantity of air available. All engines speeds from idle speed to nominal speed are possible.

Part load

Part load covers the range between no load and full load. The engine is generating an output between zero and the maximum possible torque.

Part load at idle speed

In this particular case, the governor holds the engine at idle speed. The engine generates torque output. This may extend to full load.

Lower part-load range

This is the operating range in which the diesel engine's fuel consumption is particularly economical in comparison with the gasoline engine. "Diesel knock" that was a problem on earlier diesel engines – particularly when cold – has virtually been eliminated on diesels with pre-injection.

As explained in the "Starting" section, the final compression temperature is lower at lower engine speeds and at lower loads. In comparison with full load, the combustion chamber is relatively cool (even when the engine is at operating temperature) because the energy input and therefore the temperature rise is necessarily smaller. The combustion chamber heats up relatively slowly. This is particularly true of engines with prechamber or swirl chambers because the larger surface area means that heat loss is greater.

At low loads and with pre-injection, only a few mm³ are delivered in each injection cycle. In this situation, particularly high demands are placed on the accuracy of the start of injection and injected fuel quantity. As during the starting phase, the highest combustion temperature is reached only within a small range of piston travel near TDC. Start of injection is controlled very precisely to coincide with that point.

During the ignition-lag period, only a small amount of fuel may be injected since, at the point of ignition, the quantity of fuel in the combustion chamber determines the sudden increase in pressure in the cylinder. The level of combustion noise is directly related to this pressure increase. The greater the increase in pressure, the more clearly perceptible is the noise. Pre-injection of approx. 1 mm³ of fuel virtually cancels out the ignition lag at the main injection point and thus substantially reduces combustion noise (see the chapter "Basic principles of diesel fuel injection").

Overrun

The engine is said to be overrunning when it is driven by an external force acting through the drivetrain (e.g. when descending an incline).

Steady-state operation

The engine's torque output is equal to the required torque. The engine speed is constant.

Non-steady-state operation

The engine's torque output is not equal to the required torque. The engine speed is not constant.

Transition between operating statuses

The response characteristics of an engine can be defined by means of characteristic data diagrams or maps. If, for example, the load, the engine speed or the accelerator-pedal position change, the engine's operating status changes (e.g. its speed or torque output).

The map in Figure 5 shows an example of how the engine speed changes when the accelerator-pedal position changes from 40 % to 70 % depressed. Starting from point A on the map, the new part-load point D is reached via the full-load curve (B–C). At that point, the power demand and the engine's power output are equal. The engine speed has increased from n_A to n_D.

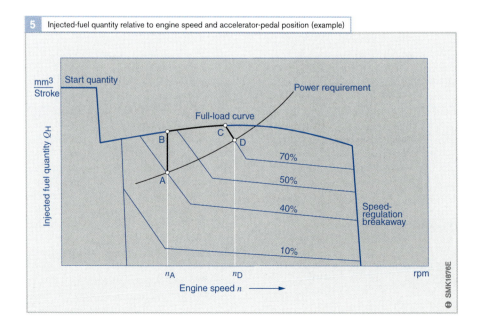

5   Injected-fuel quantity relative to engine speed and accelerator-pedal position (example)

Operating conditions

The operating conditions of a diesel engine are based on a number of process-specific circumstances.

In a diesel engine, the fuel is injected directly into the highly compressed hot air which causes it to ignite spontaneously. Therefore, and because of the heterogeneous air/fuel mixture, the diesel engine – in contrast with the gasoline engine – is not restricted by ignition limits (i.e. specific air-fuel ratios λ). Thus, with a constant air volume in the cylinder, only the fuel quantity is regulated.

The fuel-injection system thus plays a decisive role in engine operation. It is responsible for delivery of the precise amount of fuel required and "even" distribution throughout the cylinder charge – and it has to perform those tasks at all engine speeds and loads. In addition, it has to take account of the condition of the intake air in terms of pressure and temperature.

Thus, for any combination of engine operating parameters, the fuel-injection system must deliver

- The correct amount of fuel
- At the correct time
- At the correct pressure
- With the correct timing pattern and
- At the correct point in the combustion chamber

In addition to optimum air/fuel mixture considerations, determination of the correct amount of fuel to be delivered frequently requires taking account of engine or vehicle-related operating limits such as:

- Emission restrictions (e.g. smoke emission limits)
- Combustion pressure limits
- Exhaust temperature limits
- Engine speed and torque limits and
- Vehicle or engine-specific load limits

Particulate/smoke emission limits

There are prescribed statutory limits for particulate emissions and maximum exhaust smoke content. They differ according to the type of vehicle (e.g. passenger car, commercial vehicle) and from one country to another. Whereas, for cars only, the lower power band is tested, for commercial vehicles virtu-

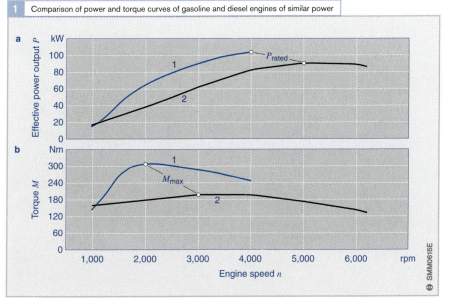

1 Comparison of power and torque curves of gasoline and diesel engines of similar power

a Effective power output P kW

b Torque M Nm

Engine speed n rpm

© SMM0615E

Fig. 1
a   Power curve
b   Torque curve

1   2.2 l 4-cylinder diesel engine with common-rail fuel injection
2   2.3 l 5-cylinder gasoline engine

M_{max} Maximum torque
P_{rated} Rated power

ally the entire output range is taken into consideration.

The largest proportion of particulate emissions is made up of soot particles (black smoke). As a large part of the air/fuel mixing process only takes place in the course of combustion, localized over-enrichment occurs and this in some cases leads to an increase in black smoke emissions even at moderate levels of excess air. The air-fuel ratio usable at the statutory full-load smoke limit is a measure of the efficiency of air utilization.

Combustion pressure limits

During the ignition process, the partially vaporized fuel mixed with the air burns under high compression at a rapid rate and with a high initial thermal-release peak (without pre-injection). This is referred to as "hard" combustion. High combustion pressure peaks are produced and this requires a relatively heavy engine. The forces generated during combustion place periodic alternating stresses on the engine components. The dimensioning and durability of the engine and drivetrain components therefore limit the permissible maximum compression pressure and consequently the amount of fuel injected.

Exhaust-gas temperature limits

The high thermal stresses placed on the engine components surrounding the hot combustion chamber, the heat resistance of the exhaust valves and of the exhaust system and cylinder head determine the maximum exhaust temperature of a diesel engine.

Engine speed limits

The fact that diesel engines operate on the basis of excess air with regulation of the injected-fuel quantity means that the power output at a constant engine speed is basically dependent solely on the amount of fuel injected. If the amount of fuel supplied to a diesel engine is increased without a corresponding increase in the load that it is working against, then the engine speed will rise. If the fuel supply is not reduced before the engine reaches a critical speed, the engine may rev itself to the point of destruction. Consequently, an engine speed limiter or governor is absolutely essential on a diesel engine.

Diesel engines that drive machinery are expected to maintain a constant speed or to keep their speed within certain upper and lower limits regardless of the load applied. For such requirements, there are variable-speed or intermediate-speed governors.

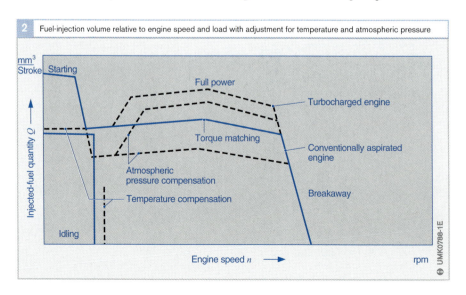

2 Fuel-injection volume relative to engine speed and load with adjustment for temperature and atmospheric pressure

On diesel engines used to drive road-going vehicles, the engine speed must be infinitely variable by the driver using the accelerator pedal. In addition, when the engine is under load or when the accelerator pedal is released, the engine speed must not be allowed to drop below the idling speed to a standstill.

The following two types of governor system are distinguished:
- Variable-speed governors which are operate across the entire engine-speed range.
- Idle-speed and maximum-speed governors which regulate only the idling and maximum speeds. The intermediate range of speeds is controlled by means of the accelerator pedal.

Taking into consideration all the requirements described, a characteristic data map can be defined for the operating range of an engine. This map (Figure 2) shows the fuel quantity in relation to the engine speed and load, together with the necessary adjustments for temperature and air-pressure variations.

Altitude and turbocharger/supercharger pressure limits

The setting of fuel-injection volumes is generally based on atmospheric pressure at sea level. In other words, the performance figures are reduced for that altitude. If the engine is operated at altitudes significantly above sea level, the fuel-injection volume must be adjusted according to the barometric altitude equation. As a general guide, it can be assumed that air density decreases by approx. 7 % per 1,000 m of altitude. In order to remain within the smoke limit, the fuel-injection volume has to be reduced accordingly.

With turbocharged/supercharged engines, the cylinder charge during dynamic operation is lower than in steady-state operation, on which the maximum injection volume is based. Therefore, as with high altitudes, the fuel volume has to be reduced according to the smaller quantity of air (full load limited by turbocharger/supercharger pressure).

Development potential

Improvements in precision regulation of fuel-injection systems and enhancements in air charge are factors that allow ever greater accuracy in complying with the limits described above. This has resulted in better specific power output of engines (Figures 3 and 4).

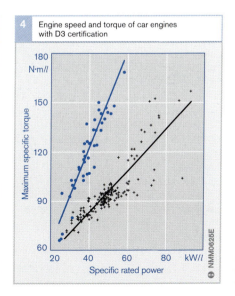

3 Development of diesel engines for mid-range cars

Engine versions
- Torque of largest engine [Nm]
- Torque of smallest engine [Nm]
- Rated power of largest engine [kW]
- Rated power of smallest engine [kW]

Year of construction (1953, 1961, 1968, 1976, 1984, 1995, 2000)

NMM0816E

4 Engine speed and torque of car engines with D3 certification

Maximum specific torque [N·m/l] vs. Specific rated power [kW/l]

NMM0826E

Fig. 4
- • Diesel engines (DI)
- + Gasoline engines

Fuel-injection system

The operating conditions referred to place exacting demands on the precision of the fuel-injection system. This can be illustrated by the following example.

The full-load injected-fuel quantity for an engine with a power output of 75 kW (102 bhp) and a specific fuel consumption of 200 g/kWh demands an overall fuel supply rate of 15 kg/h. On a four-cylinder four-stroke engine turning at an engine speed of 2,400 rpm, there are 288,000 separate injections of fuel per hour. Thus, each individual injection of fuel involves a quantity of 59 mm³. By comparison, a raindrop has a volume of approximately 30 mm³.

Even greater precision is demanded at idling speed (5 mm³ per injection) and for pre-injection (1 mm³ per injection). Even the minutest variations have a negative effect on the smooth running of the engine, noise emission and black smoke levels.

The fuel-injection system not only has to deliver precisely the right amount of fuel to suit the exact operating conditions at any particular moment, it also has to do so for each individual cylinder of a multi-cylinder engine. Furthermore, it has to prevent accuracy drift over time. The Electronic Diesel Control (EDC) system allows the injected-fuel quantity to be adjusted individually for each cylinder and thus achieves particularly smooth engine running.

The mathematically calculated injected fuel quantity serves as a guide figure for the dimensioning of a fuel-injection system. At lower engine speeds in particular, the full-load curve is limited by the engine's smoke limit and at higher speeds by the permissible maximum exhaust-gas/component temperature as well as by the maximum permissible cylinder peak pressure.

Fuel consumption

The fuel consumption of a vehicle depends on a variety of factors (e.g. driving style, route topography, tire pressure, payload, vehicle speed, electrical equipment in use, and air filter condition). In principle, the fuel consumption of diesel engines is lower than that of gasoline engines (Figure 1).

Calibrating the regulation systems

The engine, vehicle, fuel-injection and regulation systems have to be matched very precisely to one another. In this regard, a wide variety of factors must be considered. This can be illustrated by the following example.

The delivery rate of a piston pump is calculated by multiplying the area of the piston crown by the effective stroke. In port-controlled systems, pump delivery starts sooner and finishes later at higher speeds than indicated by the purely geometrical calculations, as the fuel displays inertial characteristics under dynamic flow conditions. As a result, the effective stroke under real conditions is greater than the calculated effective stroke. This "pre- and post-delivery effects" results in dynamic changes to the effective stroke and a rising or falling fuel-delivery curve. Solenoid-valve controlled fuel-injection systems also have to take account of the timing characteristics.

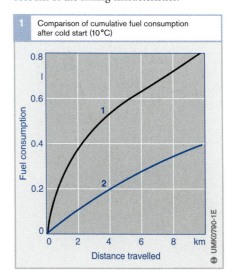

1 Comparison of cumulative fuel consumption after cold start (10 °C)

Fuel consumption

Distance travelled

UMK0790-1E

Fig. 1
1 Gasoline engine, 1.1 l, 37 kW (50 bhp)
2 Diesel engine, 1.5 l, 37 kW (50 bhp)

Combustion chambers

The shape of the combustion chamber is
one of the decisive factors in determining
the quality of combustion and therefore the
performance and exhaust characteristics of
a diesel engine. Appropriate design of combustion-chamber geometry combined with
the action of the piston can produce swirl,
squish and turbulence effects that are used
to improve distribution of liquid fuel or air
and fuel vapor inside of the combustion
chamber.

The following technologies are used:
● Undivided combustion chamber
(**direct injection (DI)** engines) and
● Divided combustion chamber
(**indirect injection (IDI)** engines)

The proportion of direct-injection engines
is increasing due to their more economical
fuel consumption (up to 20 %). The harsher
combustion noise (particularly under acceleration) can be reduced to the level of indirect-injection engines by (minimal) pre-injection. Engines with divided combustion
chambers now hardly figure at all among
new developments.

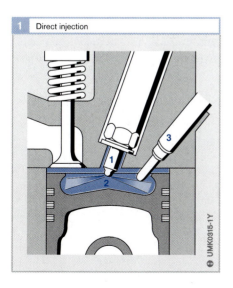

1   Direct injection

Undivided combustion chamber (direct-injection engines)

Direct-injection engines (Figure 1) have a
higher level of efficiency and operate more
economically than indirect-injection engines. Accordingly, they are used in all types
of commercial vehicles and most modern
diesel cars.

As the name suggests, the direct-injection
process involves injecting the fuel directly into
the combustion chamber, part of which is
formed by the shape of the piston crown (piston crown recess, 2). Fuel atomization, heating,
vaporization and mixing with the air must
therefore take place in rapid succession. This
places exacting demands on fuel and air delivery. During the induction and compression
strokes, the special shape of the intake port in
the cylinder head creates an air vortex inside
of the cylinder. The shape of the combustion
chamber also contributes to the air flow pattern at the end of the compression stroke (i.e.
at the moment of fuel injection). Of the combustion chamber designs used over the history
of the diesel engine, the most widely used at
present is the ω piston crown recess.

In addition to creating effective air turbulence, the technology must also ensure that
the fuel is delivered in such a way that it is
"evenly" distributed throughout the combustion chamber so as to facilitate rapid mixing.
In contrast with the indirect-injection engine
with its single-jet throttling-pintle injector,
direct-injection engines use multihole injectors (1). The positions of the jets have to be
optimized to suit the combustion chamber
design. Direct fuel injection also requires very
high injection pressures (up to 2,000 bar).
In practice, there are two types of direct
fuel injection:
● Systems in which mixture formation is
assisted by specifically created air-flow
effects and
● Systems which control mixture formation
virtually exclusively by means of fuel
injection and largely dispense with any
air-flow effects

Fig. 1
1   Multihole injector
2   ω piston recess
3   Glow plug

In the latter case, no effort is expended in creating air-turbulence effects and this is evident in smaller gas replacement losses and more effective cylinder charging. At the same time, however, far more demanding requirements are placed on the fuel-injection system with regard to nozzle positioning, number of nozzle jets and degree of atomization (achieved by small spray-hole apertures), not to mention extremely high injection pressures in order to obtain the required short injection times and atomization quality.

Divided combustion chamber (indirect injection)

For a long time, diesel engines with divided combustion chambers held an advantage over direct-injection engines in terms of noise and exhaust emissions. That was the reason why they were used in cars and light commercial vehicles. Now that high injection pressures, electronic (diesel) engine management and pre-injection are possible, however, that advantage has disappeared. As a result, indirect-injection engines are no longer used in new vehicles.

There are two types of indirect-injection system:
● The precombustion chamber system and
● The whirl-chamber system

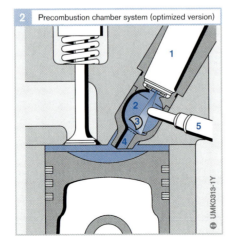

2   Precombustion chamber system (optimized version)

Fig. 2
1   Pintle nozzle
2   Precombustion
    chamber
3   Spherical pin with
    baffle surface
4   Connecting channel
5   Glow plug

Precombustion chamber system

In the precombustion chamber system, the fuel is injected into a hot precombustion chamber recessed into the cylinder head (Figure 2, Item 2). The fuel is injected through a pintle nozzle (1) at a relatively low pressure (up to 450 bar). A specially shaped baffle (3) in the center of the chamber diffuses the jet of fuel that strikes it and mixes it thoroughly with the air.

Combustion initiates inside the precombustion chamber, thereby raising the temperature and pressure and forcing the partially combusted air/fuel mixture through channels at the lower end of the precombustion chamber and into the main combustion chamber above the piston. There it mixes thoroughly with the air in the main combustion chamber so that combustion spreads and is completed.

The short ignition lag and the controlled release of energy produce a "soft" combustion effect with low levels of noise and engine stress.

A differently shaped precombustion chamber with a vaporization recess and a differently shaped and positioned baffle ("spherical pin") apply a defined degree of swirl to the air that passes from the cylinder into the precombustion chamber during the compression stroke. The fuel is injected at an angle of 5 degrees to the precombustion chamber axis in the direction of flow of the air.

So as not to disrupt the progression of combustion, the glow plug (5) is positioned on the "lee side" of the air flow. A controlled post-glow period of up to 1 minute after a cold start (dependent on coolant temperature) helps to improve exhaust-gas characteristics and reduce engine noise during the warm-up period.

The ratio of precombustion chamber volume to main combustion chamber volume is approx. $1/3$ to $2/3$.

Swirl-chamber system

With this system, combustion is also initiated in separate chamber, though in this case it accommodates almost the entire compression volume. The combustion process takes place inside a spherical or cylindrical swirl chamber with a tangentially aligned channel connecting it to the cylinder chamber (Figure 3, Item 2).

During the compression stroke, the air entering through the connecting channel is made to swirl and the fuel is injected in the swirling air flow. The nozzle jet is positioned so that the jet of fuel enters the swirling air flow perpendicular to its axis and meets a hot section of chamber wall on the opposite side of the chamber.

As soon as combustion starts, the air/fuel mixture is forced under pressure through the connecting channel into the cylinder chamber where it is turbulently mixed with the remaining air. With the swirl-chamber system, the losses due to gas flow between the main combustion chamber and the swirl chamber are less than with the precombustion chamber system because the connecting channel has a larger cross-section. This results in smaller throttle-effect losses and consequent benefits for internal efficiency and fuel consumption. However, combustion noise is louder than with the precombustion chamber system.

It is important that mixture formation takes place as completely as possible inside the swirl chamber. The shape of the swirl chamber, the alignment and shape of the fuel jet and the position of the glow plug must be carefully matched to the engine in order to obtain optimum mixture formation at all engine speeds and under all operating conditions.

Another demand is for rapid heating of the swirl chamber after a cold start. This reduces ignition lag and combustion noise as well as preventing unburned hydrocarbons (blue smoke) during the warm-up period.

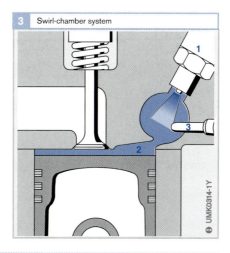

3 Swirl-chamber system

UMK0314-1Y

Fig. 3
1 Fuel injector
2 Tangential connecting channel
3 Glow plug

▶ M System

In the direct-injection system with recess-wall deposition (M system) for commercial-vehicle and fixed-installation diesel engines and multi-fuel engines, a single-jet nozzle sprays the fuel at a low injection pressure against the wall of the piston crown recess. There, it vaporizes and is absorbed by the air. This system thus uses the heat of the piston recess wall to vaporize the fuel. If the air flow inside of the combustion chamber is properly adapted, an extremely homogeneous air/fuel mixture with a long combustion period, low pressure increase and, therefore, quiet combustion can be achieved. Nevertheless, because of its less economical consumption compared with induced air-flow direct fuel injection, the M system is no longer used.

UMK0786-1Y

Diesel fuels

Diesel fuels are distilled from crude oil. They consist of a large number of different hydrocarbon compounds including n-paraffins, i-paraffins, olefins, naphthenes and aromatic compounds. They all have boiling points in the range 160...380 °C (middle distillates). Diesel fuel ignites on average at approximately 350 °C, which is very early in comparison with gasoline (500 °C) (lower limit 250 °C).

In order to cover the growing demand for diesel fuels, the refineries also add "conversion products", i.e. thermal and catalytic-cracking products. They are obtained by cracking large heavy-oil molecules.

Quality and grading criteria

The basic fuel grade is improved by the use of a series of additives, some of which have a decisive effect (see Table 2 at the end of this section).

16 grading criteria are specified by the standard EN 590 for motor vehicles which now applies throughout Europe. In many other countries around the world, the fuel standards are less stringent or in some cases nonexistent. The US standard for diesel fuels ASTM D975, for example, specifies fewer criteria and applies less strict limits to these criteria. The requirements for marine and fixed-installation engines are also much less demanding. Some of the most important grading criteria specified by EN 590 are listed in Table 1 below. It also shows the European motor manufacturers' requirements for diesel-fuel grade which are also subscribed to by Bosch. Such criteria help to keep vehicle emissions within present and future limits.

High-quality diesel fuels are characterized by the following features:
● High cetane number
● Relatively low upper boiling limit
● Narrow density and viscosity spread
● Low aromatic compounds (particularly polyaromatic compounds) content
● Low sulfur content (≤ 10 ppm)

In addition, the following characteristics are particularly important for the service life and consistent function of fuel-injection systems:
● Good lubricant qualities
● Absence of free water
● Low dirt content

The most important criteria are explained individually below.

Cetane number (CN)

The cetane number indicates the ease with which a diesel fuel ignites and is therefore of decisive importance. The higher the cetane number, the more easily combustible the fuel is.

Table 1
[1] Diesel fuel with a sulfur content of 10 ppm will be available throughout Germany from 1/1/2003 and throughout the EU from 1/1/2005.

1 Selected EN 590 grading criteria compared with the requirements of the European motor manufacturers		
Criterion	EN 590	European motor vehicle manufacturers
Cetane number	≥ 51	≥ 58
Density	820...845 kg/m³	820...840 kg/m3
Aromatic compounds content	–	≤ 20 % by vol.
Polyaromatic compounds content	≤ 11 % by vol.	≤ 1 % by vol.
Boiling point (95 %)	≤ 360 °C	≤ 340 °C
Upper boiling limit	–	≤ 350 °C
Sulfur content[1] (by mass)	≤ 350 ppm	5...10 ppm for compliance with Euro IV and V emission limits
Lubricity (HFRR)	≤ 460 μm	≤ 400 μm

The cetane number is tested using a standardized single-cylinder testing engine. The ignition lag is set for the fuel under test by means of a variable compression ratio. The engine is then run on a reference fuel made up of a mixture of cetane and α-methylnaphthalene (Figure 1) using the same compression ratio. The proportion of cetane in the mixture is altered until the same ignition lag is obtained. The proportion of cetane then gives the cetane number (for example, a mixture of 52% cetane and 48% α-methylnaphthalene has a cetane number of 52).

Paraffin fuel components have a high cetane number while aromatic compounds (chiefly cracking products) have a low cetane number; i-paraffins, olefins and naphthenes have a medium cetane number.

Ignition accelerators can be added to the fuel to improve its cetane number. All types of emission, particularly NO_x, diminish as the cetane number increases, as does the combustion noise.

Density
The energy content of diesel fuel per unit of volume increases with density. Fuels are sold by volume and delivered to the combustion chamber by fuel-injection systems on the same basis. If an engine is designed for use with a "medium-density" fuel, then if it is run on higher-density fuel (based on fuel grade), engine performance and soot emission increase; they diminish if a lower-density fuel is used. Temperature-dependent variations in fuel density are compensated for by the EDC system.
   The requirement of diesel fuel is therefore "narrow grade-based density spread". A density sensor could also provide a solution to the problem. There is a greater density spread found in fuels around the world than permitted by EN 590.

Viscosity
If the viscosity of a fuel is too low, it will lead to leakage losses in the fuel-injection system at low engine speeds in particular and therefore also to power deficiencies and hot-start problems. If the viscosity is too high, it will impair pump function and result in poor fuel atomization. Therefore, EN 590 specifies narrow tolerance limits for diesel-fuel viscosity.

Boiling range
The boiling range is the temperature range within which the fuel boils.
   A low initial boiling point makes a fuel suitable for use in cold weather but also means a lower cetane number and poor lubricant properties. A high upper boiling limit gives long-chained paraffins poor cold-starting properties but a higher cetane number.

Polyaromatic compounds with three or more rings also have a high boiling point but a low cetane number. As the polyaromatic-compound content of diesel fuel increases, more soot is produced as a byproduct of combustion.

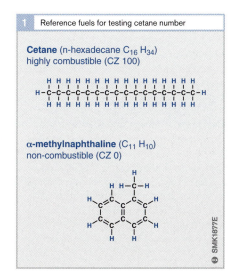

1 Reference fuels for testing cetane number

Cetane (n-hexadecane $C_{16}H_{34}$)
highly combustible (CZ 100)

α-**methylnaphthaline** ($C_{11}H_{10}$)
non-combustible (CZ 0)

© SMK1877E

Fig. 1
C   Carbon
H   Hydrogen
—   Chemical bond

With a view to avoiding poor cold-starting properties (paraffins) and high soot emissions (polyaromatic compounds), therefore, the upper limit of the boiling range should not be too high. The ACEA requirement for this property is therefore 350 °C. But although such a requirement is valuable in terms of combustion efficiency, it is offset by a lower level of crude-oil exploitation.

Cold-weather properties

At temperatures ≤ 0 °C, diesel fuels may precipitate paraffin crystals which can clog up the fuel filter. For this reason, oil companies add flow enhancers to diesel fuel in the winter to limit crystal formation so that their size still allows them to pass through the filter pores.

The previously common practice of adding gasoline or kerosene is no longer necessary and also dangerous because it lowers the flash point. In cold parts of the world, the oil industry produces winter diesel fuel with a CFPP rating (Cold Filter Plugging Point, i.e. the point at which it clogs the filter in cold weather) (e.g. at least –20 °C for Germany). For Arctic regions, the CFPP is substantially lower (as much as –44 °C).

Lubricant properties ("lubricity")

In order to reduce the sulfur content of diesel fuel, it is hydrogenated. In addition to removing sulfur, the hydrogenation process also removes the ionic fuel components that aid lubrication. After the introduction of low-sulfur diesel fuels, wear-related problems started to occur on distributor-type fuel-injection pumps which are lubricated by the fuel. The oil industry was able to fully restore the lubricant qualities, however, by adding lubricant additives.

Since 1998 lubricity has been standardized on the basis of the HFRR method (High Frequency Reciprocating Rig) (in which a steel ball is moved rapidly to and fro) by EN 590 and ISO 12 156-1 and 12156-2. A maximum permissible WSD (Wear Scar Diameter, i.e. caused by the steel ball) deter-

mined according to the HFRR method of 460 μm, is adequate to protect fuel-injection pumps. For brand new pumps, Bosch recommends the use of a diesel fuel with a WSD ≤ 400 μm.

Water in diesel fuel

Diesel fuel can absorb water in solution in varying proportions depending on temperature, e.g. 50...200 ppm (by weight) at 25...60 °C.

EN 590 permits a maximum water content of 200 mg/kg. In many countries, however, analysis of diesel fuels reveals higher water concentrations. Dissolved water does not harm the fuel-injection system. Free water, however, which cannot be dissolved in the fuel, can cause damage to fuel-lubricated injection pumps within a very short space of time and even when it is present only in very small quantities.

The presence of water in the fuel tank as a result of condensation from the air cannot be prevented. A water separator and a water sensor on the fuel filter are therefore absolutely essential. In addition, the vehicle manufacturer must design the tank ventilation system and the fuel-filler neck so as to prevent additional water from entering.

Overall contamination

Overall contamination refers to the sum total of undissolved foreign particles in the fuel such as sand, rust and undissolved organic components. EN 590 permits a maximum of 24 mg/kg. However, this figure is too high. Particularly the very hard silicates that occur in mineral dust are harmful to precision-made high-pressure fuel-injection systems. Even a fraction of the permissible overall contamination level of hard particles would produce erosive and abrasive wear (e.g. at the seats of solenoid valves). Such wear causes valve leakage which lowers the injection pressure and engine performance as well as increasing exhaust particulate emissions.

A particle size of 6...7 μm in the fuel is critical, especially considering the fact that 100 ml of fuel can contain millions of such particles. High-efficiency fuel filters that not only achieve very good filtration results but also have long replacement intervals can help to solve the problem.

Sulfur content

Diesel fuels contain varying amounts of sulfur in chemically bonded form depending on the quality of the crude oil. The sulfur is extracted from the middle distillate by hydrogenation at high pressure and temperature in the presence of a catalyst. The initial by-product of this process is hydrogen sulfide (H_2S) which is subsequently converted into pure sulfur.

Since the beginning of 2000 the EN 590 maximum limit for the sulfur content of diesel fuel has been 350 ppm. From 2005 onwards the EU (European Union) will require all diesel fuels to contain less than 10 ppm of sulfur.

Emission-control systems such as NO_X catalytic converters and particulate filters function on the basis of catalytic effects and have to be run on sulfur-free fuel ($\leq$ 10 ppm). Otherwise, instead of the NO_X and HC reactions, sulfur reactions would take place and the catalytic converter would

be to a greater or lesser degree "contaminated" for the purposes of emission elimination, and therefore incapable of performing its intended function.

Regardless of the function of the systems used for emission control in the future, sulfur dioxide (SO_2) and sulfate particle emissions can also be eliminated by the use of sulfur-free fuels.

Coking

The coking tendency of a fuel is an extremely complex process. The coking factor indicates the degree to which the fuel injectors "coke up" (resulting in restriction of flow).

Flash point

The "flash point" indicates the storage temperature at which flammable vapors are produced. For diesel fuels, it is above 55 °C (Hazard Class A III).

Additives in diesel fuel

The most important additives and their effects are listed in Table 2. Their concentration level in the fuel is generally < 1 %.

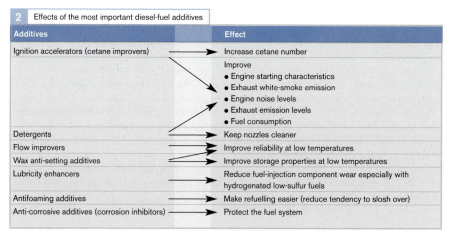

2 Effects of the most important diesel-fuel additives	
Additives	**Effect**
Ignition accelerators (cetane improvers)	Increase cetane number
	Improve
• Engine starting characteristics
• Exhaust white-smoke emission
• Engine noise levels
• Exhaust emission levels
• Fuel consumption
Detergents	Keep nozzles cleaner
Flow improvers	Improve reliability at low temperatures
Wax anti-setting additives	Improve storage properties at low temperatures
Lubricity enhancers	Reduce fuel-injection component wear especially with hydrogenated low-sulfur fuels
Antifoaming additives	Make refuelling easier (reduce tendency to slosh over)
Anti-corrosive additives (corrosion inhibitors)	Protect the fuel system

Table 2
Different additives can have similar effects. The arrows indicate the effects of each additive independently of other components.

Alternative fuels

Alternative fuels for diesel engines are fuels that are not produced from mineral oil under refinery conditions. The most important among them are alcohols and vegetable oils.

Diesel-and-water emulsions are also referred to as alternative fuels. However, they do not genuinely belong in this category.

Alcohol fuels

The alternative fuels methanol and ethanol can also be used in diesel engines. Methanol can be produced from raw materials containing carbon. In some countries (e.g. Brazil), ethanol is extracted from biomass (e.g. sugar cane).

Both of these fuels, however, have inherent disadvantages in comparison with diesel and this would demand substantial modifications to the engine design concept and the fuel-injection system.

Alcohols have poor ignition characteristics (cetane number: 3...8), a low volumetric calorific value, high evaporation heat, poor lubricating qualities, high volatility and a high tendency to produce corrosion.

Another possibility instead of using alcohol fuels in their pure forms is to mix them with diesel fuel, although it should be pointed out that they are practically immiscible without the use of additives. Such a concept therefore requires the use of large quantities of solubilizers.

Furthermore, the poor ignition characteristics demand the use of a high proportion of ignition accelerators. Ultimately, therefore, the high proportion of additives reduces the economy of these concepts.

The advantages of the use of alcohol fuels in diesel engines are their low soot and NO_x emissions.

FAME

FAME (Fatty Acid Methyl Ester) is the generic term for vegetable and animal products.

Fatty acid methyl esters are produced by transesterification of vegetable and animal fats using methanol. The best known vegetable oil methyl ester in Europe is RME (Rape seed oil Methyl Ester). There are also soya, sunflower and palm-oil methyl esters, among others.

The transesterification of the raw materials essentially improves their cold-weather characteristics, viscosity and thermal stability. Consequently, transesterified vegetable oils are rather more suitable than pure vegetable oils for use as alternative fuels for diesel engines.

Nevertheless, transesterified vegetable oils still present a large number of problems such as
- Elastomer incompatibility (leakage at seals)
- Corrosion of aluminum and zinc
- Free water in mixtures with diesel
- Insufficient oxidation stability (chemical contaminants, RME is the most suitable in this respect)
- Free glycerines (deposits, Figure 1a)
- High modulus of elasticity (excessively high injection pressures can damage the fuel-injection pump)
- High viscosity at low temperatures (high exhaust emissions), etc.

Vegetable-oil methyl esters do not offer any significant advantages with regard to emission levels. Nor do they represent a closed CO_2 cycle, as energy has to be introduced for sowing the crops, harvesting, transport and processing (more than in the case of diesel fuel).

The maximum saving of fossil fuels achieved by using RME is theoretically 65 % (50 % in practice). Thus, the only advantage that can be claimed for the sustainable fuel RME is the 65 % maximum fossil-fuel saving.

Since the end of 2000 there has been a draft European standard for FAME which is expected to come into force by the beginning of 2003. Until that time the properties of FAME remain unstandardized and the quality standards on offer in the marketplace remain widely divergent (ranging from "safe" to "fatal" for the fuel-injection system).

A common position statement on FAME issued by the fuel-injection equipment manufacturers Delphi, Stanadyne, Denso and Bosch indicates that it is likely they will only accept a maximum proportion of 5% "good quality" RME (i.e. as defined by the draft EU standard) until a standard comes into effect. Apart from that, some vehicle manufacturers have issued RME approvals (in some cases only for new specifically designed fuel-injection pumps with special seals).

Diesel-and-water emulsions

Diesel-and-water emulsions reduce soot and NO_x emissions but also lower power output relative to the proportion of water (if the injection system is set up for pure diesel). The HC emission levels increase, especially at low engine loads and/or when the engine is cold.

The companies Elf and Lubrizol have plans to sell diesel-and-water emulsions under the respective brand names "Aquazole" and "Purinox" for use in closed commercial-vehicle fleets. As far is known, these emulsions have been tested (not by Bosch) in commercial vehicles with in-line fuel-injection pumps.

The advantage quoted is that, on older vehicles, emissions can be immediately reduced for a limited period without having to take any other measures. Diesel-and-water emulsions are not suitable for more modern fuel-injection systems. For that reason, they cannot be sold on the open market. In many such systems, the fuel temperature can exceed 100 °C, meaning that the water would vaporize and subsequently condense as free water within the fuel system. Without a wa-

ter separator, that water would cause damage to the fuel-injection components. As far as is known, the extremely fine emulsion droplets measuring only a few nanometers are not necessarily removed by a water separator.

Diesel-and-water emulsions contain numerous additives such as
- Emulsifiers to stabilize the emulsion
- Anti-corrosive additives
- Anti-freeze
- Lubricant additives
- Biocides or the like for preventing the growth of micro-organisms, etc.

1 Damage to a fuel-injection pump caused by poor fuel quality

a

b

© SMK1878Y

Fig. 1
a Deposits on actuator mechanism caused by "contaminated" RME
b Bearing damage caused by free water (vehicle mileage approx. 5,600 km)

Cylinder-charge control systems

[1]) The cylinder charge is the mixture of gases trapped in the cylinder when the inlet valves are closed. It consists of the intake air and the residual burned gases from the preceding combustion cycle.

In diesel engines, both the fuel mass injected and the air mass with which it is mixed are decisive factors in determining the torque output and therefore engine performance and exhaust-gas composition. For that reason, the systems that control the cylinder-air charge [1]) have an important role to play as well as the fuel-injection system. Those cylinder-charge control systems clean the intake air and affect the flow, the density and the composition (e.g. the oxygen content) of the cylinder charge.

Overview

In order to burn the fuel, the engine requires oxygen which it extracts from the intake air. In principle, the more oxygen there is available for combustion in the combustion chamber, the greater the amount of fuel that can be injected for full load. There is thus a direct relationship between the amount of air with which the cylinder is charged and the maximum possible engine power output.

Valve and combustion-chamber design has a major effect on the efficiency of the cylinder charging process (see the chapter "Basic principles of the diesel engine"). Beyond that, the aspiration and air-intake systems have the job of conditioning the intake air and ensuring that the cylinders are properly charged.

The cylinder-charge control systems are made up of the following components (Figure 1):
- Air filter (1)
- Swirl flaps (5)
- Turbocharger/supercharger (2)
- Exhaust-gas recirculation system (4)

Most diesel engines are turbocharged or supercharged. Exhaust-gas recirculation systems are fitted on all modern diesel cars and some commercial vehicles. Systems used on cars are not transferrable to commercial vehicles.

Apart from very large, slow-running marine engines, only four-stroke engines are used nowadays. Gas exchange is thus controlled by valves operated by one or more camshafts. Systems with variable valve timing are under development.

1 Cylinder-charge control systems on a diesel engine

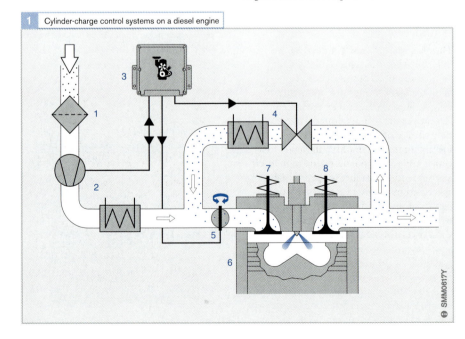

Fig. 1
1   Air filter
2   Turbocharger/
     supercharger
     with intercooler
3   Engine control unit
4   Exhaust-gas
     recirculation
     and cooler
5   Swirl flap
6   Engine cylinder
7   Inlet valve
8   Exhaust valve

SMM0617Y

Intake air filters

Intake air filters reduce the amount of particles contained in the intake air. They are generally deep-bed filters which – in contrast with surface filters – trap the particles in the internal structure of the filter element rather than obstructing their passage on the surface. Deep-bed filters with high dust retention capacities are always preferable when large flow volumes with low particle concentrations need to be efficiently filtered.

Typical air contaminants are illustrated in Figure 2. They consist of particles from both natural and artificial sources and vary widely with regard to particle size. The dust particles drawn in together with the intake air have a diameter of between 0.01 μm (mostly soot particles) and 2 mm (sand grains). Around 75 % of the particles (based on mass) are in the size range from 5 μm to 100 μm. The mass concentration in the intake air depends heavily on the environment in which the vehicle is used (e.g. motorway or dirt track). For a car over a period of ten years it may range from the extremes of a few grams to several kilograms of dust.

The air filter prevents mineral dust and particles entering the engine and the engine oil and thereby reduces the wear on components such as bearings, piston rings and cylinder walls. It also protects the sensitive air-mass meter by preventing dust being deposited on it. This might otherwise cause incorrect readings resulting in higher fuel consumption and pollutant emission levels above the allowable limits. Special high-specification air-filter element designs in combination with appropriately shaped filter housings are also capable of preventing the ingress of water in heavy rain.

Air filters which incorporate the latest technology achieve total mass filtration rates of up to 99.8 % (cars) and 99.95 % (commercial vehicles). Such figures must be capable of being maintained under all prevailing conditions including the dynamic conditions that exist in the air-intake system of an engine (pulsation). Filters of inadequate quality have greater dust passage rates under such circumstances.

The filter elements are individually designed for each engine. In that way, pressure losses can be kept to a minimum and the high filtration rates are not dependent on the flow rate. The filter elements, which may be rectangular or cylindrical, consist of a filter medium that is folded so that the maximum possible filter surface area can be accommodated within the smallest possible space. Generally cellulose-fiber based, the filter medium is compressed and impregnated to give it the required structural strength, wet rigidity and resistance to chemicals. The filter elements have to be replaced at the intervals specified by the vehicle manufacturer (for cars, every two to four or, in some cases, even every six years, i.e. every 40,000 to 60,000 km or every 90,000 km, or when the back pressure reaches 20 mbar).

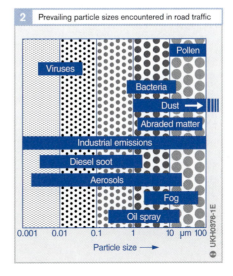

2   Prevailing particle sizes encountered in road traffic

Viruses · Pollen · Bacteria · Dust → · Abraded matter · Industrial emissions · Diesel soot · Aerosols · Fog · Oil spray

0.001   0.01   0.1   1   10 μm 100

Particle size ⟶

UKH0976-1E

3   Photograph of a filter medium made of synthetic fibers taken using an electron microscope

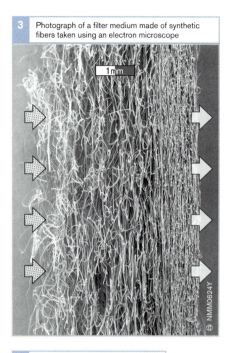

1 mm

The demands for small and highly efficient filter elements (smaller space requirements) that also offer longer servicing intervals is the driving force behind the development of innovative, new air-filter media. New air-filter media made of synthetic fibers which have substantially improved performance figures in some cases are already entering production. Figure 3 shows a photograph of a synthetic high-performance filter medium (felt) with continuously increasing density and decreasing fiber diameter across the filter section from the input side to the output side.

Better results than with purely cellulose-based media can be achieved with composite materials (e.g. paper with melt-blown layer) and special nano-fiber filter media which consist of a relatively coarse base layer made of cellulose to which ultra-thin fibers with diameters of only 30 to 40 nm are applied.

Fig. 3
The arrows indicate the direction of flow of the intake air
Source: Freudenberg Vliesstoffe KG

4   Air-intake module for a car (example)

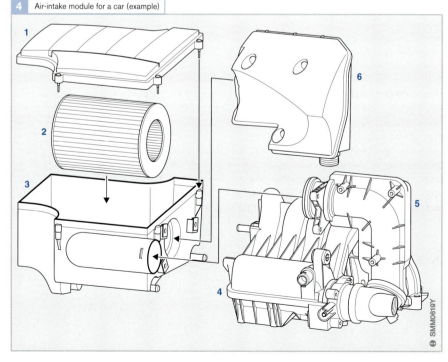

Fig. 4
1   Housing lid
2   Filter element
3   Filter housing
4   Air-intake module
5   Intake duct
6   Intake duct

New folded structures with alternately sealed channels, similar to diesel soot filters, are soon to be introduced on the market.

Conical, oval as well as stepped and trape-zoidal geometries add to the range of shapes available in order to optimize use of the space under the hood which is becoming ever more confined.

Previously, air-filter housings were almost exclusively designed as "muffler filters". Their large volume was designed for the supplementary function of reducing air intake noise. Nowadays, the two functions of filtration and engine-noise reduction are increasingly separated and the different components independently optimized. This means that the filter housing can be reduced in size. And that results in very slim filters which can be integrated in the engine trim covers while the mufflers are placed in less accessible positions inside the engine com-partment.

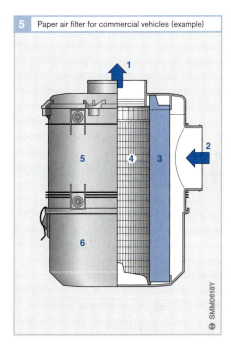

5   Paper air filter for commercial vehicles (example)

SMM0618Y

Fig. 5
1   Air outlet
2   Air inlet
3   Filter element
4   Supporting tube
5   Housing
6   Dust collector

Air filters for cars
Figure 4 shows a complete air-intake module for a car. In addition to the air-filter housing (1 and 3) with the cylindrical filter element (2), it incorporates all air-intake ducts (5 and 6) and the air-intake module (4) as well as Helmholtz resonators and lambda quarter pipes in between for noise reduction. With the aid of this type of overall system opti-mization, the individual components can be better matched to one another. This helps to comply with the ever stricter noise-output re-strictions. In this example, the engine control unit is integrated in the air-intake system so that the air flow cools the electronic circuitry.

Air filters for commercial vehicles
Figure 5 shows an easy-to-maintain and weight-optimized plastic air filter for com-mercial vehicles. In addition to having, as pre-viously mentioned, a very high filtration rate, the elements for this filter are dimensioned for servicing intervals of over 100,000 km. In countries with high levels of atmospheric dust, and on construction and agricultural machines, a pre-filter is fitted upstream of the filter element. The pre-filter filters out coarse-grained, heavy dust particles, thereby sub-stantially increasing the service life of the fine filter element. In its most simple form, it is a ring of deflector vanes which set the air flow into a rotating motion. The resulting cen-trifugal force separates out the coarse dust particles. However, only mini-cyclone pre-fil-ter batteries optimized for use in conjunction with the main filter element can properly uti-lize the potential of centrifugal separators in commercial-vehicle air filters.

Swirl flaps

The pattern of air flow inside the cylinders of a diesel engine has a fundamental effect on mixture formation. The term "swirl" refers to a circular motion of the intake air inside the cylinder. That rotating motion enables better mixing of fuel and air to be achieved. Using appropriate flaps and channels, the swirl can be regulated according to varying operating requirements. In the example shown in Figure 1, the flap (6) is closed at low engine speeds. This produces a large degree of swirl combined with sufficient air flow to the cylinder. At high speeds, the flap opens, allowing unrestricted air flow though the intake port (5). This means that the cylinder charge and the engine power output are increased at higher engine speeds. Such "intake-port shutoff" systems are currently used on some car engines.

Turbochargers and superchargers

Assisted aspiration by means of turbochargers or superchargers has been around for many years[1] on large-scale diesel engines for fixed installations and marine propulsion systems as well as on commercial vehicles. In more recent times, it has also been adopted for fast-running diesel engines in cars [2]. In contrast to a conventionally aspirated engine, the air is forced into the cylinders under pressure in a turbocharged or supercharged engine. This increases the mass of the cylinder charge and, in combination with a correspondingly greater injected fuel mass, results in a higher power yield from the same engine capacity.

The diesel engine is particularly suited to assisted aspiration as its compressed cylinder charge consists only of air rather than a mixture of fuel and air, and it can be economically combined with a supercharger/turbocharger because of its quality-based method of control.

With larger commercial-vehicle engines, a further increase in mean pressure (and therefore torque) is achieved by higher turbocharger pressures and lower compression, but is offset by poorer cold-starting characteristics.

Although, strictly speaking, the turbocharger is itself a type of supercharger, the terms turbocharger and supercharger are now generally used to distinguish between different methods of operation, so that
- the term *turbocharger* is used to refer to a supercharger driven by the flow of exhaust gas from the engine, while
- the term *supercharger* generally refers only to one that is driven directly by the engine (and usually by the crankshaft).

Volumetric efficiency
Volumetric efficiency refers to the relationship of the actual air charge trapped inside the cylinder to the theoretical air charge determined by the cylinder capacity under

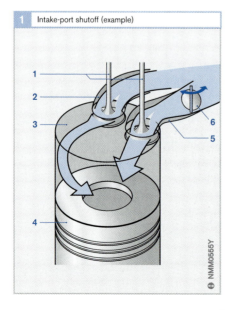

1 Intake-port shutoff (example)

1
2
3
        6
        5
4

© NMM0555Y

Fig. 1
1 Intake-valve
2 Swirl port
3 Engine cylinder
4 Piston
5 Intake-port
6 Flap

standard conditions (air pressure $p_0 =$ 1,013 hPa, temperature $T_0 = 273$ K) without assisted aspiration. For supercharged/turbocharged diesel engines, the volumetric efficiency is in the range 0.85...3.0.

Dynamic supercharging

A degree of supercharging can be achieved simply by the utilization of dynamic effects in the intake manifold. Dynamic supercharging effects of this type are less important in diesel engines than they are for gasoline engines. In diesel engines, the main emphasis of intake-manifold design is on even distribution of the air charge between all cylinders and distribution of the recirculated exhaust gas. In addition, the creation of swirl effects inside the cylinders is also of importance. At the relatively low speeds at which diesel engines run, designing the intake manifold specifically to obtain dynamic supercharging effects would require it to be extremely long. Since virtually all modern diesel engines are equipped with turbochargers, the only benefit that could be achieved would be under non steady-state operating conditions where the turbocharger has not reached full delivery pressure.

In general, the intake manifold on a diesel engine is kept as short as possible. The advantages of this are
- Improved dynamic response characteristics and
- Better control characteristics on the part of the exhaust-gas recirculation system

Turbocharging

Of the methods of assisted aspiration, the exhaust-gas-driven turbocharger is by far the most widely used. This method of assisted aspiration enables even small-capacity engines to achieve high torque and power output with a good level of engine efficiency. Turbochargers are used on engines for cars and commercial vehicles as well as on large, heavy-duty marine and locomotive engines.

Whereas the turbocharger was originally conceived as a means of improving the power-to-weight ratio, it is now increasingly used to improve the maximum torque figure at low to medium engine speeds. This is particularly true in connection with systems in which the turbocharger pressure is electronically controlled.

| 2 | Turbocharger with variable turbine geometry |

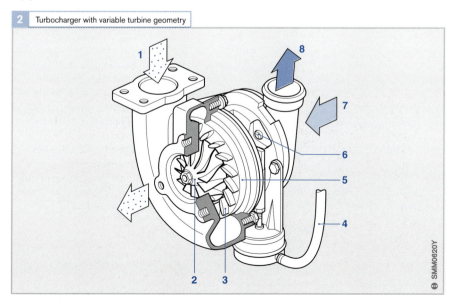

© SMM0620Y

Fig. 2
1 Exhaust inlet
2 Turbine
3 Deflector blades
4 Vacuum tube
5 Adjusting ring
6 Lubricant supply
  connection
7 Intake-air inlet
8 Intake-air outlet

Design and method of operation

The hot exhaust gas expelled under pressure from an internal-combustion engine represents a substantial loss of energy. It makes sense, therefore, to utilize some of that energy to generate pressure in the intake manifold.

The turbocharger (Figure 3) is a combination of two fluid-flow devices:
- A turbine (7) that is driven by the flow of exhaust gas and
- A compressor (2) that is directly coupled with the turbine by means of a shaft (11) and which compresses the intake air

The hot exhaust gas flows over the turbine and by so doing forces it to rotate at high speeds (in diesel engines, up to around 200,000 rpm). The inward-facing blades of the turbine divert the flow of gas into the center from where it passes out to the side (8, radial-flow turbine). The connecting shaft drives the radial-flow compressor. This is the exact reverse of the turbine: The intake air (3) is drawn in at the center of the compressor and is driven outwards by the blades of the impeller so that it is compressed (4).

On large-scale engines, axial-flow turbines are also used. In that case, the exhaust gas flows through the turbine in an axial direction. Axial-flow turbines are more efficient on such engines and are cheaper to produce than radial-flow turbines. For car and commercial-vehicle engines, the radial-flow turbine is more economical.

Because of the exhaust-gas back pressure that builds up upstream of the turbine, the engine has to work harder to expel the exhaust gas on the exhaust stroke. Nevertheless, the engine efficiency across broad areas of the characteristic-data map is greater.

For fixed-installation engines running at constant speed, the turbine and turbocharger characteristics can be tuned to a high level of efficiency and turbocharger pressure. Turbocharger design becomes more complicated when it is applied to motor-vehicle engines that do not run under steady-state conditions – because they are expected to produce high torque levels particularly when accelerating from slow speeds. Low exhaust temperatures, low exhaust-flow volumes and the inertia of the turbocharger itself all contribute

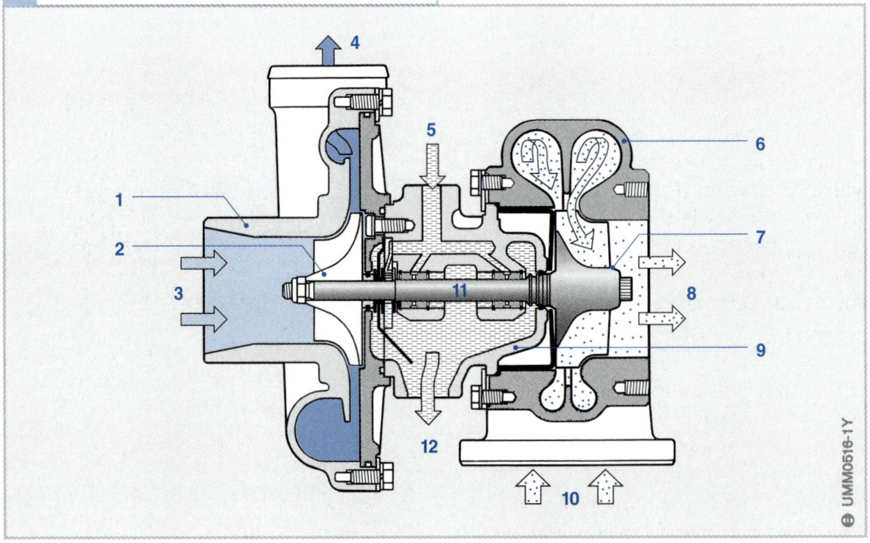

Fig. 3
1 Compressor housing
2 Centrifugal compressor
3 Intake air
4 Compressed intake air
5 Lubricant inlet
6 Turbine housing
7 Turbine
8 Exhaust outflow
9 Bearing housing
10 Exhaust inflow
11 Shaft
12 Lubricant return outlet

3 Commercial-vehicle turbocharger with twin-flow turbine

to a slow build-up of pressure in the compressor at the start of acceleration. On turbocharged car engines, this phenomenon is referred to as "turbo lag".

Because of this effect, turbochargers with a low inertial mass that respond at lower exhaust-gas flow rates have been developed especially for cars and commercial vehicles. Engine responsiveness is substantially improved by using such turbochargers – particularly at low engine speeds.

A distinction is made between two methods of turbocharging.
Constant-pressure turbocharging involves the use of an exhaust-gas accumulator upstream of the turbine to smooth out the pressure pulsations in the exhaust system. As a result, the turbine can accommodate a higher exhaust-gas flow rate at a lower pressure at high engine speeds. As the exhaust-gas back pressure that the engine is working against is lower under those operating conditions, fuel consumption is also lower. Constant-pressure turbocharging is used for large-scale marine, generator and fixed-installation engines.

Pulse turbocharging utilizes the kinetic energy of the pressure pulsations caused by the expulsion of the exhaust gas from the cylinders. Pulse turbocharging achieves higher torques at lower engine speeds. It is the principle used by turbochargers for cars and commercial vehicles. Separate exhaust manifolds are used for different banks of cylinders to prevent individual cylinders from interfering with each other during gas exchange, e.g. two groups of three cylinders on a six-cylinder engine. If twin-flow turbines – which have two outer channels – are used (Figure 3), the exhaust flows are kept separate in the turbocharger as well.

In order to obtain good response characteristics, the turbocharger is positioned as close as possible to the exhaust valves in the flow of hot exhaust gas. It therefore has to be made of highly durable materials. On ships – where hot surfaces in the engine room have

to be prevented because of the fire risk – turbochargers are water-cooled or enclosed in heat-insulating material. Turbochargers for gasoline engines, where the exhaust-gas temperatures can be 200...300 °C higher than on diesel engines, may also be water-cooled.

Designs
Engines need to be able to generate high torque even at low speeds. For that reason, turbochargers are designed for low exhaust-gas mass flow rates (e.g. full load at an engine speed of $n \leq 1,800$ rpm). To prevent the turbocharger from overloading the engine at higher exhaust-gas mass flow rates, or being damaged itself, the turbocharger pressure has to be controlled. There are three turbocharger designs which can achieve this:
● The wastegate turbocharger
● The variable-turbine-geometry turbocharger and
● The variable-inlet-valve turbocharger

Wastegate turbocharger (Figure 4)
At higher engine speeds or loads, part of the exhaust flow is diverted past the turbine by a bypass valve – the "wastegate" (5). This reduces the exhaust-gas flow passing through

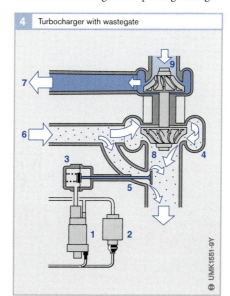

4 Turbocharger with wastegate

Fig. 4
1 Charge-pressure actuator
2 Vacuum pump
3 Pressure actuator
4 Turbocharger
5 Wastegate (bypass valve)
6 Exhaust flow
7 Intake air flow
8 Turbine
9 Centrifugal compressor

UMK1551-9Y

the turbine and lowers the exhaust-gas back pressure, thereby preventing excessive turbocharger speed.

At low engine speeds or loads, the wastegate closes and the entire exhaust flow passes through and drives the turbine.

The wastegate usually takes the form of a flap integrated in the turbine housing. In the early days of turbocharger design, a poppet valve was used in a separate housing parallel to the turbine.

The wastegate is operated by an electro-pneumatic charge-pressure actuator (1). That actuator is an electrically operated 3/2-way valve that is connected to a vacuum pump (2). In its neutral position (de-energized) it allows atmospheric pressure to act on the pressure actuator (3). The spring in the pressure actuator opens the wastegate.

If a current is applied to the charge-pressure actuator by the engine control unit, it opens the connection between the pressure actuator and the vacuum pump so that the diaphragm is drawn back against the action of the spring. The wastegate closes and the turbocharger speed increases.

The turbocharger is designed in such a way that the wastegate will always open if the control system fails. This insures that, at high engine speeds, excessive turbocharger pressure which might damage the engine or the turbocharger itself cannot be produced.

On gasoline engines, sufficient vacuum is created by the intake manifold. Therefore, unlike diesel engines, they do not require a vacuum pump. Both types of engine may also use a purely electrical wastegate actuator.

Variable-turbine-geometry (VTG) turbocharger (Figure 5)
Varying the rate of gas flow through the turbine by means of variable turbine geometry (VTG) is another method by which the exhaust-gas flow rate can be limited at high engine speeds. The adjustable deflector blades (3) alter the size of the gap through which the exhaust gas flows in order to

reach the turbine (variation of geometry). By so doing, they adjust the exhaust-gas pressure acting on the turbine in response to the required turbocharger pressure.

At low engine speeds or loads, they allow only a small gap for the exhaust gas to pass through so that the exhaust-gas back pressure increases. The exhaust-gas flow velocity through the turbine is then higher so that the turbine turns at a higher speed (a). In addition, the exhaust-gas flow is directed at the outer ends of the turbine blades. This generates more leverage which in turn produces greater torque.

At high engine speeds or loads, the deflector blades open up a larger gap for the exhaust gas to flow through with the result that the flow velocity is lower (b). Consequently, the turbocharger turns more slowly if the flow volume remains the same, or else its speed does not increase as much if the flow volume increases. In that way, the turbocharger pressure is limited.

5 Variable turbine geometry of VTG turbocharger

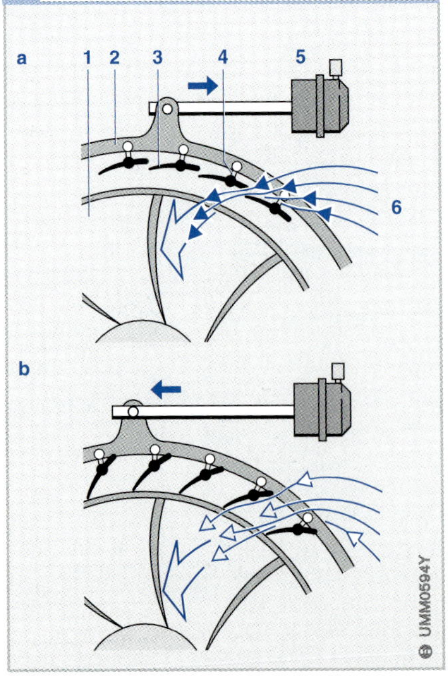

Fig. 5
a Deflector blade setting for high turbocharger pressure
b Deflector blade setting for low turbocharger pressure

1 Turbine
2 Adjusting ring
3 Deflector blade
4 Adjusting lever
5 Pneumatic actuator
6 Exhaust flow

← High flow rate
⇽ Low flow rate

The deflector blade angle is adjusted very simply by turning an adjuster ring (2). This sets the deflector blades to the desired angle by operating them either directly using adjusting levers (4) attached to the blades or indirectly by means of adjuster cams. The adjusting ring is operated by a pneumatic actuator (5) to which positive or negative pressure is applied, or alternatively by an electric motor with position feedback (position sensor). The engine control unit controls the actuator. Thus the turbocharger pressure can be adjusted to the optimum setting in response to a range of input variables.

The VTG turbocharger is fully open in its neutral position and therefore inherently safe, i.e. if the control system fails, neither the turbocharger nor the engine suffers damage as a result. There is merely a loss of power at low engine speeds.

This is the type of turbocharger most widely used on diesel engines today. It has not been able to establish itself as the preferred choice for gasoline engines because of the high thermal stresses and the higher exhaust temperatures encountered.

Variable-intake-valve turbocharger (Figure 6)
The variable-intake-valve turbocharger is used on small car engines. On this type of turbocharger, an intake slide valve (4) alters the cross-section of the inlet flow to the turbine by opening one or both of the intake ports (2, 3).

At low engine speeds or loads, only one of the intake ports is open (2). The small inlet aperture produces high exhaust-gas back pressure combined with a high exhaust-gas flow velocity, and consequently results in a high speed of rotation on the part of the turbine (1).

When the required turbocharger pressure is reached, the intake valve gradually opens the second intake port (3). The flow velocity of the exhaust gas – and therefore the turbine speed and the turbocharger pressure – then gradually reduce.

The engine control unit module controls the valve setting by means of a pneumatic actuator.

There is also a bypass channel (5) integrated in the turbine housing so that virtually the entire exhaust gas flow can be diverted past the turbine in order to obtain a very low turbocharger pressure.

6 Method of operation of variable-intake-valve turbocharger

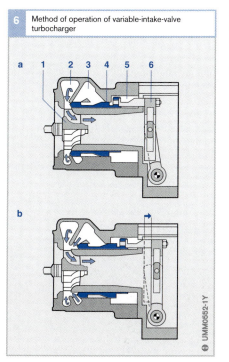

a

b

UMM0552-1Y

Fig. 6
a   Only one intake port open
b   Both intake ports open

1   Turbine
2   1st intake port
3   2nd intake port
4   Inlet slide valve
5   Bypass channel
6   Valve actuator

Advantages and disadvantages of turbocharging

Downsizing

When compared with a conventionally aspirated engine of equal power, the prime advantage of a turbocharged engine is its lighter weight and smaller dimensions. It also has better torque characteristics within the useful speed range (Figure 7). Consequently, the power output at a given speed is higher (A – B) at the same specific fuel consumption.

The same amount of power is available at a lower engine speed because of the superior torque characteristics (B – C). Thus, with a turbocharged engine, the point at which a required amount of power is produced is shifted to a position where frictional losses are lower. The result of this is lower fuel consumption (E – D).

Torque curve

At very low engine speeds, the basic torque of a turbocharged engine is similar to that of a conventionally aspirated engine. At that point, the usable energy from the exhaust-gas flow is insufficient to drive the turbine. No turbocharger pressure is generated in this way.

Under dynamic operating conditions, the torque output remains similar to that of a conventionally aspirated engine even at medium engine speeds (c). This is because of the delay in the build-up of the exhaust-gas flow. On acceleration from slow speeds, therefore, the "turbo lag" effect occurs.

On gasoline engines in particular, the turbo lag can be minimized by utilizing the dynamic supercharging effect. This improves the turbocharger's response characteristics.

On diesel engines, the use of turbochargers with variable turbine geometry provides a means of significantly reducing turbo lag.

Another design variation is the electrically assisted turbocharger which is aided by an electric motor. The motor accelerates the impeller on the compressor side of the turbocharger independently of the exhaust-gas flow through the turbine, thereby reducing turbo lag. This type of turbocharger is currently in the course of development.

The response of turbocharged engines as altitude increases is very good because the pressure differential is greater at lower atmospheric pressure. This partially offsets the lower density of air. However, the design of the turbocharger must ensure that the turbine does not over-rev in such conditions.

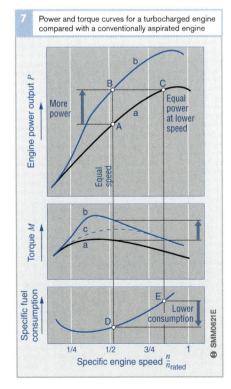

7   Power and torque curves for a turbocharged engine compared with a conventionally aspirated engine

Fig. 7

a   Conventionally aspirated engine under steady-state conditions

b   Turbocharged engine under steady-state conditions

c   Turbocharged engine under dynamic conditions

Supercharging

A supercharger consists of a compressor driven directly by the engine. The engine and the compressor are generally rigidly linked, e.g. by a belt drive system. Compared with turbochargers, superchargers are rarely used on diesel engines.

Positive-displacement supercharger

The most common type of supercharger is the positive-displacement supercharger. It is used mainly on small and medium-sized car engines. The following types of supercharger are used on diesel engines:

Positive-displacement supercharger with internal compression

With this type of supercharger, the air is compressed inside the compressor. The types used on diesel engines are the reciprocating-piston supercharger and the helical-vane supercharger.

Reciprocating-piston supercharger: This type has either a rigid piston (Figure 8) or a diaphragm (Figure 9). A piston (similar to an engine piston) compresses the air which then passes through an outlet valve to the engine cylinder.

Helical-vane supercharger (Figure 10): Two inter-meshing helical vanes (4) compress the air.

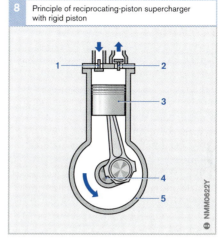

8 Principle of reciprocating-piston supercharger with rigid piston

Fig. 8
1 Inlet valve
2 Outlet valve
3 Piston
4 Drive shaft
5 Casing

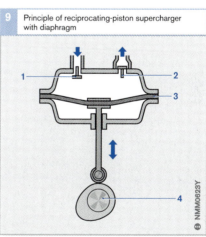

9 Principle of reciprocating-piston supercharger with diaphragm

Fig. 9
1 Inlet valve
2 Outlet valve
3 Diaphragm
4 Drive shaft

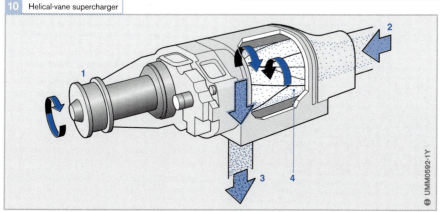

10 Helical-vane supercharger

Fig. 10
1 Drive pulley
2 Intake air
3 Compressed air
4 Helical vane

Positive-displacement supercharger without internal compression

With this type of supercharger, the air is compressed outside of the supercharger by the action of the fluid flow generated. The only example of this type to be used on diesel engines was the Rootes supercharger (Figure 11) which was fitted to some two-stroke diesels.

Rootes supercharger: Two contra-rotating rotary vanes (2) linked by gears rotate in contact with one another in similar fashion to a gear pump and in that way compress the intake air.

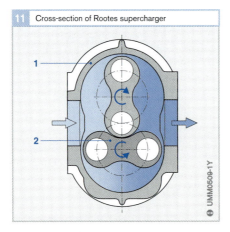

11   Cross-section of Rootes supercharger

Fig. 11
1   Housing
2   Rotary vane

12   Two-stage turbocharging (schematic)

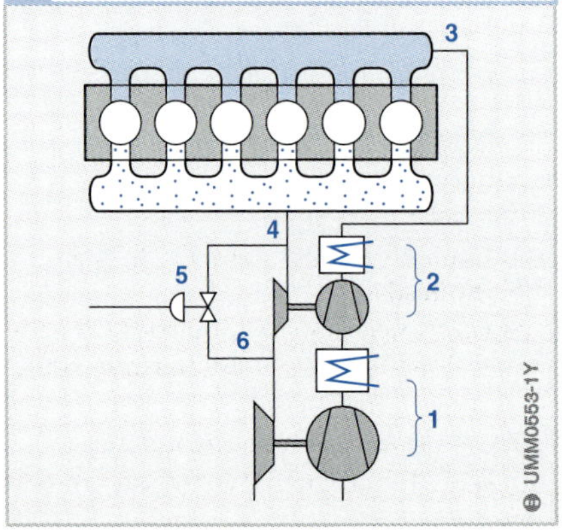

Fig. 12
1   Low-pressure stage (turbocharger with intercooler)
2   High-pressure stage (turbocharger with intercooler)
3   Intake manifold
4   Exhaust manifold
5   Bypass valve
6   Bypass pipe

Centrifugal supercharger

In addition to the positive-displacement superchargers, there are also centrifugal superchargers (centrifugal-flow compressors) in which the compressor is similar to that in a turbocharger. In order to obtain the high peripheral velocity required, they are driven via a system of gears. This type of supercharger offers good volumetric efficiency over a wide range of speeds and can be seen as an alternative to the turbocharger for small engines. Centrifugal turbochargers are rarely used on medium-sized or larger car engines.

Controlling supercharger pressure

The pressure generated by a supercharger can be controlled by means of a bypass. A proportion of the compressed air flow enters the cylinder and determines the cylinder charge. The remainder flows through the bypass and is returned to the intake side. The bypass valve is controlled by the engine control unit.

Advantages and disadvantages of supercharging

Because the supercharger is driven directly by the crankshaft, any increase in engine speed is instantaneously mirrored by an increase in compressor speed. This means that under dynamic operating conditions, higher engine torque and better response characteristics are obtained than with a turbocharger. If variable-speed gearing is used, the engine response to load changes can also be improved.

Since, however, the necessary power output for driving the compressor (approx. 10...15 kW for cars) is not available as effective engine output, those advantages are offset by a somewhat higher rate of fuel consumption than with a turbocharger. That disadvantage is mitigated if the compressor can be disconnected at low engine speeds and loads by means of a clutch operated by the engine control unit. This, on the other hand, makes the supercharger more expensive to produce. Another disadvantage of the supercharger is the greater amount of space it requires.

Multistage turbocharging

Multistage turbocharging is an improvement on single-stage turbocharging in that power limits can be significantly extended. The objective here is to improve air supply under both steady-state and dynamic operating conditions and at the same time improve the specific fuel consumption of the engine. Two methods of turbocharging have proved successful in this respect.

Sequential supercharging

Sequential supercharging involves the use of multiple turbochargers connected in parallel which successively cut in as engine load increases. Thus, in comparison with a single larger turbocharger which is geared to the engine's rated power output, two or more optimum levels of operation can be obtained. Because of the added expense of the supercharger sequencing control system, however, sequential supercharging is predominantly used on marine propulsion systems or generator engines.

Controlled two-stage turbocharging

Controlled two-stage turbocharging involves two differently dimensioned turbochargers connected in series with a controlled bypass and, ideally, two intercoolers (Figure 12, Items 1 and 2). The first turbocharger is a low-pressure turbocharger (1) and the second, a high-pressure turbocharger (2). The intake air first undergoes precompression by the low-pressure turbocharger. Consequently, the relatively small high-pressure compressor in the second turbocharger is operating at a higher input pressure with a low volumetric flow rate, so that it can deliver the required air-mass flow rate. A particularly high level of compressor efficiency can be achieved with two-stage turbocharging.

At lower engine speeds, the bypass valve (5) is closed, so that both turbochargers are working. This provides for very rapid development of a high turbocharger pressure. As engine speed increases, the bypass valve gradually opens until eventually only the low-pressure turbocharger is operating. In this way, the turbocharging system adjusts evenly to the engine's requirements. This method of turbocharging is used in automotive applications because of its straightforward control characteristics.

Electric booster

This is an additional compressor mounted upstream of the turbocharger. It is similar in design to the turbocharger's compressor but is driven by an electric motor. Under acceleration, the electric booster supplies the engine with extra air, thereby improving its response characteristics at low speeds in particular.

Intercooling

In the process of being compressed by the turbocharger, air also heats up (to as much as 180 °C). Since, under otherwise identical conditions, hot air is less dense than cold air, the higher temperature of the air has a negative effect on the cylinder charge. An intercooler between the turbocharger and the engine is therefore used to reduce the temperature of the compressed air. Intercooling consequently helps to further improve the efficiency of the cylinder charging process. It means that there is more oxygen available for combustion, with the result that a higher maximum torque and therefore greater power output is available at a given engine speed.

The lower temperature of the air entering the cylinder also reduces the temperatures generated during the compression stroke. This has a number of advantages:

- Greater thermal efficiency and therefore lower fuel consumption and soot emission on the part of diesel engines
- Reduced knocking tendency on the part of gasoline engines
- Lower thermal stresses on the cylinder block/head
- Small reduction in NO_x emissions as a result of the lower combustion temperature

Intercoolers achieve heat extraction either by cooling the air or with a separate coolant circuit.

▶ Pressure-wave superchargers

A variation of the supercharger for car engines is the pressure-wave supercharger known by the proprietary name "Comprex®". A vane rotor (2) driven by the engine rotates inside a cylindrical housing, the ends of which each have two vents (7). Specially shaped vane enclosures created by the rotor vanes insure that the pressure waves of the exhaust-gas flow (4) produce a pressure rise in the intake air flow (5). An integral governing mechanism regulates supercharger pressure according to engine requirements.

The characteristic feature of pressure-wave superchargers is the direct exchange of energy between the exhaust and intake air flows without any intermediate mechanical components. The exchange of energy takes place at the speed of sound. The system is not subject to the negative effects of turbo lag. A pressure-wave supercharger – like other types of supercharger – responds instantaneously to load changes.

If the gearing ratio between the engine and the pressure-wave supercharger is invariable, the exchange of energy is optimum only for a specific point on the power curve. But by the use of appropriate "pockets" in the ends of the housing and clever design of the vane rotor, the supercharger can be made efficient over a relatively broad operating range. In that way, the pressure-wave supercharger can achieve good supercharging characteristics for steady-state operation. It can also produce torque response characteristics that are not obtainable in the same way with other supercharging methods.

The vane rotor and exhaust pipe arrangement of a pressure-wave supercharger requires a large amount of space in comparison with other methods of supercharging. This makes it difficult to accommodate in engine compartments where space is at a premium. The necessity of balancing the exhaust-gas oscillations at all engine speeds and loads demands a very costly control system. Consequently, since an optimized turbocharger using the latest technology provides the best compromise between function and cost, this type of supercharger has failed to establish itself.

▶ Pressure-wave superchargers

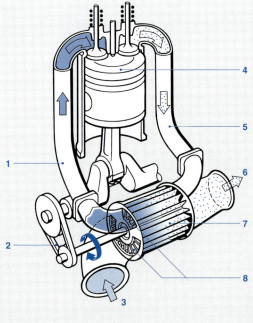

1   High-pressure intake air
2   Drive belt
3   Low-pressure intake air
4   Engine piston
5   High-pressure exhaust
6   Low-pressure exhaust
7   Vane rotor
8   Vents in housing

UMM0517-1Y

Exhaust-gas recirculation

After combustion, there is always a certain amount of burned gas left in the cylinder (internal exhaust-gas recirculation). The size of that proportion can be deliberately determined by valve timing. In addition, more exhaust gas can be diverted from the exhaust system by an exhaust-gas recirculation valve and returned (via a recirculated-exhaust-gas cooler if required) to the intake side of the engine (external exhaust-gas recirculation). Sometimes referred to as EGR (exhaust-gas recirculation), this process is a useful method for reducing NO_x emissions. The NO_x-reducing effect of exhaust-gas recirculation is based on three mechanisms:
- Reduction of the oxygen concentration in the combustion chamber
- Reduction of the amount of exhaust gas expelled and
- Reduction of temperature by virtue of the greater thermal capacity (specific heat) of the inert gases [1] H_2O and CO_2

Recirculation of cooled exhaust gas is particularly effective. The recirculation rates can be up to 50 % on cars and 5...25 % for commercial vehicles.

Addition of recirculated exhaust gas to the cylinder charge reduces its oxygen content (the excess-air factor λ diminishes). If too much exhaust gas is fed back into the cylinder, the levels of the emissions that occur as a result of insufficient air (CO, HC and soot) increase. Fuel consumption also increases if too much exhaust gas is recirculated. Consequently, precise monitoring and control of exhaust-gas recirculation is an absolute necessity.

Exhaust-gas recirculation is controlled by the exhaust-gas recirculation valve (EGR valve). In its neutral setting, it closes off a channel connecting the exhaust-gas system upstream of the turbocharger turbine to the intake system downstream of the turbocharger compressor. The EGR valve is controlled by the engine control unit in response to engine speed and load. To make the EGR valve operate precisely, it has to be designed to be resistant to deposits.

Exhaust-gas recirculation in cars

Exhaust-gas recirculation was first introduced on cars in the 1970s. Today it is used on most car diesel engines.

In accordance with statutory requirements, exhaust-gas recirculation on cars is used only within the lower speed/power band. At low loads, there is always a pressure differential between exhaust-gas back pressure and turbocharger pressure (turbocharger with wastegate or variable-turbine geometry) for exhaust-gas recirculation. The exhaust gas can therefore be recirculated by means of a valve.

Exhaust-gas recirculation in commercial vehicles

In the future exhaust-gas recirculation will also be used on commercial vehicles (heavy-duty) in order to obtain lower NO_x emissions. This will require its use across virtually the entire operating range.

Under normal circumstances, at high loads the exhaust-gas back pressure upstream of the turbocharger turbine on a commercial vehicle is lower than the turbocharger pressure downstream of the turbocharger compressor and intercooler. For this reason, in order to effect exhaust-gas recirculation, the turbocharger must be suitably modified or a VTG turbocharger that can generate the required negative pressure differential must be used. Another possibility is a flutter valve which opens whenever the pressure in the exhaust is greater than in the intake duct so that exhaust recirculates. This will be the case at high loads whenever a pressure pulse is created by the exhaust stroke of a cylinder.

Yet another alternative is the use of an adjustable venturi tube (lower pressure at the constriction point) in the bypass to the air intake. Exhaust-gas recirculation can be controlled on the basis of differential air mass using an air-mass flow meter (cars), a lambda sensor sensor or the signal from a differential pressure sensor on a venturi (commercial vehicles).

[1] Constituents of the cylinder charge that are inert, i.e. do not take part in combustion. The inert gas components do, however, influence ignition characteristics and combustion propagation.

Basic principles of diesel fuel injection

The combustion processes that take place inside a diesel engine are essentially dependent on the way in which the fuel is injected into the combustion chamber. The most important criteria are the timing and the duration of injection, the degree of atomization and the distribution of the fuel inside the combustion chamber, the timing of ignition, the mass of the fuel injected relative to crankshaft rotation, and the total amount of fuel injected relative to engine load. In order that a diesel engine and its fuel-injection system function properly, all of these variable factors must be carefully balanced.

The design of the fuel-injection system must be precisely matched to the engine concerned and its application. As a variety of factors have to be taken into account, some of which are in conflict with one another, the final design can only ever be a compromise.

The composition and conditioning of the air/fuel mixture has a fundamental effect on an engine's specific fuel consumption, torque (and therefore power output), exhaust-gas composition and combustion noise. The quality and effectiveness of the mixture formation is largely attributable to the fuel-injection system.

A number of fuel-injection variables affect mixture formation and the course of combustion inside the combustion chamber and, therefore, the engine's emission levels and power output/efficiency. They are:
- Start of injection
- Injection characteristics (injection duration and rate-of-discharge curve)
- Injection pressure
- Injection direction and
- The number of injection jets

The injection mass and the engine speed are operating parameters that determine the engine power output.

Mixture distribution

Excess-air factor λ

The excess-air factor λ was devised in order to indicate the degree to which the actual air/fuel mixture achieved in reality diverges from the theoretical (stoichiometric [1]) mass ratio. It indicates the ratio of intake air mass to required air mass for stoichiometric combustion, thus:

$$\lambda = \frac{Air\ mass}{Fuel\ mass \cdot Stoichiometric\ ratio}$$

$\lambda = 1$: The intake air mass is equal to the air mass theoretically required to burn all of the fuel injected.

$\lambda < 1$: The intake air mass is less than the amount required and therefore the mixture is rich.

$\lambda < 1$: The intake air mass is greater than the amount required and therefore the mixture is lean.

---

[1] The stoichiometric ratio indicates the air mass in kg required to completely burn 1 kg of fuel (m_L/m_K). For diesel fuel, this is approx. 14.5.

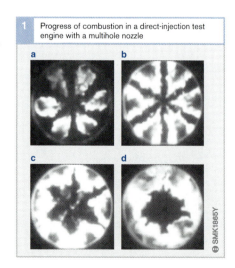

1 Progress of combustion in a direct-injection test engine with a multihole nozzle

a   b

c   d

© SMK1865Y

Lambda levels in diesel engines

Rich areas of mixture are responsible for sooty combustion. In order to prevent the formation of too many rich areas of mixture, diesel engines – in contrast to gasoline engines – have to be run with an overall excess of air.

The lambda levels for turbocharged diesel engines at full load are between $\lambda = 1.15$ and $\lambda = 2.0$. When idling and under no-load conditions, those figures rise to $\lambda > 10$.

Those excess-air factor figures represent the total masses of fuel and air in the cylinder. However, spontaneous ignition and pollutant formation are determined essentially by localized lambda levels.

Diesel engines operate with heterogeneous mixture formation and auto-ignition. It is not possible to achieve completely homogeneous mixing of the injected fuel with the air charge prior to or during combustion. Auto-ignition occurs a few degrees of crankshaft rotation after the point at which fuel injection starts (ignition lag).

Within the heterogeneous mixture encountered in a diesel engine, the localized excess-air factors can cover the entire range from $\lambda = 0$ (pure fuel) in the eye of the jet close to the injector to $\lambda = \infty$ (pure air) at the outer extremities of the spray jet. Closer examination of a single droplet of liquid fuel

reveals that around the outer zone of the droplet (vapor envelope), localized, combustible lambda levels of 0.3...1.5 occur (Figures 2 and 3). From this, it can be deduced that good atomization (large numbers of very small droplets), high levels of excess air and "moderate" motion of the air charge produce large numbers of localized zones with lean combustible lambda levels. The effect of this is that less soot and, in principle, less NO_X is produced during combustion.

Good atomization is achieved by high injection pressures (the highest currently used is over 2,000 bar). This results is a high relative velocity between the jet of fuel and the air in the cylinder which has the effect of scattering the fuel jet.

With a view to reducing engine weight and cost, the aim is to obtain as much power as possible from a given engine capacity. To achieve that aim, the engine "must" be run with a "small" air excess at high loads. But small air excesses increase emission levels. Therefore, they have to be limited, i.e. the fuel volume delivered must be precisely proportioned to match the available amount of air and the speed of the engine.

Low atmospheric pressures (e.g. at high altitudes) also require the fuel volume to be adjusted to the smaller amount of available air.

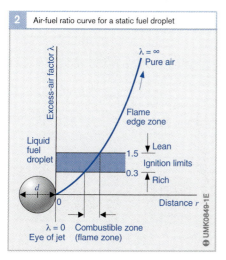

2 Air-fuel ratio curve for a static fuel droplet

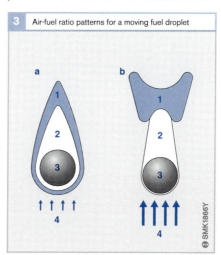

3 Air-fuel ratio patterns for a moving fuel droplet

Fig. 2
d   Droplet diameter (approx. 2...20 µm)

Fig. 3
a   Low relative velocity
b   High relative velocity

1   Flame zone
2   Vapor envelope
3   Fuel droplet
4   Air flow

Start of injection and delivery

Start of injection

The point at which injection of fuel into the combustion chamber starts has a decisive effect on the point at which combustion of the air/fuel mixture starts, and therefore on emission levels, fuel consumption and combustion noise. Consequently, injection timing plays a major role in optimizing engine performance characteristics.

The point at which injection of fuel starts is the position stated in degrees of crankshaft rotation relative to crankshaft top dead center (TDC) at which the nozzle opens and fuel starts to enter the combustion chamber.

The position of the piston relative to top dead center at that moment (as well as the shape of the intake port), determines the nature of the air flow inside the combustion chamber, and the density and temperature of the air. Accordingly, the degree of mixing of air and fuel is also dependent on start of injection. Thus, start of injection affects emissions such as soot, a product of incomplete combustion, nitrogen oxides (NO_X), unburned hydrocarbons (HC) and carbon monoxide (CO).

The start of injection requirements differ according to engine load (Figure 1). This fact demands load-dependent adjustment of the start of injection. The characteristic operating data of each engine is thus determined and stored electronically in the form of an engine data map. The engine data map plots the required start of injection points against engine load, speed and temperature. It also takes account of fuel-consumption considerations, pollutant-emission requirements and noise levels at any given power output (Figure 2).

Fig. 1
Example of an application:
α_N Optimum start of injection for emissions at no load, as NO_X emissions are lower under those conditions
α_V Optimum start of injection for emissions at full load, as HC emissions are lower under those conditions

Fig. 2
1 Cold start (<0 °C)
2 Full load
3 Medium load

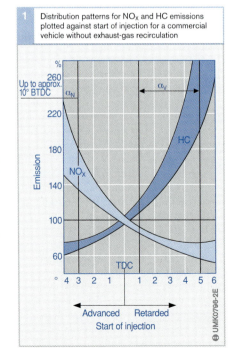

1 Distribution patterns for NO_X and HC emissions plotted against start of injection for a commercial vehicle without exhaust-gas recirculation

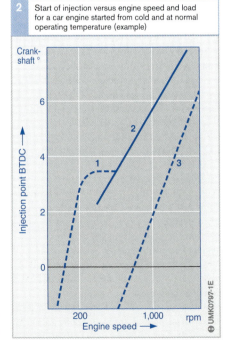

2 Start of injection versus engine speed and load for a car engine started from cold and at normal operating temperature (example)

Guide figures (EURO III)

On a diesel engine's data map, the optimum points of combustion start for low fuel consumption are in the range of 0...8° crankshaft before TDC. On that basis and based on the statutory exhaust-gas emission limits, the start of injection points are as follows:

Direct-injection car engines:
- No load: 2° crankshaft before TDC to 4° crankshaft after TDC
- Part load: 6° crankshaft before TDC to 4° crankshaft after TDC
- Full load: 6...15° crankshaft before TDC

Direct-injection commercial-vehicle engines (without exhaust-gas recirculation):
- No load: 4...12° crankshaft before TDC
- Full load: 3...6° crankshaft before TDC to 2° crankshaft after TDC

When the engine is cold, the start of injection for car and commercial-vehicle engines is 3...10° earlier. The duration of combustion at full load is 40...60° of crankshaft rotation.

Advanced start of injection

The highest final compression temperature is reached at TDC. If combustion is initiated a long way before TDC, the combustion pressure rises steeply and acts as a retarding force against the movement of the piston. The heat lost in the process diminishes the efficiency of the engine and therefore increases its fuel consumption. The steep increase in compression pressure also makes combustion much noisier.

An advanced start of injection increases the temperature in the combustion chamber. As a result, the NO_X emission levels rise while HC emissions are lower (Figure 1).

Retarded start of injection

A retarded start of injection under no-load conditions can result in incomplete combustion and therefore in the emission of unburned hydrocarbons (HC) since combustion takes place at a time when the temperature in the combustion chamber is dropping (Fig. 1).

The partially conflicting interdependence of specific fuel consumption and hydrocarbon emission levels on the one hand, and soot (black smoke) and NO_X emissions on the other, demand a trade-off combined with very tight tolerances when modifying the start of injection to suit a particular engine.

Minimizing blue and white smoke levels requires advanced start of injection and/or pre-injection when the engine is cold.

In order to keep noise and pollutant emissions at acceptable levels, a different start of injection is frequently necessary when the engine is running at part load than when it is at full power. The start-of-injection map (Figure 2) shows the inter relationship between the start of injection and engine temperature, load and speed for a car engine.

Start of delivery

In addition to start of injection, start of delivery is another aspect that is often considered. It relates to the point at which the fuel injection pump starts to deliver fuel to the injector. Since, on older fuel-injection systems and when the engine is not running, the start of delivery is easier to determine than the actual injection point, synchronization of the start of injection with the engine (particularly in the case of in-line and distributor injection pumps) is performed on the basis of the start of delivery. This is possible because there is a definite relationship between the start of delivery and the start of injection (injection lag[1])).

The time it takes for the pressure wave to travel from the high-pressure pump to the nozzle depends on the length of the pipe and produces an injection lag stated in degrees of crankshaft rotation that varies according to engine speed. The engine also has a longer ignition lag (in terms of crankshaft rotation) at higher speeds[2]. Both these effects must be compensated for – which is the reason why a fuel-injection system must be able to adjust the start of delivery/start of injection in response to engine speed, load and temperature.

[1] Time from start of fuel delivery to start of injection

[2] Time from start of injection to start of ignition

Injected-fuel quantity

The required fuel mass, m_e, in mg for an engine cylinder per power stroke is calculated using the following equation:

$$m_e = \frac{P \cdot b_e \cdot 33.33}{n \cdot z} \quad [\text{mg/stroke}]$$

where
P is the engine's power output in kW
b_e is the engine's specific fuel consumption in g/kWh
n is the engine speed in rpm and
z is the number of cylinders in the engine

The corresponding fuel volume (injected fuel quantity), Q_H, in mm³/stroke or mm³/injection cycle is then:

$$Q_H = \frac{P \cdot b_e \cdot 1,000}{30 \cdot n \cdot z \cdot \rho} \quad [\text{mm}^3/\text{stroke}]$$

Fuel density, ρ, in mg/mm³ is temperature-dependent.

It is evident from this equation that the engine's power output at a constant level of efficiency ($\eta \sim 1/b_e$) is directly proportional to the injected fuel quantity.

The mass of fuel injected by the fuel-injection system depends on the following variables:
- The fuel-metering cross-section of the nozzle
- The injection duration
- The variation over time of the pressure difference between the injection pressure and the pressure in the combustion chamber and
- The density of the fuel

At high pressures, the diesel fuel is compressible, i.e. it is, in fact, compressed. This affects the injected fuel quantity and must therefore be taken into account by the injection control system.

Variations in the injected-fuel quantity lead to fluctuations in the level of pollutant emissions and in the engine's power output. By the use of high-precision fuel-injection systems controlled by an electronic governor, the required injected fuel quantity can be delivered with a high degree of accuracy.

Figures 1 to 4
Engine:
Six-cylinder diesel commercial-vehicle engine with common-rail fuel injection

Operating conditions:
$n = 1,400$ rpm,
50 % power

In this example, the injection duration is varied by variation of the injection pressure

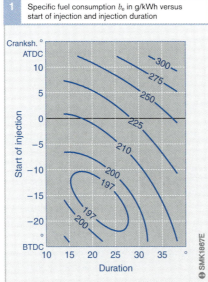

1 Specific fuel consumption b_e in g/kWh versus start of injection and injection duration

Cranksh.° ATDC — Start of injection (10, 5, 0, −5, −10, −15, −20, BTDC) — Duration (10, 15, 20, 25, 30, 35°)
Contour values: 300, 275, 250, 225, 210, 200, 197, 197, 200
SMK1867E

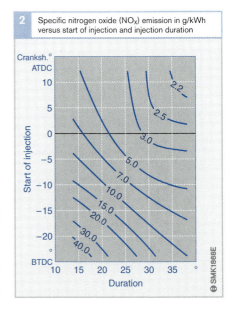

2 Specific nitrogen oxide (NO_X) emission in g/kWh versus start of injection and injection duration

Cranksh.° ATDC — Start of injection (10, 5, 0, −5, −10, −15, −20, BTDC) — Duration (10, 15, 20, 25, 30, 35°)
Contour values: 2.2, 2.5, 3.0, 5.0, 7.0, 10.0, 15.0, 20.0, 30.0, 40.0
SMK1868E

Injection characteristics

An engine's emission and fuel-consumption characteristics are very important considerations. For that reason, the following demands are placed on the fuel-injection system:
- Fuel injection must be precisely timed. Even small discrepancies have a substantial effect on fuel consumption, emission levels and combustion noise (Figures 1 to 4).
- It should be possible to vary the injection pressure as independently as possible to suit the demands of all engine - operating conditions (e.g. load, speed).
- The injection must be reliably terminated. Uncontrolled "post-injection" leads to higher emission levels.

The term "injection characteristics" refers to the pattern of the fuel quantity injected into the combustion chamber as a function of time.

Injection duration
One of the main parameters of the injection pattern is the injection duration. This refers to the period of time that the nozzle is open

and allows fuel to flow into the combustion chamber. It is specified in degrees of crankshaft or camshaft rotation, or in milliseconds. Different diesel combustion processes demand different injection durations as illustrated by the following examples (approximate figures at rated power):
- Direct-injection car engines: 32...38° of crankshaft rotation
- Indirect-injection car engines: 35...40° of crankshaft rotation and
- Direct-injection commercial-vehicle engines: 25...36° of crankshaft rotation

An injection duration of 30° of crankshaft rotation corresponds to 15° of camshaft rotation. In terms of time at an injection pump speed[1]) of 2,000 rpm, that is equal to an injection duration of 1.25 ms.

In order to minimize fuel consumption and soot emission, the injection duration must be defined on the basis of the engine operating conditions and the start of injection (Figures 1 and 4).

[1] Equal to half the engine speed on four-stroke engines

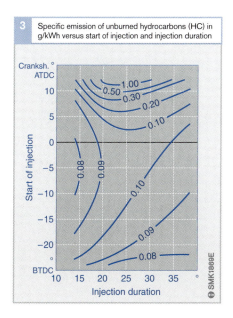

3 Specific emission of unburned hydrocarbons (HC) in g/kWh versus start of injection and injection duration

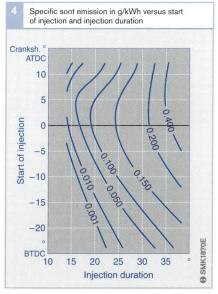

4 Specific soot emission in g/kWh versus start of injection and injection duration

Fig. 5
Adjustments aimed at low
NO$_X$ levels require starts
of injection close to TDC
at maximum load
(engines without exhaust-
gas recirculation). The
fuel delivery point is
significantly in advance
of the start of injection
and is dependent on
the injection system

1 Pre-injection (PI)
   phase
2 Main injection (MI)
   phase
3 Steep pressure
   gradient (common-
   rail system)
4 Two-stage pressure
   gradient (UPS with
   CCRS two-stage
   solenoid valve)
   (dual-spring nozzle-
   holder assemblies
   can produce a bath-
   tub needle lift curve
   [but not pressure
   gradient]. This re-
   duces combustion
   noise but not always
   soot emission levels.)
5 Gradual pressure
   gradient (conven-
   tional fuel injection)
6 Gradual pressure
   drop (in-line and
   distributor injection
   pumps)
7 Steep pressure drop
   (UIS, UPS, slightly
   less steep with
   common rail)
8 Advanced
   post-injection (PO)
9 Retarded
   post-injection
p_s Peak pressure
p_o Injector opening
   pressure
b Duration of
   combustion for main
   injection phase
v Duration of
   combustion for
   pre-injection phase
ZV Ignition lag for main
   injection phase
   without pre-injection

Injection pattern

Depending on the type of use for which the engine is intended, the following injection functions are required (Figure 5):

- *Pre-injection* (1) in order to reduce combustion noise and NO$_X$ emissions, especially on DI engines
- *Positive pressure gradient* during the main injection phase (3) in order to reduce NO$_X$ emissions on engines without exhaust-gas recirculation
- *Two-stage pressure gradient* (4) during the main injection phase in order to reduce NO$_X$ and soot emissions on engines without exhaust-gas recirculation
- *Constant high pressure* during the main injection phase (3, 7) in order to reduce soot emissions on engines with exhaust-gas recirculation
- *Post-injection* immediately following the main injection phase (8) in order to reduce soot emissions, or
- *Retarded post-injection* (9) of fuel as a reducing agent for an NO$_X$ accumulator-type catalytic converter and/or in order to raise the exhaust-gas temperature for regeneration of a particulate filter

Conventional injection pattern

With conventional fuel-injection systems, the pressure is generated continuously throughout the injection cycle by an injection pump. Thus, the speed of the pump has a direct effect on the fuel delivery rate and consequently on injection pressure.

In the case of port-controlled distributor and in-line injection pumps, the injection pattern consists exclusively of a main injection phase, i.e. without pre- or post-injection (Figure 5, Items 5 and 6).

With solenoid-valve controlled distributor injection pumps, pre-injection is also possible (1). On unit injector systems (UIS) for cars, pre-injection is currently controlled by hydromechanical means.

Pressure generation and delivery of the injected fuel quantity are interdependent by virtue of the link between the cam and the injection pump in conventional systems. This has the following consequences for the injection characteristics:

- Injection pressure increases with engine speed and injected fuel quantity (Figure 6)

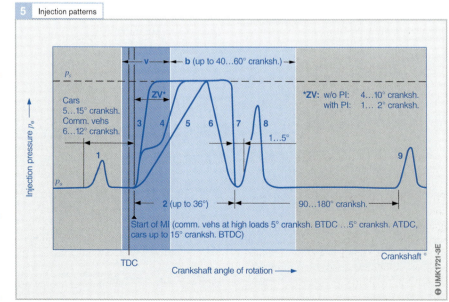

5 Injection patterns

© UMK1721-3E

- Injection pressure rises at the start of injection but drops again before the end of the injection period (as from the end of the fuel-delivery period) down to the injector closing pressure

The consequences of this are the following:
- Small injected fuel quantities are injected at low pressures and
- The injection pattern is approximately triangular, as is required for good combustion in an engine without exhaust-gas recirculation (shallow pressure gradient and therefore quiet combustion)

The determining factor for the stresses to which the components of an injection pump and its drive system are subjected is peak pressure. Peak pressure is also a measure of the quality of fuel atomization in the combustion chamber.

On indirect-injection engines (precombustion or swirl-chamber engines), throttling-pintle nozzles are used which produce a single jet of fuel and determine the shape of the injection pattern. This type of nozzle controls the outlet cross-section as a function of the needle lift. This produces a gradual increase in pressure and consequently, "quiet combustion".

Pre-injection
The pressure curve of an engine without pre-injection (Figure 7a) shows only a shallow gradient leading up to TDC in keeping with the compression. The gradient then rises steeply from the start of combustion. That rapid rise in pressure is the cause of the noisier combustion encountered on diesel engines without pre-injection.

Pre-injection enables a less abrupt rise in combustion pressure to be achieved. The ignition lag of the main injection quantity is very short. The pattern of combustion is affected in such a way that combustion noise, fuel consumption and – depending on the type of combustion – NO_X and HC emissions are reduced.

Pre-injection involves the injection of a small quantity of fuel (1...4 mm³) in advance of the main injection phase in order to "precondition" the combustion chamber. This has the following effects:
- The ignition lag of the main-injection phase is shortened and
- The combustion pressure gradient is less steep (Figure 7b)

Depending on the timing of the main injection phase and the gap between the pre-injection and main-injection phases, the specific fuel consumption will vary.

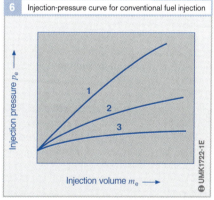

6 Injection-pressure curve for conventional fuel injection

Injection pressure p_e →

Injection volume m_e →

1
2
3

UMK1722-1E

Fig. 6
1 High engine speeds
2 Medium engine speeds
3 Low engine speeds

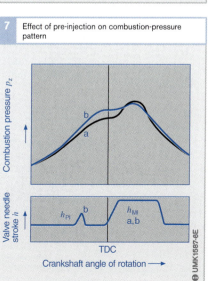

7 Effect of pre-injection on combustion-pressure pattern

Combustion pressure p_z →

Valve needle stroke h →

b
a

h_{PI} b h_{MI} a,b

TDC

Crankshaft angle of rotation →

UMK1687-8E

Fig. 7
a Without pre-injection
b With pre-injection

h_{PI} Needle lift during pre-injection
h_{MI} Needle lift during main injection

Post-injection

Retarded post-injection

Post-injection can be employed as a means of delivering a measured amount of reducing agent for a particular type of NO_X catalytic converter. The post-injection phase follows the main-injection phase during the ignition or exhaust stroke at a point up to 200° crankshaft after TDC. It introduces a precisely measured amount of fuel into the exhaust gas.

In contrast with the pre-injection and main-injection phases, the fuel injected is not burned but is merely vaporized by the heat of the exhaust gas. The resulting mixture of fuel and exhaust gas is expelled through the exhaust ports into the exhaust-gas system during the exhaust stroke. The fuel in the exhaust gas acts as a reducing agent for nitrogen oxides in suitable NO_X catalytic converters. As a result, the NO_X emission levels are moderately reduced.

Another means of reducing NO_X emissions is the NO_X accumulator-type catalytic converter (see chapter "Emission control systems").

Retarded post-injection can also be used to raise the exhaust temperature in an oxidation-type catalytic converter in order to assist regeneration on the part of a particulate filter.

Retarded post-injection can lead to thinning of the engine oil by the diesel fuel. It is therefore essential that the injection system is designed in consultation with the engine manufacturer.

Advanced post-injection

The common-rail fuel-injection system can perform post-injection immediately following the main-injection phase independently of any post-injection for an NO_X catalytic converter or particulate filter. In this case, the fuel is injected while combustion is still in progress. In that way, soot particles are re-burned and soot emissions can be reduced by 20...70%.

Camshaft-driven injection systems that are capable of post-injection are also under development.

Post-injection and dead volumes

Unintended post-injection has a particularly undesirable effect. Post-injection occurs when the nozzle momentarily re-opens after closing and allows "poorly conditioned" fuel to escape into the cylinder at a late stage in the combustion process. This fuel is not completely burned, or may not be burned at all, with the result that it is released into the exhaust gas as unburned hydrocarbons. Rapidly closing nozzles with a sufficiently high closing pressure and a low static pressure in the supply line can prevent this undesirable effect.

Fuel retained in the nozzle on the cylinder side of the needle-seal seats has a similar effect to post-injection. That dead volume runs into the cylinder after the combustion process has finished and also partially escapes into the exhaust gas. This fuel component similarly increases the level of unburned hydrocarbons in the exhaust gas (Figure 8). Sac-less nozzles in which the injection orifices are drilled into the needle-seal seats have the smallest dead volume.

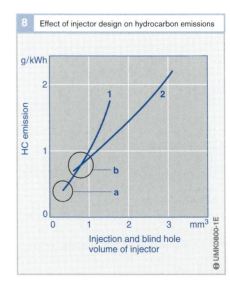

8   Effect of injector design on hydrocarbon emissions

Axis: g/kWh — HC emission (values 0, 1, 2)
Curves labeled 1 and 2; circles labeled a and b.
Injection and blind hole volume of injector (mm³): 0, 1, 2, 3

UMK0800-1E

Fig. 8
a   Injector without blind hole
b   Injector with micro-blind hole

1   Engine with 1 l/cylinder
2   Engine with 2 l/cylinder

Timing characteristics of fuel-injection systems

Taking as its example a radial-piston distributor injection pump (VP 44), Figure 9 illustrates how the cam on the cam ring initiates delivery of fuel by the pump and the fuel ultimately exits from the nozzle. It shows that the pressure and injection patterns vary greatly between the pump and the nozzle and are determined by the characteristics of the components that control injection (cam, pump, high-pressure valve, fuel line and nozzle). For that reason, the fuel-injection system must be precisely matched to the engine.

In all fuel-injection systems in which the pressure is generated by a pump piston (in-line injection pumps, unit injectors and unit pumps) the characteristics are similar. The common-rail system on the other hand behaves entirely differently.

Detrimental volume in conventional injection systems

The term "detrimental volume" refers to the volume of fuel in the high-pressure side of the fuel-injection system for an individual nozzle. This is made up of the high-pressure side of the fuel-injection pump, the high-pressure fuel lines and the nozzle.

Every time fuel is injected, the detrimental volume is pressurized and depressurized. As a result, compression losses occur and a fuel injection lag is produced. The fuel volume inside the pipes is compressed by the dynamic processes generated by the pressure wave.

The greater the detrimental volume, the poorer the hydraulic efficiency of the fuel-injection system. A major consideration when developing a fuel-injection system is therefore to keep the detrimental volume as small as possible. The unit injector system has the smallest detrimental volume.

In order to guarantee consistency of control for the benefit of the engine, the detrimental volume must be equal for all cylinders.

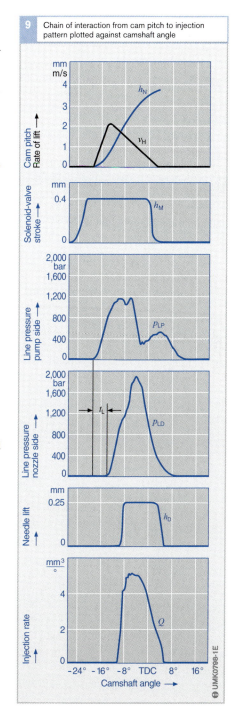

9 Chain of interaction from cam pitch to injection pattern plotted against camshaft angle

Fig. 9
Example of radial-piston distributor injection pump (VP 44) at full load without pre-injection

t_L Fuel transit time in pipe

Injection characteristics of common-rail system

A high-pressure pump generates the fuel-rail pressure independently of the injection cycle. The fuel-rail pressure remains virtually constant for the entire injection cycle (Figure 10). Because of the almost uniform delivery pattern, the high-pressure pump can be significantly smaller and designed for a lower peak drive torque.

Short pipes join the fuel rail to the injectors. Since the control unit controls the injectors, start of injection and end of injection are infinitely variable in engine applications. Multiple pre- and post-injection phases are possible.

For a given system pressure, the injected fuel quantity is proportional to the length of time that the injector valve is open and entirely independent of the engine or pump speed (time-based injection system).

Thus, start of injection, duration and pressure can be individually regulated to suit all engine operating points and optimized to the engine's operating requirements. They are controlled by the crankshaft-position/time-based system of the electronic diesel control (EDC) system.

Injection pressure

The process of fuel injection uses the pressure in the fuel system to induce the flow of fuel through the injector jets. A high fuel-system pressure results in a high rate of fuel outflow at the nozzle. Fuel atomization is caused by the collision of the turbulent jet of fuel with the air inside the combustion chamber. Therefore, the higher the relative velocity between fuel and air, and the higher the density of the air, the more finely the fuel is atomized. By clever dimensioning of the high-pressure fuel line, the injection pressure at the nozzle can be higher than in the fuel-injection pump.

Direct-injection (DI) engines

In diesel engines with direct injection, the speed of movement of the air inside the combustion chamber is relatively slow as it only moves as a result of its mass inertia (i.e. the air "attempts" to maintain the velocity at which it enters the cylinder; swirl effect). This effect is assisted by the movement of the piston. The degree of swirl increases as the piston approaches TDC.

Fig. 10
p_r Fuel-rail pressure
p_o Nozzle-opening
       pressure

Fig. 11
Direct-injection engine,
engine speed 1,200 rpm,
mean pressure 16.2 bar

p_e Injection pressure
α_S Start of injection
       after TDC

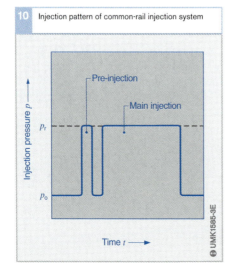

10   Injection pattern of common-rail injection system

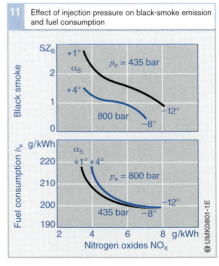

11   Effect of injection pressure on black-smoke emission and fuel consumption

In a direct-injection engine, fuel is injected into the combustion chamber at high pressure. Injection pressures as high as 2,000 bar enable smoke and particulate emissions to be substantially reduced. Modern direct-injection systems now generate full-load peak pressures of 1,000...2,050 bar for car engines and 1,000...1,800 bar for commercial vehicles.

However, peak pressure is available only at the higher engine speeds (except with the common-rail system). Overall, a good torque curve combined with low smoke emission demands a high injection pressure when the engine is under maximum load at low speeds. Based on those conditions, the target figures at maximum torque for cars and commercial vehicles are in the range 800...1,400 bar.

Indirect-injection (IDI) engines

Indirect-injection engines, in which rising combustion pressure propels the air/fuel mixture out of the swirl/precombustion chamber, have high rates of air flow in the swirl/precombustion chamber and in the channel connecting it to the main combustion chamber. In this type of engine, no advantage is gained by increasing injection pressure above about 450 bar.

Injection direction and number of injection jets

Direct-injection engines

Diesel engines with direct injection generally have hole-type nozzles with between 4 and 10 injection orifices (most commonly 6...8 injection orifices, see chapter "Nozzles") arranged as centrally as possible. The injection direction is very precisely matched to the combustion chamber. Divergences of the order of only 2 degrees from the optimum injection direction lead to a detectable increase in black-smoke emission and fuel consumption.

Indirect-injection engines

Indirect-injection engines use pintle nozzles with only a single injection jet. The nozzle injects the fuel into the precombustion or swirl chamber in such a way that the glow plug is just within the injection jet. The injection direction is matched precisely to the combustion chamber. Inaccuracies in injection direction result in poorer utilization of combustion air and therefore to an increase in black smoke and hydrocarbon emissions.

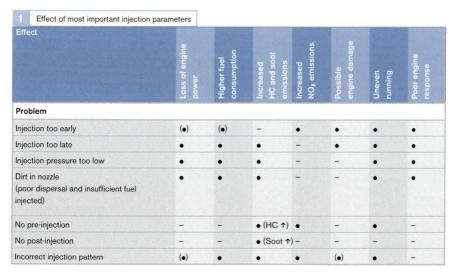

1 Effect of most important injection parameters							
Effect / **Problem**	Loss of engine power	Higher fuel consumption	Increased HC and soot emissions	Increased NOₓ emissions	Possible engine damage	Uneven running	Poor engine response
Injection too early	(•)	(•)	–	•	•	•	•
Injection too late	•	•	•	–	•	•	•
Injection pressure too low	•	•	•	–	–	•	•
Dirt in nozzle (poor dispersal and insufficient fuel injected)	•	•	•	–	–	•	•
No pre-injection	–	–	• (HC ↑)	•	–	•	–
No post-injection	–	–	• (Soot ↑)	–	–	–	–
Incorrect injection pattern	(•)	•	•	•	(•)	•	–

Table 1
This table illustrates how greatly fuel-injection parameters affect engine characteristics. Only a well matched and precisely functioning fuel-injection system can guarantee quiet, low-emission and economical diesel-engine operation.

Overview of diesel fuel-injection systems

Diesel engines are characterized by high fuel economy. Since the first volume-production fuel-injection pump was introduced by Bosch in 1927, fuel-injection systems have experienced a process of continual advancement.

Diesel engines are used in a wide variety of design for many different purposes (Figure 1 and Table 1), for example
- To drive mobile power generators (up to approx. 10 kW/cylinder)
- As fast-running engines for cars and light commercial vehicles (up to approx. 50 kW/cylinder)
- As engines for construction-industry and agricultural machinery (up to approx. 50 kW/cylinder)
- As engines for heavy trucks, omnibuses and tractor vehicles (up to approx. 80 kW/cylinder)
- To drive fixed installations such as emergency power generators (up to approx. 160 kW/cylinder)
- As engines for railway locomotives and ships (up to 1,000 kW/cylinder)

Requirements

Ever stricter statutory regulations on noise and exhaust-gas emissions and the desire for more economical fuel consumption continually place greater demands on the fuel-injection system of a diesel engine.

Basically, the fuel-injection system is required to inject a precisely metered amount of fuel at high pressure into the combustion chamber in such a way that it mixes effectively with the air in the cylinder as demanded by the type of engine (direct or indirect-injection) and its present operating status. The power output and speed of a diesel engine is controlled by means of the injected fuel volume as it has no air intake throttle.

Mechanical control of diesel fuel-injection systems is being increasingly displaced by Electronic Diesel Control (EDC) systems. All new diesel-injection systems for cars and commercial vehicles are electronically controlled.

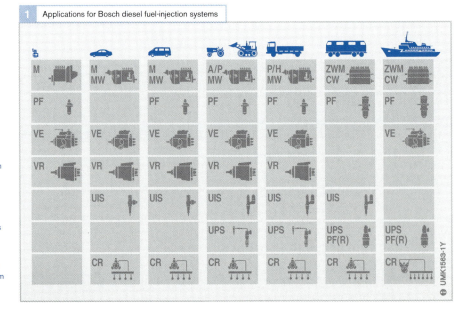

1 Applications for Bosch diesel fuel-injection systems

Fig. 1
M, MW,
A, P, H,
ZWM,
CW In-line fuel-injection pumps of increasing size
PF Discrete fuel-injection pumps
VE Axial-piston pumps
VR Radial-piston pumps
UPS Unit pump system
UIS Unit injector system
CR Common-rail system

Table 1   Properties and characteristic data of the most important fuel-injection systems for diesel engines

Type (Fuel-injection system)	Type of use P: Cars and light commercials N: Trucks and buses O: Off-road vehicles [1] S: Ships/trains	Injected volume per stroke/injection cycle mm³	Max. permissible pressure at jet bar (0.1 MPa)	Pre-injection / Post-injection PI / PO	Control method h: Hydraulic m: Mechanical em: Electromechanical Mv: Solenoid valve	Direct injection / Indirect injection DI / IDI	Number of cylinders	Max. rated speed rpm	Max. power output per cylinder kW
In-line injection pumps									
M	P, O	60	550	–	m, em	IDI	4 … 6	5,000	20
A	O	120	750	–	m	DI/IDI	2 … 12	2,800	27
MW[8]	P, N, O	150	1,100	–	m	DI	4 … 8	2,600	36
P3000	N, O	250	950	–	m, em	DI	4 … 12	2,600	45
P7100	N, O	250	1,200	–	m, em	DI	4 … 12	2,500	55
P8000	N, O	250	1,300	–	m, em	DI	6 … 12	2,500	55
P8500	N, O	250	1,300	–	m, em	DI	4 … 12	2,500	55
H1	N	240	1,300	–	em	DI	6 … 8	2,400	55
H1000	N	250	1,350	–	em	DI	5 … 8	2,200	70
P10	S, O	800	1,200	–	m, em, h	DI/IDI	6 … 12	2,400	140
ZW (M)	S, O	900	950	–	m, em, h	DI/IDI	4 … 12	2,400	160
P9	S, O	1,200	1,200	–	m, em, h	DI/IDI	6 … 12	2,000	180
CW	S, O	1,500	1,000	–	m, em, h	DI/IDI	6 … 10	1,800	200
Axial-piston pumps									
VE..F	P	70	350	–	m	IDI	3 … 6	4,800	25
VE..F	P	70	1,250	–	m	DI	4 … 6	4,400	25
VE..F	N, O	125	800	–	m	DI	4, 6	3,800	30
VP37 (VE..EDC)	P	70	1,250	–	em[7]	DI	3 … 6	4,400	25
VP37 (VE..EDC)	O	125	800	–	em[7]	DI	4, 6	3,800	30
VP30 (VE..MV)	P	70	1,400	PI	Mv[7]	DI	4 … 6	4,500	25
VP30 (VE..MV)	O	125	800	PI	Mv[7]	DI	4, 6	2,600	30
Radial-piston pumps									
VP44 (VR)	P	85	1,900	PI	Mv[7]	DI	4, 6	4,500	25
VP44 (VR)	N	175	1,500	–	Mv[7]	DI	4, 6	3,300	45
Discrete/cylinder-pump systems									
PF(R)…	O	13 … 120	450 … 1,150	–	m, em	DI/IDI	Any	4,000	4 … 30
PF(R)… large-scale diesel	P, N, O, S	150 … 18,000	800 … 1,500	–	m, em	DI/IDI	Any	300 … 2,000	75 … 1,000
UIS P1	P	60	2,050	PI	Mv	DI	5[2, 2a]	4,800	25
UIS 30	N	160	1,600	–	Mv	DI	8[2]	4,000	35
UIS 31	N	300	1,600	–	Mv	DI	8[2]	2,400	75
UIS 32	N	400	1,800	–	Mv	DI	8[2]	2,400	80
UPS 12	N	180	1,600	–	Mv	DI	8[2]	2,400	35
UPS 20	N	250	1,800	–	Mv	DI	8[2]	3,000	80
UPS (PF..MV)	S	3,000	1,600	–	Mv	DI	6 … 20	1,000	450
Common-rail injection systems									
CR 1st generation	P	100	1,350	PI, PO[3]	Mv	DI	3 … 8	4,800[4]	30
CR 2nd generation	P	100	1,600	PI, PO[5]	Mv	DI	3 … 8	5,200	30
CR	N, S	400	1,400	PI, PO[6]	Mv	DI	6 … 16	2,800	200

Table 1

[1] Fixed-installation engines, construction and agricultural machinery
[2] Larger numbers of cylinders are also possible with two control units
[2a] EDC 16 and above: 6 cylinders
[3] PI up to 90° BTDC, PO possible
[4] Up to 5,500 rpm when overrunning
[5] PI up to 90° BTDC, PO up to 210° ATDC
[6] PI up to 30° BTDC, PO possible
[7] Electrohydraulic injection timing adjustment using solenoid valve
[8] This type of pump is no longer used with new systems

Designs

The function of a fuel-injection system for a diesel engine is to inject the fuel into the combustion chamber at high pressure, in the precise quantity required, and at precisely the right moment.

The nozzle projects into either the swirl/precombustion chamber or the main combustion chamber, depending on the type of engine. The nozzle opens – if it is not externally controlled – at a specific opening pressure that is set to suit the engine and the fuel-injection system. It closes when the fuel pressure drops. The essential difference between the different types of fuel-injection system is the method by which they generate the fuel pressure.

Because of the high pressures involved, all individual components are made to high-precision tolerances from high-strength materials. All components must be precisely matched to one another.

Electronic control concepts enable systems to perform a variety of supplementary functions (e.g. active surge damping, smooth-running control, cruise control and boost-pressure control).

In-line fuel-injection pumps
Standard in-line fuel-injection pumps (Type PE)

In-line fuel-injection pumps (Figure 1) have a separate pump unit consisting of a cylinder (1) and plunger (4) for each cylinder of the engine. The pump plunger is moved in the delivery direction (in this case upwards) by the camshaft (7) integrated in the injection pump and driven by the engine, and is returned to its starting position by the plunger spring (5). The individual pump units are generally arranged in-line (hence the name in-line fuel-injection pump).

The stroke of the plunger is invariable. The point at which the top edge of the plunger closes off the inlet port (2) on its upward stroke marks the beginning of the pressure generation phase. This point is referred to as the start of delivery. The plunger continues to move upwards. The fuel pressure therefore increases, the nozzle opens and fuel is injected into the combustion chamber.

When the helix (3) of the plunger clears, the inlet port, fuel can escape and pressure is

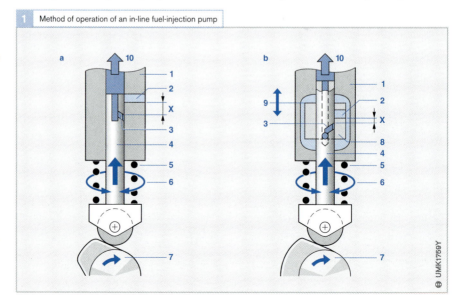

1   Method of operation of an in-line fuel-injection pump

© UMK1759Y

2 Method of operation of port-controlled axial-piston distributor injection pump

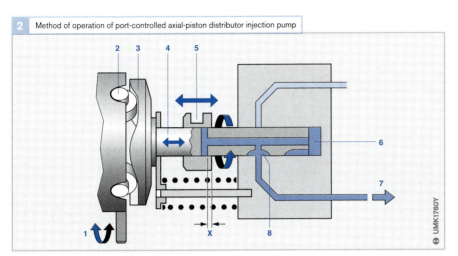

Fig. 2
1   Injection timing
    adjustment range
    on roller ring
2   Roller
3   Cam plate
4   Axial piston
5   Control sleeve
6   High-pressure
    chamber
7   Fuel outflow to
    nozzle
8   Metering slot

X   Effective stroke

lost. The nozzle closes and fuel injection ceases.

Plunger travel between the points at which the inlet port is closed and opened is termed the effective stroke (X).

The position of the helix can be altered by means of a control rod (6). This alters the effective stroke and therefore the injected-fuel quantity. The control rod is controlled by a mechanical governor or an electrical actuator mechanism.

Control-sleeve in-line fuel-injection pump

This type of in-line fuel-injection pump differs from a conventional type by virtue of a control sleeve (8) which slides over the pump plunger. It allows plunger lift to port closing – that is the distance travel by the plunger before it closes off the inlet port – to be altered by means of an actuator shaft (9). This changes the start of delivery.

The control-sleeve in-line fuel-injection pump thus has an additional degree of independent control in comparison with a Type PE standard in-line fuel-injection pump – it allows the start of injection to be varied independently of engine speed.

Distributor injection pumps

Distributor injection pumps have only one pump unit that serves all cylinders (Figures 2 and 3). A vane pump forces the fuel into the high-pressure chamber (6). High-pressure generation is performed by an axial piston (Figure 2, Item 4) or several radial pistons (Figure 3, Item 4). A rotating central distributor piston opens and closes metering slots (8) and spill ports, thereby distributing the fuel to the individual cylinders of the engine (7). The injection duration can be varied by means of a control sleeve (Figure 2, Item 5) or a high-pressure solenoid valve (Figure 3, Item 5).

Axial-piston distributor pumps

A rotating cam plate (Figure 2, Item 3) is driven by the engine. The number of cam lobes on the underneath of the cam plate is equal to the number of cylinders in the engine. They travel over rollers (2) on the roller ring and thus cause the distributor piston to describe a rotating as well as a lifting action. In the course of each rotation of the drive shaft, the piston accordingly completes a number of strokes equal to the number of engine cylinders to be supplied.

3   Method of operation of solenoid-valve controlled radial-piston distributor injection pump

In a Type VE port-controlled axial-piston distributor pump with mechanical governor or electronically controlled actuator mechanism, a control sleeve (5) determines the effective stroke, thereby controlling the injected-fuel quantity.

The timing device can vary the pump's start of delivery by turning the roller ring (1).

Radial-piston distributor injection pump

Instead of the cam plate used on the axial-piston pump, high-pressure is generated by a radial-piston pump with a cam ring (Figure 3, Item 3) and two to four radial pistons (4). Radial-piston pumps can generate higher injection pressures than axial-piston pumps. However, they have to be capable of withstanding greater mechanical stresses.

The cam ring is rotated by the timing device (1). With radial-piston distributor pumps, the start of injection and start of delivery are always controlled by solenoid valve.

Solenoid-valve controlled distributor injection pumps

With this type of distributor injection pump, an electronically controlled high-pressure solenoid valve (5) meters the injected-fuel quantity and varies the start of injection. When the solenoid valve is closed, pressure can build up in the high-pressure chamber (6). When it is open, the fuel escapes so that no pressure build-up occurs and therefore fuel injection is not possible. One or two electronic control units (pump control unit and engine control unit) generate the control and regulation signals.

Discrete cylinder systems

Type PF discrete injection pumps

Type PF discrete injection pumps are used primarily on marine engines, locomotive engines, construction machinery and small-scale engines. They are also suitable for use with high-viscosity heavy oils.

They operate in the same way as Type PE in-line fuel-injection pumps. But – in common with all discrete cylinder systems – discrete fuel-injection pumps do not have their own camshaft (externally driven). The cams which drive the discrete fuel-injection pumps are on the same camshaft that operates the engine valvegear. On large-scale engines, the hydro-mechanical or electronic control system is mounted directly on the engine block. It adjusts the injected-fuel quantity by means of a linkage integrated in the engine. Due to the fact that the pump is linked directly to the engine camshaft, it is not possible to vary fuel-injection timing by adjusting the camshaft. In this case, an adjustable roller can be used to provide an adjustment range of a few degrees. Control by solenoid valves is also possible.

Unit injector system (UIS)
In a unit injector system (UIS), the fuel-injection pump and nozzle form a single unit (Figure 4). There is a unit injector for each cylinder fitted in the cylinder head. It is actuated either directly by a tappet or indirectly by a rocker arm driven by the engine camshaft.

Since there are no high-pressure fuel lines, a significantly higher fuel-injection pressure (as much as 2,050 bar) is possible than is achievable with in-line or distributor pumps.

The fuel-injection parameters are calculated by an electronic control unit and controlled by opening and closing the high-pressure solenoid valve (3).

Unit pump system (UPS)
The modular unit pump system (UPS) operates on the same principle as the unit injector system (Figure 5). In contrast with the unit injector system, however, the nozzle-and-holder assembly (2) and the fuel-injection pump are linked by a short high-pressure line (3) specifically designed for the system components. This separation of high-pressure generation and nozzle-and-holder assembly allows for more straightforward attachment to the engine. There is one unit pump assembly (fuel-injection pump, fuel line and nozzle-and-holder assembly) for each cylinder of the engine. The unit pump assemblies are driven by the engine camshaft (6).

As with the unit injector system, the unit pump system uses an electronically controlled fast-switching high-pressure solenoid valve (4) to regulate injection duration and start of injection.

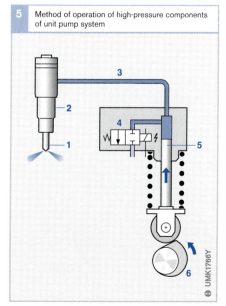

4	Method of operation of high-pressure components of unit injector system

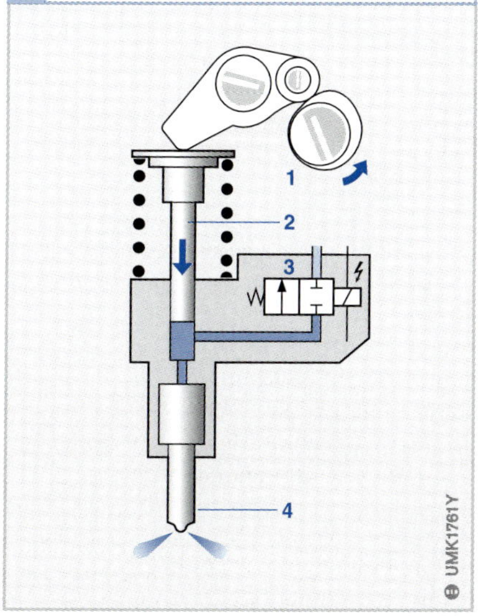

5	Method of operation of high-pressure components of unit pump system

UMK1781Y

UMK1766Y

Fig. 4
1 Drive cam
2 Pump plunger
3 High-pressure solenoid valve
4 Nozzle

Fig. 5
1 Nozzle
2 Nozzle-and-holder assembly
3 High-pressure fuel line
4 High-pressure solenoid valve
5 Pump plunger
6 Drive cam

Common-rail (CR) system

In the common-rail accumulator fuel-injection system, the functions of pressure generation and fuel injection are separated (Figure 6). The injection pressure is largely independent of engine speed or injected-fuel quantity, and is generated and controlled by a high-pressure pump (1). The pressure is held in a pressure accumulator, the "fuel rail" (2).

This system thus offers the greatest degree of flexibility in the choice of fuel-injection parameters.

There is a nozzle (4) fitted in each cylinder of the engine. Fuel injection is effected by opening and closing the high-pressure solenoid valve (3). Start of injection and injected-fuel quantity are calculated by an electronic control unit.

Fig. 6
1   High-pressure pump
2   Fuel rail (pressure
    accumulator)
3   High-pressure
    solenoid valve
4   Nozzle
5   Nozzle

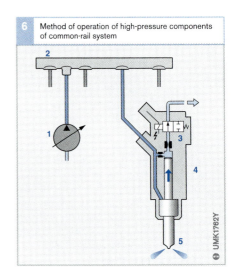

6   Method of operation of high-pressure components of common-rail system

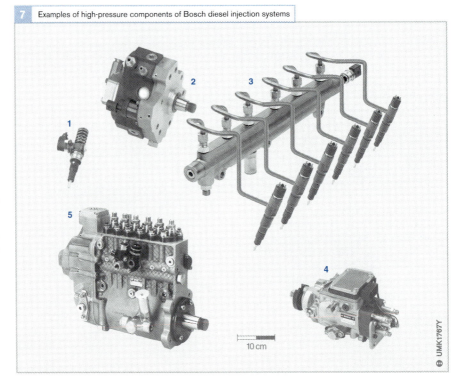

7   Examples of high-pressure components of Bosch diesel injection systems

10 cm

Fig. 7
1   Unit injector Type
    P1 (cars)
2   Common-rail high-
    pressure pump Type
    CP3 (commercial
    vehicles)
3   Fuel rail and nozzles
    (common-rail system
    for commercial
    vehicles)
4   Distributor injection
    pump Type VP30
    (cars)
5   Control-sleeve
    in-line fuel-injection
    pump Type RP39
    (commercial
    vehicles)

▶ History of diesel fuel injection

Development by Bosch of a fuel-injection system for diesel engines started in 1922. The technological omens were good: Bosch had experience with internal-combustion engines, its production systems were highly advanced and, above all, expertise developed in the production of lubrication pumps could be utilized. Nevertheless, this step was still a substantial risk for Bosch as there were still many difficulties to be overcome.

The first volume-production fuel-injection pumps appeared in 1927. At the time, the level of precision of the product was unmatched. They were small, light, and enabled diesel engines to run at higher speeds. These in-line fuel-injection pumps were used on commercial vehicles from 1932 and in cars from 1936. Since that time, the technological advancement of the diesel engine and its fuel-injection systems has continued unabated.

In 1962, the distributor injection pump with automatic timing device developed by Bosch gave the diesel engine an additional boost. More than two decades later, many years of intensive development work at Bosch culminated in the arrival of the electronically controlled diesel fuel-injection system.

The pursuit of ever more precise metering of minute volumes of fuel delivered at exactly the right moment coupled with the aim of increasing the injection pressure is a constant challenge for developers. This has led to many more innovations in the design of fuel-injection systems (see graphic).

In terms of fuel consumption and energy efficiency, the compression-ignition engine remains the benchmark.

New fuel-injection systems have helped to further exploit its potential. In addition, engine performance has been continually improved while noise and exhaust-gas emissions have been consistently lowered.

▶ Milestones in diesel fuel injection

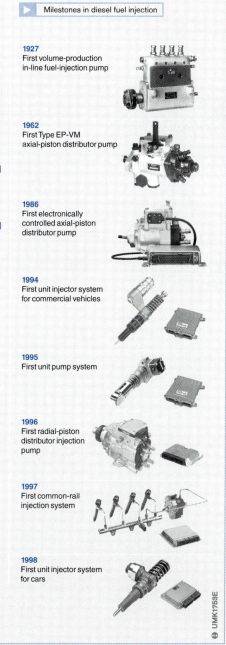

1927
First volume-production
in-line fuel-injection pump

1962
First Type EP-VM
axial-piston distributor pump

1986
First electronically
controlled axial-piston
distributor pump

1994
First unit injector system
for commercial vehicles

1995
First unit pump system

1996
First radial-piston
distributor injection
pump

1997
First common-rail
injection system

1998
First unit injector system
for cars

UMK1763E

Fuel supply system (low-pressure stage)

The job of the fuel supply system is to store the fuel required, to filter it and to supply it to the fuel-injection installation at a specific supply pressure under all operating conditions. For some applications, the fuel return flow is also cooled.

The essential components of the fuel supply system are as follows:
- The fuel tank (Figure 1, Item 1)
- The preliminary filter (except UIS and cars) (2)
- The control unit cooler (optional) (3)
- The presupply pump (optional, and may be inside the fuel tank on cars) (4)
- The fuel filter (5)
- The main presupply pump (low pressure) (6)
- The pressure-control valve (overflow valve) (7)
- The fuel cooler (optional) (9)
- The low-pressure fuel lines

Some of those components may be integrated in a single assembly (e.g. presupply pump and pressure limiter). In axial and radial-piston distributor injection pump systems, and in the common-rail system, the presupply pump is integrated in the high-pressure pump.

Fuel tank

The fuel tank stores the fuel. It has to be corrosion-resistant and leakproof to a pressure equivalent to double the system pressure and at least 30 kPa (0.3 bar). Any gauge pressure must be relieved automatically by suitable vents or safety valves. When the vehicle is negotiating corners, inclines or bumps, fuel must not escape past the filler cap or leak out of the pressure-relief vents or valves. The fuel tank must be fitted in a position where it is sufficiently distant from the engine to ensure that fuel will not ignite in the event of an accident.

Fuel lines

The fuel lines for the low-pressure stage can be either metal lines or flexible, fire-resistant lines with braided steel armor. They must be routed so as to avoid contact with moving components that might damage them and in such a way that any leak fuel or evaporation cannot collect or ignite. The function of the fuel lines must not be impaired by twisting of the chassis, movement of the engine or any other similar effects. All parts that carry fuel must be protected from levels of heat likely to have a negative effect on the operation of the system. On busses, fuel lines must not be routed through the passenger compartment or cockpit and the fuel system must not be gravity-fed.

1 Fuel-supply components (low-pressure stage)

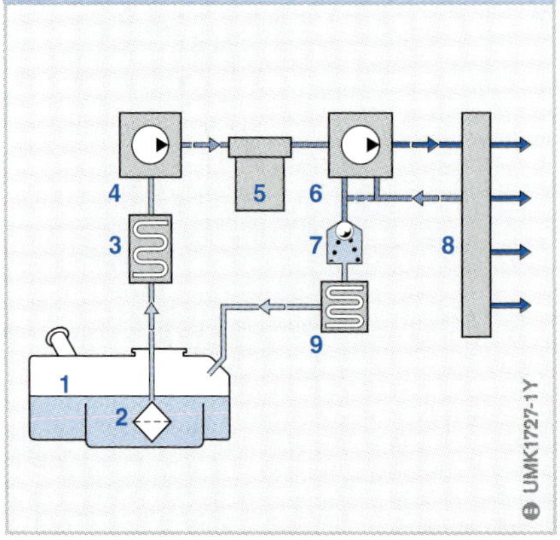

Fig. 1
1 Fuel tank
2 Preliminary filter
3 Control unit cooler
4 Presupply pump with non-return valve
5 Fuel filter
6 Main presupply pump
7 Pressure-control valve (UIS, UPS)
8 Fuel-distribution line (UIS, cars)
9 Fuel cooler (UIS, UPS, CR)

Diesel fuel filter

The job of the diesel fuel filter is to reduce contamination of the fuel by suspended particles. It therefore ensures that the fuel meets a minimum purity standard before it passes through components in which wear is critical. The fuel filter must also be capable of accumulating an adequate quantity of particles in order that servicing intervals are sufficiently long. If a filter clogs up, the fuel delivery quantity is restricted and the engine performance then dwindles.

The high-precision fuel-injection equipment used on diesel engines is sensitive to even minute amounts of contamination. High levels of protection against wear are therefore demanded in order to ensure that the desired levels of reliability, fuel consumption and exhaust-gas emissions are maintained over the entire life of the vehicle (1,000,000 km in the case of commercial vehicles). Consequently, the fuel filter must be designed to be compatible with the fuel-injection system with which it is used.

For cases where particularly exacting demands are placed on wear protection and/or maintenance intervals, there are filter systems consisting of a preliminary filter and a fine filter.

Design variations
The following functions are used in combination:

Preliminary filter for presupply pump
The preliminary filter (Figure 1, Item 2) is generally a strainer-type filter with a mesh size of 300 μm that is used in addition to the fuel filter proper (5).

Main filter
Easy-change filters (Figure 2) with spiral vee-shaped or wound filter elements (3) are widely used. They are screw-mounted to a filter console. In some cases, two filters connected in parallel (greater accumulation capacity) or in series (multistage filter to increase filtration rate, or fine filter with

preliminary filter) may be used. The replaceable-element filter is also becoming increasingly popular.

Water separator
Fuel may contain emulsified or free water (e.g. condensation caused by temperature change) which must be prevented from entering the fuel-injection equipment.

Because of the different surface tensions of fuel and water, water droplets form on the filter element (coalescence). They then collect in the water accumulation chamber (8). Free water can be removed by the use of a discrete water separator in which water droplets are separated out by centrifugal force. Conductivity sensors are used to monitor the water level.

Fuel preheating
Preheating of the fuel prevents clogging of the filter pores by paraffin crystals in cold weather. The most common methods use an electric heater element, the engine coolant or recirculated fuel to heat the fuel supply.

Manual priming pumps
These are used to prime and vent the system after the filter has been changed. They are generally integrated in the filter cover.

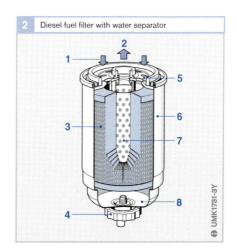

2   Diesel fuel filter with water separator

Fig. 2
1 Inlet
2 Outlet
3 Filter element
4 Water drain plug
5 Cover plate
6 Housing
7 Supporting tube
8 Water accumulation
   chamber

UMK1781-3Y

Fuel-supply pump

The fuel-supply pump in the low-pressure stage (the so-called presupply pump) is responsible for maintaining an adequate supply of fuel to the high-pressure components. This applies
● irrespective of operating state,
● with a minimum of noise,
● at the necessary pressure, and
● throughout the vehicle's complete service life.

In the axial-piston and radial-piston distributor pumps, a vane-type pump is used as the presupply pump and is integrated directly in the injection pump.

In the UIS/UPS injection systems, the presupply pump draws fuel out of the vehicle's fuel tank and continuously delivers the correct quantity (fuel for injection and fuel for purging) in the direction of the high-pressure injection system (60...200 l/h, 300...700 kPa). Many pumps bleed themselves automatically so that starting is possible even when the tank has been run dry before filling again.

There are three designs:
● Electric fuel pump (as used in passenger cars)
● Mechanically driven gear-type fuel pumps and
● Tandem fuel pumps (passenger-car UIS)

Electric fuel pump EKP

The electric fuel pump (Figs. 1 and 2) is only used in passenger cars and light commercial vehicles. Within the system-monitoring framework, in addition to fuel delivery it is also responsible for cutting off the supply of fuel if this is necessary in an emergency.

Electric fuel pumps are available as in-line or in-tank versions. In-line pumps are fitted to the vehicle's body platform outside the fuel tank in the fuel line between tank and fuel filter. In-tank pumps on the other hand are mounted in the fuel tank itself using a special mounting which usually also incorporates a suction-side strainer, a fuel-level indicator, a swirl pot which acts as a fuel reservoir, and electrical and hydraulic connections to the outside.

Starting with the engine cranking process, the electric fuel pump runs continuously independent of engine speed. This means that it permanently delivers fuel from the fuel tank and through a fuel filter to the fuel-injection system. Excess fuel flows back to the tank through an overflow valve.

A safety circuit is provided to prevent the delivery of fuel should the ignition be on with the engine stopped.

An electric fuel-supply pump is comprised of three function elements inside a common housing:

Pumping element (Fig. 1, Pos. A)
There are a variety of different pumping-element versions available depending upon the fuel pump's particular field of application. Roller-cell pumps (RZP) are normally used for diesel applications.

The roller-cell pump (Fig. 2) is a positive-displacement pump consisting of an eccentrically located base plate (4) in which a slotted rotor (2) is free to rotate. There is a movable roller in each slot (3) which, when the rotor rotates, is forced outwards against the outside roller path and against the driving flanks of the slots by centrifugal force and the pressure of the fuel. The result is that the

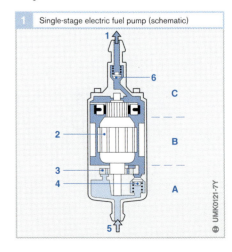

1   Single-stage electric fuel pump (schematic)

UMK0121-7Y

Fig. 1
A Pumping element
B Electric motor
C End cover

1 Pressure end
2 Motor armature
3 Pumping element
4 Pressure limiter
5 Suction end
6 Non-return valve

rollers now act as rotating seals, whereby a chamber is formed between the rollers of adjacent slots and the roller path. The pumping effect is due to the fact that once the kidney-shaped intake opening (1) has closed, the chamber volume reduces continuously.

Electric motor (Fig. 1, Pos. B)

The electric motor comprises a permanent-magnet system and an armature (2). Design is determined by the required delivery quantity at the given system pressure. The electric motor is permanently flushed by fuel so that it remains cool. This design permits high motor performance without the necessity for complicated sealing elements between pumping element and electric motor.

End cover (Fig. 1, Pos. C)

The end cover contains the electrical connections as well as the pressure-side hydraulic connection. A non-return valve (6) is incorporated to prevent the fuel lines emptying once the fuel pump has been switched off. Interference-suppression units can also be fitted in the end cover.

Gear-type fuel pump

This gear-type fuel pump (Fig. 3) is used to supply the fuel-injection modules of the single-plunger injection systems (passenger cars) and of the Common Rail System (for passenger cars, commercial vehicles, and off-road vehicles). It is directly attached to the engine, and in the case of Common Rail is integrated in the high-pressure pump. Common forms of drive are via coupling, gearwheel, or toothed belt.

The main components are two counter-rotating gearwheels which mesh with each other when rotating, whereby fuel is trapped in the chambers formed between the gear teeth and transported from the intake (suction) side (1) to the outlet (pressure) side (3). The line of contact between the rotating gearwheels provides the seal between the suction and pressure ends of the pump, and prevents fuel flowing back again.

The delivery quantity is practically proportional to engine speed, so that it is necessary to reduce the delivery quantity by a suction throttle at the inlet (suction) end, or limit it by an overflow valve at the outlet (pressure) end.

The gear-type fuel pump is maintenance-free. In order to bleed the fuel system before the first start, or when the tank has been driven "dry", a hand pump can be installed directly on the gear pump or in the low-pressure lines.

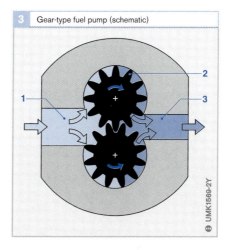

2   Roller-cell pump (schematic)

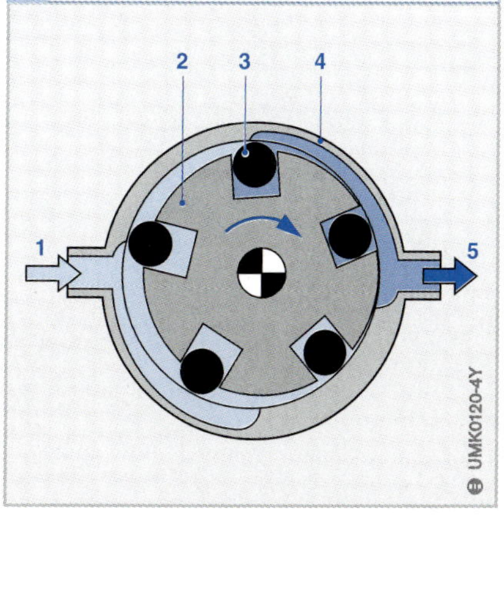

UMK0120-4Y

3   Gear-type fuel pump (schematic)

UMK1569-2Y

Fig. 2
1   Suction (intake) end
2   Slotted rotor
3   Roller
4   Base plate
5   Pressure (outlet) end

Fig. 3
1   Suction (intake) end
2   Drive gear
3   Pressure (outlet) end

Vane-type pump with separating vanes

In the version of this pump used with the passenger-car UIS (Fig. 4), two separating vanes are pressed by springs (3) against a rotor (1). When the rotor rotates, volume increases at the intake (suction) end (2) and fuel is drawn into two chambers. With continued rotation, chamber volumes decrease, and fuel is forced out of the chambers at the outlet (pressure) end (5). This pump delivers fuel even at very low rotational speeds.

Tandem pump

The tandem pump used on the passenger-car UIS is a unit comprising the fuel pump (Fig. 5) and the vacuum pump for the brake booster. It is attached to the engine's cylinder head and driven by the engine's camshaft. The fuel pump itself is either a vane-type pump with separating vanes or a gear pump (3), and even at low speeds (cranking speeds) delivers enough fuel to ensure that the engine starts reliably. The pump contains a variety of valves and throttling orifices:

Suction throttling orifice (6): Essentially, the quantity of fuel delivered by the pump is proportional to the pump's speed. The pump's maximum delivery quantity is limited by the suction throttling orifice so that not too much fuel is delivered.

Overpressure valve (7): This is used to limit the maximum pressure in the high-pressure stage.

Throttling bore (4): Vapor bubbles in the fuel-pump outlet are eliminated in the fuel-return throttling bore (1).

Bypass (12): If there is air in the fuel system (for instance if the vehicle has been driven until the fuel tank is empty), the low-pressure pressure-control valve remains closed. The air is forced out of the fuel system through the bypass by the pressure of the pumped fuel.

Thanks to the ingenious routing of the pump passages, the pump's gearwheels never run dry even when the fuel tank is empty. When restarting after filling the tank, therefore, this means that the pump draws in fuel immediately.

The fuel pump is provided with a connection (8) for measuring the fuel pressure in the pump outlet.

Distributor tube

The passenger-car UIS is provided with a distributor tube which, as its name implies, distributes the fuel to the unit injectors. This form of distribution ensures that the individual injectors all receive the same quanti-

Fig. 4

1 Rotor
2 Inlet (suction) side
3 Spring,
4 Separating vane
5 Outlet (pressure) side

Fig. 5

1 Return to tank
2 Entry from the fuel tank
3 Pumping element (gearwheel)
4 Throttling bore
5 Filter
6 Suction throttling orifice
7 Overpressure valve
8 Connection for pressure measurement
9 Outlet to the injector
10 Return from the injector
11 Non-return valve
12 Bypass

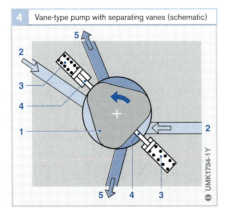

4 Vane-type pump with separating vanes (schematic)

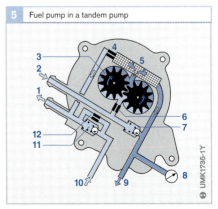

5 Fuel pump in a tandem pump

ties of fuel at the same temperature, and smooth engine running is the result. In the distributor tube, fuel flowing to the unit injectors mixes with fuel flowing back from them in order to even out the temperature.

Low-pressure pressure-control valve

The pressure-control valve (Fig. 1) is an overflow valve installed in the fuel return of the UIS and UPS systems. Independent of operating status, it provides for adequate operating pressure in the respective low-pressure stages so that the pumps are always well filled with a consistently even charge of fuel. The accumulator plunger (5) opens at a "snap-open pressure" of 3...3.5 bar, so that the conical seat (7) releases the accumulator volume (6). Only very little leakage fuel can escape through the gap seal (4). The spring (3) is compressed as a function of the fuel pressure, so that the accumulator volume changes and compensates for minor pressure fluctuations.

When pressure has increased to 4...4.5 bar, the gap seal also opens and the flow quantity increases abruptly.

The valve closes again when the pressure drops. Two threaded elements, each with a different spring seat, are available for preliminary adjustment of opening pressure.

ECU cooler

On commercial vehicles, ECU cooling must be provided if the ECU for the UIS or UPS systems is mounted directly on the engine. In such cases, fuel is used as the cooling medium. It flows past the ECU in special cooling channels and in the process absorbs heat from the electronics.

Fuel cooler

Due to the high pressures in the injectors for the passenger-car UIS, and some Common Rail systems (CRS), the fuel heats up to such an extent that in order to prevent damage to fuel tank and level sensor it must be cooled down before returning. Fuel flowing back from the injectors passes through the fuel cooler (heat exchanger, Fig. 2, Pos. 3) and transfers heat energy to the coolant in the fuel-cooling circuit. This is separated from the engine-cooling circuit (6) since at normal engine temperatures the engine coolant is too hot to absorb heat from the fuel. In order that the fuel-cooling circuit can be filled and temperature fluctuations compensated for, the fuel-cooling circuit is connected to the engine-cooling circuit near the equalizing reservoir. Connection is such that the fuel-cooling circuit is not adversely affected by the engine-cooling circuit which is at a higher temperature.

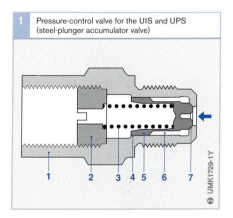

1 Pressure-control valve for the UIS and UPS (steel-plunger accumulator valve)

UMK1729-1Y

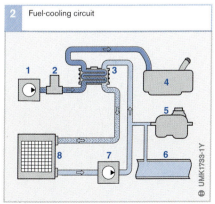

2 Fuel-cooling circuit

UMK1733-1Y

Fig. 1
1   Valve body
2   Threaded element
3   Spring
4   Gap seal
5   Accumulator plunger
6   Accumulator volume
7   Conical seat

Fig. 2
1   Fuel pump
2   Fuel-temperature sensor
3   Fuel cooler
4   Fuel tank
5   Equalizing reservoir
6   Engine-cooling circuit
7   Coolant pump
8   Auxiliary cooler

Supplementary valves for in-line fuel-injection pumps

In addition to the overflow valve, electronically controlled in-line fuel-injection pumps also have an electric shutoff valve (Type ELAB) or an electrohydraulic shutoff device (Type EHAB).

Overflow valve

The overflow valve is fitted to the pump's fuel-return outlet. It opens at a pressure (2...3 bar) that is set to suit the fuel-injection pump concerned and thereby maintains the pressure in the fuel gallery at a constant level A valve spring (Figure 1, Item 4) acts on a spring seat (2) which presses the valve cone (5) against the valve seat (6). As the pressure, p_i in the fuel-injection pump rises, it pushes the valve seat back, thus opening the valve. When the pressure drops, the valve closes again. The valve seat has to travel a certain distance before the valve is fully open. The buffer volume thus created evens out rapid pressure variations, which has a positive effect on valve service life.

Type ELAB electric shutoff valve

The Type ELAB electric shutoff valve acts as a redundant (i.e. duplicate) back-up safety device. It is a 2/2-way solenoid valve which is screwed into the fuel inlet of the in-line fuel-injection pump (Figure 2). When not energized, it cuts off the fuel supply to the pump's fuel gallery. As a result, the fuel-injection pump is prevented from delivering fuel to the nozzles even if the actuator mechanism is defective, and the engine cannot overrev. The engine control unit closes the electric shutoff valve if it detects a permanent governor deviation or if a fault in the control unit's fuel-quantity controller is detected.

When it is energized (i.e. when the status of Terminal 15 is "Ignition on"), the electromagnet (Figure 2, Item 3) draws in the solenoid armature (4) (12 or 24 V, stroke approx. 1.1 mm). The sealing cone seal (7) attached to the armature then opens the channel to the inlet passage (9). When the engine is switched off using the starter switch ("ignition switch"), the supply of electricity to the solenoid coil is also disconnected. This causes the magnetic field to collapse so that the compression spring (5) pushes the armature and the attached sealing cone back against the valve seat.

Fig. 1
1 Sealing ball
2 Spring seat
3 Sealing washer
4 Valve spring
5 Valve cone
6 Valve seat
7 Hollow screw housing
8 Fuel return

p_i Pump fuel gallery pressure

Fig. 2
1 Electrical connection to engine control unit
2 Solenoid valve housing
3 Solenoid coil
4 Solenoid armature
5 Compression spring
6 Fuel inlet
7 Plastic sealing cone
8 Constriction plug for venting
9 Inlet passage to pump
10 Connection for overflow valve
11 Housing (ground)
12 Mounting-bolt eyes

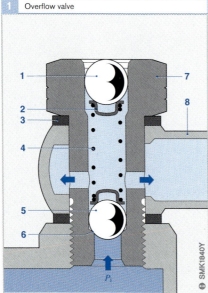

1 Overflow valve

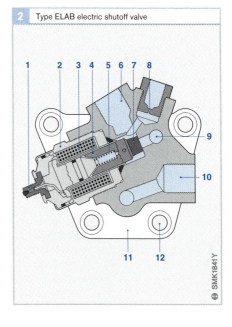

2 Type ELAB electric shutoff valve

Type EHAB electrohydraulic shutoff device

The Type EHAB electrohydraulic shutoff device is used as a safety shutoff for fuel-injection pumps with relatively high fuel gallery pressures. In such cases, the capabilities of the Type ELAB electric shutoff valve are insufficient. With high fuel-gallery pressures and in the absence of any special compensating devices, it can take up to 10 s for the pressure to drop sufficiently for fuel injection to stop. The electrohydraulic shutoff device thus ensures that fuel is drawn back out of the fuel-injection pump by the pre-supply pump. Thus, when the valve is de-energized, the fuel gallery pressure in the fuel-injection pump is dissipated much more quickly and the engine can be stopped within a period of no more than 2 s. The electrohydraulic shutoff device is mounted directly on the fuel-injection pump. The EHAB housing also incorporates an integrated fuel-temperature sensor for the electronic governing system (Figure 3, Item 8).

Normal operation setting (Figure 3a)

As soon as the engine control unit activates the electrohydraulic shutoff device ("Ignition on"), the electromagnet (6) draws in the solenoid armature (5, operating voltage 12 V). Fuel can then flow from the fuel tank (10) via the heat exchanger (11) for cold starting and the preliminary filter (3) to port A. From there, the fuel passes through the right-hand valve past the solenoid armature to port B. This is connected to the presupply pump (1) which pumps the fuel via the main fuel filter (2) to port C of the electrohydraulic shutoff device. The fuel then passes through the open left-hand valve to port D and finally from there to the fuel-injection pump (12).

Reversed-flow setting (Figure 3b)

When the ignition is switched off, the valve spring (7) presses the solenoid armature back to its resting position. The intake side of the presupply pump is then connected directly to the fuel-injection pump's inlet passage so that fuel flows back from the fuel gallery to the fuel tank. The right hand valve opens the connection between the preliminary filter and main fuel filter, allowing fuel to return to the fuel tank.

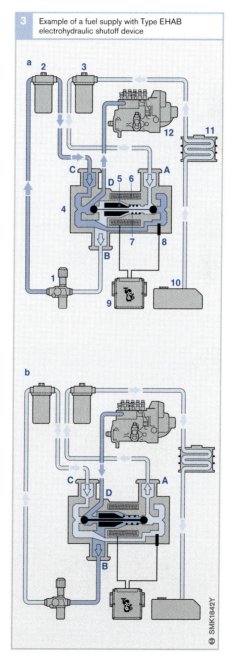

3 Example of a fuel supply with Type EHAB electrohydraulic shutoff device

SMK1842Y

Fig. 3
a Normal operation setting
b Reversed-flow/ emergency shutoff setting

1 Presupply pump
2 Main fuel filter
3 Preliminary filter
4 Type EHAB electrohydraulic shutoff device
5 Solenoid armature
6 Electromagnet
7 Valve spring
8 Fuel-temperature sensor
9 Engine control unit
10 Fuel tank
11 Heat exchanger
12 Fuel-injection pump

A...D valve ports

Overview of in-line fuel-injection pump systems

No other fuel-injection system is as widely used as the in-line fuel-injection pump – the "classic" diesel fuel-injection technology. Over the years, this system has been continually refined and adapted to suit its many areas of application. As a result, a large variety of different versions are still in use today. The particular strength of these pumps is their rugged durability and ease of maintenance.

Areas of application

The fuel-injection system supplies the diesel engine with fuel. To perform that function, the fuel-injection pump generates the necessary fuel pressure for injection and delivers the fuel at the required rate. The fuel is pumped through a high-pressure fuel line to the nozzle, which injects it into the engine's combustion chamber. The combustion processes in a diesel engine are primarily dependent on the quantity and manner in which the fuel is introduced into the combustion chamber. The most important criteria in that regard are
● The timing and duration of fuel injection
● The dispersal of fuel throughout the combustion chamber
● The point at which ignition is initiated
● The volume of fuel injected relative to crankshaft rotation and
● The total volume of fuel injected relative to the desired power output of the engine

The in-line fuel-injection pump is used all over the world in medium-sized and heavy-duty trucks as well as on marine and fixed-installation engines. It is controlled either by a mechanical governor, which may be combined with a timing device, or by an electronic actuator mechanism (Table 1, next double page).

In contrast with all other fuel-injection systems, the in-line fuel-injection pump is lubricated by the engine's lubrication system. For that reason, it is capable of handling poorer fuel qualities.

Types

Standard in-line fuel-injection pumps
The range of standard in-line fuel-injection pumps currently produced encompasses a large number of pump types (see Table 1, next double page). They are used on diesel engines with anything from 2 to 12 cylinders and ranging in power output from 10 to 200 kW per cylinder (see also Table 1 in the chapter "Overview of diesel fuel-injection systems"). They are equally suitable for use on direct-injection (DI) or indirect-injection (IDI) engines.

Depending on the required injection pressure, injected-fuel quantity and injection duration, the following versions are available:
● Type M for 4...6 cyl. up to 550 bar
● Type A for 2...12 cyl. up to 750 bar
● Type P3000 for 4...12 cyl. up to 950 bar
● Type P7100 for 4...12 cyl. up to 1,200 bar
● Type P8000 for 6...12 cyl. up to 1,300 bar
● Type P8500 for 4...12 cyl. up to 1,300 bar
● Type R for 4...12 cyl. up to 1,150 bar
● Type P10 for 6...12 cyl. up to 1,200 bar
● Type ZW(M) for 4...12 cyl. up to 950 bar
● Type P9 for 6...12 cyl. up to 1,200 bar
● Type CW for 6...10 cyl. up to 1,000 bar
The version most commonly fitted in commercial vehicles is the Type P.

Control-sleeve in-line fuel-injection pump
The range of in-line fuel-injection pumps also includes the control-sleeve version (Type H), which allows the start-of-delivery point to be varied in addition to the injection quantity. The Type H pump is controlled by a Type RE electronic controller which has two actuator mechanisms. This arrangement enables the control of the start of injection and the injected-fuel quantity with the aid of two control rods and thus makes the automatic timing device superfluous. The following versions are available:
● Type H1 for 6...8 cyl. up to 1,300 bar
● Type H1000 for 5...8 cyl. up to 1,350 bar

Design

Apart from the in-line fuel-injection pump, the complete diesel fuel-injection system (Figures 1 and 2) comprises
- A fuel pump for pumping the fuel from the fuel tank through the fuel filter and the fuel line to the injection pump
- A mechanical governor or electronic control system for controlling the engine speed and the injected-fuel quantity
- A timing device (if required) for varying the start of delivery according to engine speed
- A set of high-pressure fuel lines corresponding to the number of cylinders in the engine and
- A corresponding number of nozzle-and-holder assemblies

In order for the diesel engine to function properly, all of those components must be matched to each other.

Control

The operating parameters are controlled by the injection pump and the governor which operates the fuel-injection pump's control rod. The engine's torque output is approximately proportional to the quantity of fuel injected per piston stroke.

Mechanical governors

Mechanical governors used with in-line fuel-injection pumps are centrifugal governors. This type of governor is linked to the accelerator pedal by means of a rod linkage and an adjusting lever. On its output side, it operates the pump's control rod. Depending on the type of use, different control characteristics are required of the governor:
- The Type RQ maximum-speed governor limits the maximum speed.
- The Type RQ and RQU minimum/maximum-speed governors also control the idle speed in addition to limiting the maximum speed.

1 Fuel-injection system with mechanically governed standard in-line fuel-injection pump

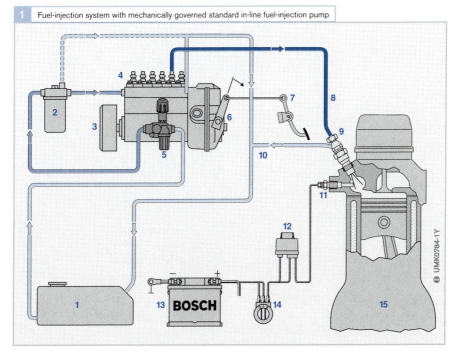

Fig. 1
1 Fuel tank
2 Fuel filter with overflow valve (option)
3 Timing device
4 In-line fuel-injection pump
5 Fuel pump (mounted on injection pump)
6 Governor
7 Accelerator pedal
8 High-pressure fuel line
9 Nozzle-and-holder assembly
10 Fuel-return line
11 Type GSK glow plug
12 Type GZS glow plug control unit
13 Battery
14 Glow plug/starter switch ("ignition switch")
15 Diesel engine (IDI)

- The Type RQV, RQUV, RQV..K, RSV and RSUV variable-speed governors also control the intermediate speed range.

Timing devices

In order to control start of injection and compensate for the time taken by the pressure wave to travel along the high-pressure fuel line, standard in-line fuel-injection pumps use a timing device which "advances" the start of delivery of the fuel-injection pump as the engine speed increases. In special cases, a load-dependent control system is employed. Diesel-engine load and speed are controlled by the injected-fuel quantity without exerting any throttle action on the intake air.

Electronic control systems

If an electronic control system is used, there is an accelerator-pedal sensor which is connected to the electronic control unit. The control unit then converts the accelerator-position signal into a corresponding nominal control-rack travel while taking into account the engine speed.

An electronic control system performs significantly more extensive functions than the mechanical governor. By means of electrical measuring processes, flexible electronic data processing and closed-loop control systems with electrical actuators, it enables more comprehensive response to variable factors than is possible with the mechanical governor.

Electronic diesel control systems can also exchange data with other electronic control systems on the vehicle (e.g. Traction Control System, electronic transmission control) and can therefore be integrated in a vehicle's overall system network.

Electronic control of diesel engines improves their emission characteristics by more precise metering of fuel delivery.

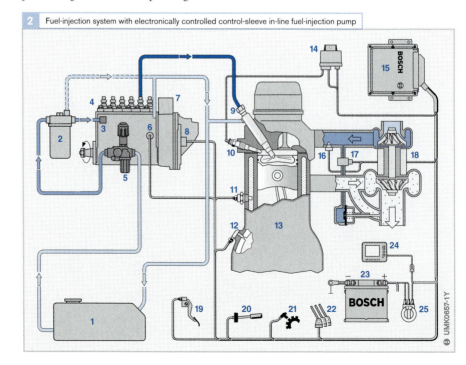

2   Fuel-injection system with electronically controlled control-sleeve in-line fuel-injection pump

UMK0657-1Y

1   Areas of application for the most important in-line fuel-injection pumps and their governors

Area of application	Cars	Fixed-installation engines	Commercial vehicles	Construction and agricultural machinery	Railway locomotives	Ships
Pump type						
Standard in-line fuel-injection pump Type M	●	–	–	●	–	–
Standard in-line fuel-injection pump Type A	–	●	–	●	–	–
Standard in-line fuel-injection pump Type MW[1]	–	–	●	●	–	–
Standard in-line fuel-injection pump Type P	–	●	●	●	●	●
Standard in-line fuel-injection pump Type R[2]	–	–	●	●	●	●
Standard in-line fuel-injection pump Type P10	–	●	–	●	●	●
Standard in-line fuel-injection pump Type ZW(U)	–	–	–	–	●	●
Standard in-line fuel-injection pump Type P9	–	●	–	●	●	●
Standard in-line fuel-injection pump Type CW	–	–	–	–	●	●
Control-sleeve in-line fuel-injection pump Type O	–	–	●	–	–	–
Governor type						
Minimum/maximum speed governor Type RSF	●	–	–	●	–	–
Minimum/maximum speed governor Type RQ	–	–	●	●	–	–
Minimum/maximum speed governor Type RQU	–	–	–	–	–	●
Variable-speed governor Type RQV	–	●	●	●	–	–
Variable-speed governor Type RQUV	–	–	–	–	●	●
Variable-speed governor Type RQV..K	–	–	●	–	–	–
Variable-speed governor Type RSV	–	●	–	●	–	–
Variable-speed governor Type RSUV	–	–	–	–	–	●
Type RE (electric actuator mechanism)	●	–	●	–	–	–

Table 1
[1] This type of pump is no longer used with new systems.
[2] Same design as Type P but for heavier duty.

3   Examples of in-line fuel-injection pumps

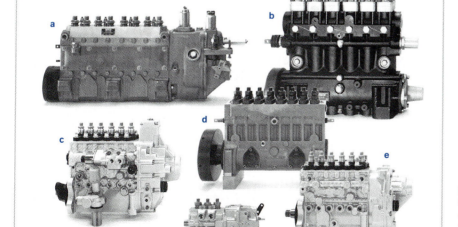

NMK1813Y

20 cm

Fig. 3
Pump types:
a   ZWM (8 cylinders)
b   CW (6 cylinders)
c   H (control-sleeve type) (6 cylinders)
d   P9/P10 (8 cylinders)
e   P7100 (6 cylinders)
f   A (3 cylinders)

Presupply pumps for in-line fuel-injection pumps

The presupply pump's job is to supply the in-line fuel-injection pump with sufficient diesel fuel under all operating conditions. In addition, it "flushes" the fuel-injection pump with fuel to cool it down by extracting heat from the fuel and returning it through the overflow valve to the fuel tank. In addition to the presupply pumps described in this section, there are also multifuel and electric presupply pumps. In certain relatively rare applications, the in-line fuel-injection pump can be operated without a presupply pump in a gravity-feed fuel-tank system.

Applications

In applications where there is an insufficient height difference or a large distance between the fuel tank and the fuel-injection pump, a presupply pump (Bosch type designation FP) is fitted. This is normally flange-mounted on the in-line fuel-injection pump. Depending on the conditions in which the engine is to be used and the specifics of the engine design, various fuel line arrange-

ments are required. Figures 1 and 2 illustrate two possible variations.

If the fuel filter is located in the immediate vicinity of the engine, the heat radiated from the engine can cause bubbles to form in the fuel lines. In order to prevent this, the fuel is made to circulate through the fuel-injection pump's fuel gallery so as to cool the pump. With this line arrangement, the excess fuel flows through the overflow valve (6) and the return line back to the fuel tank (1).

If, in addition, the ambient temperature in the engine compartment is high, the line arrangement shown in Figure 2 may also be used. With this system, there is an overflow restriction (7) on the fuel filter through which a proportion of the fuel flows back to the fuel tank during normal operation, taking any gas or vapor bubbles with it. Bubbles that form inside the fuel-injection pump's fuel gallery are removed by the excess fuel that escapes through the overflow valve (6) to the fuel tank. The presupply pump must therefore be dimensioned to be able to deliver not only the fuel volume

Fig. 1
1 Fuel tank
2 Presupply pump
3 Fuel filter
4 In-line fuel-injection
   pump
5 Nozzle-and-holder
   assembly
6 Overflow valve

— Supply line
– – Return line

Fig. 2
1 Fuel tank
2 Presupply pump
3 Fuel filter
4 In-line fuel-injection
   pump
5 Nozzle-and-holder
   assembly
6 Overflow valve
7 Overflow restriction

— Supply line
– – Return line

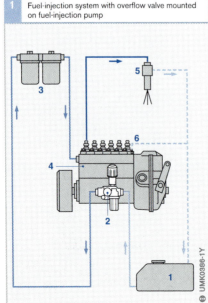

1 Fuel-injection system with overflow valve mounted on fuel-injection pump

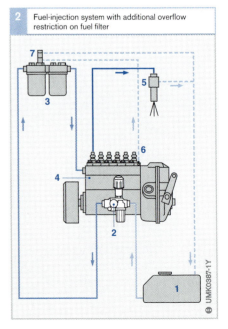

2 Fuel-injection system with additional overflow restriction on fuel filter

required by the fuel-injection pump but also the volume that "bypasses" the fuel-injection pump and returns to the fuel tank.

The following criteria determine the choice of presupply pump:
- The type of fuel-injection pump
- The delivery rate
- The line routing arrangement and
- The available space in the engine compartment

Design and method of operation

A presupply pump draws the fuel from the fuel tank and pumps it under pressure through the fuel filter and into the fuel gallery of the fuel-injection pump (100 ... 350 kPa or 1 ... 3.5 bar). Presupply pumps are generally mechanical plunger pumps that are mounted on the fuel-injection pump (or in rare cases on the engine).

The presupply pump is then driven by an eccentric (Figure 3, Item 1) on the fuel-injection pump or engine camshaft (2).

Depending on the fuel delivery rate required, presupply pumps may be single or double-action designs.

Single-action presupply pumps

Single-action presupply pumps (Figures 3 and 4) are available for fuel-injection pump sizes M, A, MW and P. The drive cam or eccentric (Figure 3, Item 1) drives the pump plunger (5) via a push rod (3). The piston is also spring-loaded by a compression spring (7) which effects the return stroke.

The single-action presupply pump operates according to the throughflow principle as follows. The cam pitch on the push rod moves the pump plunger and its integrated suction valve (8) against the force of the compression spring. In the process, the suction valve is opened by the lower pressure created in the fuel gallery (4, Figure 3a).

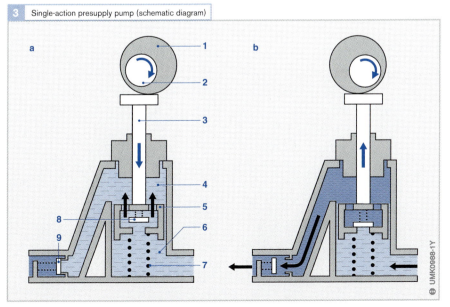

3 Single-action presupply pump (schematic diagram)

a
b

© UMK0988-1Y

Fig. 3
a Cam pitch
b Return stroke

1 Drive eccentric
2 Fuel-injection pump camshaft
3 Push rod
4 Pressure chamber
5 Pump plunger
6 Fuel gallery
7 Compression spring
8 Suction valve
9 Delivery valve

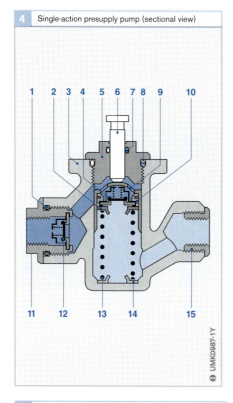

4   Single-action presupply pump (sectional view)

1   2   3   4   5   6   7   8   9   10

11   12   13   14   15

UMK0987-1Y

Fig. 4
1 Sealing ring
2 Spring seat
3 Pump housing
   (aluminum)
4 Suction valve
5 Roller-tappet shell
6 Push rod
7 Sealing ring
8 Sealing ring
9 Pump plunger
10 Spacer ring
11 Pressure port
12 Delivery valve
13 Compression spring
14 Spring seat
15 Suction port

As a result, the fuel passes into the chamber between the suction valve and the delivery valve (9). When the pump performs its return stroke under the action of the compression spring, the suction valve closes and the delivery valve opens (Figure 3b). The fuel then passes under pressure along the high-pressure line to the fuel-injection pump.

Double-action presupply pumps
Double-action presupply pumps (Figure 5) offer a higher delivery rate and are used for fuel-injection pumps that serve larger numbers of engine cylinders and which consequently must themselves provide greater delivery quantities. This type of presupply pump is suitable for Type P and ZW fuel-injection pumps. As with the single-action version, the double-action presupply pump is driven by a cam or eccentric.

In the double-action plunger pump, fuel is delivered to the fuel-injection pump on both the cam-initiated stroke and the return stroke, in other words there are two delivery strokes for every revolution of the camshaft.

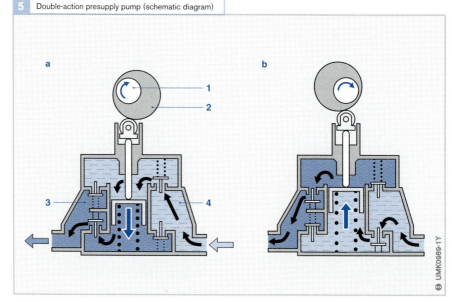

5   Double-action presupply pump (schematic diagram)

a   b

1
2

3   4

UMK0988-1Y

Fig. 5
a   Cam pitch
b   Return stroke

1   Fuel-injection pump
    camshaft
2   Drive eccentric
3   Pressure chamber
4   Fuel gallery

Manual priming pumps

The priming pump is usually integrated in the presupply pump (Figure 6, Item 1). However, it can also be fitted in the fuel line between the fuel tank and the presupply pump. It performs the following functions:
- Priming the suction side of the fuel-injection installation prior to initial operation
- Priming and venting the system after repairs or servicing and
- Priming and venting the system after the fuel tank has been run dry

The latest version of the Bosch priming pump replaces virtually all previous designs. It is backwardly compatible and can therefore be used to replace pumps of older designs. It no longer has to be released or locked in its end position. Consequently, it is easy to operate even in awkward positions.

The priming pump also contains a non-return valve which prevents the fuel flowing back in the wrong direction.

For applications in which the pump has to be fireproof, there is a special version with a steel body.

Preliminary filter

The preliminary filter protects the presupply pump against contamination from coarse particles. In difficult operating conditions, such as where engines are refueled from barrels, it is advisable to fit an additional strainer-type filter inside the fuel tank or in the fuel line to the presupply pump.

The preliminary filter may be integrated in the presupply pump (Figure 6, Item 2), mounted on the presupply pump intake or connected to the intake passage between the fuel tank and the presupply pump.

Gravity-feed fuel-tank system

Gravity-feed fuel-tank systems (which operate without a presupply pump) are generally used on tractors and very small diesel engines. The arrangement of the tank and the fuel lines is such that the fuel flows through the fuel filter to the fuel-injection pump under the force of gravity.

With smaller height differences between the fuel tank and the fuel filter or fuel-injection pump, larger-bore lines are better suited to providing an adequate flow of fuel to the fuel-injection pump. In such systems, it is useful to fit a stopcock between the fuel tank and the fuel filter. This allows the fuel inlet to be shut off when carrying out repairs or maintenance so that the fuel tank does not have to be drained.

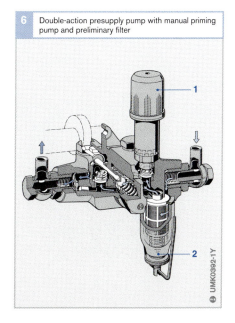

6 Double-action presupply pump with manual priming pump and preliminary filter

UMK0392-1Y

Fig. 6
1 Manual priming pump
2 Preliminary filter

Type PE standard in-line fuel-injection pumps

In-line fuel-injection pumps are among the classics of diesel fuel-injection technology. This dependable design has been used on diesel engines since 1927. Over the years they have been continuously refined and adapted to suit their many areas of application. In-line fuel-injection pumps are designed for use on fixed-installation engines, commercial vehicles, and construction and agricultural machinery. They enable high power outputs per cylinder on diesel engines with between 2 and 12 cylinders. When used in conjunction with a governor, a timing device and various auxiliary components, the in-line fuel-injection pump offers considerable versatility. Today in-line fuel-injection pumps are no longer produced for cars.

The power output of a diesel engine is determined essentially by the amount of fuel injected into the cylinder. The in-line fuel-injection pump must precisely meter the amount of fuel delivered to suit every possible engine operating mode.

In order to facilitate effective mixture preparation, a fuel-injection pump must deliver the fuel at the pressure required by the combustion system employed and in precisely the right quantities. In order to achieve the optimum balance between pollutant emission levels, fuel consumption and combustion noise on the part of the diesel engine, the start of delivery must be accurate to within 1 degree of crankshaft rotation.

In order to control start of delivery and compensate for the time taken by the pressure wave to travel along the high-pressure delivery line, standard in-line fuel-injection pumps use a timing device (Figure 1, Item 3) which "advances" the start of delivery of the fuel-injection pump as the engine speed increases (see chapter "Governors for in-line fuel-injection pumps"). In special cases, a load-dependent control system is employed. Diesel-engine load and speed are controlled by varying the injected fuel quantity.

A distinction is made between standard in-line fuel-injection pumps and control-sleeve in-line fuel-injection pumps.

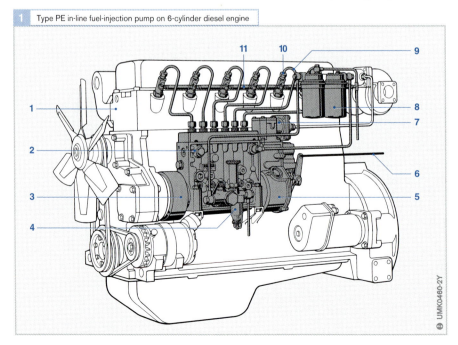

1   Type PE in-line fuel-injection pump on 6-cylinder diesel engine

Fig. 1
1   Diesel engine
2   Standard in-line
    fuel-injection pump
3   Timing device
4   Presupply pump
5   Governor
6   Control lever with
    linkage to accelerator
7   Manifold-pressure
    compensator
8   Fuel filter
9   High-pressure
    delivery line
10 Nozzle-and-holder
    assembly
11 Fuel-return line

UMK0460-2Y

Fitting and drive system

In-line fuel-injection pumps are attached directly to the diesel engine (Figure 1). The engine drives the pump's camshaft. On two-stroke engines, the pump speed is the same as the crankshaft speed. On four-stroke engines, the pump speed is half the speed of the crankshaft – in other words, it is the same as the engine camshaft speed.

In order to produce the high injection pressures required, the drive system between the engine and the fuel-injection pump must be as "rigid" as possible.

There is a certain amount of oil inside the fuel-injection pump in order to lubricate the moving parts (e.g. camshaft, roller tappets, etc.). The fuel-injection pump is connected to the engine lube-oil circuit so that oil circulates when the engine is running.

Design and method of operation

Type PE in-line fuel-injection pumps have an internal camshaft that is integrated in the aluminum pump housing (Figure 2, Item 14). It is driven either via a clutch unit or a timing device or directly by the engine. Pumps of this type with an integrated camshaft are referred to by the type designation PE.

Above each cam on the camshaft is a roller tappet (13) and a spring seat (12) for each cylinder of the engine. The spring seat forms the positive link between the roller tappet and the pump plunger (8). The pump barrel (4) forms the guide for the pump plunger. The two components together form the pump-and-barrel assembly.

---

2 Type PE in-line fuel-injection pump for 6-cylinder diesel engine

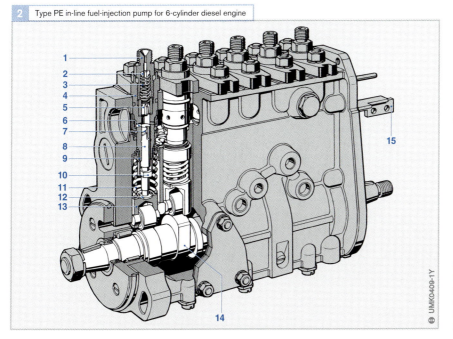

Fig. 2
1 Pressure-valve holder
2 Filler piece
3 Pressure-valve spring
4 Pump barrel
5 Delivery-valve cone
6 Intake and control port
7 Helix
8 Pump plunger
9 Control sleeve
10 Plunger control arm
11 Plunger spring
12 Spring seat
13 Roller tappet
14 Camshaft
15 Control rack

Design of the pump-and-barrel assembly

In its basic form, a pump-and-barrel assembly consists of a pump plunger (Figure 3, Item 9) and a pump barrel (8). The pump barrel has one or two inlet passages that lead from the fuel gallery (1) into the inside of the cylinder. On the top of the pump-and-barrel assembly is the delivery-valve holder (5) with the delivery-valve cone (7). The control sleeve (3) forms the connection between the pump plunger and the control rack (10). The control rack moves inside the pump housing – under the control of the governor as described in the chapter "Governors for in-line fuel-injection pumps" – so as to rotate the positively interlocking "control-sleeve-and-piston" assembly by means of a ring gear or linkage lever. This enables precise regulation of the pump delivery quantity.

The plunger's total stroke is constant. The effective stroke, on the other hand, and therefore the delivery quantity, can be altered by rotating the pump plunger.

In addition to a vertical groove (Figure 4, Item 2), the pump plunger also has a helical channel (7) cut into it. The helical channel is referred to as the helix (6).

For injection pressures up to 600 bar, a single helix is sufficient, whereas higher pressures require the piston to have two helixes on opposite sides. This design feature prevents the units from "seizing" as the piston is no longer

Fig. 3
1 Fuel gallery
2 Control-sleeve gear
3 Control sleeve
4 Cover plate
5 Pressure-valve holder
6 Pressure-valve body
7 Delivery-valve cone
8 Pump barrel
9 Pump plunger
10 Control rack
11 Plunger control arm
12 Plunger spring
13 Spring seat
14 Adjusting screw
15 Roller tappet
16 Camshaft

Fig. 4
a Single-port plunger-and-barrel assembly
b Two-port plunger-and-barrel assembly

1 Inlet passage
2 Vertical groove
3 Pump barrel
4 Pump plunger
5 Control port (inlet and return lines)
6 Helix
7 Helical channel
8 Ring groove for lubrication

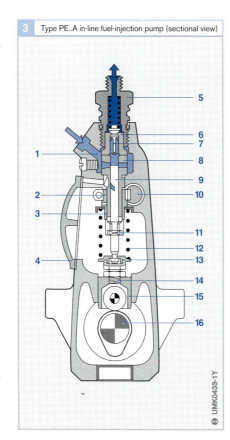

3   Type PE..A in-line fuel-injection pump (sectional view)

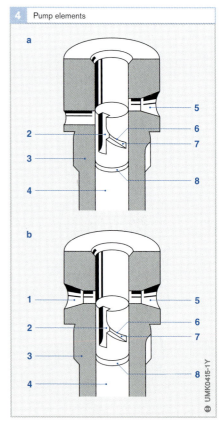

4   Pump elements

forced sideways against the cylinder wall by the injection pressure.

The cylinder then has one or two bores for fuel supply and return (Figure 4).

The pump plunger is such an exact fit inside the pump barrel that it provides a leakproof seal even at extremely high pressures and at low rotational speeds. Because of this precise fit, pump plungers and barrels can only be replaced as a complete plunger-and-barrel assembly.

The injected fuel quantity possible is dependent on the charge volume of the pump barrel. The maximum injection pressures vary between 400 and 1,350 bar at the nozzle depending on the pump design.

The relative angular positions of the cams on the pump camshaft are such that the injection process is precisely synchronized with the firing sequence of the engine.

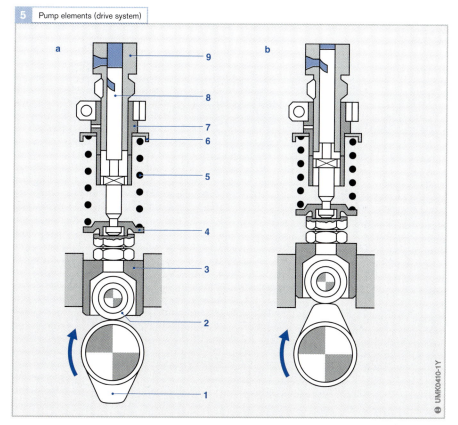

5 Pump elements (drive system)

a

b

UMK0410-1Y

Fig. 5
a BDC position
b TDC position

1 Cam
2 Tappet roller
3 Roller tappet
4 Lower spring seat
5 Plunger spring
6 Upper spring seat
7 Control sleeve
8 Pump plunger
9 Pump barrel

Method of operation of plunger-and-barrel assembly (stroke phase sequence)

The rotation of the camshaft is converted directly into a reciprocating motion on the part of the roller tappet and consequently into a similar reciprocating action on the part of the pump plunger.

The delivery stroke, whereby the piston moves towards its "top dead center" (TDC), is assumed by the action of the cam. A compression spring performs the task of returning the plunger to "bottom dead center" (BDC). It is dimensioned to keep the roller in contact with the cam even at maximum speed, as loss of contact between roller and cam, and the consequent impact of the two surfaces coming back into contact, would

inevitably cause damage to both components in the course of continuous operation.

The plunger-and-barrel assembly operates according to the overflow principle with helix control (Figure 6). This is the principle adopted on Type PE in-line fuel-injection pumps and Type PF single-plunger fuel-injection pumps.

When the pump plunger is at bottom dead center (BDC) the cylinder inlet passages are open. Under pressure from the presupply pump, fuel is able to flow through those passages from the fuel gallery to the plunger chamber. During the delivery stroke, the pump plunger closes off the inlet passages. This phase of the plunger lift is referred to as

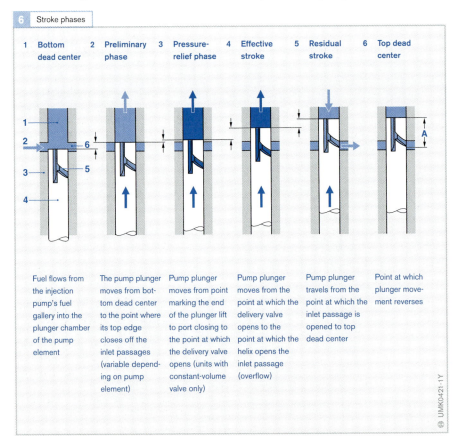

6   Stroke phases

1 Bottom dead center	2 Preliminary phase	3 Pressure-relief phase	4 Effective stroke	5 Residual stroke	6 Top dead center
Fuel flows from the injection pump's fuel gallery into the plunger chamber of the pump element	The pump plunger moves from bottom dead center to the point where its top edge closes off the inlet passages (variable depending on pump element)	Pump plunger moves from point marking the end of the plunger lift to port closing to the point at which the delivery valve opens (units with constant-volume valve only)	Pump plunger moves from the point at which the delivery valve opens to the point at which the helix opens the inlet passage (overflow)	Pump plunger travels from the point at which the inlet passage is opened to top dead center	Point at which plunger movement reverses

Fig. 6
1 Plunger chamber
2 Fuel inlet
3 Pump barrel
4 Pump plunger
5 Helix
6 Fuel return

A Total stroke

UMK0421-1Y

7 Fuel-delivery control

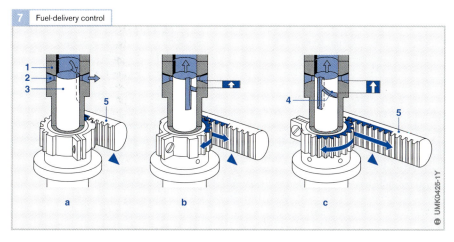

Fig. 7
a Zero delivery
b Partial delivery
c Maximum delivery

1 Pump barrel
2 Inlet passage
3 Pump plunger
4 Helix
5 Geared control rack

the preliminary phase. As the delivery stroke continues, fuel pressure increases and causes the delivery valve at the top of the plunger-and-barrel assembly to open. If a constant-volume valve is used (see section "Delivery valves") the delivery stroke also includes a retraction-lift phase. Once the delivery valve has opened, fuel flows along the high-pressure line to the nozzle for the duration of the effective stroke. Finally, the nozzle injects a precisely metered quantity of fuel into the combustion chamber of the engine.

Once the pump plunger's helix releases the inlet passage again, the effective stroke is complete. From this point on, no more fuel is delivered to the nozzle as, during the residual stroke, the fuel can escape through the vertical groove from the plunger chamber back into the fuel gallery so that pressure in the plunger-and-barrel assembly breaks down.

After the piston reaches top dead center (TDC) and starts to move back in the opposite direction, fuel flows through the vertical groove from the fuel gallery to the plunger chamber until the helix closes off the inlet passage again. As the plunger continues its return stroke, a vacuum is created inside the pump barrel. When the inlet passage is opened again, fuel then immediately flows into the plunger chamber. At this point, the cycle starts again from the beginning.

Fuel-delivery control

Fuel delivery can be controlled by varying the effective stroke (Figure 7). This is achieved by means of a control rack (5) which twists the pump plunger (3) so that the pump plunger helix (4) alters the point at which the effective delivery stroke ends and therefore the quantity of fuel delivered.

In the final zero-delivery position (a), the vertical groove is directly in line with the inlet passage. With the plunger in this position, the pressure chamber is connected to the fuel gallery through the pump plunger for the entire delivery stroke. Consequently, no fuel is delivered. The pump plungers are placed in this position when the engine is switched off.

For partial delivery (b), fuel delivery is terminated depending on the position of the pump plunger.

For maximum delivery (c), fuel delivery is not terminated until the maximum effective stroke is reached, i.e. when the greatest possible delivery quantity has been reached.

The force transfer between the control rack and the pump plunger, see Figure 7, takes place by means of a geared control rack (PE..A and PF pumps) or via a ball joint with a suspension arm and control sleeve (Type PE..M, MW, P, R, ZW(M) and CW pumps).

Pump unit with leakage return channel

If the fuel-injection pump is connected to the engine lube-oil circuit, leakage fuel can result in thinning of the engine oil under certain circumstances. Assemblies with a leakage return channel to the fuel gallery of the fuel-injection pump largely avoid this problem. There are two designs:

● A ring groove (Figure 8a, Item 3) in the plunger collects the leakage fuel and returns it to the fuel gallery via other specially located grooves (2) in the piston.
● Leakage fuel flows back to the fuel gallery via a ring groove in the pump barrel (Figure 8b, Item 4) and a hole (1).

Pump plunger design variations

Special requirements such as reducing noise or lowering pollutant emissions in the exhaust gas make it necessary to vary the start of delivery according to engine load. Pump plungers that have an upper helix (Figure 9, Item 2) in addition to the lower helix (1) allow load-dependent variation of start of delivery. In order to improve the starting characteristics of some engines, special pump plungers with a starting groove (3) are used. The starting groove – an extra groove cut into the top edge of the plunger – only comes into effect when the plunger is set to the starting position. It retards the start of delivery by 5...10° in terms of crankshaft position.

Fig. 8
a   Version with ring groove in plunger
a   Version with ring groove in barrel

1   Leakage return bore
2   Leakage-return slots
3   Ring groove in pump plunger
4   Ring groove in pump barrel

Fig. 9
a   Helix at bottom
b   Helix at top and bottom
c   Helix at bottom and starting groove

1   Bottom helix
2   Top helix
3   Starting groove
4   Start-quantity limitation groove

8 Pump elements with leakage return channel

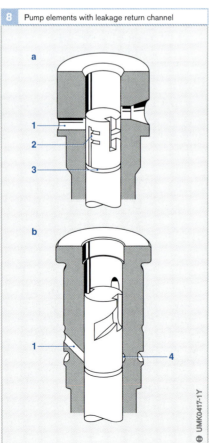

UMK0417-1Y

9 Pump plunger design variations

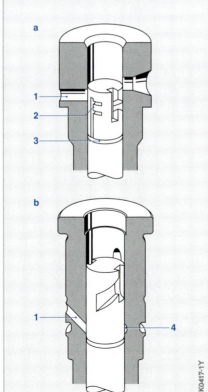

UMK0418-1Y

Cam shapes

Different combustion-chamber geometries and combustion methods demand different fuel-injection parameters. In other words, each individual engine design requires an individually adapted fuel-injection process. The piston speed (and therefore the length of the injection duration) depends on the cam pitch relative to the camshaft angle of rotation. For this reason, there are various different cam shapes according to the specifics of the application. In order to improve injection parameters such as the "rate-of-discharge curve" and "pressure load", special cam shapes can be designed by computer.

The trailing edge of the cam can also be varied (Figure 10): There are symmetrical cams (a), cams with asymmetric trailing edge (b) and reversal-inhibiting cams (c) which make it more difficult for the engine to start rotating in the wrong direction.

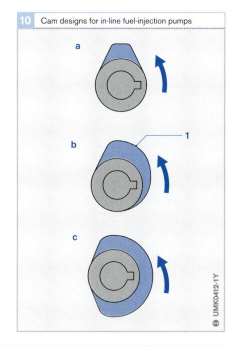

10 Cam designs for in-line fuel-injection pumps

Fig. 10
a Symmetrical cam
b Asymmetrical cam
c Reversal-inhibiting cam

1 Trailing edge

▶ **History of in-line fuel-injection pumps**

No other diesel fuel-injection system can look back on a history as long as the Bosch in-line fuel-injection pump. The very first examples of this famously reliable design came off the production line in Stuttgart as long ago as 1927.

Although the basic method of operation has remained the same, pump and governor design has been continuously adapted and improved to meet new demands. The arrival of electronic diesel control in 1987 and the control-sleeve in-line fuel-injection pump in 1993 opened up new horizons.

Sales figures show that, for a wide range of applications, the in-line fuel-injection pump is far from reaching its "sell-by date" even today. In 2001 roughly 150,000 Type P and Type H pumps left the Bosch factory in Homburg.

▼ Type PE..A in-line fuel-injection pump

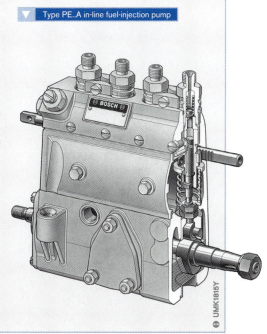

Delivery valve

The delivery valve is fitted between the plunger-and-barrel assembly and the high-pressure delivery line. Its purpose is to isolate the high-pressure delivery line from the plunger-and-barrel assembly. It also reduces the pressure in the high-pressure delivery line and the nozzle chamber following fuel injection to a set static pressure. Pressure reduction causes rapid and precise closure of the nozzle and prevents undesirable fuel dribble into the combustion chamber.

In the course of the delivery stroke, the increasing pressure in the plunger chamber lifts the delivery-valve cone (Figure 11, Item 3) from the valve seat (4) in the delivery-valve body (5). Fuel then passes through the delivery-valve holder (1) and into the high-pressure delivery line to the nozzle. As soon as the helix of the pump plunger brings the injection process to an end, the pressure in the plunger chamber drops. The delivery-valve cone is then pressed back against the valve seat by the valve spring (2). This isolates the space above the pump plunger and the high-pressure side of the system from one another until the next delivery stroke.

Constant-volume valve without return-flow restriction

In a constant-volume valve (Bosch designation GRV), part of the valve stem takes the form of a "retraction piston" (Figure 12, Item 2). It fits into the valve guide with a minimum degree of play. At the end of fuel delivery, the retraction piston slides into the valve guide and shuts off the plunger chamber from the high-pressure delivery line. This increases the space available to the fuel in the high-pressure delivery line by the charge volume of the retraction piston. The retraction volume is dimensioned precisely to suit the length of the high-pressure delivery line, which means that the latter must not be altered.

In order to achieve the desired fuel-delivery characteristics, torque-control valves are used in some special cases. They have a retraction piston with a specially ground pintle (6) on one side.

Constant-volume valve with return-flow restriction

A return-flow restriction (Bosch designation RDV or RSD) may also be used in addition to the constant-volume valve. Its purpose is to dampen and render harmless returning pressure waves that are produced when the nozzle

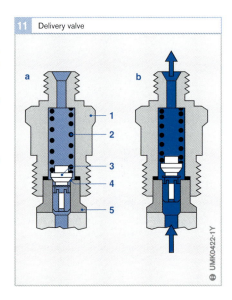

11   Delivery valve

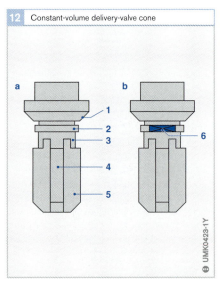

12   Constant-volume delivery-valve cone

closes. This reduces or entirely eliminates wear effects and cavitation in the plunger chamber. It also prevents undesirable secondary injection.

The return-flow restriction is integrated in the upper part of the delivery-valve holder (Figure 13), in other words between the constant-volume valve and the nozzle. The valve body (4) has a small bore (3) the size of which is dimensioned to suit the application so as to achieve, firstly, the desired flow restriction and, secondly, to prevent reflection of pressure waves as much as possible. The valve opens when fuel is flowing in delivery direction. The delivery flow is therefore not restricted. For pressures up to approx. 800 bar, the valve body shaped like a disk. For higher pressures it is a guided cone.

Pumps with return-flow throttle valves are "open systems", i.e. during the plunger lift to port closing and retraction lift, the static pressure in the high-pressure delivery line is the same as the internal pump pressure. Consequently, this pressure must be at least 3 bar.

Constant-pressure valve
The constant-pressure valve (Bosch designation GDV) is used on fuel-injection pumps

with high injection pressures (Figure 14). It consists of forward-delivery valve (consisting of delivery valve, 1, 2, 3) and a pressure-holding valve for the return-flow direction (consisting of 2, 5, 6, 7 and 8) which is integrated in the delivery-valve cone (2). The pressure-holding valve maintains a virtually constant static pressure in the high-pressure delivery line between fuel-injection phases under all operating conditions. The advantages of the constant-pressure valve are the prevention of cavitation and improved hydraulic stability which means more precise fuel injection.

During the delivery stroke, the valve acts as a conventional delivery valve. At the end of the delivery stroke, the ball valve (7) is initially open and the valve acts like a valve with a return-flow restriction. Once the closing pressure is reached, the compression spring (5) closes the return-flow valve, thereby maintaining a constant pressure in the fuel line.

However, correct functioning of the constant-pressure valve demands greater accuracy of adjustment and modifications to the governor. It is used for high-pressure fuel-injection pumps (upwards of approx. 800 bar) and for small, fast-revving direct-injection engines.

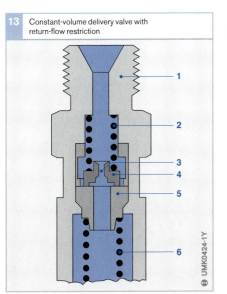
13 Constant-volume delivery valve with return-flow restriction

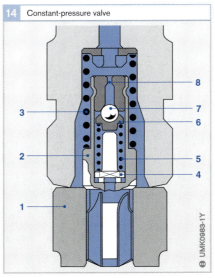
14 Constant-pressure valve

Fig. 13
1 Pressure-valve holder
2 Valve spring
3 Flow throttle
4 Valve body (disk in this case)
5 Valve holder
6 Pressure-valve spring

Fig. 14
1 Delivery-valve support
2 Delivery-valve cone
3 Pressure-valve spring
4 Filler piece
5 Compression spring (pressure-holding valve)
6 Spring seat
7 Ball
8 Flow throttle

Design variations

The range of power outputs for diesel engines with in-line fuel-injection pumps extends from 10 to 200 kW per cylinder. Various pump design variations allow such a wide range of power outputs to be accommodated. The designs are grouped into series whose engine output ranges overlap to some degree. Pump sizes A, M, MW and P are produced in large volumes (Figure 1).

There are two different designs of the standard in-line fuel-injection pump:
- The open-type design of the Type M and A pumps with a cover plate at the side and
- The closed-type design of the Type MW and P pumps in which the plunger-and-barrel assemblies are inserted from the top

For even higher per-cylinder outputs, there are the pump sizes P10, ZW, P9 and CW.

There are two ways in which the plunger-and-element assemblies can be supplied with fuel (Figure 2):
    With the *longitudinal scavenging* (a), fuel flows from one plunger-and-barrel assembly to the next *in sequence*.

With the *crossflow scavenging* (b), the plunger-and-barrel assemblies are supplied individually from a *common supply channel*. In this way, the fuel-delivery termination pressure does not affect the adjacent cylinder. This achieves tighter quantity tolerances and more precise fuel proportioning.

Fig. 2
a Longitudinal
   scavenging
b Crossflow
   scavenging
   (Type P-8000 pump)

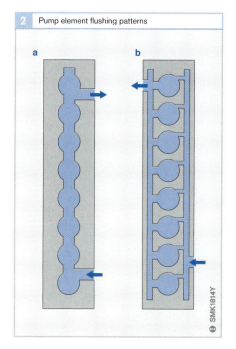

2 Pump element flushing patterns

a   b

1 Comparison of in-line fuel-injection pump sizes (sectional view)

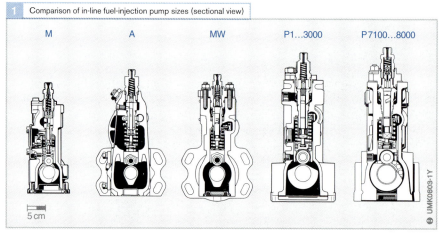

M   A   MW   P1...3000   P7100...8000

5 cm

1978 diesel speed records

In April 1978 the experimental Mercedes-Benz C111-III set nine world speed records, some of which still stand today, and eleven international class records. Some of those records had previously been held by gasoline-engine cars.

The average speed of the record attempts was approximately 325 kph. The highest speed reached was measured at 338 kph. The average fuel consumption was only 16 *l*/100 km.

These considerable achievements were made possible primarily by the highly streamlined plastic body. Its aerodynamic drag coefficient of 0.195 was sensationally low for the time.

The car was powered by a 3-liter, five-cylinder in-line diesel engine with a maximum power output of 170 kW (230 bhp). That meant that it was twice as powerful as its standard production counterpart. The maximum torque of 401 Nm was produced at 3,600 rpm. This performance was made possible by a turbocharger and an intercooler.

NMM0598Y

Engine compartment of the Mercedes-Benz C111-III

At the engine's nominal speed, the turbocharger was rotating at 150,000 rpm.

Precise fuel delivery and metering was provided by a Bosch Type PE...M in-line fuel-injection pump

NMM0599Y

Size M fuel-injection pumps

The size M in-line fuel-injection pump (Figures 3 and 4) is the smallest of the Series PE pumps. It has a light-metal (aluminum) body that is attached to the engine by means of a flange.

The size M pump is an open-type in-line fuel-injection pump which has a cover plate on the side and the base. On size M pumps, the peak injection pressure is limited by the pump to 400 bar.

After removal of the side cover plate, the delivery quantities of the plunger-and-barrel assemblies can be adjusted and matched to one another. Individual adjustment is effected by moving the position of the clamp blocks (Figure 4, Item 5) on the control rack (4). When the fuel-injection pump is running, the control rack is used to adjust the position of the pump plungers and, as a result, the delivery quantity within design limits. On the size M pump, the control rack consists of a round steel rod that is flatted on one side. Fitted over the control rack are the slotted clamp blocks. Together with its control sleeve, the lever (3), which is rigidly attached to the control sleeve, forms the mechanical link with the corresponding clamp block. This arrangement is referred to as a rod-and-lever control linkage.

The pump plungers sit directly on top of the roller tappets (6). LPC adjustment is achieved by selecting tappet rollers of different diameters.

The size M pump is available in 4, 5 and 6 cylinder versions, and is suitable for use with diesel fuel only.

Fig. 4

1   Delivery valve
2   Pump barrel
3   Control-sleeve
     lever arm
4   Control rack
5   Clamp block
6   Roller tappet
7   Camshaft
8   Cam

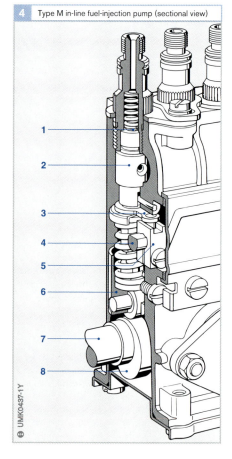

3   Type M in-line fuel-injection pump (external view)

UMK0436-1Y

4   Type M in-line fuel-injection pump (sectional view)

1
2
3
4
5
6
7
8

UMK0437-1Y

Size A fuel-injection pumps

The size A in-line fuel-injection pump (Figures 5 and 6) is the next size up from the size M pump and offers larger delivery quantities as a result.

It has a light-metal housing and can be either flange-mounted to the engine or attached by means of a cradle mounting.

On the size A fuel-injection pump, which is also an open-type design, the pump barrel (Figure 6, Item 2) is inserted directly into the aluminum body from above. It is pressed by the pressure-valve holder against the pump housing via the pressure-valve support. The sealing pressures, which are considerably higher than the hydraulic delivery pressures, must be withstood by the pump housing. For this reason, the peak pressure for a size A pump is internally limited to 600 bar.

In contrast with the size M pump, the size A pump has an adjusting screw (7) for setting the plunger lift to port closing. This simplifies the process of adjusting the basic setting. The adjusting screw is screwed into the roller tappet and fixed by a locking nut.

Another difference with the size M pump is the rack-and-pinion control linkage instead of the rod-and-lever arrangement. This means that the control rack is replaced by a rack (4). Clamped to the control sleeve (5) there is a control-sleeve gear. By loosening the clamp bolt, each control sleeve can be rotated relative to its control-sleeve gear in order to equalize the delivery quantities between individual plunger-and-barrel assemblies.

With this design of pump, all adjustments must be carried out without the pump running and with the housing open. A cover plate is positioned on the side of the pump housing and provides access to the valve-spring chamber.

Size A pumps are available in versions for up to 12 cylinders and, in contrast with the size M models, are suitable for multifuel operation.

5 Type A in-line fuel-injection pump (external view)

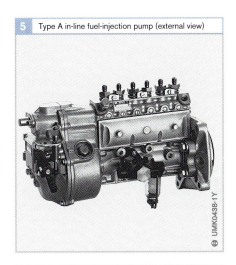

UMK0438-1Y

6 Type A in-line fuel-injection pump (sectional view)

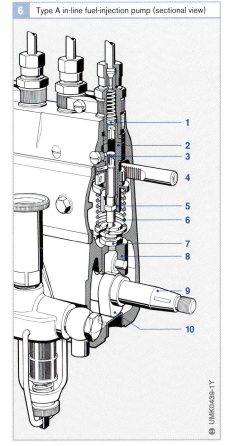

UMK0439-1Y

Fig. 6
1 Delivery valve
2 Pump barrel
3 Pump plunger
4 Control rack
5 Control sleeve
6 Plunger spring
7 Adjusting screw
8 Roller tappet
9 Camshaft
10 Cam

Size MW fuel-injection pumps

For higher pump outputs, the size MW in-line fuel-injection pump was developed (Figures 7 and 8).

The MW pump is a closed-type in-line fuel-injection pump which has a peak pressure limited to 900 bar, it is a lightweight metal design similar to the smaller models, and is attached to the engine by a baseplate, flange or cradle mounting.

Its design differs significantly from that of the Series M and A pumps. The main distinguishing feature of the MW pump is the barrel-and-valve assembly that is inserted into the pump housing from above. The barrel-and-valve assembly is assembled outside the housing and consists of the pump barrel (Figure 8, Item 3), the delivery valve (2) and the pressure-valve holder. On the MW pump, the pressure-valve holder is screwed directly into the top of the longer pump barrel. Shims or spacers of varying thicknesses are fitted between the pump housing and the barrel-and-valve assembly to achieve LPC adjustment. The uniformity of fuel delivery between the barrel-and-valve assemblies is adjusted by rotating the barrel-and-valve assembly from the outside. To achieve this, the flange (1) is provided with slots. The position of the pump plunger is not altered by this adjustment.

The MW pump is available with the various mounting options in versions for up to 8 cylinders. It is suitable for diesel fuel only. MW pumps are no longer used for new engine designs.

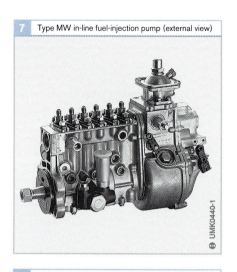

7 Type MW in-line fuel-injection pump (external view)

UMK0440-1

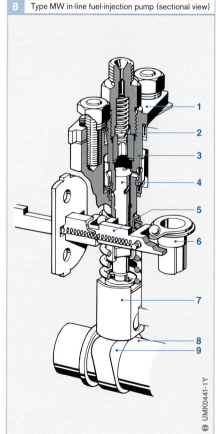

8 Type MW in-line fuel-injection pump (sectional view)

1
2
3
4
5
6
7
8
9

UMK0441-1Y

Fig. 8

1   Pump unit mounting flange
2   Delivery valve
3   Pump barrel
4   Pump plunger
5   Control rack
6   Control sleeve
7   Roller tappet
8   Camshaft
9   Cam

Size P fuel-injection pump

The size P in-line fuel-injection pump was similarly developed for higher pump outputs (Figures 9 and 10). Like the MW pump, it is a closed-type fuel-injection pump and is attached to the engine by its base or by a flange. On size P pumps for peak internal pressures of up to 850 bar, the pump barrel (Figure 10, Item 4) is inside an additional flange bushing (3) in which there is an internal thread for the pressure-valve holder. With this design, the sealing forces do not act on the pump housing. LPC adjustment on the P pump takes place in the same way as on the MW pump.

In-line fuel-injection pumps with low injection pressures use conventional fuel gallery flushing whereby the fuel passes through the fuel galleries of the individual barrel-and-valve assemblies one after the other from the fuel inlet to the return outlet, traveling along the pump longitudinal axis (longitudinal scavenging). On size P pumps of the type P 8000, which are designed for injection pressures at the pump of 1,150 bar, this flushing method inside the pump would result in a significant temperature difference in fuel temperature (as much as 40 °C) between the first and the last cylinder. Consequently, different quantities of energy would be injected into the individual combustion chambers of the engine (the energy density of the fuel decreases with increasing temperature and the associated increase in volume). For this reason, this type of fuel-injection pump has crossflow scavenging (i.e. at right angles to the pump longitudinal axis) whereby the fuel galleries of the individual barrels are isolated from one another by flow throttles and are flushed in parallel with fuel at virtually identical temperatures.

The P-type pump is produced in versions for up to 12 cylinders and is suitable both for diesel-only and for multifuel operation.

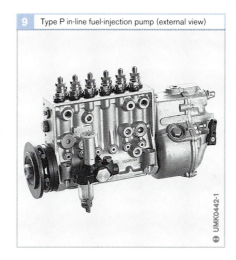

9   Type P in-line fuel-injection pump (external view)

UMK0442-1

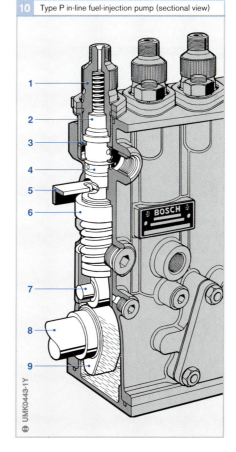

10   Type P in-line fuel-injection pump (sectional view)

UMK0443-1Y

Fig. 10

1   Pressure-valve holder
2   Delivery valve
3   Flange bushing
4   Pump barrel
5   Control rack
6   Control sleeve
7   Roller tappet
8   Camshaft
9   Cam

Size P10 fuel-injection pump

The size P10 in-line fuel-injection pump is the smallest of the models described below for larger diesel engines such as are used for off-road applications, fixed installations, construction and agricultural machinery, specialized vehicles, railway locomotives and ships. It is mounted on the engine by means of a baseplate.

The peak injector pressure is limited to approx. 1,200 bar.

The closed-type light-metal body (Figure 12, Item 13) holds the barrel-and-flange elements that are inserted from the top. They consist of a pump barrel (5), a constant-pressure valve and a pump plunger (12). They are held in position by stud bolts (3). A pressure-valve holder (1) seals the constant-pressure valve. As a result, the pump housing is not subjected to sealing stresses. Fitted directly in the pump barrels are impact-deflector screws (4) which protect the pump housing from damage caused by high-energy cutoff jets at the end of the delivery stroke. On the control sleeve (8) there are two link arms with thin cylindrical end lugs which locate in mating slots on the control rack (6).

For balancing the delivery quantity between plunger-and-barrel assemblies, the pump barrels have slotted mounting holes on their flanges. This allows the pump barrels to be suitably adjusted before they are tightened in position. The LPC is adjusted by inserting shims or spacers (2) of varying thicknesses between the pump barrels and the pump housing. To make them easier to replace, the shims are slotted so that they can be inserted from the side.

In order to remove a roller tappet (10) when servicing the pump, the corresponding pump barrel must first be removed. The spring seat (7) above the plunger spring (9) is then pressed downwards. A retaining spring (11) holding the spring seat then releases it. The spring seat, control sleeve, plunger spring, pump plunger and roller tappet can then be removed from above.

Fig. 12
1 Constant-pressure valve socket
2 Shims
3 Stud bolts
4 Impact-deflector screw
5 Pump barrel with mounting flange
6 Control rack
7 Spring seat
8 Control sleeve
9 Plunger spring
10 Roller tappet
11 Spring ring
12 Pump plunger
13 Housing
14 Camshaft

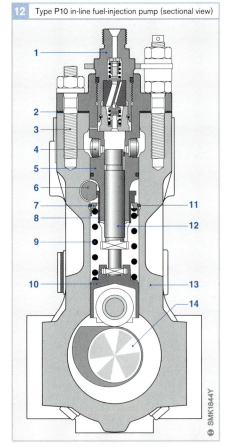

11 Type P10 in-line fuel-injection pump (external view)

12 Type P10 in-line fuel-injection pump (sectional view)

To refit these components, the plunger spring is compressed using the spring seat and the retaining spring which is snapped into position in the pump housing using a special device.

The camshaft runs on roller elements in the pump housing at each end. In order to obtain a high degree of rigidity, it is also supported by one or two half-shell plain bearings.

The size P10 fuel- injection pump is connected to the engine lube-oil circuit. A throttle bore determines the rate of oil flow. The fuel galleries of the individual plunger-and-barrel assemblies are interconnected and fuel circulates through the pump in a longitudinal direction (longitudinal scavenging). The presupply pump is usually either a gear pump driven by the engine or an electric fuel pump. For effective supply of the fuel-injection pump (and therefore efficient pump cooling), its delivery rate is several times the required fuel quantity.

Size P10 fuel-injection pumps are produced in versions for 6, 8 and 12 cylinders. The standard design is for diesel fuel only, with a special version available for multifuel operation.

Size P9 fuel-injection pump

The size P9 in-line fuel-injection pump is more or less identical in design to the P10 pump. However, it is somewhat larger and therefore positioned between the ZW and CW models.

The P9 fuel-injection pump has a closed-type light-metal housing. As with the P10, the peak nozzle pressure is limited to approx. 1,200 bar. It is attached to the engine by means of a cradle mounting. It is produced in versions for 6, 8 and 12 cylinders. The pump delivery quantity is controlled by a hydraulic or electromechanical governor provided by the engine manufacturer.

Size ZW fuel-injection pump

The size ZW in-line fuel-injection pump (Figure 13) has an open-style light-metal housing. The pump is attached to the engine by means of a cradle mounting. The peak nozzle pressure is limited to 950 bar.

The pressure-valve holder (Figure 14 overleaf, Item 1) screwed into the pump housing (18) provides the seal between the delivery valve and the pump barrel (2) as well as transmitting the hydraulic forces from the plunger. A fixing bolt (14) holds the pump barrel in position.

Two hardened impact-deflector screws (3) fitted in the pump housing opposite the control ports for each cylinder protect the pump housing from damage caused by the high-energy cutoff jet at the end of the delivery stroke.

The delivery quantity is controlled by means of a control rack in the form of a rack (4). This meshes with the control-sleeve gear that is clamped to the control sleeves (6).

For balancing the delivery quantities of the individual plunger-and-barrel assemblies, the clamp bolts (15) are loosened. Each control-sleeve gear can then be rotated relative to its control sleeve. The clamp bolts are then retightened.

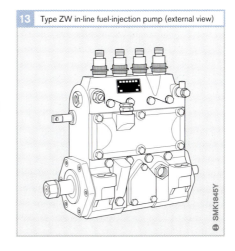

13 Type ZW in-line fuel-injection pump (external view)

SMK1845Y

LPC adjustment takes place by fitting or replacing the LPC disk (9) or a screw in the roller tappet (10).

For the purposes of removing the camshaft (11), the roller tappets can be held at their upper limit of travel by a retaining screw (17) fitted in the side of the pump housing. The camshaft runs on roller elements. For larger numbers of cylinders, there may also be one or two half-shell plain bearings in addition.

The presupply pump used may be a reciprocating piston pump which is flange-mounted on the side of the fuel-injection pump or a separate ring-gear pump or electric fuel pump. The fuel-injection pump is lubricated by the engine lube-oil circuit.

Size ZW fuel-injection pumps are available for engines with 4...12 cylinders. They are suitable for operation with diesel fuel. Fuel-injection pumps with the designation ZW(M) are designed for multifuel operation.

Size CW fuel-injection pump

The size CW in-line fuel-injection pump completes the top end of the Bosch in-line fuel-injection pumps range. The typical area of application for this model is on heavy-duty and relatively slow-revving marine engines and off-highway power units with nominal speeds of up to 1,800 rpm and power outputs of up to 200 kW per cylinder.

Even the 6-cylinder version of this fuel-injection pump with its closed-style pump housing made of nodulized cast iron weighs around 100 kg – this is roughly the weight of medium-sized car engine.

The pump is attached to the engine by eight bolts through its base.

The peak injection pressure is limited to approx. 1,000 bar.

The sealing and retention forces of the pump barrels with their plunger diameters of up to 20 mm are transferred to the pump housing by means of four strong clamp bolts (Figure 15, Item 1).

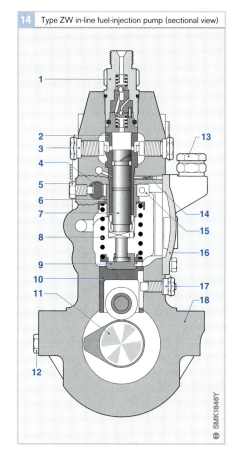

14 Type ZW in-line fuel-injection pump (sectional view)

SMK1846Y

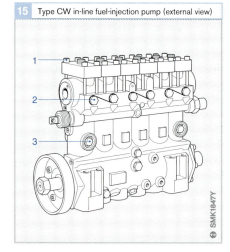

15 Type CW in-line fuel-injection pump (external view)

SMK1847Y

Fig. 14
1 Constant-pressure valve socket
2 Pump barrel
3 Impact-deflector screw
4 Control rack
5 Control rack guide screw
6 Control sleeve
7 Pump plunger
8 Plunger spring
9 LPC disk
10 Roller tappet
11 Camshaft
12 Oil-level checking plug
13 Oil filler plug
14 Pump-unit fixing screw
15 Clamp bolt
16 Cover plate
17 Retaining screw
18 Pump housing

Fig. 15
1 Clamp bolt
2 Impact-deflector screw
2 Screw cap

The control rack is in the form of a rack. Balancing of the delivery quantity between plunger-and-barrel assemblies is achieved with the aid of small orifices in the side of the pump housing. They are sealed by screw caps (3). LPC adjustment is by inserting shims of varying thicknesses between the roller tappets and the pump plungers.

Fuel supply to the fuel-injection pump is provided by a gear pump driven by the engine or an electric fuel pump.
   The fuel-injection pump is controlled by a hydraulic or electromechanical governor provided by the engine manufacturer.
   The pump is produced in 6, 8 and 10-cylinder versions and is suitable for use with diesel fuel.

In-line fuel-injection pumps for special applications
In addition to their use with internal combustion engines, there are a number of specialized applications in which in-line fuel-injection pumps (e.g. driven by an electric motor) are employed. Those include applications in the
● Chemical industry
● Textiles industry
● Machine-tool industry and
● Plant engineering industry

Fuel-injection pumps used in these areas are referred to as *press pumps*. They are mainly Type P and Type ZW(M) designs. Type PE single-plunger fuel-injection pumps without their own camshaft may also be used.
   The applications listed above require the delivery or finely and evenly atomized injection of fluids in very small but precisely metered quantities at high pressures. They frequently also demand the ability to vary the delivery quantity quickly, smoothly and as easily as possible.

The fluids pumped must not chemically attack the pump materials (aluminum, copper, steel, perbunane, nylon) to any discernible degree nor contain any solid, i.e. abrasive, components as this is the only way in which premature wear of the pump elements can be prevented. Where necessary, the fluids must be thoroughly filtered before they enter the press pumps. Depending on the fluids involved, special components (e.g. non-corroding compression springs, treated-surface fuel galleries, special seals) may need to be fitted to the press pumps.

High-viscosity fluids must be delivered to the press pump under sufficiently high pressure or made less viscous before passing through the filter by being heated (to max. 80 °C).
   The viscosity limits for pumped fluids are $v = 7.5 \cdot 10^{-5}$ m²/s; or with a higher fuel-gallery pressure of up to 2 bar $v = 38 \cdot 10^{-5}$ m²/s.
   The fluid pumped should enter the fuel gallery at a pressure of up to 2 bar – depending on viscosity. This can be achieved by a presupply pump mounted on the press pump, a sufficient static head of pressure or a pressurized fluid reservoir.

Delivery capacities are measured using standard commercially available diesel fuels. If fluids of differing viscosities are used, delivery capacities may vary. Precise determination of the maximum delivery quantity is only possible using the actual fluid pumped and in situ in the actual installation.

The permissible *delivery pressure* also depends on whether the pump is operated intermittently or continuously. For Type ZW(M) press pumps, the maximum permissible pressure may be as much as 1,000 bar under certain circumstances (consultation required). If there is a possibility that a peak pressure above the maximum permissible limit may occur during operation, then a safety valve must be fitted in the high-pressure line.

Type PE in-line fuel-injection pumps for alternative fuels

Some specially designed diesel engines can also be run on "alternative" fuels. For such applications, modified versions of the MW and P-type pumps are used.

Multifuel operation

Multifuel engines can be run not only on diesel fuel but also on petrol, paraffin or kerosene. The changeover from one type of fuel to another requires adjustments to the fuel metering system in order to prevent large differences in power output. The most important fuel properties are boiling point, density and viscosity. In order that those properties can be balanced against one another to optimum effect, design modifications to the fuel-injection equipment and the engine are necessary.

Because of the low boiling points of alternative fuels, the fuel has to circulate more rapidly and under greater pressure through the fuel gallery of the fuel-injection pump. There is a special presupply pump available for this purpose.

With low-density fuels (e.g. petrol), the full-load delivery quantity is increased with the aid of a reversible control-rod stop.

In order to prevent leakage losses with low-viscosity fuels, the pump elements have a leakage trap that takes the form of two ring grooves in the pump barrel (see section "Pump unit with leakage return channel"). The upper groove is connected to the fuel gallery by a bore. The fuel that leaks past the plunger during the delivery stroke expands into this groove and flows through the bore back into the fuel gallery.

The lower groove has an inlet passage for the sealing oil. Oil from the engine lube-oil circuit is forced under pressure into this groove via a fine filter. At normal operating speeds, this pressure is greater than the fuel pressure in the fuel gallery, thereby reliably sealing the pump element. A non-return valve prevents crossover of fuel into the lubrication system if the oil pressure drops below a certain level at idle speeds.

Running on alcohol fuels

Suitably modified and equipped in-line fuel-injection pumps can also be used on engines that run on the alcohol fuels methanol or ethanol. The necessary modifications include:
- Fitting special seals
- Special protection for the surfaces in contact with the alcohol fuel
- Fitting non-corroding steel springs and
- Using special lubricants

In order to supply an equivalent quantity of energy, the delivery quantity has to be 2.3 times higher than for diesel fuel in the case of methanol and 1.7 times greater with ethanol. In addition, greater rates of wear must be expected on the delivery-valve and nozzle-needle seats than with diesel fuel.

Running on organic fuels (FAME[1]))

For use with FAME, the fuel-injection pump has to be modified in a similar manner to the changes required for alcohol fuels.

RME[2]) is one of the varieties of FAME frequently used. With *unmodified fuel-injection pumps*, the present maximum allowable proportion of RME that may be added to the diesel fuel is 5 % based on the draft European standard of 2000. If higher proportions or poorer fuel qualities are used, the fuel-injection system may become clogged or damaged. In future there may be other types of FAME that are used either in pure form or as an additive to diesel fuel ($\leq 5\%$).

A definitive standard for FAME is currently in preparation. It will be required to precisely define fuel properties, stability and maximum permissible levels of contamination. Only by such means can trouble-free operation of the fuel-injection system and the engine be ensured.

[1]) FAME: Fatty Acid Methyl Ester, i.e. animal or vegetable oil

[2]) RME: Rape-oil Methyl Ester

Operating in-line fuel-injection pumps

In order to operate correctly, a fuel-injection pump must be correctly adjusted, vented to remove all air, connected to the engine lube-oil circuit and its start of delivery must be synchronized with the engine. Only in this way is it possible to obtain the optimum balance between engine fuel consumption and performance and the ever stricter statutory regulations for exhaust-gas emission levels. Consequently a fuel-injection pump test bench is indispensable (see chapter "Service technology").

Venting

Air bubbles in the fuel impair the proper operation of the fuel-injection pump or disable it entirely. The system should therefore always be vented after replacing the filter or any other repair or maintenance work on the fuel-injection pump. While the system is in operation, air is reliably expelled via the overflow valve on the fuel filter (continuous venting). On fuel-injection pumps without an overflow valve, a flow throttle is used.

Lubrication

Fuel-injection pumps and governors are connected to the engine lube-oil circuit. Then the fuel-injection pump is maintenance-free.

On pumps that are attached to the engine through the base or by a cradle mounting, the oil returns to the engine through a lube-oil return (Figure 1). If the fuel-injection pump is flange-mounted to the engine at its end face, the oil can return directly through the camshaft bearing or special oil bores.

The oil level check takes place at the same time as the regular engine oil changes specified by the engine manufacturer and is performed by removing the oil check plug on the governor. Fuel-injection pumps and governors with separate oil systems have their own dipsticks for checking the oil level.

Shutting down

If the engine, and therefore the fuel-injection pump, is taken out of service for a long period, no diesel fuel may remain inside the fuel-injection pump. Resinification of the diesel fuel would occur, causing the pump plungers and delivery valves to stick and possibly corrode. For this reason, a proportion of up to 10 % of a reliable rust-inhibiting oil is added to the diesel fuel in the fuel tank and the fuel is then circulated through the fuel-injection pump for 15 minutes. The same proportion of rust-inhibiting oil is added to the lubricant in the fuel-injection pump's camshaft housing.

New fuel-injection pumps with a "p" in their identification code have been factory-treated with an effective anticorrosive.

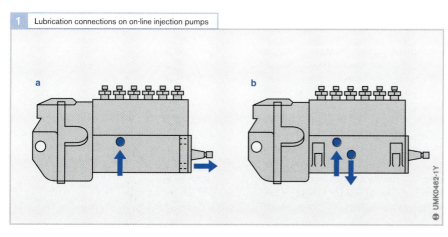

1 Lubrication connections on on-line injection pumps

a

b

Fig. 1
a Return line via bearing at driven end
b Return via return line

Governors and control systems for in-line fuel-injection pumps

A diesel fuel-injection pump must reliably supply the engine with precisely the right amount of fuel at exactly the right time under all operating conditions, in all operating statuses and at all engine loads. Even with the control rack in a fixed position, the engine would not maintain an absolutely constant speed. Effective operation of the fuel-injection pump therefore requires a mechanical centrifugal governor or an electronic control system.

The fuel-injection pump delivers precisely metered amounts of fuel at high pressure to the nozzles so that it is injected into the engine's combustion chamber. The fuel-injection system has to ensure that fuel is injected
- In precisely metered quantities according to engine load
- At precisely the right moment
- For a precisely defined length of time and
- In a manner compatible with the combustion method used

It is the job of the fuel-injection pump and governor to ensure that these requirements are met.

The characteristic features of *mechanical governors* are their durability and ease of maintenance. The main topic of this chapter is an examination of the various types of governor and adjustment mechanisms.
An *Electronic Diesel Control* (Electronic Diesel Control, EDC) performs a substantially more comprehensive range of tasks than a mechanical governor. The system of electrical actuators for the EDC system is described at the end of this chapter. The structure of the system is described in a separate chapter.
In the past *pneumatic governors* were also used for smaller fuel-injection pumps. They utilize the intake-manifold pressure (see next page). Because of today's greater demands with regard to control quality, however, the pneumatic governor is no longer produced and therefore not described in any greater detail in this manual.

Open and closed-loop control

Control systems are systems in which one or more input variables (reference variables and disturbance values) govern one or more output variables (Figure 1).

Open-loop control
In an open-loop control system (Figure 1a), the effects of control commands are not monitored (open-control loop). This method is used for proportioning the start quantity, for example.

Closed-loop control
The distinguishing feature of a closed-loop control system (Figure 1b) is the circular nature of the control sequence. The actual value of the controlled variable is constantly compared with the setpoint value. As soon as a discrepancy is detected, an adjustment is made to the settings of the actuators. The advantage of closed-loop control is that external disturbance values on the control process can be detected and compensated for (e.g. changes in engine load). Closed-loop control is used for the engine idle speed, for example.

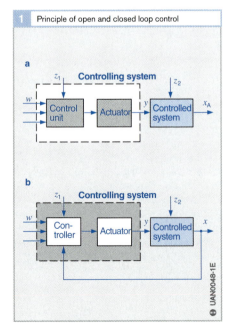

1   Principle of open and closed loop control

a

Controlling system

Control unit → Actuator → Controlled system

b

Controlling system

Controller → Actuator → Controlled system

UAN0048-1E

Fig. 1
a   Open control loop
b   Closed control loop

w Reference variable(s)
x Controlled variable (closed loop)
x_A Controlled variable (open loop)
y Manipulated variable(s)
z_1, z_2 Disturbance values

▶ History of the governor

"Anyone who thinks a diesel engine is a crude machine that will tolerate crude solutions is mistaken!" [1]

A large degree of sensitivity and precision is needed to obtain and maintain the very best performance from a diesel engine.

The specific method by which a diesel engine was governed was originally left to the engine manufacturers themselves. However, in order to do away with the need for a drive system running off the engine, they started to demand fuel-injection pumps with ready-mounted governors.

At the end of the 1920s Bosch took up that new challenge and, as a result of some outstanding engineering work, had a centrifugal idling and maximum-speed governor in volume production by 1931. A variation of that design followed shortly in the guise of a variable-speed governor that was in great demand for tractors and marine engines.

For smaller, faster running diesel engines in motor vehicles, on the other hand, a centrifugal governor did not seem suitable. It wasn't until the pneumatic governor was conceived that new impetus was introduced: "The control rack is attached to a leather diaphragm and the depression in the intake manifold, which is dependent on engine speed, alters the position of the diaphragm and adjusts the delivery quantity according to the position of the control flap" (see illustration). [2]

In the post-war years, an enormous variety of improved designs were used such as floating-pivot governors (1946 to 1948), governors with external springs (1955 onwards) and governors with vibration dampers.

Additional attachments for matching the full-load delivery quantity to the desired engine torque curve were also used as well as devices for automatically adjusting the start quantity.

Today, electronics are as important in diesel fuel injection as in any other branch of technology. An optimized diesel engine controlled by an electronic control system is now virtually taken for granted.

▼ Pneumatic governor
Illustration taken from the publication "Bosch und der Dieselmotor" issued in 1950

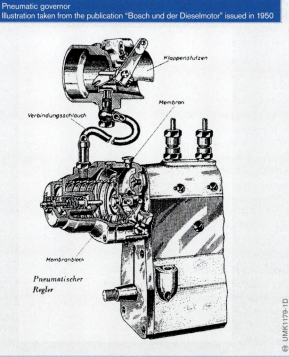

Klappenstutzen

Membran

Verbindungsschlauch

Membranblock

Pneumatischer Regler

UMK1179-1D

[1] Georg Auer; "Der Widerspenstigen Zähmung"; *Diesel-Report;* Robert Bosch GmbH; Stuttgart, 1977/78

[2] Friedrich Schildberger; *Bosch und der Dieselmotor;* Stuttgart, 1950

Action of the governor/control system

All in-line fuel-injection pumps have a pump element consisting of a pump barrel (Figure 1, Item 8) and plunger (9) for each engine cylinder. The quantity of fuel injected can be altered by rotating the pump plunger (see chapter "Type PE standard in-line fuel-injection pumps"). The governor/control system adjusts the position of all pump plungers simultaneously by means of the control rack (7) in order to vary the injected fuel quantity between zero and maximum delivery quantity. The control rack travel, s, is proportional to the injected fuel quantity and therefore to the torque produced by the engine.

The helix on the pump plunger can be of various types. Where there is only a bottom helix, fuel delivery always starts at the same point of plunger lift but ends at a variable point dependent on the angle of rotation of the piston. Where there is a top helix, the start of delivery can also be varied. There are also designs which incorporate both a top and bottom helix.

Definitions

No load

No load refers to all engine operating statuses in which the engine is overcoming only its own internal friction. It is not producing any torque output. The accelerator pedal may be in any position. All speed ranges up to and including breakaway speed are possible.

Idle

The engine is said to be idling when it is running at the lowest no-load speed. The accelerator pedal is not depressed. The engine is not generating any output torque. It is overcoming only the internal friction. Some sources refer to the entire no-load range as idling. The upper no-load speed (breakaway speed) is then called the upper idle speed.

Full load

At full load (wide-open throttle, WOT), the accelerator pedal is fully depressed. Under steady-state conditions, the engine is generating its maximum possible torque. Under non steady-state conditions (limited by turbocharger/supercharger pressure) the engine

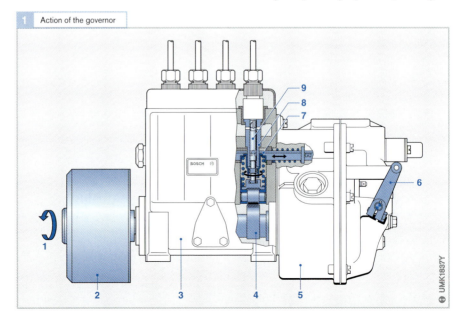

1   Action of the governor

Fig. 1
1   Pump drive
2   Timing device
3   Pump housing
4   Camshaft
5   Governor housing
6   Control lever
7   Control rack
8   Pump barrel
9   Pump plunger

develops the maximum possible (lower) full-load torque with the quantity of air available. All engine speeds from idle speed to nominal speed are possible. At breakaway speed the governor/control system automatically reduces the injected fuel quantity and therefore engine torque.

Part load
Part load covers the range between no load and full load. The engine is generating an output between zero and the maximum possible torque.

Part load at idle speed
In this particular case, the governor holds the engine at idle speed. The engine is generating torque output. This may extend to full load.

Overrunning
The engine is said to be overrunning when it is driven by an external force acting through the drivetrain (e.g. when descending an incline).

Steady-state operation
The engine's torque output is equal to the required torque. The engine speed is constant.

Non steady-state operation
The engine's torque output is not equal to the required torque. The engine speed is not constant.

Indices
The indices used in the diagrams and equations in the rest of this chapter have the following meanings:
I   Idle
n   No load
v   Full load
u   Minimum figure
o   Maximum figure

Some examples:
n_{nu} Minimum no-load speed (= idle speed n_I)
n_n Any no-load speed
n_{no} Maximum no-load speed
n_v Any full-load speed
n_{vo} Maximum full-load speed (nominal speed)

Proportional response of the governor

Every engine has a torque curve that corresponds to its maximum load capacity. For every engine speed there is a corresponding maximum torque.

If the load is removed from the engine without the position of the control lever being altered, the engine speed must not be allowed to increase by more than a permissible degree specified by the engine manufacturer (e.g. from full-load speed n_v to no-load speed, n_n, Fig. 2). The increase in engine speed is proportional to the change in engine load, i.e. the greater the amount by which the engine load is reduced, the greater the increase in engine speed. Hence the terms proportional response and proportional characteristics in connection with governors. The proportional response of the governor generally relates to the maximum full-load speed. That is equivalent to the nominal speed.

The proportional response δ is calculated as follows:

$$\delta = \frac{n_{no} - n_{vo}}{n_{vo}}$$

or as a percentage thus:

$$\delta = \frac{n_{no} - n_{vo}}{n_{vo}} \cdot 100\,\%$$

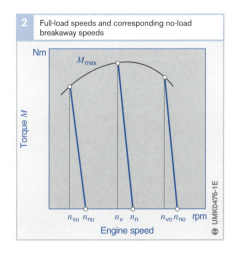

2   Full-load speeds and corresponding no-load breakaway speeds

Nm
M_{max}

Torque M

$n_{vu}\ n_{nu}$ n_v n_n $n_{vo}\ n_{no}$ rpm
Engine speed

UMK0475-1E

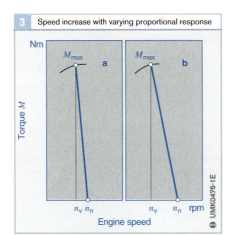

UMK0476-1E

Fig. 3
a   Small proportional response
b   Large proportional response

UMK0477-1E

Fig. 4
Curve for varying speeds set by means of the control lever

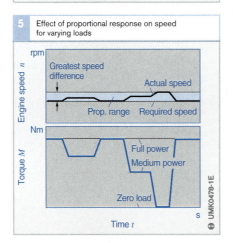

UMK0478-1E

where
n_{no} Upper no-load speed
n_{vo} Upper full-load speed
As the pump speed on four-stroke engines is equivalent to half the engine speed, both the pump speed and the engine speed can be used in the calculation.

Example (pump speeds):
$n_{no} = 1{,}000$ rpm, $n_{vo} = 920$ rpm

$$\delta = \frac{1{,}000 - 920}{920} \cdot 100\% = 8.7\%$$

Figure 5 shows the effect of the proportional response based on a practical example: At a constant set speed, the actual speed varies when the engine load alters (e.g. variations in gradient) within the proportional response range.

In general, a greater proportional response allows the achievement of more stable characteristics on the part of the entire control loop consisting of governor, engine and driven machine or vehicle. On the other hand, the proportional response is limited by the operating conditions.

Examples of proportional response:
● Approx. 0...5 % for power generators
● Approx. 6...15 % for vehicles

Purpose of the governor/ control system

The basic task of any governor/control system is to prevent the engine from exceeding the maximum revving speed specified by the engine manufacturer. Since the diesel engine always operates with excess air because the intake flow is not restricted, it would "overrev" if there were no means of limiting its maximum speed. Depending upon the type of governor/control system, its functions may also include holding the engine speed at specific constant levels such as idling or other speeds within a specific band or the entire range between idling and maximum speed. There may be also be other tasks beyond those mentioned, in which case the capabilities of an electronic control system are substantially more extensive than those of a mechanical governor.

The governor/control system is also required to perform control tasks such as
- Automatic enabling or disabling of the greater fuel delivery quantity required for starting (start quantity)
- Variation of the full-load delivery quantity according to engine speed (torque control)
- Variation of the full-load delivery quantity according to turbocharger and atmospheric pressure

Some of those tasks necessitate additional equipment which will be explained in more detail at a later stage.

Maximum speed control function

When the engine is running at maximum full-load speed, n_{vo}, it must not be allowed to exceed the maximum no-load speed, n_{no}, when the load is removed in accordance with the permissible proportional response (Figure 6). The governor/control system achieves this by moving the control rack back towards the stop setting.

The range $n_{vo} - n_{no}$ is referred to as the maximum speed cutoff range. The greater the proportional response of the governor, the greater the increase in speed from n_{vo} to n_{no}.

Intermediate-speed regulation

If the task so requires (e.g. on vehicles with PTO drives), the governor/control system can also hold the engine at specific constant speeds between idle speed and maximum speed according to the proportional response (Figure 7). When the intermediate speed regulation function is active, the engine speed, n, thus only varies according to load and within the engine's power band between n_v (at full load) and n_n (at no load).

Idle-speed regulation

The diesel engine's speed can also be controlled at the lower end of the speed range (Figure 8) – the idle speed range. Without a governor or control system, the engine would either slow down to a standstill or overrev uncontrollably when not under load.

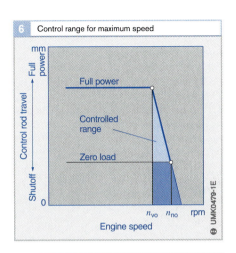

6   Control range for maximum speed

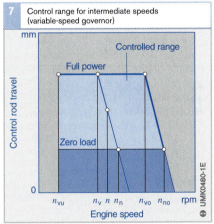

7   Control range for intermediate speeds (variable-speed governor)

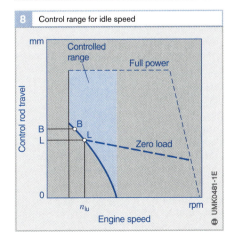

8   Control range for idle speed

When the control rack returns to position B from the starting position after the engine has been started from cold, the engine's internal friction levels are still relatively high. The quantity of fuel required to keep the engine running is therefore somewhat greater and the engine speed somewhat lower than that represented by the idle speed setting L.

As the engine warms up, the internal friction of the engine decreases as do the resistance levels of the external units such as the alternator, air compressor, fuel-injection pump, etc. that are driven by the engine. Consequently, the engine speed gradually increases and the control rack eventually returns to the position L. The engine is then at the idle speed for normal operating temperature.

Torque control

Torque control enables optimum utilization of the combustion air available in the cylinder. Torque control is not a true governor/control system function but is one of the control functions allocated to the governor/control system. It is calibrated for the full-load delivery capacities, i.e. the maximum quantity of fuel delivered within the engine's power output band and combustible without the production of smoke.

Conventionally aspirated engines

The fuel requirement of a non-turbocharged diesel engine generally decreases as the engine speed increases (lower relative air throughput, thermal limits, changes in mixture formation parameters). By contrast, the fuel delivery quantity of a Bosch fuel-injection pump increases with engine speed over a specific range assuming the control rack setting remains unchanged (throttle effect of the pump unit control port). Too much fuel injected into the cylinder, on the other hand, produces smoke or may cause the engine to overheat. The quantity of fuel injected must therefore be adjusted to suit the fuel requirement (Figure 9).

Governors/control systems with a torque control function move the control rack a specified distance towards the stop setting in the torque control range (Figure 10). Thus, as the engine speed increases (from n_1 to n_2),

Fig. 9
a   Engine fuel requirement
b   Full-load delivery quantity without torque control
c   Torque-matched full-load delivery quantity

Fig. 11
a   With torque control
b   Without torque control

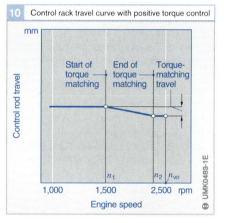

9   Characteristic curves for fuel requirement and delivery quantity

UMK0482-1E

10   Control rack travel curve with positive torque control

UMK0483-1E

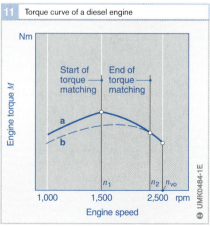

11   Torque curve of a diesel engine

UMK0484-1E

the fuel delivery quantity decreases (positive torque control), and when the engine speed decreases (from n_2 to n_1) the fuel delivery quantity increases.

Torque control mechanisms vary in design and arrangement from one governor/control system to another. Details are provided in the descriptions of the individual governors/control systems.

Figure 11 shows the torque curves of diesel engines with and without torque control. The maximum torque is obtained without exceeding the smoke limit across the entire engine speed range.

Turbocharged engines
In engines with high-compression turbochargers, the full-load fuel requirement at lower engine speeds increases so much that the inherent increase in delivery quantity of the fuel-injection pump is insufficient. In such cases, torque control must be based on engine speed or turbocharger pressure and effected by means of the governor/control system or the manifold pressure compensator alone or by the two in conjunction depending on the circumstances.

This type of torque control is referred to as negative. That means that the delivery quantity increases more rapidly as engine speed rises (Figure 12). This is in contrast with the usual positive torque control whereby the injection quantity is reduced as engine speed increases.

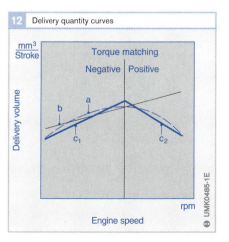

12 Delivery quantity curves

$\frac{mm^3}{Stroke}$

Torque matching

Negative | Positive

Delivery volume

a

b

c_1 c_2

rpm

Engine speed

UMK0485-1E

Types of governor/control system

Continually increasing demands with regard to exhaust-gas emissions, fuel economy and engine smoothness and performance are the defining characteristics of diesel engine development. Those demands are reflected in the requirements placed on the fuel-injection system and in particular the governor or control system.

The various different control tasks required result in the following types of governor:
- *Maximum-speed governors*
  only limit the engine's maximum speed.
- *Minimum/maximum-speed governors*
  also control the idle speed in addition to limiting the maximum speed. They do not control the intermediate range. The injected fuel quantity in that range is controlled by means of the accelerator pedal. This type of governor is used primarily on motor vehicles.
- *Variable-speed governors*
  limit not only the minimum and maximum speeds but also control the intermediate speed range.
- *Combination governors*
  are, as their name suggests, a combination of minimum/maximum-speed governors and variable-speed governors.
- *Generator-engine governors*
  are for use on engines that drive power generators designed to comply with DIN 6280 or ISO 8528.

Mechanical governors
The mechanical governors used with in-line fuel-injection pumps are also referred to as centrifugal governors because of the flyweights they employ. This type of governor is linked to the accelerator pedal by means of a rod linkage and a control lever (Figure 1 overleaf).

Timing device
In order to control the start of injection and compensate for the time taken by the pressure wave to travel along the high-pressure delivery line, a timing device is used to "advance" the start of delivery of the fuel-injection pump as the engine speed increases.

Fig. 12
a   Engine fuel requirement
b   Full-load delivery quantity without torque control
c   Torque-matched full-load delivery quantity
c_1 Negative torque control
c_2 Positive torque control

Electronic control systems

The Electronic Diesel Control EDC satisfies the greater demands placed on modern control systems. It enables electronic sensing of engine parameters and flexible electronic data processing. Closed control loops with electrical actuators offer not only more effective control functions in comparison with mechanical governors but also have additional capabilities such as smooth-running control. In addition, Electronic Diesel Control provide the facility for data exchange with other electronic systems such as the transmission control system and permit comprehensive electronic fault diagnosis. The subsystems and components of the EDC system for in-line fuel-injection pumps are described in the chapter "Electronic Diesel Control EDC".

Figures 1 and 2 show schematic diagrams of the control loops for mechanical governors and electronic control systems. Detailed illustrations of the control loops for standard in-line fuel-injection pumps and control-sleeve in-line fuel-injection pumps are displayed on the next double page.

Advantages of electronic control systems

The use of an electronically controlled fuel-injection system offers the following advantages:
- The extensive range of functions and available data enables the achievement of optimum engine response across the entire operating range.
- Clear separation of individual functions: Governor characteristics and fuel rate-of-discharge curves are no longer interdependent; consequently there is wider scope for adaptation to individual applications.
- Greater capacity for manipulating variables that previously could not be included in the equation with mechanical systems (e.g. compensation for fuel temperature, controlling idle speed independently of engine load).
- Higher levels of control precision and consistency throughout engine life by diminishing tolerance effects.
- Improved engine response characteristics: The large volume of stored data (e.g. engine data maps) and parameters allows optimization of the engine-and-vehicle combination.

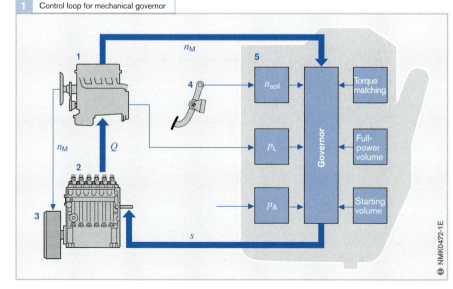

1   Control loop for mechanical governor

Fig. 1

1   Diesel engine
2   In-line fuel-injection pump
3   Timing device
4   Accelerator pedal
5   Governor

n_{req} Required engine speed
n_M Engine speed
p_A Atmospheric pressure
p_L Turbocharger pressure
Q Injected fuel quantity
s Control rack travel

- More extensive range of functions: Additional functions such as cruise control and intermediate-speed regulation can be implemented without major complications.
- Interaction with other electronic systems on the vehicle provides the potential for making future vehicles generally easier to use, more economical, more environmentally friendly and safer (e.g. electronic transmission control EGS, traction control system TCS).
- Substantial reduction of space requirements because mechanical attachments to the fuel-injection pump are no longer required.
- Versatility and adaptability: Data maps and stored parameters are programmed individually when the control unit reaches the end of the production line at Bosch or the engine/vehicle manufacturer. This means that a single control unit design can be used for several different engine or vehicle models.

Safety concept

For safety reasons, a compression spring moves the control rack back to the "zero delivery" position whenever the electrical actuators are disconnected from the power supply.

Self-monitoring: The Electronic Diesel Control EDC incorporates functions for monitoring the sensors, actuators and the microcontroller in the control unit. Additional safety is provided by extensive use of redundant backup. The diagnostic system provides the facility for obtaining a read-out of recorded faults on a compatible tester or, on older systems, using a diagnostic lamp.

Substitute functions: The system incorporates an extensive array of substitute functions. If, for example, the engine speed sensor fails, the signal from terminal W on the alternator can be used as a substitute for the speed sensor signal. If important sensors fail, a warning lamp lights up.

Fuel shutoff function: In addition to the fuel shutoff function of the control rack when in stop setting, a solenoid valve in the fuel supply line shuts off the fuel supply when disconnected from the power supply. This separate electric or electrohydraulic shutoff valve also shuts off the fuel supply if, for example, the fuel quantity control mechanism fails, thus stopping the engine.

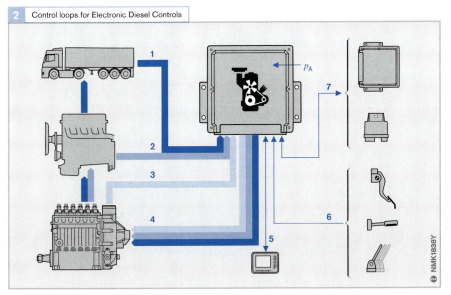

Fig. 2
1 Vehicle sensors (e.g. for road speed)
2 Engine sensors (e.g. for engine temperature)
3 Injection system sensors (e.g. for start of delivery)
4 Control signals
5 Diagnosis interface
6 Accelerator pedal and desired-value generators (switches)
7 Data communication (e.g. for glow-plug control)

P_A Atmospheric pressure

Control loop configurations

Engine starting, idling, performance, soot emissions and response characteristics are decisively affected by the injected fuel quantity. Accordingly, there are data maps for starting, idling, full load, accelerator characteristics, smoke emission limitation and pump characteristics stored on the control unit.

The control rack travel is used as a substitute variable for injected fuel quantity. For engine response characteristics, an RQ or RQV control characteristic familiar from mechanical governors can be specified.

A pedal-travel sensor detects the driver's torque/engine speed requirements as indicated by the accelerator pedal (Figure 3). The control unit calculates the required injected fuel quantity (required fuel-injection pump control rack setting), taking account of the stored data maps and the current sensor readings. The required control rack setting is then the reference variable for the control loop. A position control circuit in the control unit detects the actual position of the control rack, and thus the required adjustment, and provides for rapid and precise adjustment of the control rack position.

There are control functions for maintaining various engine speeds: idle speed, a fixed intermediate speed, e.g. for PTO drives, or a set speed for the cruise control function.

Control loop for injected fuel quantity
Based on the calculated required setting, the control unit sends electrical signals to the control rack actuation system for the fuel-injection pump. The required injected fuel quantity specified by the control unit is set using the position control loop: The control unit specifies a required control-rack travel and receives a feedback signal indicating the

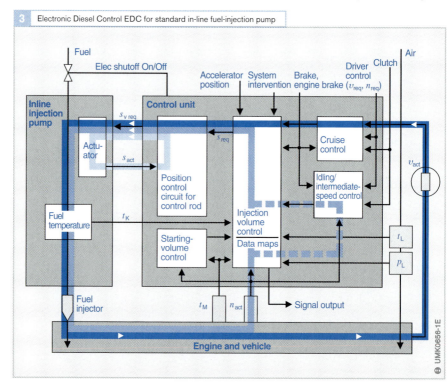

3   Electronic Diesel Control EDC for standard in-line fuel-injection pump

Fig. 3

n_{act} Actual engine speed
n_{req} Required engine speed
p_L Turbocharger pressure
s_{act} Actual control rack travel
s_{req} Required control rack travel
$s_{v\,req}$ Control rack positioning signal
t_K Fuel temperature
t_L Air temperature
t_M Engine temperature
v_{act} Actual road speed
v_{req} Required road speed

actual control rack position from the control rack position sensor. To complete the control loop, the control unit repeatedly recalculates the adjustment required to achieve the required fuel-injection pump setting, thereby continually correcting the actual setting to match the required setting.

Start of delivery control loop
Control-sleeve in-line fuel-injection pumps have a means of adjusting start of delivery as well as the mechanism for adjusting the injected fuel quantity (Figure 4).

The start of delivery is also adjusted by means of a closed control loop. A needle-motion sensor in one of the nozzle holders signals to the control unit the actual point in time at which injection takes place. Using this information in conjunction with stored data, the control unit then calculates the actual crankshaft position at which injection

starts. Next, it compares the actual start of delivery with the calculated required start of delivery. A signal control circuit in the control unit then operates the timing device on the fuel-injection pump, thus bringing the actual start of delivery into line with the required setting.

Because the timing device actuation mechanism is "structurally rigid", there is no need for an adjustment travel feedback sensor. Structurally rigid means that the lines of action of solenoid and spring always have a precise intersection point so that the travel of the solenoid is proportional to the signal current. That is equivalent to feedback within a closed control loop.

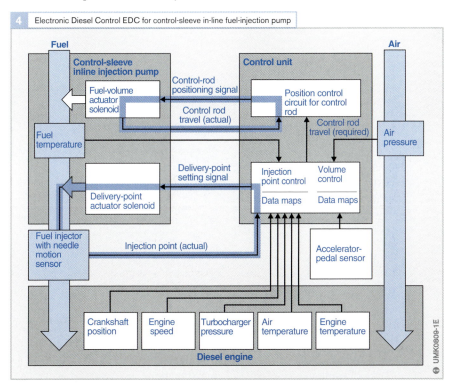

4   Electronic Diesel Control EDC for control-sleeve in-line fuel-injection pump

Overview of governor types

Governor type designations

The governor type designation is shown on the identification plate. It indicates the essential features of the governor (e.g. design type, governed speed range, etc.). Figure 3 details the individual components of the governor type designation.

Maximum-speed governors

Maximum-speed governors are intended for diesel engines that drive machinery at their nominal speed. For such applications, the governor's job is merely to hold the engine at its maximum speed; control of idle speed and start quantity are not required. If the engine speed rises above the nominal speed, n_{vo}, because the load decreases, the governor shifts the control rack towards the stop setting, i.e. the control rack travel is shortened and the delivery quantity reduced (Figure 1). Engine speed increase and control rack travel decrease follow the gradient A – B. The maximum no-load speed, n_{no}, is reached when the engine load is removed entirely. The difference between n_{no} and n_{vo} is determined by the proportional response of the governor.

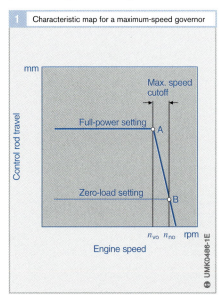

1 Characteristic map for a maximum-speed governor

mm

Control rod travel

Max. speed cutoff

Full-power setting
A

Zero-load setting
B

n_{vo} n_{no} rpm

Engine speed

UMK0486-1E

Minimum/maximum-speed governors

Diesel engines for motor vehicles frequently do not require engine speeds between idling and maximum speed to be governed. Within this range, the fuel-injection pump's control rack is directly operated by the accelerator pedal under the control of the driver so as to obtain the required engine torque. At idle speed, the governor ensures that the engine does not cut out; it also limits the engine's maximum speed. The governor's characteristic map (Figure 2) shows the following: When the engine is cold, it is started using the start quantity (A). At this point, the driver has fully depressed the accelerator pedal.

If the driver releases the accelerator, the control rack returns to the idle speed setting (B). While the engine is warming up, the idle speed fluctuates along the idle speed curve and finally comes to rest at the point L. Once the engine has warmed up, the maximum start quantity is not generally required when the engine is restarted. Some engines can even start with the control rack actuating lever (accelerator pedal) in the idling position.

An additional device, the temperature-dependent start quantity limiter, can be used to limit the start quantity when the engine is warm even if the accelerator pedal is fully depressed. If the driver fully depresses the accelerator pedal when the engine is running, the control rack is moved to the full-load setting. The engine speed increases as a result and when it reaches n_1, the torque control function comes into effect, i.e. the full-load delivery quantity is slightly reduced. If the engine speed continues to increase, the torque control function ceases to be effective at n_2. With the accelerator pedal fully depressed, the full-load volume continues to be injected until the maximum full-load speed, n_{vo}, is reached. Upwards of n_{vo}, the maximum speed limiting function comes into effect in accordance with the proportional response characteristics so that a further small increase in engine speed results in the control rack travel backing off so as to reduce the delivery quantity. The maximum no-load speed, n_{no}, is reached when the engine load is entirely removed.

2   Characteristic map for a minimum/maximum-speed governor with torque control

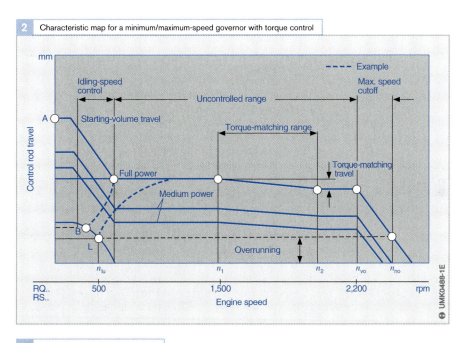

3   Bosch governor type designations

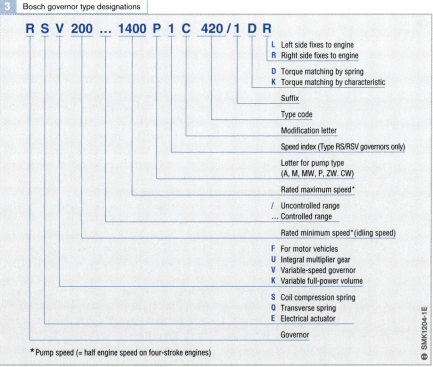

L	Left side fixes to engine
R	Right side fixes to engine
D	Torque matching by spring
K	Torque matching by characteristic
	Suffix
	Type code
	Modification letter
	Speed index (Type RS/RSV governors only)
	Letter for pump type (A, M, MW, P, ZW. CW)
	Rated maximum speed*
/	Uncontrolled range
...	Controlled range
	Rated minimum speed* (idling speed)
F	For motor vehicles
U	Integral multiplier gear
V	Variable-speed governor
K	Variable full-power volume
S	Coil compression spring
Q	Transverse spring
E	Electrical actuator
	Governor

* Pump speed (= half engine speed on four-stroke engines)

Fig. 3
With combination governors, multiple speeds are specified (e.g. 300/900...1,200).

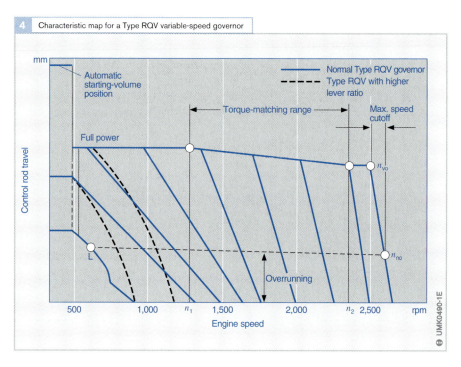

4   Characteristic map for a Type RQV variable-speed governor

When the engine is overrunning (e.g. coasting on a descent) the engine speed may increase further but the control rack travel is backed off to zero.

Variable-speed governors

Vehicles that have to maintain a specific speed (e.g. agricultural tractors, roadsweepers, ships) or have a PTO drive that requires the engine speed to be kept at a constant level (e.g. tank pumps, fire-engine ladders) are fitted with variable-speed governors.

This type of governor controls not only the idling and maximum speeds but also intermediate speeds independently of engine load. The desired speed is set by means of the control lever. The governor characteristic map (Figure 4) shows the following: the start quantity setting for cold starting, the full-load control characteristic with torque control between n_1 and n_2 up to the maximum speed cutoff band from the maximum full-load speed along the gradient n_{vo}, n_{no}.

The remaining curves show the cutoff characteristics for intermediate speeds. They reveal an increase in the proportional response as speed decreases. The curves shown as broken lines apply to vehicles whose PTO drives operate within the lower engine speed range. As the load increases, the engine speed does not dip as sharply as with normal governors (shallower curves). This is achieved by the use of a higher transmission ratio for the control lever.

Combination governors

If the normal proportional response of a Type RQV or RQUV variable-speed governor at the upper or lower end of the adjustment range is too great for the intended application, and control of intermediate speeds is not required, then the governor movement is designed as a combination governor. With such an arrangement, torque control is not possible in the uncontrolled range of the maximum-speed governor component. On this characteristic map (Figure 5), the uncontrolled stage is in the lower speed band, while the controlled

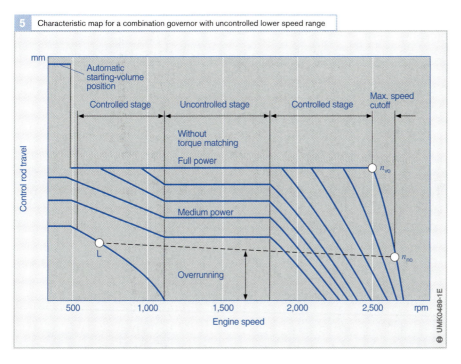

5 Characteristic map for a combination governor with uncontrolled lower speed range

stage is in the upper speed band. A different type of governor operates in the lower speed band as a variable-speed governor (downward-gradient curves), after which follows an uncontrolled band (horizontal sections of the curves) extending to the maximum speed cut-off band. In both cases, the horizontal sections of the curves represent the control rack travel for varying part-load control lever settings. The lines descending from the full-load curve represent the speed-regulation breakaway characteristics for varying set intermediate speeds. The combination governor differs in design from a variable-speed governor simply by virtue of the use of different governor springs.

Generator governor

For engines driving power generators, German regulations require that governors conform to DIN 6280 (see tables overleaf). Bosch centrifugal governors can be used for design classes 1, 2 and 3. The conditions for design class 4, to which units with a 0 % proportional response belong, usually require the use of an

electronic control system. A characteristic map for a generator governor is shown in Figure 6. If parallel operation is not required, the speed setting can be fixed, i.e. a straightforward maximum-speed governor can be used.

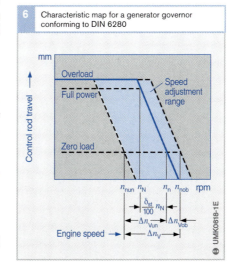

6 Characteristic map for a generator governor conforming to DIN 6280

1	Governor types			
Type	**Function**	**Actuator mechanism**	**Pump size**	**Torque control**
RQ	Minimum/maximum-speed governor or maximum-speed governor	flyweights	A, MW, P	Positive
RQ	Generator governor	flyweights	A, MW, P	None
RQU	Minimum/maximum-speed governor or maximum-speed governor	flyweights[1]	ZW, P9, P10	Positive
RS	Minimum/maximum-speed governor	flyweights	A, MW, P	Positive
RSF	Minimum/maximum-speed governor	flyweights	M	Negative/positive
RQV	Variable-speed or combination governor	flyweights	A, MW, P	Positive
RQUV	Variable-speed governor	flyweights[1]	ZW, P9, P10	Positive
RQV..K	Variable-speed governor	flyweights	A, MW, P	Negative/positive
RSV	Variable-speed governor	flyweights	A, M, MW, P	Positive
RSUV	Variable-speed governor	flyweights[1]	P	Positive
RE	Any characteristic	Electromagnet	A, MW, P	Negative/positive

Table 1
[1] With transmission
     ratio for slow-running
     engines

2	Operating limits for design classes						
No.	**Description**	**Symbol**	**Unit**	**Design class**			
				1	**2**	**3**	**4**
4.2.4	Static speed difference or proportional response	δ_{st}	%	8	5	3	STA
4.2.5	Speed fluctuation range	ν_n	%	–	1.5	0.5	STA
4.2.1	Lower speed setting range	$\delta \cdot n_{Vun}$	%	$-(2.5+\delta_{st})$	$-(2.5+\delta_{st})$	$-(2.5+\delta_{st})$	STA
4.2.2	Upper speed setting range	$\delta \cdot n_{Vob}$	%	+2.5	+2.5	+2.5	STA
4.1.6	Frequency regulation time STA	t_{fzu}, t_{fab}	s	–	5	3	STA

Table 2
Applies only to power
generator applications
Excerpt from DIN 6280,
Part 3

STA Subject to
     agreement

No.	Description	Symbol	Definition
3	**Speed definitions**		
4.1	Nominal speed	n_N	The engine speed corresponding to the rated frequency of the generator and to which the generator rated output relates.
4.3	Zero-output speed	n_n	Steady-state speed of the engine under no load. Associated figures for rated-output and intermediate output speeds relate to an unchanged speed setting.
4.7	Minimum variable zero-output speed	n_{nun}	Minimum steady-state engine speed under no load that can be set on the speed setting device or governor.
4.8	Maximum variable zero-output speed	n_{nob}	Maximum steady-state engine speed under no load that can be set on the speed setting device or governor.
4.9	Speed setting range	Δn_V	Range between set minimum and maximum zero-output speeds; the figure for the speed range is obtained by adding the figures for the upper and lower speed setting ranges as per sections 4.9.1 and 4.9.2.
4.9.1	Lower speed setting range	Δn_{Vun}	Range between set minimum zero-output speed and the zero-output speed that results from removal of the engine load at the rated output speed without alteration of the speed setting.
		δn_{Vun}	$\Delta n_{Vun} = n_n - n_{nun}$
			The difference between the two speeds is expressed as a percentage of the nominal speed
			$$\delta n_{Vun} = \frac{(n_n - n_{nun})}{n_N} \cdot 100$$
4.9.2	Upper speed setting range	Δn_{Vob}	Range between set maximum zero-output speed and the zero-output speed that results from removal of the engine load at the rated output speed without alteration of the speed setting.
			$\Delta n_{Vob} = n_{nob} - n_n$
		δn_{Vob}	The difference between the two speeds is expressed as a percentage of the nominal speed.
			$$\delta n_{Vob} = \frac{(n_{nob} - n_n)}{n_N} \cdot 100$$
5.1	Static speed difference or proportional response	δ_{St}	Ratio of speed difference between zero-output speed, n_n, and nominal speed, n_N, expressed as a percentage of the nominal speed.
			$$\delta_{St} = \frac{(n_n - n_N)}{n_N} \cdot 100$$

Table 3
Applies only to power generator applications
Excerpt from DIN 6280, Part 4

Mechanical governors

The Bosch centrifugal mechanical governor is mounted on the fuel-injection pump. The pump's control rack is connected by a rod linkage to the governor. The control lever on the governor housing forms the link to the accelerator pedal.

There are two possible governor movement designs with centrifugal governors:
- Type RQ and RQV governors: The governor springs are fitted inside the flyweights. The two flyweights then each act directly on a set of springs, which are dimensioned for the relevant nominal speed and corresponding proportional response.
- Type RSV, RS and RSF: The centrifugal force acts via a lever system on the governor spring that is external to the two flyweights. The two flyweights then act via the sliding bolt on the tensioning lever to which the governor spring is attached and acting in the opposite direction.

On the Type RSV governor (variable-speed governor) the driver sets the desired engine speed by tensioning the governor spring by means of the control lever. On the Type RS/RSF governor (minimum/maximum-speed governor) the governor spring setting for the high-idle speed is fixed and cannot be altered by means of the accelerator pedal. The governor springs in the movements are chosen so that the spring force and the centrifugal force are in equilibrium at the desired speed. When that speed is exceeded, the greater centrifugal force of the weights moves the control rack by means of a lever system and reduces the pump's delivery quantity.

Type RQ minimum/maximum-speed governor
Design
The governor hub is driven via a vibration damper by the fuel-injection pump camshaft (Figure 3, Items 14 and 18). Mounted on the hub are the two flyweights (17) with their bell cranks (13). A set of springs (16) is fitted inside each flyweight. The bell cranks convert the radial travel of the flyweights into an axial movement on the part of the sliding bolt (12) which the latter transmits to the sliding block (10). The sliding block, which is guided in a straight line by the guide pin (11), combines with the variable-fulcrum lever (5) to form the link between the flyweight speed-sensing element and the control rack (7). The lower end of the variable-fulcrum lever locates in the sliding block. Inside the variable-fulcrum lever is a sliding-block guide. The movable pivot block is attached radially to a linkage lever which is connected to the control lever (2) on the same axis. The control lever is operated either manually or via a rod linkage by the accelerator pedal. Movement of the control lever moves the position of the guide block so that the variable-fulcrum lever pivots around the clevis pin on the sliding block. When the governor comes into action, the pivot of the variable-fulcrum lever on the guide block. The sliding block thus varies the transmission ratio of the variable-fulcrum lever. Consequently, there is sufficient force to move the control rack even at idle speeds when the centrifugal force is relatively small. The spring sets inside the

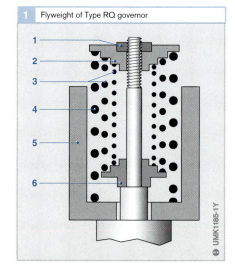

1 Flyweight of Type RQ governor

Fig. 1
1 Adjusting nut
2 Outer spring seat
3 Maximum-speed springs
4 Idle-speed spring
5 Flyweight
6 Inner spring seat

2   Type RQ minimum/maximum-speed governor

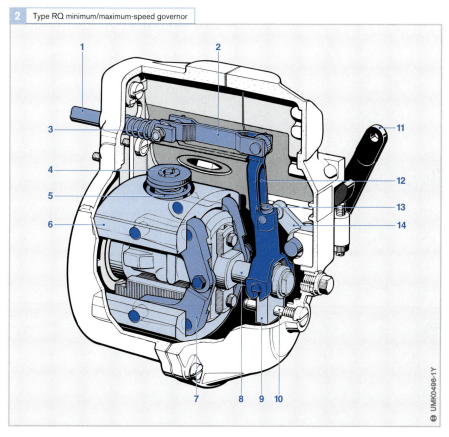

UMK0498-1Y

Fig. 2
1   Control rack
2   Link fork
3   Play compensating
    spring
4   Adjusting nut
5   Governor spring
6   Flyweight
7   Bell crank
8   Sliding bolt
9   Sliding block
10   Guide pin
11   Control lever
12   Variable-fulcrum
    lever
13   Guide block
14   Linkage lever

3   Type RQ minimum/maximum-speed governor in stop position

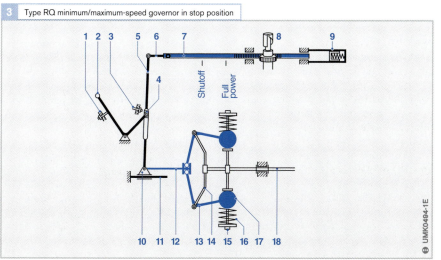

UMK0494-1E

Fig. 3
1   Shutoff stop
2   Control lever
3   Full-load stop
4   Guide block
5   Variable-fulcrum lever
6   Link fork
7   Control rack
8   Pump plunger
9   Control-rod stop
    (buffered)
10   Sliding block
11   Guide pin
12   Sliding bolt
13   Bell crank
14   Governor hub
15   Adjusting nut
16   Governor spring
17   Flyweight
18   Fuel-injection pump
    camshaft

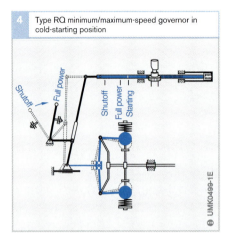

4 Type RQ minimum/maximum-speed governor in cold-starting position

UMK0499-1E

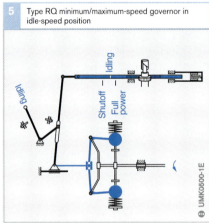

5 Type RQ minimum/maximum-speed governor in idle-speed position

UMK0500-1E

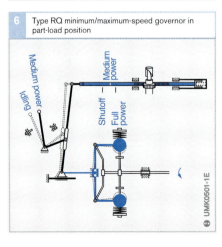

6 Type RQ minimum/maximum-speed governor in part-load position

UMK0501-1E

flyweights (Figure 1) generally consist of three concentrically arranged helical springs, namely the idle-speed spring (4) and the two maximum-speed springs (3).

Starting the engine

The manual for the vehicle specifies the required accelerator position when starting the engine.

A fully depressed accelerator pedal provides the required start quantity for starting the engine from cold at cold outside temperatures. If the engine is already warm, the injected fuel quantity obtained when the control lever is in the idle-speed position is usually sufficient for starting. In this situation, fully depressing the accelerator pedal would merely result in unnecessary emission of smoke (Figures 4 and 7).

Operating characteristics

Idle speed
After the engine has started and the control lever (accelerator pedal) has been released, the latter returns to the idle speed position. The control rack then also returns to the idle-speed position as dictated by the now active governor (Figure 5).

The idle speed of an engine is defined as the lowest speed at which it will reliably continue to run when not under load. The only resistance overcome by the engine is that created by its internal friction and the auxiliary assemblies driven by it such as the alternator, fuel-injection pump, radiator fan, etc. In order to be able to overcome that minimal level of resistance, it requires a specific quantity of fuel. That is provided when the control lever is in a position that corresponds to the specified idle-speed setting.

Intermediate speeds
When the engine is operating under load (i.e. between no load and full load, Figure 6) and the accelerator pedal is depressed, the engine accelerates. As a result, the flyweights in the governor move outwards. The immediate response of the governor is to prevent an increase in engine speed.

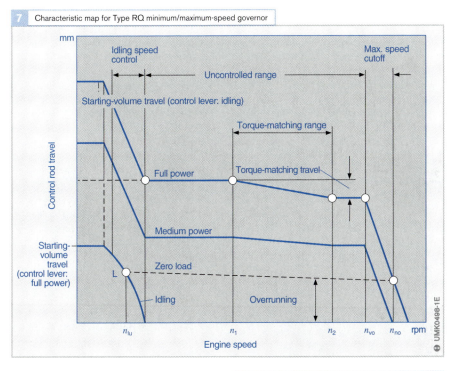

7 Characteristic map for Type RQ minimum/maximum-speed governor

mm

Idling speed control

Uncontrolled range

Max. speed cutoff

Starting-volume travel (control lever: idling)

Torque-matching range

Control rod travel

Full power

Torque-matching travel

Medium power

Starting-volume travel (control lever: full power)

Zero load

L

Idling

Overrunning

n_{lu} n_1 n_2 n_{vo} n_{no} rpm

Engine speed

UMK0498-1E

Fig. 7
n_{lu} Minimum idle speed
n_{vo} Maximum full-load speed
n_1 Start of torque-control phase
n_2 End of torque-control phase
n_{no} Maximum no-load speed

But as soon as the speed rises above idle by only a small amount, the flyweights come into contact with the spring seats against which the maximum-speed springs are acting and remain in that position until the engine reaches its maximum speed – because the maximum-speed springs are not overcome by the centrifugal force until the engine is about to exceed its nominal speed. Consequently, the governor is ineffective between idle speed and maximum speed. In that range, the position of the control rack, and therefore the torque output of the engine, is determined solely by the driver. The torque control phase within that range is described below.

Torque control
On a Type RQ governor, the torque control mechanism is inside the flyweights; to be precise: in between the inner spring seat (Figure 8, Item 9) and the maximum-speed springs (3). The torque-control spring (7) is

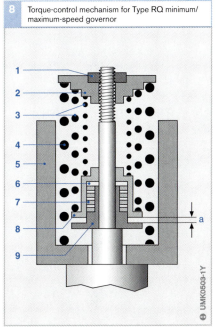

8 Torque-control mechanism for Type RQ minimum/maximum-speed governor

1
2
3
4
5
6
7
8
9

a

UMK0503-1Y

Fig. 8
1   Adjusting nut
2   Outer spring seat
3   Maximum-speed springs
4   Idle-speed spring
5   Flyweight
6   Shim
7   Torque-control spring
8   Spring retainer
9   Inner spring seat

a   Torque-control travel

inside a spring retainer (8) against the outside of which the two maximum-speed springs rest. It therefore comes into effect before the maximum-speed springs. The distance between the inner spring seat and the spring retainer is the torque-control travel (a). It can be adjusted by means of shims. The point n_1 at which torque control comes into effect is dependent on the engine's fuel characteristic curve.

At a point slightly before maximum speed (n_2), the torque-control spring is compressed to such an extent that the inner spring seat and the spring retainer are touching. Without torque-control springs, the governor is ineffective between idle speed and maximum speed. Due to the "give" provided by the torque-control springs, the flyweights can move outwards by the distance represented by the torque-control travel at speeds between n_1 and n_2 and thus move the control rack the corresponding distance towards the stop setting (positive torque control).

Maximum speed
The maximum-speed limiting function comes into operation when the engine exceeds nominal speed, n_{vo}. Depending on the control lever position, therefore, that may take place at full or part load (Figure 9). As soon as the maximum-speed limiting function is active, the position of the control

rack is no longer controlled solely by the driver but also by the governor. The maximum-speed limiting travel of the flyweights is designed so as to bring about speed-regulation breakaway between maximum full-load speed and maximum no-load speed.

Type RQU minimum/maximum-speed governor
Design
The Type RQU governor is designed for controlling very low speeds. It is equipped with a multiplier gear (transmission ratio 1:1.5 to 1:3.7 depending on requirements) between the fuel-injection pump camshaft that provides the drive and the governor hub (Figures 10 and 20). The Type RQU governor was developed for Type ZW, P9 and P10 fuel-injection pumps that are used on larger and generally slower-running diesel engines.

As with the Type RQV governor, the linkage lever on the Type RQU model is a two-part construction that runs in a plate cam.

Operating characteristics
The method of operation and operating characteristics are the same as the Type RQ governor.

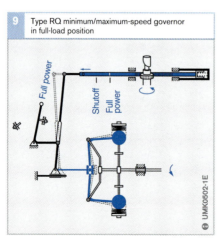

9 Type RQ minimum/maximum-speed governor in full-load position

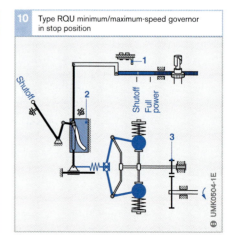

10 Type RQU minimum/maximum-speed governor in stop position

Fig. 10
1 Full-load stop
2 Plate cam
3 Multiplier gear

Type RQ and RQU maximum-speed governors

Design

The design of the maximum-speed governor differs from that of the minimum/maximum-speed governor essentially by virtue of the fact that the idle-speed stage is omitted.

Operating characteristics

In operation, the maximum-speed governor behaves in the same way as the maximum-speed stage of the Type RQ/RQU minimum/maximum-speed governor.

Maximum speed

The speed-regulation breakaway function comes into operation when the engine exceeds maximum full-load speed. The maximum-speed limiting travel of the flyweights is designed so as to bring about maximum-speed breakaway between maximum full-load speed and maximum no-load speed.

Type RQV variable-speed governor

Design

The Type RQV governor is similar in design to the Type RQ model. The governor springs are fitted inside the flyweights. However, the flyweights move continuously outwards as engine speed increases within the specified control range (Figure 11). Every control lever position corresponds to a specific speed at which speed-regulation breakaway begins. The movement of the control lever (1) is transmitted via the two-piece linkage lever (2) and the guide block (4) to the variable-fulcrum lever(5) and thence to the control rack (8). The pivot of the variable-fulcrum lever is movable along the sliding-block guide; it also runs in a plate cam (3) fixed to the governor housing so that the transmission ratio of the variable-fulcrum lever also changes. The sliding bolt (12), which forms the link between the flyweight speed-sensing element and the variable-fulcrum lever, is sprung against pressure and tension (drag spring).

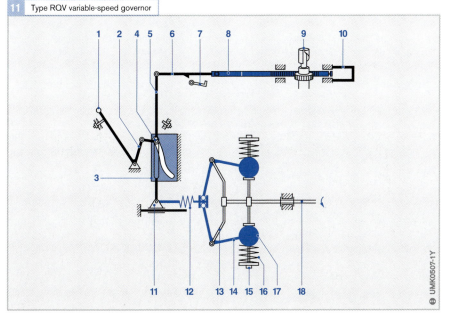

11   Type RQV variable-speed governor

UMK0507-1Y

Fig. 11
1 Control lever
2 Linkage lever
3 Plate cam
4 Guide block
5 Variable-fulcrum lever
6 Link fork
7 Full-load stop (automatic)
8 Control rack
9 Pump plunger
10 Start-quantity stop
11 Sliding block
12 Sliding bolt with drag spring
13 Governor hub
14 Bell crank
15 Adjusting nut
16 Governor spring
17 Flyweight
18 Fuel-injection pump camshaft

As with the Type RQ governor, the spring sets fitted inside the flyweights generally consist of three concentric helical springs. The outer spring acts as the idle-speed control spring (Figure 12, Item 4); it is held between the flyweight (5) and the adjusting nut (1) for setting the initial spring tension. After completing the short idle-speed travel (idle-speed stage), the flyweight comes into contact with the spring seat, and the inner springs that are fitted between the spring seat and the adjusting nut come into effect.

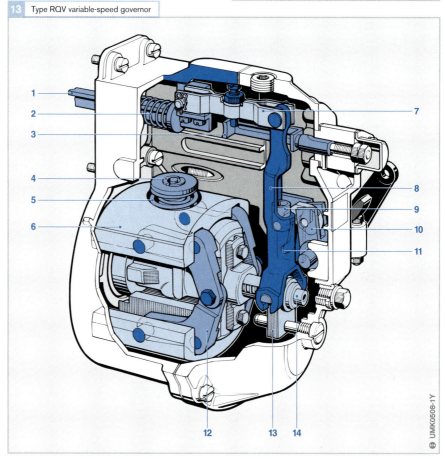

12 Flyweight for Type RQV variable-speed governor

1
2
3
4
5
a
UMK0509-1Y

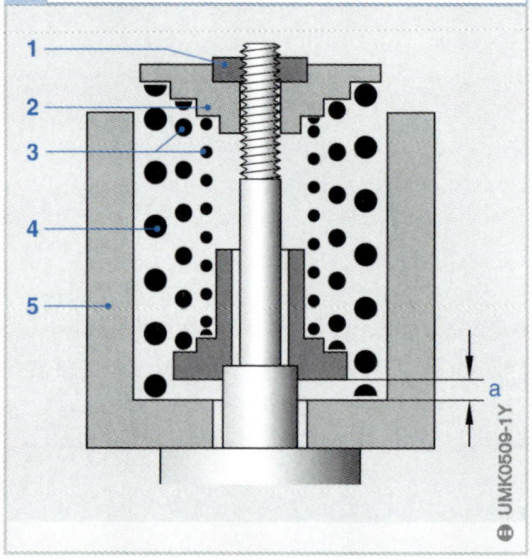

Fig. 12
1 Adjusting nut
2 Spring seat
3 Maximum-speed springs
4 Idle-speed spring
5 Flyweight

a Idle-speed travel

13 Type RQV variable-speed governor

1
2
3
4
5
6
7
8
9
10
11
12
13
14
UMK0508-1Y

Fig. 13
1 Control rack
2 Play compensating spring
3 Full-load stop
4 Adjusting nut
5 Governor spring
6 Flyweight
7 Link fork
8 Variable-fulcrum lever
9 Guide block
10 Linkage lever
11 Plate cam
12 Bell crank
13 Sliding block
14 Sliding bolt (with drag spring)

Starting the engine

When applying the start quantity, the Type RQV governor operates in the same way as the Type RQ apart from the following difference:

If the driver fully depresses the accelerator pedal the first time the engine is started, when it reaches idle speed, unlike the Type RQ, the governor does not back off to the full-load setting. Instead, the control rack remains in the start quantity position until the engine reaches maximum speed. Only once the speed-regulation breakaway function has first come into action does the full-load stop drop into its normal operating position (Figure 14).

Operating characteristics

Idle speed (Figure 15)

After the engine has started and the control lever (accelerator pedal) has been released, the latter returns to the idle-speed position. The control rack then also returns to the idle-speed position as dictated by the now active governor (position L in Figure 17 overleaf).

Intermediate speeds (Figure 16)

If a load is applied to or removed from the engine at any speed set by the control lever (accelerator pedal), the variable-speed governor maintains the speed setting by increasing or reducing the fuel delivery quantity within the limits determined by the proportional response.

Example: The driver moves the control lever from the idle-speed position to a position that corresponds to a desired vehicle speed. The movement of the control lever is transmitted to the variable-fulcrum lever via the linkage lever. The transmission ratio of the variable-fulcrum lever is variable, and at a position just above the idle-speed setting is such that even a relatively small amount of control-lever or centrifugal-weight travel is enough to move the control rack to the set full-load position (distance L – B in Figure 17); consequently, a fixed (i.e. unbuffered) control-rod stop must exist.

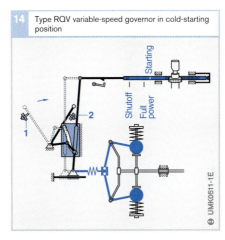

14 Type RQV variable-speed governor in cold-starting position

UMK0511-1E

Fig. 14
1 Stop setting stop
2 Maximum-speed stop

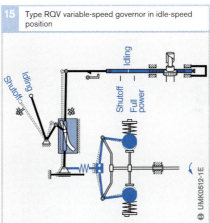

15 Type RQV variable-speed governor in idle-speed position

UMK0512-1E

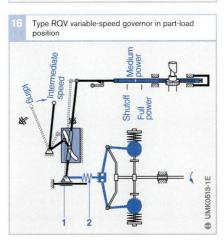

16 Type RQV variable-speed governor in part-load position

UMK0513-1E

Fig. 16
1 Sliding block
2 Sliding bolt with drag spring

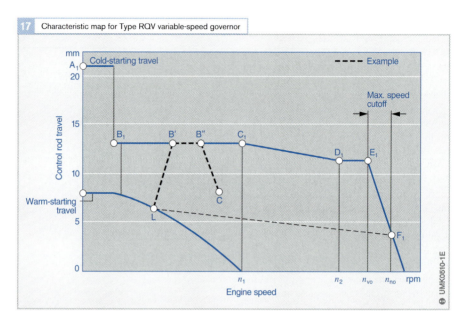

17   Characteristic map for Type RQV variable-speed governor

Further movement of the control lever applies tension to the drag spring. The control rack remains at "full load" so that the engine rapidly increases speed (Figure 17, distance B' – B"). The flyweights move outwards but the control rack remains at "full load" until the drag spring is no longer under tension. Only then do the flyweights start to act on the variable-fulcrum lever, and the control rack is then moved towards the stop setting. Consequently, the fuel delivery quantity is reduced again and the engine's speed is limited. The engine speed limit corresponds to the setting of the control lever and the position of the flyweights (distance B" – C).

Thus, under normal operation, every control lever position corresponds to a specific range of engine speeds, provided the engine is not overloaded or is driven by the road wheels (overrunning, e.g. on a descent). If the load on the engine increases, e.g. due to an uphill gradient, the speed of the engine and governor drops. As a result, the flyweights move inwards and move the control rack towards the "full load" setting, thereby holding the engine at a constant speed as determined by the control lever position and the propor-

tional response. If, however, the gradient (which equates to the load) is such that even when the control rack is moved right up to the "full load" stop the engine speed still drops, the flyweights retract even further and move the sliding bolt to the left. The flyweights are thus attempting to move the control rack further in the same direction beyond the stop setting. Since, however, the control rack is already in contact with the "full load" stop and cannot therefore move any further in that direction. The drag spring is tensioned. This means that the engine is overloaded. Under such circumstances, the driver will have to change down to a lower gear. When traveling downhill, precisely the opposite will apply. The engine is driven and accelerated by the road wheels. As a result, the flyweights move outwards and the control rack is moved towards the stop setting. If the engine speed continues to increase (control rack has reached the stop setting), the drag spring is extended in the other direction.

The governor response described above applies in principle to all control lever positions if the engine load or speed alters for any reason to such a degree that the control rack comes into contact with either of its end positions.

Torque control
Torque control comes into effect between n_1 and n_2 (Figure 17) along the line $C_1 - D_1$ under full load. On the Type RQV governor, the torque-control mechanism is fitted in a special control-rod stop or a torque-control bar that replaces the normal link fork (for a detailed description refer to the section "Control-rod stops").

Maximum speed (Figure 18)
When the engine exceeds the maximum full-load speed, full-load speed regulation $E_1 - F_1$ (Figure 17) comes into operation. The flyweights move outwards and the control rack moves towards the stop setting. The maximum no-load speed, n_{no}, is reached when the engine load is entirely removed.

Type RQUV variable-speed governor
Design
The Type RQUV variable-speed governor is used to control very low engine speeds such as are encountered with marine engines. It is a variation of the Type RQV governor.

It is available with various multiplier gear ratios (roughly in the range of 1:1.5 to 1:3.7) between the fuel-injection pump camshaft that provides the drive and the governor hub (Figure 19). The transmission ratio of the variable-fulcrum lever is similar to that on the Type RQV governor. For this reason, this model also has a plate cam (Figure 20).

The Type RQUV governor is used with Type ZW, P9 and P10 fuel-injection pumps.

Operating characteristics
The method of operation and operating characteristics are the same as those of the Type RQV governor except that there is no start quantity function.

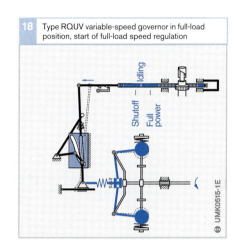

18 Type RQUV variable-speed governor in full-load position, start of full-load speed regulation

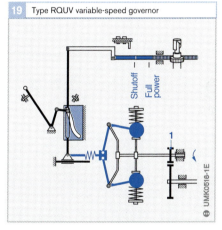

19 Type RQUV variable-speed governor

Fig. 19
1 Multiplier gear

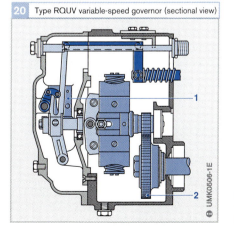

20 Type RQUV variable-speed governor (sectional view)

Fig. 20
1 Plate cam
2 Multiplier gear

Type RQV..K variable-speed governor

Design

The Type RQV..K governor (Figures 21 and 22) has basically the same flyweight speed-sensing element as the Type RQV governor with the governor springs fitted inside the flyweights. The essential difference is the method by which the torque-control function is achieved. Whereas with all other governors, torque control consists basically of reducing the fuel delivery quantity when engine speed is increasing at full-power, the Type RQV..K governor can either increase or decrease the delivery quantity at full load.

Starting the engine

As previously explained in connection with the Type RQ governor, the required starting procedure is specified by the engine manufacturer. If the cold-start quantity is required, the governor control lever must be set to the high idle speed (Figure 23).

The rocker then tips over and slides under the full-load stop, and the control rack moves to the start-quantity position, A_1 (Figure 26, next double page). There is a stop for the start quantity on the fuel-injection pump. When the starter motor is switched on, the fuel-injection pump delivers the fuel quantity necessary for starting (the start quantity) to the nozzle and then to the combustion chamber. On the Type RQV..K governor, the start quantity can also be varied according to temperature by means of a temperature-dependent stop.

Operating characteristics

Idle speed (Figure 24)

Once the engine has started, the control lever is returned to the idle-speed position.

The spring-loaded rocker then slides back from under the full-load stop and returns to the idle-speed position. The engine then runs at idle speed.

Fig. 21
1 Control lever
2 Linkage lever
3 Plate cam
4 Guide block
5 Variable-fulcrum lever
6 Rocker
7 Plate-cam compression spring
8 Sprung link
9 Control rack
10 Full-load stop
11 Pump plunger
12 Start-quantity stop
13 Sliding block
14 Guide lever
15 Sliding bolt
16 Bell crank
17 Adjusting nut
18 Governor spring
19 Flyweight
20 Fuel-injection pump camshaft

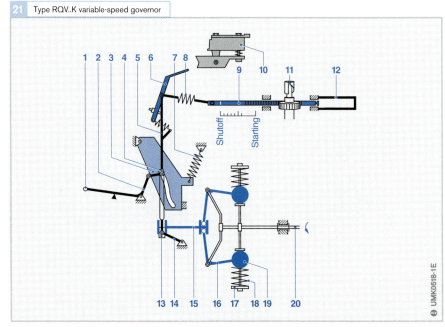

21	Type RQV..K variable-speed governor

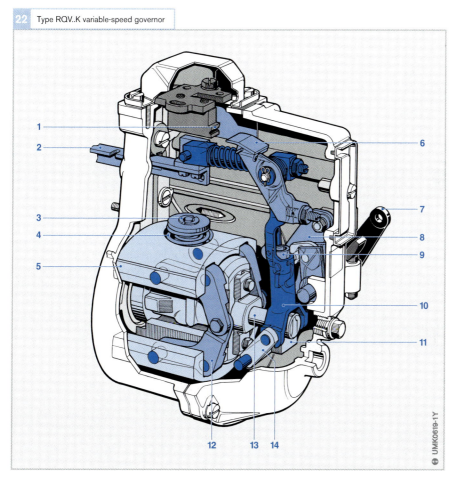

22 Type RQV..K variable-speed governor

Fig. 22
1 Full-load stop with
   cam track
2 Control rack
3 Adjusting nut
4 Governor spring
5 Flyweight
6 Rocker
7 Control lever
8 Plate cam
9 Guide block
10 Variable-fulcrum
   lever
11 Sliding block
12 Bell crank
13 Sliding bolt
14 Guide lever

UMK0619-1Y

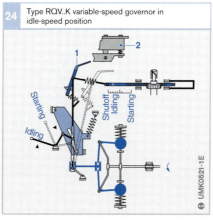

23 Type RQV..K variable-speed governor in
        cold-starting position

UMK0620-1E

24 Type RQV..K variable-speed governor in
        idle-speed position

UMK0621-1E

Fig. 23
1 Start-quantity stop

Fig. 24
1 Rocker
2 Full-load stop with
   cam track
   (adjustable)

Intermediate speeds
The group of curves at point B, for example, show the possibilities for regulating intermediate speeds.

Low intermediate speed and full-load quantity (Figure 25)
If, for example, the control lever is moved from the idle-speed position to the maximum-speed position, the guide block moves along the cam track in the guide plate and simultaneously downwards in guide of the variable-fulcrum lever. At the same time, the variable-fulcrum lever pivots to the right around the clevis pin on the sliding block and pushes the control rack via the link towards the full-load position. The delivery quantity increases and the engine speed rises.

The flyweights move outwards and the sleeve moves to the right. This moves the variable-fulcrum lever and the guide lever so that the rocker slides along the cam track of the full-load stop (path A – B on the characteristic map, Figure 26).

If the control lever is moved forwards, the rocker comes into contact with the cam track of the full-load stop and the plate cam lifts away from its stop on the governor body against the action of the compression spring.

Torque control
On the RQV..K governor, the torque-control function take place by the fact that the rocker attached to the top end of the variable-fulcrum lever follows the cam track of the full-load stop which mirrors the fuel requirement pattern of the engine. The link between the variable-fulcrum lever and the control rack transmits the movement to the control rack. As a result, a full-load delivery quantity corresponding to the desired torque curve is obtained.

Depending on the shape of the cam, the delivery quantity may be increased or decreased. The full-load stop can be moved along its longitudinal axis in order to adjust the delivery quantity.

Medium intermediate speed with torque control and full-load volume (Figure 27)
If the engine speed increases further, the flyweights move further outwards and the rocker slides along the cam track of the full-load stop. Up to the point at which the curve changes direction at B, torque control takes the form of an increase in the full-load delivery quantity in response to an increase in engine speed (negative torque control); after the changeover point, it involves a reduction in full-load delivery quantity (positive torque control, path B – C on the characteristic map, Figure 26).

Maximum speed (Figure 28)
At the end of the torque-control phase at the art of the speed regulation, the plate cam rests back against the stop on the governor body.

If the engine speed continues to increase, the maximum-speed limiting (speed-regulation breakaway) function comes into action. The flyweights move further outwards and the sliding sleeve moves further to the right. Consequently, the variable-fulcrum lever pivots around the guide block and its upper end moves to the left. This moves the con-

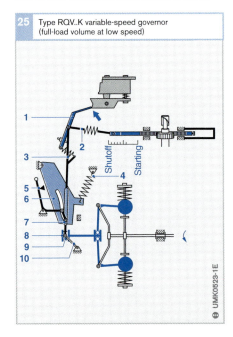

25 Type RQV..K variable-speed governor (full-load volume at low speed)

Fig. 25
Start of negative
torque-control phase
1 Rocker
2 Sprung link
3 Variable-fulcrum
   lever
4 Compression spring
5 Control lever
6 Plate cam
7 Guide block
8 Sliding sleeve
9 Sliding block
10 Guide lever

26 Characteristic map for Type RQV..K variable-speed governor

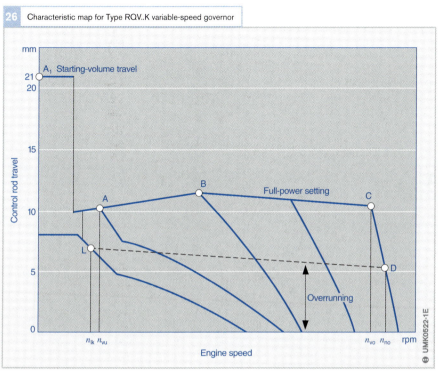

27 Type RQV..K variable-speed governor
(full-load volume at medium speed)

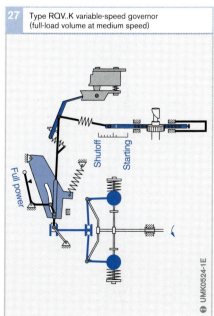

28 Type RQV..K variable-speed governor
(maximum full-load speed)

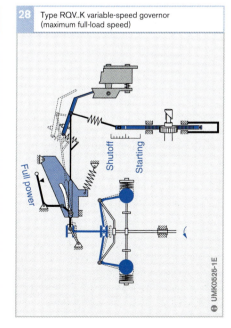

Fig. 27
Torque-control
changeover point

Fig. 28
End of positive
torque-control phase
(broken line: speed-reg.
breakaway)

trol rack towards the stop setting (path C – D on the characteristic map, Figure 26).

Thus, under normal operation, every control lever position corresponds to a specific range of engine speeds, provided the engine is not overloaded or is driven by the road wheels, e.g. when overrunning on a descent.

If the load on the engine increases, e.g. due to an uphill gradient, the speed of the engine and governor drops. As a result, the flyweights move inwards and move the control rack towards the "full load" setting, thereby holding the engine at a constant speed as determined by the control lever (or accelerator) position. If, however, the load (which equates to the gradient) is such that, even when the control rack is moved right up to the "full load" stop, the engine speed still drops, the flyweights retract even further and move the sliding sleeve further towards the "full load" position.

Since, however, the control rack cannot move any further in that direction as it is already in contact with the stop, the lower end of the variable-fulcrum lever moves to the left against the resistance of the plate-cam compression spring and moves the plate cam away from its stop.

When descending an incline, the reverse situation applies. The engine is driven and accelerated by the road wheels. As a result, the flyweights move outwards and the control rack is moved towards the stop setting. If the engine speed then continues to increase (control rack reaches the stop setting), the sprung link that connects the variable-fulcrum lever to the control rack "gives". If the driver brakes the vehicle or changes into a higher gear, the link retracts to its normal length again.

The governor response described above applies in principle to all control lever positions if the engine load or speed alters for any reason to such a degree that the control rack comes into contact with either of its end positions.

Type RSV variable-speed governor
Design
The Type RSV variable-speed governor has a different design from the comparable Type RQV model. It has only one governor spring, which is a pivoted design and external to the flyweights (Figure 29, Item 12 and Figure 30, Item 16). When the speed is set by means of the control lever, the position and tension of the spring alters in such a way that, when the engine is running at the desired speed, the turning force acting on the tensioning lever is held in equilibrium by the opposing turning force produced by the flyweights. All changes to the control lever position and to the position of the flyweights are transmitted to the control rack via the governor linkage mechanism.

The starting spring (Figure 30, Item 5) that hooks into the top end of the variable-fulcrum lever pulls the control rack (2) into the starting position, thus automatically setting the fuel-injection pump to the start quantity. The full-load stop (20) and torque-control mechanism (19) are integrated in the governor. The auxiliary idle-speed spring (17) and adjusting screw (13) fitted in the governor housing serve the function of stabilizing the idle speed.

The governor spring is connected at one end to the tensioning lever (18) and at the other to the rocker (8). The screw (7) on the rocker allows the active lever arm of the governor spring to adjust in relation to the tensioning-lever pivot. This enables the proportional response to be varied within certain limits without having to change the governor spring – one of the advantages of the Type RSV governor. Lighter flyweights are available for higher engine speeds.

29 Type RSV variable-speed governor

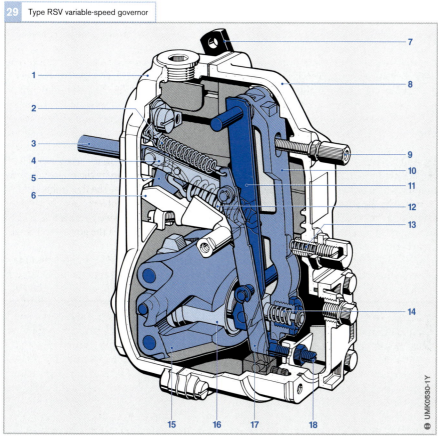

Fig. 29
1 Governor housing
2 Starting spring
3 Control rack
4 Rigid link
5 Rocker
6 Swiveling lever
7 Control lever
8 Governor cover
9 Stop setting/
   low-idle stop
10 Tensioning lever
11 Guide lever
12 Governor spring
13 Auxiliary idle-speed
   spring
14 Torque-control
   spring
15 Flyweight
16 Guide bushing
17 Variable-fulcrum
   lever
18 Full-load stop

Fig. 30
1 Pump plunger
2 Control rack
3 Maximum-speed
   stop
4 Control lever
5 Starting spring
6 Swiveling lever
7 Adjusting screw
8 Rocker
9 Fuel-injection pump
   camshaft
10 Governor hub
11 Flyweight
12 Sliding bolt
13 Stop setting/
   low-idle stop
14 Guide lever
15 Variable-fulcrum
   lever
16 Governor spring
17 Auxiliary idle-speed
   spring
18 Tensioning lever
19 Torque-control
   spring
20 Full-load stop

30 Type RSV variable-speed governor

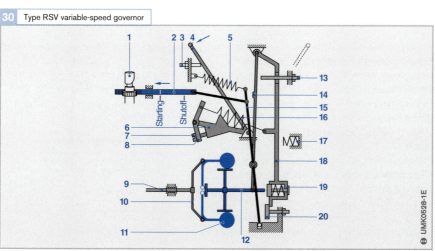

Starting the engine

On the Type RSV governor, the control rack is always at point A (characteristic map, Figure 32) when the engine is not running regardless of the position of the control lever. For this reason, the Type TAS temperature-dependent start-quantity stop is always recommended for this type of governor.

Operating characteristics

Idle speed (Figure 31)

The control lever is resting against the low-idle stop. Consequently, there is virtually no tension on the governor spring (3) and it is in an almost vertical position. The spring action is very soft so that the flyweights start to move outwards at a very low speed. As a result, the sliding bolt (7) moves to the right and, with it, the guide lever. The latter pivots the variable-fulcrum lever (4) to the right so that the control rack moves towards the stop setting to the idle-speed position L (characteristic map, Figure 32). The tensioning lever (6) comes into contact with the auxiliary idle-speed spring which helps to regulate the idle speed.

Low intermediate speeds (Figure 33)

Even a relatively small movement of the control lever beyond the idle-speed position is enough to move the control rack from its initial position (point L on Figure 32) to the full-load position (point B' on Figure 32). The fuel-injection pump then delivers the full-load volume and the engine speed increases (path B' – B").

As soon as the centrifugal force exceeds the force of the governor spring at the corresponding position of the control lever, the flyweights move outwards and move the guide bushing, sliding bolt, variable-fulcrum lever and control rack back to a lower delivery quantity (point C on Figure 32). The engine speed then stops rising and is held constant by the governor assuming other conditions remain the same.

Torque control

On governors with a torque-control mechanism, as soon as the tensioning lever comes into contact with the full-load stop, the torque-control spring is continuously compressed (path D – E on Figure 32) as engine speed increases, thereby backing off the guide lever, variable-fulcrum lever and control rack towards the stop setting to adjust the fuel delivery quantity by the appropriate amount, i.e. the torque-control travel.

Maximum speed (Figure 34)

If the control lever is moved to the maximum-speed stop, the governor basically operates in the same way as described for low intermediate speeds. The one difference, however, is that the swiveling lever fully tensions the governor spring. The governor spring therefore pulls the tensioning lever against the full-load stop with greater force and thus the control rack as well. The engine speed increases and the centrifugal force rises continuously.

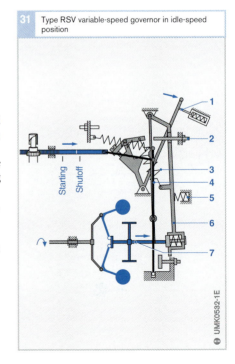

31 Type RSV variable-speed governor in idle-speed position

Starting   Shutoff

1
2
3
4
5
6
7

© UMK0532-1E

Fig. 31

1   Low-idle stop
2   Stop setting stop
3   Governor spring
4   Variable-fulcrum lever
5   Auxiliary idle-speed spring
6   Tensioning lever
7   Sliding bolt

32   Characteristic map for Type RSV variable-speed governor

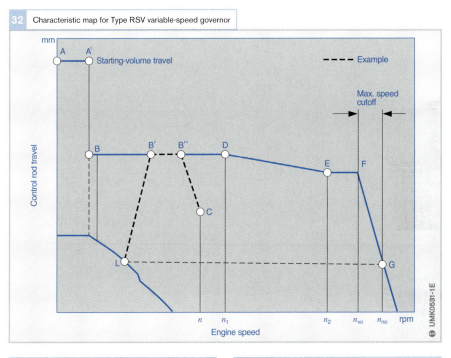

33   Type RSV variable-speed governor (full-load at low speed)

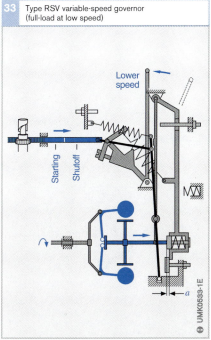

34   Type RSV variable-speed governor (no load from full load)

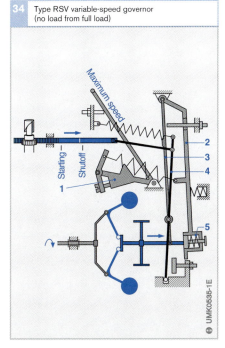

Fig. 33
Start of torque-control phase
a Torque-control travel

Fig. 34
1   Swiveling lever
2   Tensioning lever
3   Guide lever
4   Variable-fulcrum lever
5   Torque-control spring

When it reaches the maximum full-load speed, n_{vo}, the centrifugal force overcomes the force of the governor spring and the tensioning lever is moved to the right. The sliding bolt and guide lever along with the control rack to which the variable-fulcrum lever is linked move towards the stop setting (path W – G on Figure 32) until an appropriately reduced injected fuel quantity is obtained for the new load conditions.

The maximum no-load speed, n_{no}, is reached if the engine load is entirely removed.

Stopping the engine
Stopping the engine using the control lever (Figure 35)
Engines with governors that do not a have a dedicated shutoff device are stopped by moving the governor control lever to the stop setting. The lug on the swiveling lever (arrow in Figure 35, Item 1) then presses against the guide lever. The latter pivots to the right taking the variable-fulcrum lever and consequently the control rack with it to the stop setting. Because the sliding bolt is relieved of the tension of the governor spring, the flyweights move outwards.

Stopping the engine using the shutoff lever (Figure 36)
On governors which have a dedicated shutoff mechanism, the control rack can be set to the stop setting by operating the shutoff lever (2).

Moving the shutoff lever to the stop setting causes the variable-fulcrum lever to pivot around the pivot point C on the guide lever so that its top end moves to the right. The rigid link then pulls the control rack to the stop setting. A compression spring (not illustrated) moves the shutoff lever back to its original position when it is released.

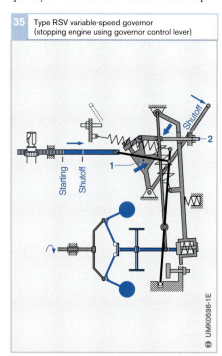

35 Type RSV variable-speed governor (stopping engine using governor control lever)

36 Type RSV variable-speed governor (stopping engine using shutoff lever)

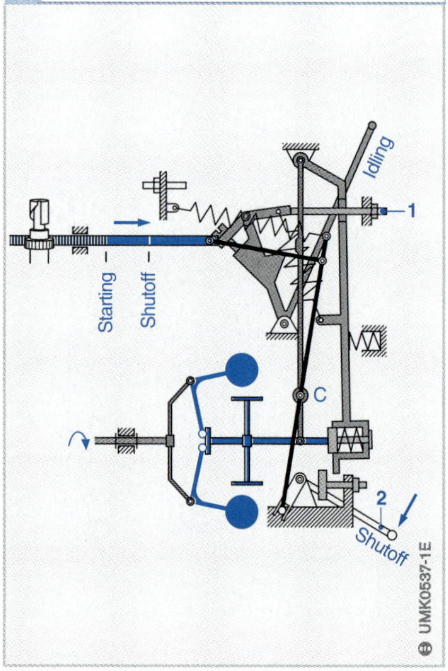

UMK0536-1E

UMK0537-1E

Fig. 35
1   Lug on swiveling lever
2   Shutoff stop

Fig. 36
1   Low-idle stop
2   Shutoff lever

37 Type RSUV variable-speed governor

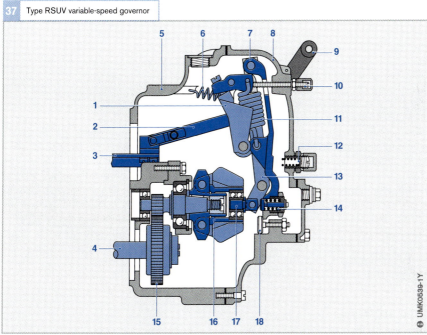

Fig. 37
1   Swiveling lever
2   Rigid link
3   Control rack
4   Fuel-injection pump camshaft
5   Governor housing
6   Starting spring
7   Tensioning lever
8   Governor cover
9   Control lever
10 Stop setting/ low-idle stop
11 Governor spring
12 Auxiliary idle-speed spring
13 Guide lever
14 Torque-control spring
15 Multiplier gear
16 Guide bushing
17 Sliding bolt
18 Full-load stop

38 Type RSUV variable-speed governor in start position

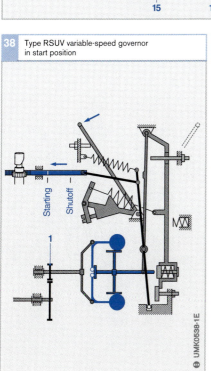

Type RSUV variable-speed governor

Design

The Type RSUV variable-speed governor is used to control very low engine speeds such as are encountered with slow-running marine engines. It differs in design from the Type RSV governor essentially by virtue of the multiplier gear fitted between the fuel-injection pump camshaft that provides the drive and the governor hub (Figure 37). A number different of transmission ratios are possible (Type A 1:3, 1:2; Type B 1:1.36, 1:1.86; Type Z 1:2.2, 1:2.6).

Type RSUV variable-speed governors are used in conjunction with Type P in-line fuel-injection pumps.

Operating characteristics

The method of operation and operating characteristics are identical with those of the Type RSV governor.

Figure 38 shows the Type RSUV variable-speed governor in the maximum-speed position.

Fig. 38
1   Multiplier gear

Type RS minimum/maximum-speed governor

Design

The Type RS minimum/maximum-speed governor (derived from the Type RSV variable-speed governor) has minimal control lever leverage. The control lever, which on the Type RSV governor tensions the governor spring and is therefore used to set the engine speed, is held in the maximum-speed position by an adjustable stop on the governor cover. It is also possible to set an intermediate speed (e.g. on vehicles with a PTO drive). The shut-off lever of the Type RSV governor acts as an accelerator lever on the Type RS governor and is therefore operated in the reverse direction (Figure 39).

Starting the engine

For starting the engine, the accelerator lever (Figure 42, Item 12) is pivoted towards the full-load position. It then presses the sliding bolt (14) via the variable-fulcrum lever (10) and guide lever (8) against the spring retainer (11) in which the idle-speed spring (11.3) sets the control rack (2) to the start-quantity position.

Operating characteristics

Idle speed

The flyweights move outwards even at low speeds. As a result, the sliding bolt moves to the right and, with it, the guide lever. The guide lever pivots the variable-fulcrum lever to the right so that the control rack moves towards the stop setting to the idle-speed position L (characteristic map, Figure 41). In addition, the sliding bolt presses against the spring retainer in which there is an idle-speed spring as well as the torque-control spring. The low-idle stop screw and auxiliary idle-speed spring found on the Type RSV governor are not present in this case.

Maximum speed

If the engine speed exceeds the maximum full-load speed, the control rack moves towards the stop setting (path E – F, Figure 41). The maximum no-load speed is reached if the engine load is entirely removed.

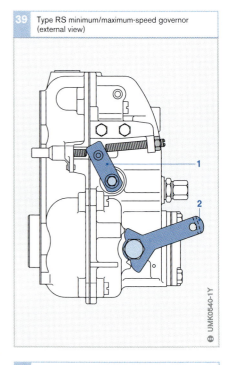

39 Type RS minimum/maximum-speed governor (external view)

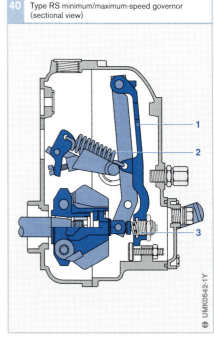

40 Type RS minimum/maximum-speed governor (sectional view)

41 Characteristic map for Type RS minimum/maximum-speed governor

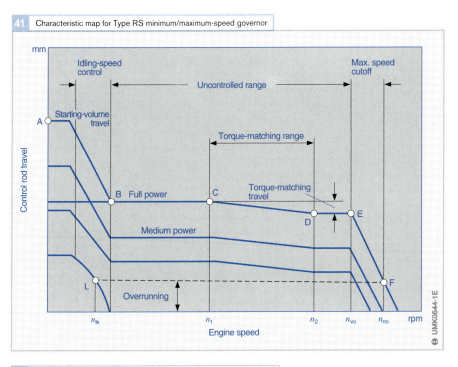

42 Type RS minimum/maximum-speed governor in cold-starting position

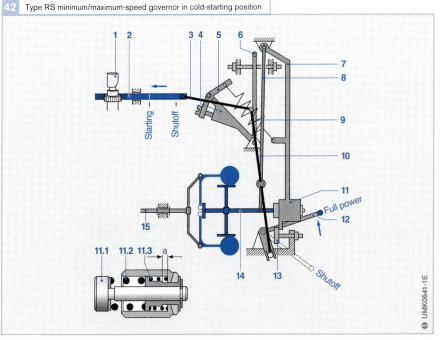

Fig. 42
1 Pump plunger
2 Control rack
3 Rigid link
4 Swiveling lever
5 Rocker
6 Control lever
7 Swiveling lever
8 Guide lever
9 Governor spring
10 Variable-fulcrum lever
11 Spring retainer
11.1 Pressure pin
11.2 Torque-control spring
11.3 Idle-speed spring
12 Accelerator lever
13 Full-load stop
14 Sliding bolt
15 Fuel-injection pump camshaft

a Idle-speed stage

Type RSF minimum/maximum-speed governor

Design

The Type RSF centrifugal mechanical governor was designed specifically as a minimum/ maximum-speed governor for motor-vehicle engines with Type M in-line diesel fuel-injection pumps. It is suitable for road-going vehicles (cars and commercials) which only require limitation of the minimum and maximum speeds. Within the uncontrolled intermediate-speed range, the fuel-injection pump's control rack is directly operated by the accelerator pedal under the control of the driver so as to obtain the required engine torque (Figure 43).

The Type RSF governor meets demanding requirements in respect of governor characteristics, ease of operation and driver convenience. It is intended primarily for fast-revving diesel engines in cars. In addition, it offers the facility for combination with compensating mecha-

nisms and is easily adjustable. The governor design can be divided into two sections: the governor movement and the actuator mechanism (Figure 44).

Governor movement stage 1 (idle speed)
The force originates from the flyweights (22) and is transmitted via the sliding sleeve (20) and the guide lever (9) to the idle-speed spring (12) and the auxiliary idle-speed spring (14) – both are leaf springs.

Governor movement stage 2
(up to full-load speed regulation)
After the idle-speed travel has been completed, the force is transmitted from the sliding sleeve (20) via the torque-control spring retainer (18) and the tensioning lever (16) to the governor spring (17).

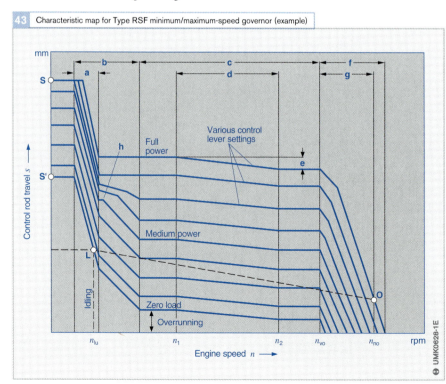

43   Characteristic map for Type RSF minimum/maximum-speed governor (example)

When the flyweights move outwards, the sliding sleeve is moved along its central axis. Apart from in the idle-speed, full-load torque-control and breakaway range, the sliding sleeve is stationary and the injected fuel quantity required to obtain the necessary engine output is set by moving the control lever of the actuator mechanism.

At the pivot B, the guide lever (9) is attached to the sliding sleeve. In addition, the guide lever and the tensioning lever (16) pivot around point A.

Actuator mechanism
The desired engine speed is set by means of the control lever (6) which acts via the linkage lever (5) and the reverse-transfer lever (11) on the variable-fulcrum lever (13) which in turn transmits the movement to the sprung link (2) and the control rack (4) of the fuel-injection pump.

The sprung link compensates for excess travel on the part of the variable-fulcrum lever. Like the guide lever, the reverse-transfer lever is also attached by a pivoting joint at point B to the sliding sleeve, and is also attached by another connecting pin to the variable-fulcrum lever (13).

The lower anchor point of the variable-fulcrum lever can be moved by means of the full-load adjusting screw (19) in order to vary the full-load injected fuel quantity. It also acts as a spring buffer for the variable-fulcrum lever so that the excess sliding-sleeve travel can be accommodated if the engine is over-revving.

The pivot shaft of the stop lever (3) passes through the governor housing and is attached on the outside to a shutoff lever (1) that can be used to stop the engine. In that case, the stop lever moves the control rack to the stop setting.

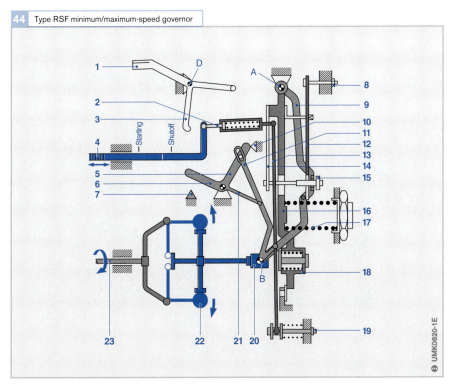

44 Type RSF minimum/maximum-speed governor

Fig. 44
1 Shutoff lever
2 Sprung link
3 Stop lever
4 Control rack
5 Linkage lever (internal)
6 Control lever (external)
7 Full-load stop
8 Adjusting screw for idle speed
9 Guide lever
10 Low-idle stop
11 Reverse-transfer lever
12 Idle-speed spring
13 Variable-fulcrum lever
14 Auxiliary idle-speed spring
15 Adjusting screw for auxiliary idle-speed spring
16 Tensioning lever
17 Governor spring
18 Spring retainer (torque control)
19 Full-load adjusting screw
20 Sliding sleeve
21 Auxiliary idle-speed spring shutoff
22 Flyweight
23 Fuel-injection pump camshaft

Starting the engine

The required setting for starting the engine is specified by the engine manufacturer. As a rule, the engine can be started without pressing the accelerator pedal. Only when there is a combination of cold weather and cold engine is the control lever (6) set to the full-load stop (7) – a fixed stop on the governor housing corresponding to a fully depressed accelerator (Figure 45). The reverse-transfer lever (11) pivots around point B, thereby moving the variable-fulcrum lever (13) towards the start-quantity position. As a result, the control rack (4) moves to the start-quantity position so that the engine receives the necessary fuel quantity for starting. Rapid speed-regulation breakaway from the start-quantity position is made possible by the fact that when the control lever is in full-load position, the auxiliary idle-speed spring (14) is lifted away from the guide lever (9) by a shutoff arm (21).

Operating characteristics

Idle speed (Figure 46)

When the accelerator is released after the engine has started, a compression spring (not illustrated) moves the control lever (6) back to the idle-speed position. This means that the linkage lever (5) is in contact with the low-idle stop screw (10).

While the engine is warming up, the idle speed fluctuates along the idle speed curve and finally comes to rest at the point L (Figure 43). As the engine speed increases, the flyweights (22) move outwards and push the sliding sleeve (20) to the right. During the idle-speed stage, the control rack (4) is moved by the action of the sliding sleeve transmitted via the reverse-transfer lever (11) and the variable-fulcrum lever (13) towards the stop setting. At the same time, the movement of the sliding sleeve causes the guide lever (9) to pivot around point A and press against the idle-speed spring (12), the tension of which (and therefore the idle speed) can be preset by means of the adjusting screw (8). At a certain speed, the guide lever also comes into contact with the adjusting nut for the auxiliary idle-speed spring (14).

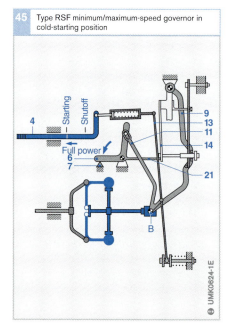

45   Type RSF minimum/maximum-speed governor in cold-starting position

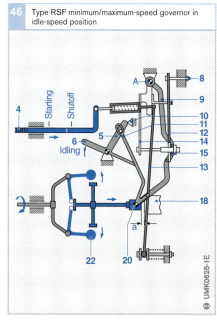

46   Type RSF minimum/maximum-speed governor in idle-speed position

Intermediate speeds

After passing beyond the idle-speed travel (a), the sliding sleeve (20) comes into contact with the torque-control spring retainer (18). In the uncontrolled range between idle speed and maximum speed, the flyweights (22) do not change their position, apart from the small amount of travel for torque control, until the maximum speed is reached. The control-rod position, and therefore the injected fuel quantity, is set directly by moving the control lever (6), i.e. the driver varies the delivery quantity (e.g. in order to increase vehicle speed or to negotiate an uphill gradient) by means of the accelerator pedal (control lever position is between the idle-speed and maximum-speed stops. If the accelerator pedal is fully depressed, the control rack moves to the full-load position.

Torque control

When the torque-control function is active, the full-load delivery quantity is reduced if the engine speed exceeds n_1 because the force of the flyweights acting on the sliding sleeve (20) is greater than the force of the torque-control spring in the spring retainer (18). The torque-control spring "gives" so that the control rack (4) shifts by the torque-control travel if the speed continues to increase. At the speed n_2 the torque-control phase comes to an end. The Type RSF governor may also incorporate a mechanism for negative as well as positive torque control. In this case, the control rack position is controlled by combination of springs.

High-idle speed (Figure 47)

With the accelerator pedal fully depressed, the full-load volume continues to be injected until the maximum full-load speed, n_{vo}, (breakaway speed) is reached. If the engine speed continues to increase beyond the maximum full-load speed, the force of the flyweights (22) is enough to overcome the force of the governor spring (17). Full-load speed regulation then comes into effect. The engine speed then increases a little further, the control rack is pushed back towards the stop setting and as a result the fuel delivery quantity is reduced. The point at which start of speed regulation takes effect depends on the tension of the governor spring. The maximum no-load speed, n_{no}, is reached when the engine load is entirely removed. When the engine is overrunning, e.g. if the vehicle is traveling downhill, the engine is accelerated by the road wheels. Under such conditions, no fuel is injected (overrun fuel cutoff).

Stopping the engine

Manual operation of the shutoff lever (1) moves the control rack (4) by means of the stop lever (3) to the stop setting. Fuel delivery is shut off and, therefore, the engine stopped. The engine can also be stopped by means of a pneumatically operated shutoff valve (refer to the section "Calibration devices").

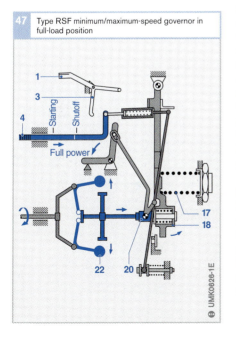

47   Type RSF minimum/maximum-speed governor in full-load position

Fig. 47
(Only those components involved in the governing function are illustrated)
1   Shutoff lever
3   Stop lever
4   Control rack
17   Governor spring
18   Spring retainer (torque control)
20   Sliding sleeve
22   Flyweight

Calibration devices

Control-lever stops

On every governor there are stops for the minimum and maximum control-lever deflection. If, for example, the driver fully depresses the accelerator, the control lever is brought into contact with an adjustable stop screw. Adjusting the screw alters
- The control-lever deflection, i.e. the injected fuel quantity, on a minimum/maximum-speed governor
- The maximum speed on a variable-speed governor

The stop screw is factory-adjusted and sealed; tampering with it voids the manufacturer's warranty.

The other stop is normally used to adjust the idle speed. This stop may be sprung or rigid.

Rigid stop

With a rigid stop (Figure 1) the fuel-injection equipment must incorporate a separate device for stopping the engine.

Sprung stop

If a sprung stop is used (Figures 2 and 3) the stop setting is reached by pressing the lever past the stop position against the force of the spring.

If necessary, the minimum stop can be set to "shutoff", but in this case there must be a low-idle stop elsewhere on the engine.

Stops for intermediate fuel volumes or engine speeds

Stops for intermediate control lever settings can be fitted as an option.

Depending on governor type, either a "reduced-delivery stop" for setting a lower full-load delivery quantity, or an "intermediate-speed stop" for setting an engine speed below nominal speed can be used (Figure 4).

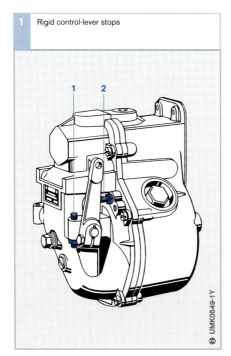

1 Rigid control-lever stops

Fig. 1
1   Stop for idle speed (or shutoff)
2   Stop for full-load volume on minimum/maximum-speed governor or for nominal speed on variable-speed governor

Fig. 2
1   Lever
2   Stop lever
3   Control-lever shaft
4   Clamping screw

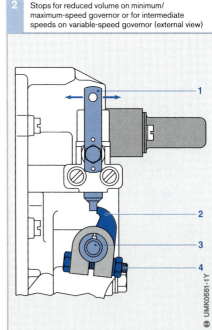

2 Stops for reduced volume on minimum/maximum-speed governor or for intermediate speeds on variable-speed governor (external view)

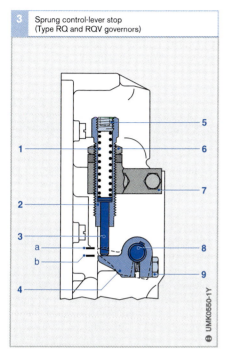

3 Sprung control-lever stop (Type RQ and RQV governors)

Control-rod stops

Apart from the stops for idle speed/shutoff, full-load volume/maximum speed (present on every governor for limiting control-lever movement) a special stop is required to limit control rack travel at full load or when starting from cold.

There are also full-load stops for performing specific compensating functions. Control-rod stops may be fitted on the fuel-injection pump or on the governor. A selection of the possible variations is described below in more detail.

Rigid start-quantity stop

The rigid start-quantity stop is used primarily on Type RQ governors with low idle-speed settings (Figure 5). When the engine is running, the excess fuel for starting is backed off by the governor so that it does not have an adverse effect (emission of smoke).

Fig. 3
a   Shutoff
b   Idle speed

1   Spring
2   Threaded sleeve
3   Pin
4   Stop lever
5   Screw cap
6   Locking nut
7   Fixing bracket
8   Control-lever shaft
9   Clamping screw

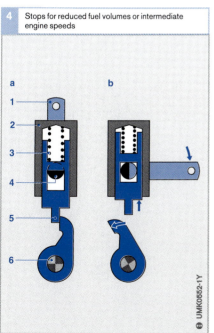

4 Stops for reduced fuel volumes or intermediate engine speeds

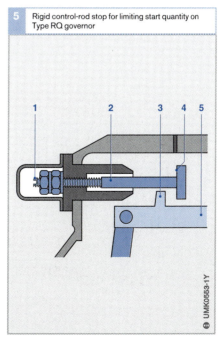

5 Rigid control-rod stop for limiting start quantity on Type RQ governor

Fig. 4
a   Locked
b   Released

1   Lever
2   Housing
3   Spring
4   Switching shaft
5   Stud
6   Control-lever shaft

Fig. 5
1   Excess starting fuel adjusting screw
2   Stop pin
3   Stop lug
4   Start-quantity limitation
5   Link fork

Sprung start-quantity stop for Type RQ governor

When the engine is started (accelerator fully depressed), the stop pin is moved against the resistance of the spring to the set start-quantity position. The spring in the stop acts against the idle-speed spring and thus initiates early return of the control rack from the start-quantity position (Figure 6). That means that if the engine is accelerated rapidly from idle speed, partial application of the start quantity is prevented.

Automatic full-load stop

When the engine is not running, the governor springs in the flyweights act via the sliding bolt (Figure 8, Item 13) to overcome the rocker spring (12). The rocker (9) pushes the stop strap (8) with the full-load stop (7) downwards (position shown in gray). If the accelerator is fully depressed when the engine is started, the control rack (6) can be moved to the start-quantity position.

After the engine has started, the sliding bolt is drawn back from the rocker (arrow) by the action of the flyweights. For the same reason, the control rack moves back from the start-quantity position to a lower quantity setting. Consequently, the rocker spring pivots the rocker so that its long arm moves back upwards (position shown in blue). The full-load stop once again prevents the control rack moving past the full-load position by catching against the lug on the link fork (4).

Stop with external torque-control mechanism for Type RQV governor

This external stop provides the facility for adjusting the full-load control-rod position and the torque-control settings (starting point, characteristic and travel). Torque control is effected by the interaction between the governor drag spring and torque-control spring (Figure 7) and requires that the springs are precisely matched to one another.

If there is also a tension spring for enabling the start quantity, the rocker (i.e. speed-dependent enabling) is omitted (Figure 9).

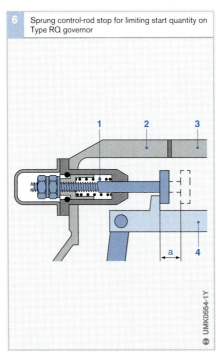

6   Sprung control-rod stop for limiting start quantity on Type RQ governor

Fig. 6
1 Spring
2 Governor cover
3 Governor housing
4 Control-rod link fork

a Start-quantity stop travel

Fig. 7
Torque-control spring overcomes drag spring
1 Torque-control spring
2 Control rack
3 Drag spring

a Torque-control travel

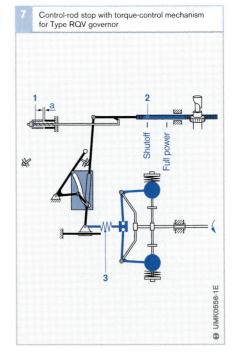

7   Control-rod stop with torque-control mechanism for Type RQV governor

8 Automatic full-load control-rod stop for Type RQV governor

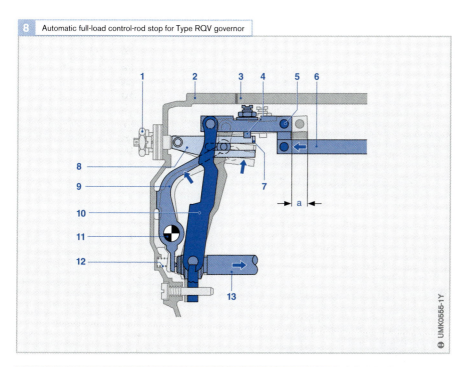

Fig. 8
Position shown in gray:
start quantity enabled
Position shown in blue:
full-load quantity setting

1   Full-load quantity
    adjuster
2   Governor cover
3   Governor housing
4   Stop lug
5   Link fork
6   Control rack
7   Full-load stop
8   Stop strap
9   Rocker
10 Variable-fulcrum
    lever
11 Control-lever shaft
12 Rocker spring
13 Sliding bolt

a   Start-quantity stop
    travel

9 Control-rod stop for Type RQV governor with lever for excess starting fuel and torque-control mechanism

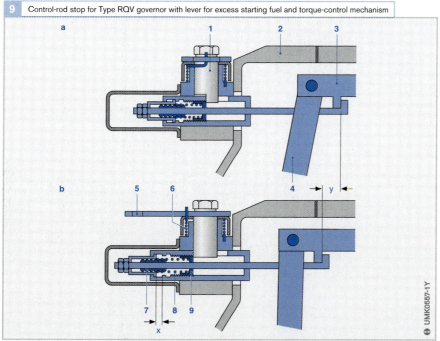

Fig. 9
a   Start-quantity
    position
b   Full-load setting with
    torque control

1   Locking pin
2   Governor cover
3   Link fork
4   Variable-fulcrum lever
5   Start-quantity lever
6   Lever compression
    spring
7   Threaded sleeve
8   Torque-control
    spring
9   Adjusting screw

x   Torque-control travel
y   Start-quantity stop
    travel

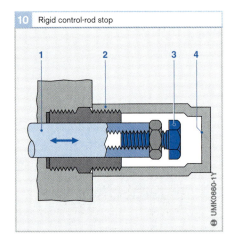

10   Rigid control-rod stop

UMK0860-1Y

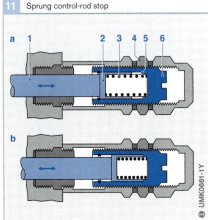

11   Sprung control-rod stop

UMK0861-1Y

Stop with internal torque-control mechanism for Type RQV governor

The control-rod stop with internal torque-control mechanism (Figure 12) for Type RQV governors protrudes only approximately 25 % of the length of the stop with external torque-control mechanism. Designed for situations where space is limited, this stop allows adjustment of the point at which torque control starts and the torque-control travel, but not the torque-control rate.

Pump-mounted stops

The full-load volume is generally adjusted on the governor. However, there are also rigid and sprung control-rod stops for mounting on the drive input side of the fuel-injection pump. They normally set the maximum permissible start quantity, and in a few cases the full-load volume as well.

Rigid version

A rigid stop set to the excess fuel for starting as shown in Figure 10 can be used in place of the governor-mounted stop shown in Figure 5. A rigid stop set to the full-load position will, by definition, not permit excess fuel for starting.

Sprung version

A pump-mounted sprung control-rod stop as shown in Figure 11 can be used in place of the governor-mounted stop shown in Figure 6; its function is identical.

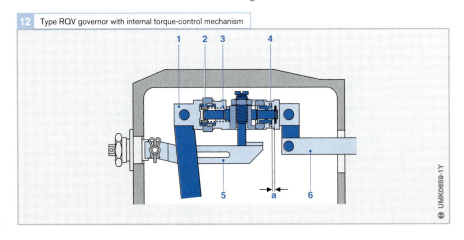

12   Type RQV governor with internal torque-control mechanism

UMK0859-1Y

Type LDA manifold-pressure compensator

Usage

On turbocharged engines, the full-load volume is set on the basis of the turbocharger pressure. However, the turbocharger pressure is lower at lower engine speeds and the mass of the cylinder charge therefore smaller. Consequently, the full-load volume must be adjusted in proportion to the smaller amount of air. The Type LDA manifold-pressure compensator reduces the full-load delivery quantity at lower engine speeds from a specific (selectable) turbocharger pressure onwards. There are versions of the manifold-pressure compensator for fitting on the fuel-injection pump as well as on the governor (top or rear). The version described below is intended for fitting on the Type RSV governor (Figures 13, 14 and 15).

Design and method of operation

The design of all such control-rod stops is essentially the same. In between the compensator housing, which screws onto the top of the governor, and its cover there is a diaphragm which forms an airtight seal (Figure 13, Item 3). In the compensator cover, there is a connection via which the manifold (turbocharger) pressure p_L acts on the diaphragm. A compression spring (4) acts on the diaphragm in the opposite direction from below. The other end of the compression spring is seated on a guide sleeve (5)

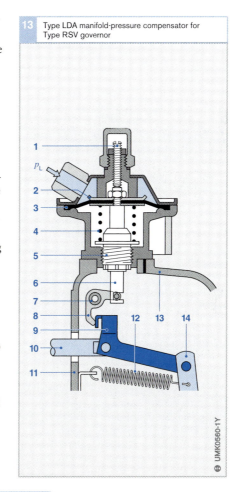

13 Type LDA manifold-pressure compensator for Type RSV governor

Fig. 13
1 Grub screw
2 Diaphragm disc
3 Diaphragm
4 Spring
5 Guide bushing
6 Pin
7 Setting shaft
8 Bell crank
9 Rigid link
10 Control rack
11 Governor housing
12 Starting spring
13 Governor cover
14 Variable-fulcrum lever

p_L Manifold pressure

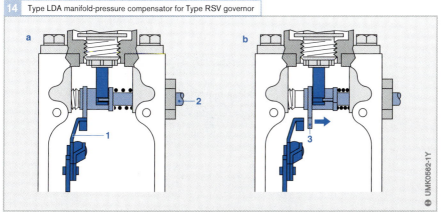

14 Type LDA manifold-pressure compensator for Type RSV governor

Fig. 14
a Normal operation setting
b Position of link relative to bell crank for starting

1 Rigid link
2 Setting shaft
3 Bell crank

which screws into the governor housing. The initial tension of the spring can thus be varied within certain limits.

The diaphragm is attached to a pin (6) that has a transverse slot cut into its lower end. A stud on the end of the bell crank (7) locates in that slot. When the manifold-pressure compensator is fitted on the governor, minor adjustments can be made by means of the grub screw (1). When the charge-air pressure is acting on the diaphragm, the pin is moved against the force of the compression spring. The maximum pin travel occurs when the charge-air pressure is at its highest. The pin acts via the bell crank, which pivots around a bell crank mounted inside the governor housing, and the rigid link, and thus ultimately on the control rack (10) of the fuel-injection pump. When the charge-air pressure drops, the control rack is moved towards the stop setting.

A version of the manifold-pressure compensator for Type RQV governors is shown in Figure 16.

   In order that the control rack can be set to the start-quantity position for starting the engine, the bell crank can be disengaged from contact with the rigid link by lateral movement of the setting shaft (Figure 14). This can be effected manually either by means of a cable-operated mechanism or a rod linkage;

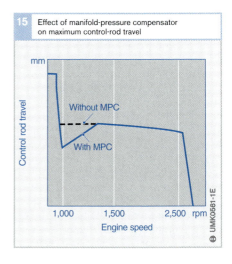

15 Effect of manifold-pressure compensator on maximum control-rod travel

Control rod travel (mm)

Without MPC

With MPC

1,000   1,500   2,500 rpm
Engine speed

UMK0561-1E

there are also governor designs where the setting shaft is operated by an electromagnet that is only energized during the starting sequence. A thermostatic switch can cut off the power supply to the solenoid if the start quantity is not required due to the temperature of the engine.

   The Type HSV hydraulic start-quantity locking device is another variation on the same theme that is operated by the engine-oil pressure. In this case, the oil pressure generated when the engine is started locks the excess fuel for starting. The hydraulic start-quantity locking device is screwed onto the side of the governor housing.

16 Type LDA manifold-pressure compensator for Type RQV governor

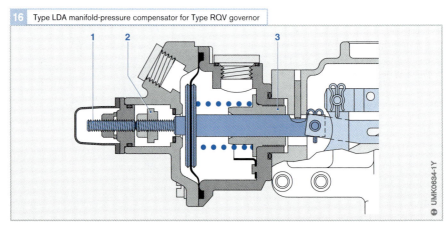

    1   2   3

Fig. 16
Adjusters for:
1   Suction quantity
2   Press-charge
    fuel-delivery quantity
3   Cutin point

UMK0634-1Y

Type ADA altitude-pressure compensator

Usage

Engines that are used at widely varying altitudes require adjustment of the injected fuel quantity to take account of the reduced mass of the cylinder charge upwards of a specific altitude.

The Type ADA altitude-pressure compensator (Figures 17 and 18) reduces the injected fuel quantity in response to increasing altitude (diminishing atmospheric pressure). On Type RQ(V) and RSF governors is fitted on the governor cover.

Design and method of operation

On the Type RQV governor, the altitude-pressure compensator consists of a barometric capsule (3) that is fitted vertically inside an outer housing and can be adjusted to a specific altitude setting by means of an adjusting screw (1) and an opposing spring-loaded pin (5). As the altitude increases, the barometric capsule expands.

The spring-loaded pin resting against the underside of the barometric capsule and the fork (4) screwed onto the end of the pin transmit the expansion and contraction of the barometric capsule to the pivoted cam disc (8). The cam disc acts on the pin connected to the stop strap. The cam disc pivots downwards. The pin attached to the stop strap moves the control rack towards the stop setting and the fuel delivery quantity is reduced. If the barometric capsule contracts again due to a reduction altitude, the delivery quantity increases again. The cam disc can be adjusted in the horizontal plane by means of a screw in order to set the full-load volume.

The arrangement and design in the case of the Type RSF governor are similar. In this case, changes in altitude are transmitted to the fuel-injection pump control rack by a spring-loaded pin and a series of connected levers. A similar design is also used on the Type RQ governor.

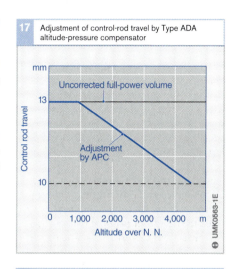

17   Adjustment of control-rod travel by Type ADA altitude-pressure compensator

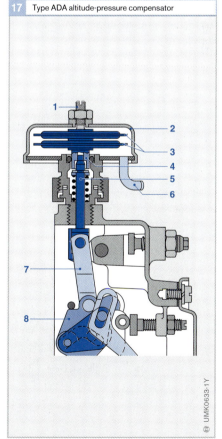

17   Type ADA altitude-pressure compensator

Fig. 18
1   Adjusting screw
2   Cap
3   Barometric capsules
4   Clevis
5   Spring-loaded pin
6   Connection to outside (for detecting atmospheric pressure)
7   Compensating link
8   Cam disc

Type ALDA absolute manifold-pressure compensator

Usage

The charge-air pressure of the turbocharged engine is measured relative to the ambient atmospheric pressure, the effect of which is felt most strongly when there are significant changes in altitude. If atmospheric pressure and manifold pressure are added, the absolute pressure is obtained.

Design and method of operation

The absolute manifold-pressure compensator on the Type RSF governor also has barometric capsules that can be adjusted for different altitudes and which are subjected to the absolute pressure via a connection to the engine's intake manifold (Figure 19, Item 1).

The barometric capsules respond to changes in pressure by expanding or contracting, thereby adjusting the injected fuel quantity by acting on a system of levers connected to the control rack.

Type PLA pneumatic idle-speed increase

Usage

The fuel volume required by a diesel engine when idling diminishes as engine temperature increases.

The temperature-dependent idle speed increase on the Type RSF governor (Figure 20) increases the engine's idle speed when it is cold, thus helping the engine to warm up more quickly. It also prevents the engine from dying if auxiliary equipment such as power steering, air conditioning, etc. cuts in while the engine is still cold. Once the engine has reached a certain temperature, it ceases to operate.

Design and method of operation

A temperature-dependent vacuum acts on the diaphragm (3) in the vacuum unit. The diaphragm moves a sliding bolt (2) which varies the tension on the idle-speed spring (1). This causes the governor linkage to move the control rack to a higher fuel-quantity setting.

Fig. 19
1. Connection to engine intake manifold (absolute pressure detection)
2. Adjusting screw
3. Pressure capsule
4. Barometric capsules
5. Compensating linkage
6. Plate cam

Fig. 20
1. Idle-speed spring
2. Sliding bolt
3. Diaphragm
4. Vacuum connection
5. Vacuum unit
6. Compression spring

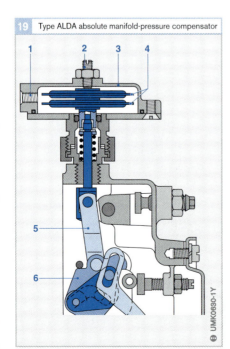

19   Type ALDA absolute manifold-pressure compensator

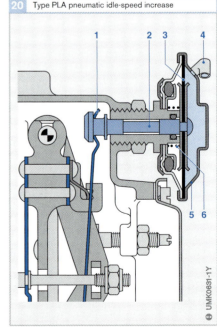

20   Type PLA pneumatic idle-speed increase

Type ELR electronic idle-speed control system

Usage

Instead of normal pneumatic idle-speed increase, the Type RSF can be combined with an electronic idle-speed control system for more demanding applications.

Design and method of operation

The electronic idle-speed control system consists of
- an electronic control unit, and
- an actuator solenoid.

The electronic control unit adjusts the idle speed by means of the actuator solenoid in response to changes in temperature and engine-load conditions. As shown in Figure 21, the actuator solenoid is mounted on the Type RSF governor cover in such a way that the energized solenoid armature can augment the force of the idle-speed spring and thus increase the idle speed.

Type ARD surge damping

Usage

Pulsations caused by sudden load changes can be largely eliminated by the use of surge damping on the Type RSF governor.

Design and method of operation

Surge damping consists of an electronic control unit, an engine speed sensor and an actuator solenoid.

The electronic control unit reads and analyzes the signals from the engine-speed sensor. In order to prevent vehicle judder caused by bucking oscillations, it operates the actuator solenoid (Figure 21, Item 2) of the Type RSF governor in such a way that it moves the lower anchor point of the variable-fulcrum lever to a less extended position in response to the oscillations. As a result, the injected fuel quantity is reduced accordingly, thereby counteracting the bucking oscillations.

21 Type RSF governor with electronic idle-speed control and active surge damping

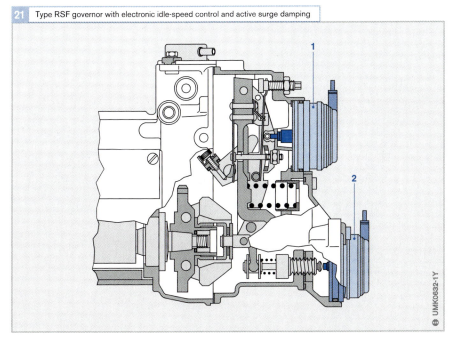

Fig. 21
1 Actuator solenoid for electronic idle-speed control
2 Actuator solenoid for active surge damping

Type TAS temperature-compensating start-quantity stop

Usage

On many engines, a greater start quantity is only required at low ambient temperatures when the engine is also cold. For environmental safety reasons, unnecessary injection of excess fuel for starting should be avoided. The temperature-compensating start-quantity stop ensures that the quantity delivered when starting does not exceed the required amount as specified by the engine manufacturer. This device is available for virtually all governor types.

Design and method of operation
With the aid of an expansion element

(Figure 22) that responds to ambient temperature or a temperature-controlled electromagnet (Figure 23), the start quantity is limited when hot-starting the engine by limiting the control rack travel according to ambient temperature.

Depending on the fitting constraints on the fuel-injection pump and the type of governor, the following types of expansion element/electromagnet are used:
1. If there is sufficient space on the drive input side of the pump, the expansion element acts directly on the control rack (Figure 22). The illustration shows the hot-start position in which the stop pin is pressed against the action of a spring by the sliding bolt of the ex-

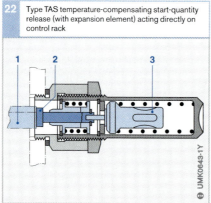

22 Type TAS temperature-compensating start-quantity release (with expansion element) acting directly on control rack

1
2
3

UMK0643-1Y

Fig. 22
1   Control rack
2   Stop pin
3   Expansion element

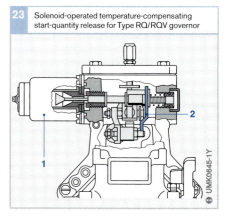

23 Solenoid-operated temperature-compensating start-quantity release for Type RQ/RQV governor

2

1

UMK0645-1Y

Fig. 23
1   Electromagnet
2   Catch

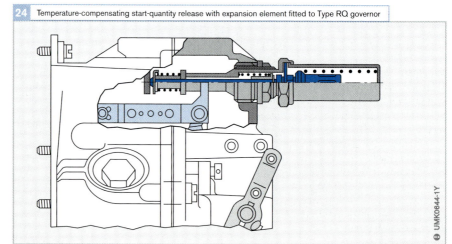

24 Temperature-compensating start-quantity release with expansion element fitted to Type RQ governor

UMK0644-1Y

pansion element so that only limited control rack travel is possible. With this arrangement, the hot-start control-rod travel is equal to or greater than the full-load control-rod travel.
2. If the device is fitted on the governor (Figure 24), in the case of a Type RQ governor the expansion element is mounted on the governor cover. In combination with the action of the springs, the expansion element has the effect of reducing control-rack travel when starting the engine from warm. In this case, the hot-start control-rod travel is equal to or greater than the full-load control rack travel.
3. In the case of elements with a starting groove, the following rule applies to cold/hot starting:
- Cold starting: higher fuel volume and retarded injection.
- Hot starting: normal fuel volume, injection not retarded.

A standard feature of Type RQ/RQV governors is also the solenoid-operated starting-volume release which can be controlled according to temperature. When starting the engine from cold, movement of the catch (Figure 23, Item 2) clears the way for the control rack to move to the start-quantity position. When the engine is hot, the solenoid is switched off so that the catch engages and the hot-start quantity is the same as the full-load volume.
4. On Type RQ/RQV governors with governor-mounted manifold-pressure compensator, the temperature-dependent start quantity setting is effected by the expansion element acting via a lever arrangement in the governor that allows a start quantity or unassisted-aspiration quantity setting according to whether the engine is being started from cold or hot (Figure 27).
5. For special operating conditions, the Type RQ governor can also be fitted with an expansion cartridge (temperature-compensating link, Figure 26).
6. On the Type RSV governor with control-rod stop or manifold-pressure compensator, the excess fuel for starting can be enabled by a temperature-controlled electromagnet (Figure 25). When starting the engine from cold, the solenoid moves the catch (1), thus clearing the way for the control rack to move to the cold-starting position.

25 Solenoid-operated temperature-compensating start-quantity release for Type RSV governor with control-rod stop or manifold-pressure compensator

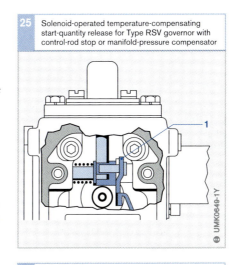

Fig. 25
1   Catch

26 Temperature-compensating link for Type RQ governor

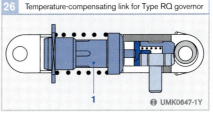

Fig. 26
1   Expansion sensor

27 Mechanical start-quantity locking device with expansion element for Type RQ/RQV governor with manifold-pressure compensator

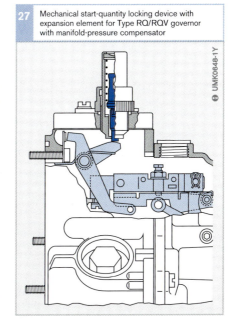

Stabilizer

Usage

The stabilizer is best suited to use on governors for engines that drive power generators in order to stabilize borderline or marginally unstable systems and to reduce proportional response on stable systems. However, it is not intended as a means of shortening the transient recovery time or reducing dynamic proportional response.

Design and method of operation

The stabilizer is hydraulically operated. It consists of a plunger (Figure 28, Item 7) that fits very tightly in the stabilizer body (6) bolted onto the governor cover. The plunger chamber is connected to an oil reservoir (5) by a channel with a variable throttle bore. A spring attached to the plunger connects at its opposite end to the tensioning lever on the Type RSV governor, or to the sliding bolt on the Type RQV governor in such a way that there is no free play between the components. The oil reservoir is connected to the engine lube-oil circuit and is designed so that no air can enter the plunger chamber at the angles of inclination normally encountered.

If the speed of the engine changes or fluctuates, the movement of the flyweights is damped by a "transient action" spring. This increases the dynamic proportional response. Once the engine speed has settled again under the operating conditions, the auxiliary spring is deactivated again, i.e. the static proportional response is not altered by the stabilizer. If the flyweights move inwards or outwards, the stabilizer spring is either extended or compressed. The resistance of the spring combines with the resistance of the governor springs and thus temporarily produces a greater proportional response, which has a stabilizing effect on the control loop as a whole. As the other end of the spring is connected to the hydraulic piston, the piston is moved until the force of the stabilizer spring is equalized. The damping effect of the stabilizer depends on the spring constant of the stabilizer spring (choice of springs) and the setting of the throttle screw between the

28 Type RSV governor with stabilizer

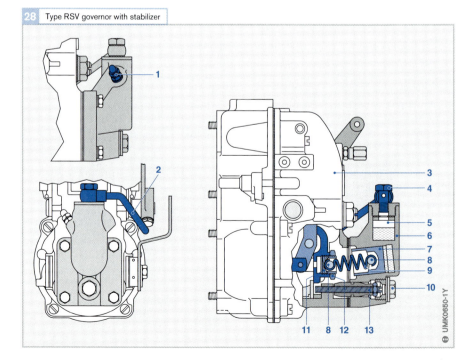

Fig. 28

1 Throttle screw
2 Oil supply line
3 Governor cover
4 Hollow screw with input throttle
5 Oil reservoir
6 Housing
7 Plunger
8 Retaining pin
9 Stabilizer spring
10 Screw cap
11 Threaded sleeve
12 Hexagonal nut
13 Full-load adjusting screw

plunger chamber and the oil reservoir. Since proper functioning requires that there is no air whatsoever in the plunger chamber, the stabilizer incorporates an automatic vent function. When it is first operated or after it has been idle, it requires a short startup phase before it is fully functional.

Type PNAB pneumatic shutoff device

In order to stop the engine, the "ignition" key is turned to the "off" position. The vacuum produced by a separate vacuum pump then acts on the diaphragm in the pneumatic shut-off device on the Type RSF governor (Figure 29). As a result, the rod connected to the diaphragm moves the control rack (5) to the stop setting.

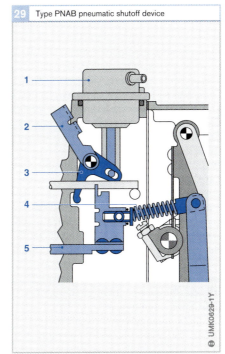

29 Type PNAB pneumatic shutoff device

Fig. 29
1   Pneumatic shutoff
     device
2   Shutoff lever for
     manual operation
3   Stop lever
4   Sprung link
5   Control rack

30 Type RQ/RQV governor with stabilizer

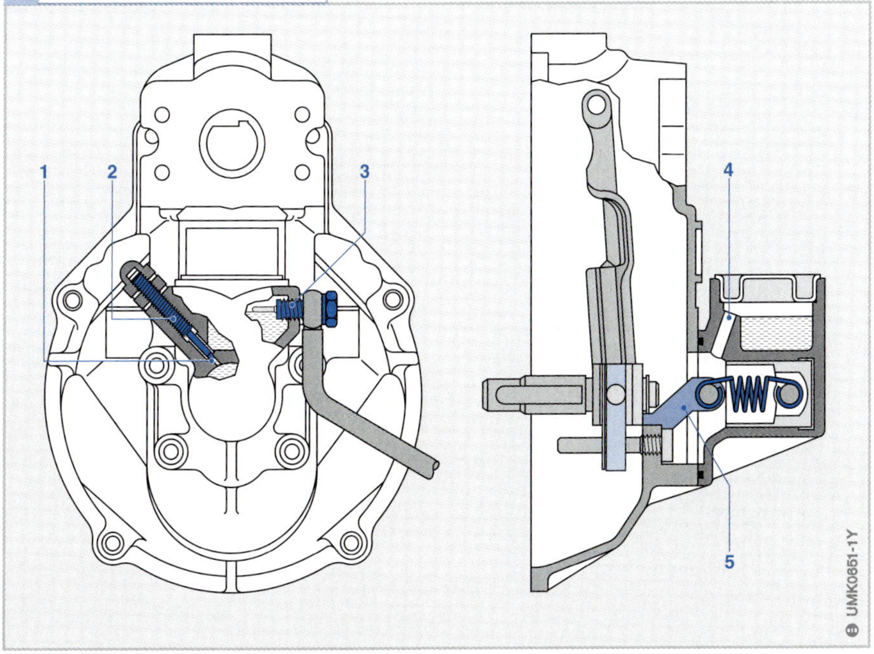

Fig. 30
1   Throttle bore
2   Adjustable throttle
     screw
3   Oil inlet
4   Oil overflow
5   Connecting link

Timing devices

The start of delivery (Figure 1, DP) represents the point at which fuel delivery by the fuel-injection pump commences. The timing of the start of delivery depends on the variables "injection lag" (IL) and "ignition lag" (CL) which are dependent on the operating status of the engine. The *injection lag* refers to the time delay between the start of delivery (DP) and the start of injection (IP), i.e. the time at which the nozzle opens and starts injecting fuel into the combustion chamber. The *ignition lag* is the time that elapses between the start of delivery and the combustion start (CP). The combustion start defines the point when air-and-fuel mixture ignites. It can be varied by altering the start of delivery.

Start of delivery, start of injection and combustion start are specified in degrees of crankshaft rotation relative to crankshaft top dead center (TDC).

Engine-speed related adjustment of start of delivery on an in-line fuel-injection pump is best achieved by means of a timing device.

Functions

Strictly speaking, based on its function, the timing device should really be called a delivery start adjuster, as it actually varies the start of delivery directly. It transmits the drive torque for the fuel-injection pump and simultaneously performs its adjustment function. The torque required to drive the fuel-injection pump depends on the pump size, the number of cylinders, the injected fuel quantity, the injection pressure, the plunger diameter, and the cam shape used. The drive torque has a retroactive effect on the timing characteristics which must be taken into account in the design as well as the work capacity.

Design

On in-line fuel-injection pumps, the timing device is mounted directly on the injection-pump camshaft. There are basically two types of design – open and closed.

A *closed-type timing device* has its own oil supply outside the housing which is independent of the engine lube-oil circuit.

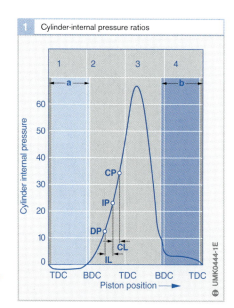

1 Cylinder-internal pressure ratios

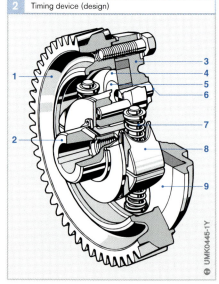

2 Timing device (design)

In the case of the *open-type design*, the timing device is connected directly to the engine lube-oil circuit. Its housing is bolted to a gear wheel. Inside the housing, the adjusting and balancing eccentrics are able to rotate around their respective bearings. They are held by a pin that is rigidly attached to the timing adjuster housing. The advantages of the open-type design are the smaller space requirements, more effective lube-oil supply and lower cost.

Method of operation

The link between the input and output sides of the of the timing device is formed by the nested pairs of eccentrics (Figures 2 and 3).

The larger eccentrics – the adjusting eccentrics (4) – fit inside the bearing plate (9) that is bolted to the gear wheel that forms the drive input side (1). Fitted inside the adjusting eccentrics are the balancing eccentrics (5). The latter are held by the adjusting eccentrics and the hub pins (6).

The hub pins are attached directly to the hub that forms the drive output side (2). The flyweights (8) locate in the adjusting eccentrics by means of flyweight bolts and are held in their resting position (Figure 3a) by compression springs (7).

The higher the engine speed – and therefore the speed of the timing device – the further outwards the flyweights move against the action of the compression springs. As a result, the relative position of the input and output sides of the timing adjuster alters by the angle α. Consequently, the engine and pump camshafts are offset by that angle relative to one another and the start of delivery is thus "advanced".

Sizes

By their external diameter and width, the size of the timing device determines the possible mass of the flyweights, the center of gravity separation and the available centrifugal-weight travel. Those three criteria are also the major factors in determining the working capacity and type of application of the timing device.

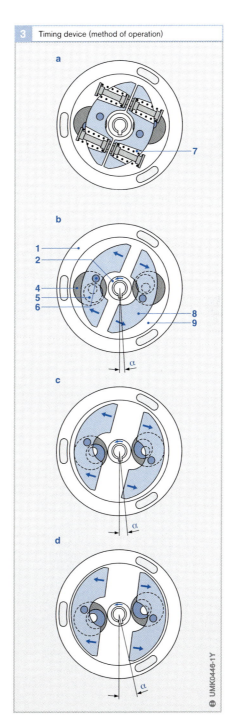

3 Timing device (method of operation)

UMK0446-1Y

Fig. 3
a   Resting position
b   Position at low
    speed
c   Position at medium
    speed
d   Position at high
    speed

1   Drive input side
2   Drive output side
    (hub)
4   Adjusting eccentric
5   Balancing eccentric
6   Hub pin
7   Compression spring
8   Flyweight
9   Bearing plate

α Advance angle

Electric actuator mechanism

Fuel-injection systems with Electronic
Diesel Control EDC use an electric actuator
mechanism mounted directly on the fuel-injection pump instead of a mechanical governor. The electrical actuator is controlled by
the engine control unit or ECU (electronic
control unit). The control unit calculates the
required control signals on the basis of the
input data from the sensors and desired-
value generators and using stored programs
and characteristic data maps. For example, it
may be programmed with an RQ or RQV
control characteristic for the purposes of engine response.

A semi-differential short-circuit-ring sensor
signals the position of the control rack to the
engine control unit so that a closed control
loop is formed. The sensor is also called a rack
travel sensor.

Design and method of operation
The injected fuel quantity is determined – as
with in-line fuel-injection pumps with mechanical governors – by the control rack position and the pump speed.

The linear magnet of the actuator mechanism (Figure 1, Item 4) moves the fuel-injection pump's control rack (1) against the
action of the compression spring (2). When
the magnet is de-energized, the spring pushes
the control rack back to the stop setting, thereby
cutting off the fuel supply to the engine. As the
effective control current increases, the magnet
draws the solenoid armature (5) to a higher
injected fuel quantity setting. Thus, varying
the effective signal current provides a means
of infinitely varying the control-rack travel between zero and maximum injected fuel quantity.

The control signal is not a direct-current
signal but a pulse-width modulation signal
(PWM signal, Figure 2). This is a square-
wave signal with a constant frequency and a

1   Actuator mechanism for Electronic Diesel Control EDC

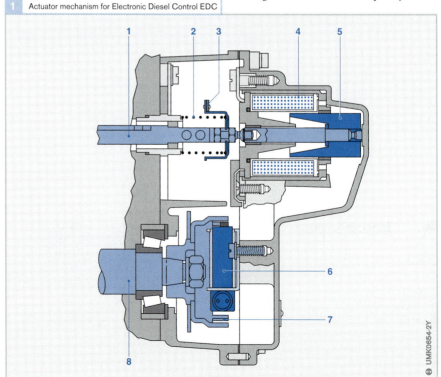

Fig. 1
1   Control rack
2   Compression spring
3   short-circuiting ring
    for rack-travel sensor
4   Linear magnet
5   Solenoid armature
6   Speed sensor
7   Sensor ring for
    speed sensor or
    marker for start of
    delivery
8   Fuel-injection pump
    camshaft

variable pulse duration. The size of the cutin current is always the same. The effective current, which determines the excursion of the armature in the actuator mechanism, depends on the ratio of the pulse duration to the pulse interval. A short pulse duration produces a low effective current, and a long pulse duration a high effective current. The frequency of the signal is chosen to suit the actuator mechanism. This method of control avoids interference problems which low currents might otherwise be susceptible to.

Control-sleeve actuator mechanism

Control-sleeve in-line fuel-injection pumps also have a setting shaft (Figure 3, Item 3) for the start of delivery as well as the control rack for the injected fuel quantity (5) (see also the chapter "Control-sleeve in-line fuel-injection pumps"). This shaft is rotated by an additional actuator mechanism (1) by way of a

ball joint (2). A low effective signal current produces a small amount of shaft travel and therefore a retarded start of delivery. As the signal current increases, the start of delivery is shifted towards an "advanced" setting.

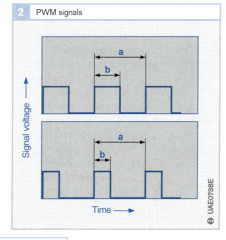

2 PWM signals

Signal voltage ⟶

a
b

a
b

Time ⟶

UAE0738E

Fig. 2
a   Fixed frequency
b   Variable pulse
    duration

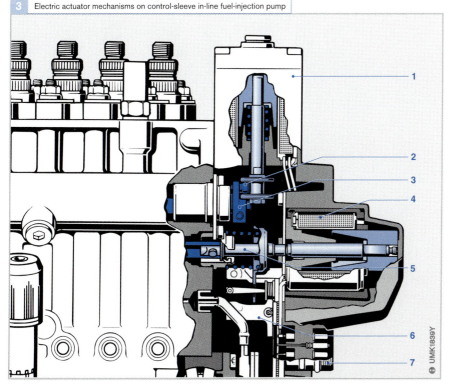

3 Electric actuator mechanisms on control-sleeve in-line fuel-injection pump

1

2

3

4

5

6

7

UMK1839Y

Fig. 3
1   Control-sleeve
    actuator mechanism
    (start of delivery
    actuator mechanism)
2   Ball joint
3   control-collar shaft
4   Linear magnet
    of control-rack travel
    actuator mechanism
5   control rack
6   Control-track travel
    sensor
7   Connector

Control-sleeve in-line fuel-injection pumps

The reduction of harmful exhaust-gas emissions is a subject to which commercial-vehicle producers are paying increasing attention. On commercial diesel engines, high fuel-injection pressures and optimized start of delivery make a major contribution here. This has led to the development of a new generation of high-pressure in-line fuel-injection pumps – control-sleeve in-line fuel-injection pumps (Figure 1). This type is capable of varying not only the injected fuel quantity but also the start of delivery independently of engine speed. In comparison with standard in-line fuel-injection pumps, therefore, it offers an additional independently variable fuel-injection parameter. Control-sleeve in-line fuel-injection pumps are always electronically controlled.

The control-sleeve in-line fuel-injection pump is a component of the electric actuation system with which the start of delivery and the injected fuel quantity can be independently varied in response to a variety of determining factors (see chapter "Electronic Diesel Control EDC"). This method of control makes it possible to
- Minimize harmful exhaust-gas emissions
- Optimize fuel consumption under all operating conditions
- Precise fuel metering and
- Effective improvement of the starting and warm-up phases

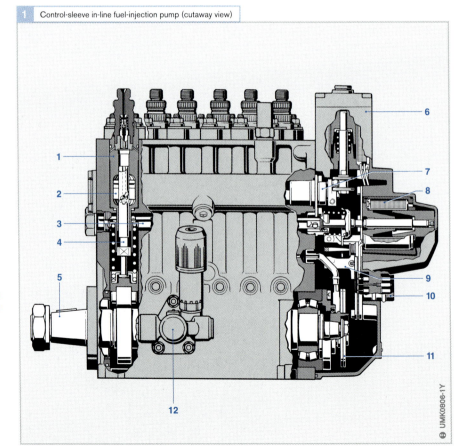

1 Control-sleeve in-line fuel-injection pump (cutaway view)

Fig. 1
1 Pump barrel
2 Control sleeve
3 Control rack
4 Pump plunger
5 Camshaft
   (connection to
   engine)
6 Start of delivery
   actuator mechanism
7 Control-sleeve shaft
8 Actuator solenoid
   for control-rack
   travel
9 Control-rack travel
   sensor
10 Connector
11 Disc for preventing
   fuel delivery which
   is also part of the
   oil-return pump
12 Presupply pump

UMK0806-1Y

A "rigid" pump-mounted timing device designed to cope with high torques is no longer required.

There are two designs of control-sleeve in-line fuel-injection pump:
- The Type H1 for 6...8 cylinders and up to 1,300 bar at the nozzle and
- The Type H1000 which offers a higher delivery rate for 5...8 cylinders and up to 1,350 bar at the nozzle for engines with greater fuel-quantity requirements

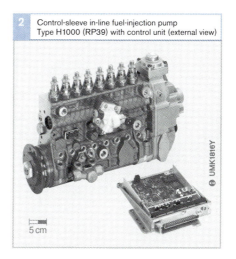

2 Control-sleeve in-line fuel-injection pump Type H1000 (RP39) with control unit (external view)

5 cm

UMK1816Y

Design and method of operation

The control-sleeve in-line fuel-injection pump differs in design from a standard type by virtue of a control sleeve (Figure 3, Item 4) which slides over the pump plunger. In all other aspects, it is the same.

The control sleeve, which slides over the pump plunger (1) inside a recess (2) in the pump barrel, provides the facility for varying the preliminary phase of the delivery stroke in order to alter the start of delivery and consequently the start of injection. In comparison with a standard in-line fuel-injection pump, this provides a second variable fuel-injection parameter that can be electronically controlled.

A control sleeve in each pump barrel incorporates the conventional spill port (3). A control-sleeve shaft with control-sleeve levers (6) which engage in the control sleeves changes the positions of all control sleeves at the same time. Depending on the position of the control sleeve (up or down), the start of delivery is advanced or retarded relative to the position of the cam. The injected fuel quantity is then controlled by the helix as on standard in-line fuel-injection pumps.

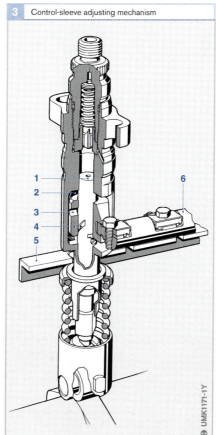

3 Control-sleeve adjusting mechanism

UMK1171-1Y

Fig. 3
1 Pump plunger
2 Recess for control sleeve
3 Spill port
4 Control sleeve
5 Control rack (injected fuel quantity)
6 Control-sleeve shaft

Fig. 4
a Bottom dead center
b Start of delivery
c End of delivery
d Top dead center

1 Delivery valve
2 Plunger chamber
3 Pump barrel
4 Control sleeve
5 Helix
6 Control port
  (start of delivery)
7 Pump plunger
8 Plunger spring
9 Roller tappet
10 Drive cam
11 Spill port

h_1 Plunger lift to port
  closing
h_2 Effective stroke
h_3 Residual travel

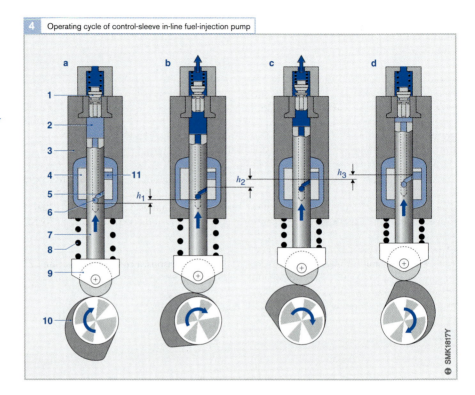

4 Operating cycle of control-sleeve in-line fuel-injection pump

Start of delivery

As soon as the pump plunger (Figure 4b, Item 7) has completed the preliminary phase (h_1) of the delivery stroke, the control sleeve (4) closes off the control port (6) in the pump plunger. From this point on, the pressure inside the plunger chamber (2) increases and fuel delivery begins.

The point at which fuel delivery, and therefore fuel injection, begins is altered by moving the control sleeve vertically relative to the pump plunger. When the control sleeve is closer to the piston top dead center, the plunger lift to port closing is longer and the start of delivery is therefore later. When the control sleeve is closer to the piston's bottom dead center position, the plunger lift to port closing is shorter and the start of injection is earlier.

The cam shape used determines the delivery velocity and the fuel-delivery rate (theoretical amount of fuel delivered per degree of cam rotation) as well as the injection pressure.

Spill

The piston's effective delivery stroke (h_2) ends when the helix (Figure 4c, Item 5) in the pump plunger overlaps the spill port (11) in the control sleeve and allows pressure to escape. Rotating the pump plunger by means of the control rack changes the point at which this occurs and, therefore, the quantity of fuel delivered in the same way as on a standard in-line fuel-injection pump.

Electronic control system

From the input data received from the sensors and desired-value generators described in the chapter "Electronic Diesel Control EDC", the control unit (Figure 5, Item 5) calculates the required fuel-injection pump settings. It then sends the appropriate electrical signals to the actuator mechanisms for start of delivery (1) and injected fuel quantity (4) on the fuel-injection pump.

Controlling start of delivery

Start of delivery is adjusted by means of a closed control loop. A needle-motion sensor in one of the nozzle holders (generally on no. 1 cylinder) signals to the control unit the actual point at which injection occurs. This information is used to determine the actual start of injection in terms of crankshaft position. This can then be compared with the setpoint value and the appropriate adjustment made by sending a current signal to the electrical start of delivery actuator mechanism.

The start of delivery actuator mechanism is "structurally rigid". For this reason, a separate travel feedback sensor can be dispensed with. Structurally rigid means that the lines of action of solenoid and spring always have a definite point of intersection. This means that the forces are always in equilibrium. Thus, the travel of the linear solenoid is proportional

to the signal current. This is equivalent to feedback within a closed control loop.

Controlling injected fuel quantity

The required injected fuel quantity calculated by the microcontroller in the control unit is set using the position control loop: The control unit specifies a required control-rack travel and receives a signal indicating the actual control-rack travel from the control-rack travel sensor (3). The control unit repeatedly recalculates the adjustment needed to achieve the required actuator mechanism setting, thereby continuously correcting the actual setting to match the setpoint setting (closed control loop).

For safety reasons, a compression spring (2) moves the control rack back to the "zero delivery" position whenever the actuator mechanism is de-energized.

---

5   Control-sleeve in-line fuel-injection pump Type H1 (RP43) with control unit

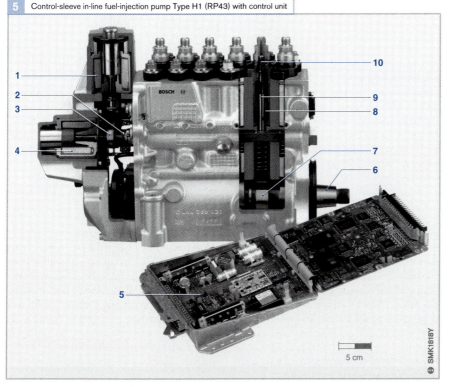

5 cm

⊕ SMK1818Y

Fig. 5
1 Fuel delivery actuator mechanism
2 Compression spring
3 Control-rack travel sensor
4 Control-rack actuator mechanism (injected fuel quantity)
5 ECU
6 Connection to engine
7 Camshaft
8 Control sleeve
9 Pump plunger
10 Delivery valve

Overview of distributor fuel-injection pump systems

The combustion processes that take place inside a diesel engine are essentially dependent on the way in which the fuel is delivered by the fuel-injection system. The fuel-injection pump plays a decisive role in that connection. It generates the necessary fuel pressure for fuel injection. The fuel is delivered via high-pressure fuel lines to the nozzles, which in turn inject it into the combustion chamber. Small, fast-running diesel engines require a high-performance fuel-injection system capable of rapid injection sequences, that is also light in weight and compact in dimensions. Distributor injection pumps meet those requirements. They consist of a small, compact unit comprising the fuel pump, high-pressure fuel-injection pump and control mechanism.

Areas of application

Since its introduction in 1962, the axial-piston distributor injection pump has become the most widely used fuel-injection pump for cars. The pump and its control system have been continually improved over that period. An increase in the fuel-injection pressure was required in order to achieve lower fuel consumption and exhaust-gas emissions on engines with direct injection. A total of more than 45 million distributor injection pumps were produced by Bosch between 1962 and 2001. The available designs and overall system configurations are accordingly varied.

Axial-piston distributor pumps for engines with indirect injection (IDI) generate pressures of as much as 350 bar (35 MPa) at the nozzle. For direct-injection (DI) engines, both axial-piston and radial-piston distributor injection pumps are used. They produce pressures of up to 900 bar (90 MPa) for slow-running engines, and up to 1,900 bar (190 MPa) for fast-running diesels.

The mechanical governors originally used on distributor injection pumps were succeeded by electronic control systems with electrical actuator mechanisms. Later on, pumps with high-pressure solenoid valves were developed.

Apart from their compact dimensions, the characteristic feature of distributor injection pumps is their versatility of application which allows them to be used on cars, light commercial vehicles, fixed-installation engines, and construction and agricultural machinery (off-road vehicles).

The rated speed, power output and design of the diesel engine determine the type and model of distributor injection pump chosen. They are used on engines with between 3 and 6 cylinders.

Axial-piston distributor pumps are used on engines with power outputs of up to 30 kW per cylinder, while radial-piston types are suitable for outputs of up to 45 kW per cylinder.

Distributor injection pumps are lubricated by the fuel and are therefore maintenance-free.

Designs

Three types of distributor injection pump are distinguished according to the method of fuel-quantity control, type of control system and method of high-pressure generation (Figure 1).

Method of fuel-quantity control
Port-controlled injection pumps
The injection duration is varied by means of control ports, channels and slide valves. A hydraulic timing device varies the start of injection.

Solenoid-valve-controlled injection pumps
A high-pressure solenoid valve opens and closes the high-pressure chamber outlet, thereby controlling start of injection and injection duration. Radial-piston distributor injection pumps are always controlled by solenoid valves.

Method of high-pressure generation

Type VE axial-piston distributor pumps
These compress the fuel by means of a piston which moves in an axial direction relative to the pump drive shaft.

Type VR radial-piston distributor pumps
These compress the fuel by means of several pistons arranged radially in relation to the pump drive shaft. Radial-piston pumps can produce higher pressures than axial-piston versions.

Type of control system

Mechanical governor
The fuel-injection pump is controlled by a governor linked to levers, springs, vacuum actuators, etc.

Electronic control system
The driver signals the desired torque output/engine speed by means of the accelerator pedal (sensor). Stored in the control unit are data maps for starting, idling, full load, accelerator characteristics, smoke limits and pump characteristics.

Using that stored information and the actual values from the sensors, specified settings for the fuel-injection pump actuators are calculated. The resulting settings take account of the current engine operating status and the ambient conditions (e.g. crankshaft position and speed, charge-air pressure, temperature of intake air, engine coolant and fuel, vehicle road speed, etc.). The control unit then operates the actuators or the solenoid valves in the fuel-injection pump according to the required settings.

The EDC (Electronic Diesel Control) system offers many advantages over a mechanical governor:

- Lower fuel consumption, lower emissions, higher power output and torque by virtue of more precise control of fuel quantity and start of injection.
- Lower idling speed and ability to adjust to auxiliary systems (e.g. air conditioning) by virtue of better control of engine speed.
- Greater sophistication (e.g. active surge damping, smooth-running control, cruise control).
- Improved diagnostic functions.
- Additional control functions (e.g. preheating function, exhaust-gas recirculation, charge-air pressure control, electronic engine immobilisation).
- Data exchange with other electronic control systems (e.g. traction control system, electronic transmission control) and therefore integration in the vehicle's overall control network.

1   Types of distributor injection pump

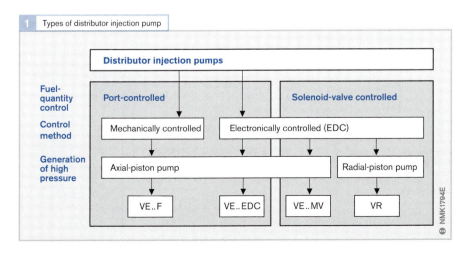

Helix and port-controlled systems

Mechanically controlled distributor injection pumps

Mechanical control is used only on axial-piston distributor pumps. This arrangement's assets consist of low manufacturing cost and relatively simple maintenance.

Mechanical rotational-speed control monitors the various operating conditions to ensure high quality in mixture formation. Supplementary control modules adapt start of delivery and injected-fuel quantity to various engine operating statuses and load factors:

● Engine speed
● Engine load
● Engine temperature
● Charge-air pressure and
● Barometric pressure

In addition to the fuel-injection pump (Fig. 1, 4), the diesel fuel-injection system includes the fuel tank (11), the fuel filter (10), the presupply pump (12), the nozzle-and-holder assembly (8) and the fuel lines (1, 6 and 7). The nozzles and their nozzle-and-holder assemblies are the vital elements of the fuel-injection system. Their design configuration has a major influence on the spray patterns and the rate-of-discharge curves. The solenoid-operated shutoff valve (5) (ELAB) interrupts the flow of fuel to the pump's plunger chamber [1] when the "ignition" is switched off.

A Bowden cable or mechanical linkage (2) relays driver commands recorded by the accelerator pedal (3) to the fuel-injection pump's controller. Specialized control modules are available to regulate idle, intermediate and high-idle speed along. The VE..F series designation stands for "Verteilereinspritzpumpe, fliehkraftgeregelt", which translates as flyweight-controlled distributor injection pump.

[1] The operating concept is reversed on marine engines. On these power-plants, the ELAB shutoff closes under current.

Fig. 1

1 Fuel supply line
2 Linkage
3 Accelerator pedal
4 Distributor injection pump
5 Solenoid-operated shutoff valve (ELAB)
6 High-pressure fuel line
7 Fuel-return line
8 Nozzle-and-holder assembly
9 Sheathed-element glow plug, Type GSK
10 Fuel filter
11 Fuel tank
12 Fuel presupply pump (installed only with extremely long supply lines or substantial differences in relative elevations of fuel tank and fuel-injection pump)
13 Battery
14 Glow-plug and starter switch ("ignition switch")
15 Glow control unit, Type GZS
16 Diesel engine (IDI)

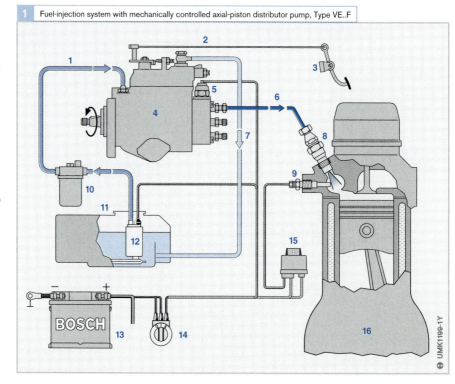

1 Fuel-injection system with mechanically controlled axial-piston distributor pump, Type VE..F

Electronically controlled distributor injection pumps

Electronic Diesel Control (EDC) supports a higher level of functionality than that provided by mechanical control systems. Electric measuring combines with the flexibility contributed by electronic data processing and closed-loop control featuring electric actuators to embed additional operational parameters in the control process.

Figure 2 illustrates the components in a fully-equipped fuel-injection system featuring an electronically-controlled axial-piston distributor pump. Some individual components may not be present in certain applications or vehicle types. The system consists of four sectors:

- Fuel supply (low-pressure circuit)
- Fuel-injection pump
- Electronic Diesel Control (EDC) with system modules for sensors, control unit and final controlling elements (actuators), and
- Peripherals (e.g. turbocharger, exhaust-gas recirculation, glow-plug control, etc.)

The solenoid-controlled actuator mechanism in the distributor injection pump (rotary actuator) replaces the mechanical controller and its auxiliary modules. It employs a shaft to shift the control collar's position and regulate injected fuel quantity. As in the mechanical pump, control collar travel is employed to vary the points at which the port is opened and closed. The ECU uses the stored program map and instantaneous data from the sensors to define the default value for the solenoid actuator position in the fuel-injection pump.

An angle sensor (such as a semidifferential short-circuiting ring sensor) registers the actuator mechanism's angle. This serves as an indicator of control collar travel and this information is fed back to the ECU.

A pulse-controlled solenoid valve compensates for fluctuations in the pump's internal pressure arising from variations in engine speed by shifting the timing device to modify start of delivery.

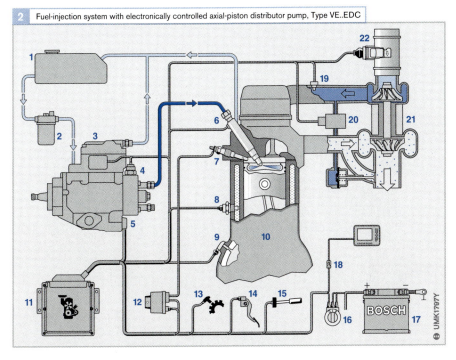

2   Fuel-injection system with electronically controlled axial-piston distributor pump, Type VE..EDC

Solenoid-valve-controlled systems

Solenoid-valve controlled fuel-injection systems allow a greater degree of flexibility with regard to fuel metering and variation of injection start than port-controlled systems. They also enable pre-injection, which helps to reduce engine noise, and individual adjustment of injection quantity for each cylinder.

Engine management systems that use solenoid-valve controlled distributor injection pumps consist of four stages (Figure 1):
- The fuel supply system (low-pressure stage)
- The high-pressure stage including all the fuel-injection components
- The Electronic Diesel Control (EDC) system made up of sensors, electronic control unit(s) and actuators and
- The air-intake and exhaust-gas systems (air supply, exhaust-gas treatment and exhaust-gas recirculation)

Control-unit configuration
Separate control units
First-generation diesel fuel-injection systems with solenoid-valve controlled distributor injection pumps (Type VE..MV [VP30], VR [VP44] for DI engines and VE..MV [VP29] for IDI engines) require two electronic control units (ECUs) – an engine ECU (Type MSG) and a pump ECU (Type PSG). There were two reasons for this separation of functions: Firstly, it was designed to prevent the overheating of certain electronic components by removing them from the immediate vicinity of pump and engine. Secondly, it allowed the use of short control leads for the solenoid valve. This eliminates interference signals that may occur as a result of very high currents (up to 20 A).

While the pump ECU detects and analyzes the pump's internal sensor signals for angle of rotation and fuel temperature in order to adjust start of injection, the engine ECU

---

1   Components of an engine ECU with solenoid-valve-controlled distributor injection pumps

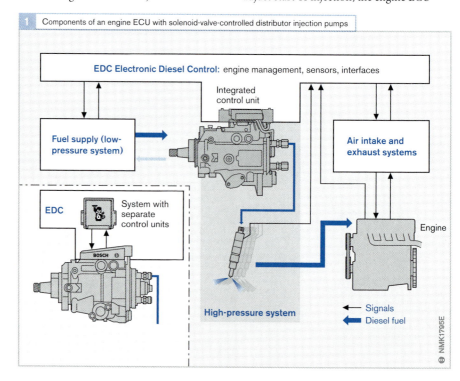

EDC Electronic Diesel Control: engine management, sensors, interfaces

Integrated control unit

Fuel supply (low-pressure system)

Air intake and exhaust systems

EDC   System with separate control units

BOSCH

Engine

High-pressure system

Signals
Diesel fuel

NMK1795E

processes all engine and ambient data signals from external sensors and uses them to calculate the actuator adjustments on the fuel-injection pump.

The two ECUs communicate over a CAN interface.

Integrated ECU

Heat-resistant printed-circuit boards designed using hybrid technology allow the integration of the engine ECU in the pump ECU on second-generation solenoid-valve-controlled distributor injection pumps. The use of integrated ECUs permits a space-saving system configuration.

Exhaust-gas treatment

There are various means employed for improving emissions and user-friendliness. They include such things as exhaust-gas recirculation, control of injection pattern (e.g. the use of pre-injection) and the use of higher injection pressures. However, in order to meet the increasingly stringent exhaust-gas regulations, some vehicles will require additional exhaust-gas treatment systems.

A number of exhaust-gas treatment systems are currently under development. It is not yet clear which of them will eventually become established. The most important are dealt with in a separate chapter.

---

2 Example of a diesel fuel-injection system with solenoid-valve-controlled radial-piston distributor injection pump and separate control units for engine and pump ECUs

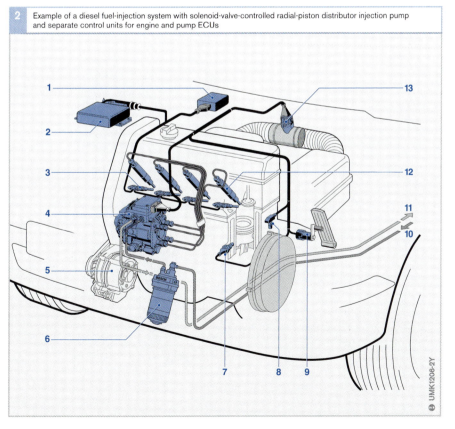

UMK1206-2Y

Fig. 2
1 Type GZS glow control unit
2 Type MSG engine ECU
3 Type GSK sheathed-element glow plug
4 Type VP44 radial-piston distributor injection pump with Type PSG5 pump ECU
5 Alternator
6 Fuel filter
7 Engine-temperature sensor (in cooling system)
8 Crankshaft speed sensor
9 Pedal travel sensor
10 Fuel inlet
11 Fuel return
12 Nozzle-and-holder assembly
13 Air-mass meter

System diagram

Figure 3 shows an example of a diesel fuel-injection system using a Type VR radial-piston distributor injection pump on a four-cylinder diesel engine (DI). That pump is fitted with an integrated engine and pump ECU. The diagram shows the full-configuration system. Depending on the nature of the application and the type of vehicle, certain components may not be used.

For the sake of clarity, the sensors and desired-value generators (A) are not shown in their fitted locations. One exception to this is the needle-motion sensor (21).

The CAN bus in the "Interfaces" section (B) provides the means for data exchange with a wide variety of systems and components such as

- The starter motor
- The alternator
- The electronic immobilizer
- The transmission-shift control system
- The traction control system (ASR) and
- The electronic stability program (ESP)

The instrument cluster (12) and the air conditioner (13) can also be connected to the CAN bus.

Fig. 3

Engine, engine ECU and high-pressure fuel-injection components

16 Fuel-injection pump drive
17 Type PSG16 integrated engine/pump ECU
18 Radial-piston distributor injection pump (VP44)
21 Nozzle-and-holder assembly with needle-motion sensor (cylinder no. 1)
22 Sheathed-element glow plug
23 Diesel engine (DI)
M Torque

A   Sensors and desired-value generators

1 Pedal-travel sensor
2 Clutch switch
3 Brake contacts (2)
4 Vehicle-speed control operator unit
5 Glow-plug and starter switch ("ignition switch")
6 Vehicle-speed sensor
7 Crankshaft-speed sensor (inductive)
8 Engine-temperature sensor (in coolant system)
9 Intake-air temperature sensor
10 Boost-pressure sensor
11 Hot-film air mass-flow sensor (intake air)

B   Interfaces

12 Instrument cluster with signal output for fuel consumption, rotational speed, etc.
13 Air-conditioner compressor and operator unit
14 Diagnosis interface
15 Glow control unit
CAN **C**ontroller **A**rea **N**etwork (onboard serial data bus)

C   Fuel supply system (low-pressure stage)

19 Fuel filter with overflow valve
20 Fuel tank with preliminary filter and presupply pump (preliminary pump is only required with long fuel pipes or large height difference between fuel tank and fuel-injection pump)

D   Air supply system

24 Exhaust-gas recirculation positioner and valve
25 Vacuum pump
26 Control valve
27 Exhaust-gas turbocharger with VTG (variable turbine geometry)
28 Charge-pressure actuator

E   Exhaust-gas treatment

29 Diesel-oxidation catalytic converter (DOC)

3 Diesel fuel-injection system with Type VP44 solenoid-valve-controlled radial-piston distributor injection pump and Type PSG16 integrated engine and pump ECU

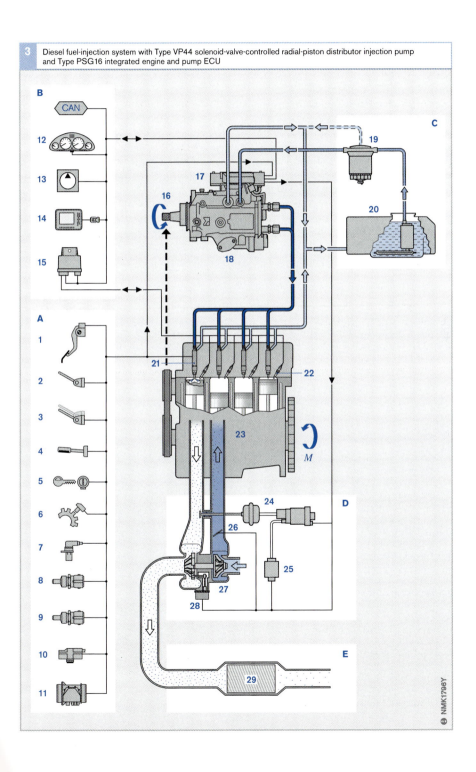

Helix and port-controlled distributor injection pumps

The helix-and-port distributor injection pump is always an axial-piston unit. As the design relies on a single high-pressure element to serve all of the engine's cylinders, units can be extremely compact. Helices, ports and collars modulate injected-fuel quantities. The point in the cycle at which fuel is discharged is determined by the hydraulic timing device. Mechanical control modules or an electric actuator mechanism (refer to section on "Auxiliary control modules for distributor injection pumps") provide flow control. The essential features of this injection-pump design are its maintenance friendliness, low weight and compact dimensions.

This type of pump makes up our VE series. This design replaced the EP/VA series pumps in 1975. In the intervening years it has undergone a range of engineering advances intended to adapt it to meet growing demands. The electric actuator mechanism's advent in 1986 (Fig. 2) started a major expansion in the VE distributor pump's performance potential. In the period up to mid-2002 roughly

42 million VE pumps were manufactured at Bosch. Every year well over a million of these ultra-reliable pumps emerge from assembly lines throughout the world.

The fuel-injection pump pressurizes the diesel fuel to prepare it for injection. The pump supplies the fuel along the high-pressure injection lines to the nozzle-and-holder assemblies that inject it into the combustion chambers.

The shape of the combustion process in the diesel engine depends on several factors, including injected-fuel quantity, the method used to compress and transport the fuel, and the way in which this fuel is injected in the combustion chamber. The critical criteria in this process are:
- The timing and duration of fuel injection
- The distribution pattern in the combustion chamber
- The point at which combustion starts
- The quantity of fuel injected for each degree of crankshaft travel and
- The total quantity of fuel supplied relative to the engine's load factor

1 Series VE...F mechanically-controlled distributor injection pump on a 4-cylinder diesel engine

Fig. 1
1 Pump drive
2 Fuel inlet
3 Accelerator pedal linkage
4 Fuel return
5 High-pressure fuel line
6 Nozzle-and-holder assembly

2 Series VE...EDC axial-piston distributor injection pump with electric actuator mechanism

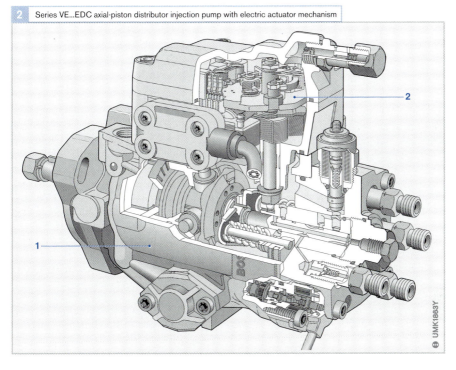

Fig. 2
1   Axial-piston
    distributor
    injection pump
2   Electric actuator
    mechanism

Applications and installation

Fast-turning diesels with limited displacement are one of the applications for helix and port-controlled distributor injection pumps. The pumps furnish fuel in both direct-injection (DI) and prechamber (IDI) powerplants.

The application and the fuel-injection pump's configuration are defined by such factors as nominal speed, power output and design of the individual diesel engine. Distributor injection pumps are fitted in passenger cars, commercial vehicles, construction and agricultural machinery, ships and stationary powerplants to produce power of up to 30 kW per cylinder.

These distributor injection pumps are available with high-pressure spill ports for engines of 3...6 cylinders. The maximum injected-fuel quantity is 125 mm³ per stroke. Requirements for injection pressure vary according to the specific engine's individual demand (DI or IDI). These pressures reach levels of 350...1,250 bar.

The distributor injection pump is flange-mounted directly on the diesel engine (Fig. 1). Motive force from the crankshaft is transferred to the pump by toothed belt, pinion, ring gear, or a chain and sprocket. Regardless of the arrangement selected, it ensures that the pump remains synchronized with the movement of the pistons in the engine (positive coupling).

On the 4-stroke diesel engine, the rotational speed of the pump is half that of the crankshaft. Expressed another way: the pump's rotational speed is the same as that of the camshaft.

Distributor injection pumps are available for both clockwise and counterclockwise rotation[1]. While the injection sequences varies according to rotational direction, the injection sequence always matches the geometrical progression of the delivery ports.

[1]   Rotational direction as viewed from the pump drive side

In order to avoid confusion with the designations of the engine's cylinders (cylinder no. 1, 2, 3, etc.) the distributor pump's delivery ports carry the alphabetic designations A, B, C, etc. Example: on a four-stroke engine with the firing order 1–3–4–2, the correlation of delivery ports to cylinders is A-1, B-3, C-4 and D-2.

The high-pressure lines running from the fuel-injection pump to the nozzle-and-holder assemblies are kept as short as possible to ensure optimized hydraulic properties. This is why the distributor injection pump is mounted as close as possible to the diesel engine's cylinder head.

The distributor injection pump's lubricant is fuel. This makes the units maintenance-free.
    The components and surfaces in the fuel-injection pump's high-pressure stage and the nozzles are both manufactured to tolerances of just a few thousandths of a millimeter. As a result, contamination in the fuel can have a negative impact on operation. This consideration renders the use of high-quality fuel essential, while a special fuel filter, custom-designed to meet the individual fuel-injec-

tion system's requirements, is another factor. These two elements combine to prevent damage to pump components, delivery valves and nozzles and ensure trouble-free operation throughout a long service life.

Diesel fuel can absorb 50...200 ppm water (by weight) in solution. Any additional water entering the fuel (such as moisture from condensation) will be present in unbound form. Should this water enter the fuel-injection pump, corrosion damage will be the result. This is why fuel filters equipped with a water trap are vital for the distributor injection pump. The water collected in the trap must be drained at the required intervals. The increasing popularity of diesel engines in passenger cars has resulted in the need for an automatic water level warning system. This system employs a warning lamp to signal that it is time to drain the collected water.
    Both the fuel-injection system and the diesel engine in general rely on consistently optimized operating parameters to ensure ideal performance. This is why neither fuel lines nor nozzle-and-holder assemblies should be modified during service work on the vehicle.

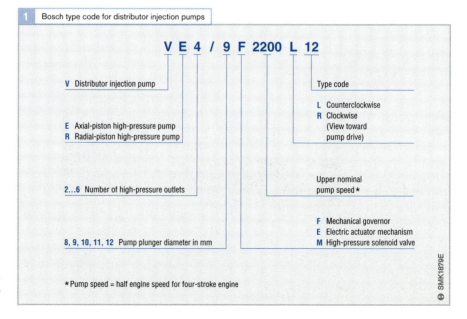

1   Bosch type code for distributor injection pumps

Fig. 1
This data code is affixed to the distributor injection pump housing for precise identification

Design

The distributor injection pump consists of the following main assembly groups (Fig. 2):

Low-pressure stage (7)
The vane-type supply pump takes in the diesel fuel and pressurizes the inner chamber of the pump. The pressure-control valve controls this internal pressure (3...4 bar at idle, 10...12 bar at maximum rpm). Air is discharged through the overflow valve. It also returns fuel in order to cool the pump.

High-pressure pump with distributor (8)
High pressure in the helix and port-controlled distributor injection pump is generated by an axial piston. A distributor slot in the pump's rotating plunger distributes the pressurized fuel to the delivery valves (9). The number of these valves is the same as the number of cylinders in the engine.

Control mechanism (2)
The control mechanism regulates the injection process. The configuration of this mechanism is the most distinctive feature of the helix-and-port distributor pump. Here the operative distinction is between:
- The mechanical governor assembly, with supplementary control modules and switches as needed and
- The electric actuator mechanism (VE...EDC), which is controlled by the engine ECU

Both can be equipped with a solenoid-operated shutoff valve (ELAB) (4). This solenoid device shuts off the fuel-injection system by isolating the high-pressure from the low-pressure side of the pump.
    Pump versions equipped with a mechanical governor also include a mechanical shutoff device which is integrated in the governor cover.

Hydraulic timing device (10)
The timing device varies the point at which the pump starts to deliver fuel.

---

2   Component assemblies in axial-piston distributor injection pump (schematic)

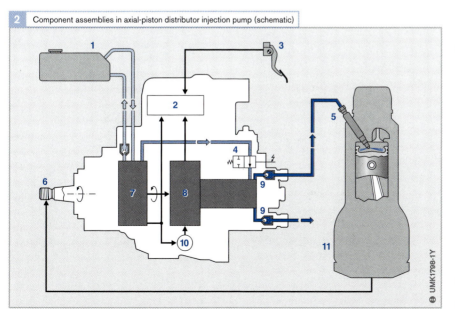

Fig. 2
1 Fuel supply (low-pressure)
2 Control mechanism
3 Accelerator pedal
4 Solenoid-operated shutoff valve (ELAB)
5 Nozzle-and-holder assembly
6 Pump drive
7 Low-pressure stage (vane-type supply pump with pressure-control valve and overflow throttle valve)
8 High-pressure pump with fuel rail
9 Delivery valve
10 Hydraulic timing device
11 Diesel engine

Fuel flow and control lever

The diesel engine powers an input shaft mounted in two plain bearings in the pump's housing (Fig. 4, Pos. 2). The input shaft supports the vane-type supply pump (3) and also drives the high-pressure pump's cam plate (6). The cam plate's travel pattern is both rotational and reciprocating; its motion is transferred to the plunger (11).

On injection pumps with mechanical control, the input shaft drives the control assembly (9) via a gear pair (4) with a rubber damper.

At the top of the control assembly is a control-lever shaft connected to the external control lever (1) on the governor cover. This control-lever shaft intervenes in pump operation based on the commands transmitted to it through the linkage leading to the accelerator pedal. The governor cover seals the top of the distributor injection pump.

Fuel supply

The fuel-injection pump depends on a continuous supply of pressurized bubble-free fuel in the high-pressure stage. On passenger cars and light trucks, the difference in the elevations of fuel-injection pump and fuel tank is usually minimal, while the supply lines are short with large diameters. As a result, the suction generated by the distributor injection pump's internal vane-type supply pump is usually sufficient.

Vehicles with substantial elevations differences and/or long fuel lines between fuel tank and fuel-injection pump need a pre-supply pump to overcome resistance in lines and filters. Gravity-feed fuel-tank operation is found primarily in stationary power-plants.

3 Component assemblies and their functions (cutaway)

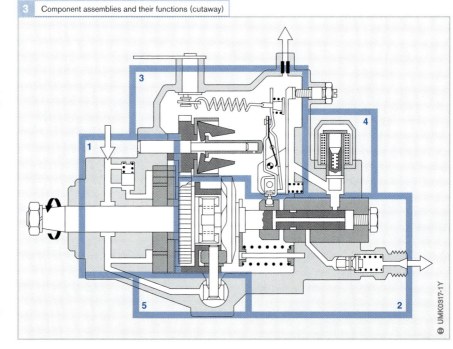

4   Series VE...F axial-piston distributor injection pump with mechanical governor

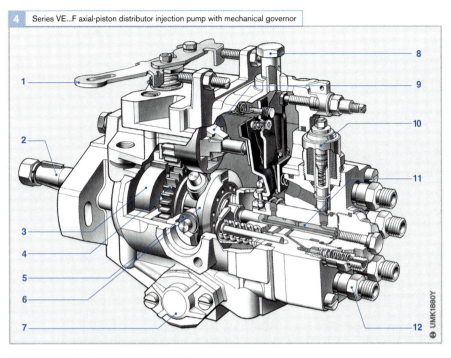

Fig. 4
1   Flow-control lever (linked to accelerator pedal)
2   Input shaft
3   Vane-type supply pump
4   Governor drive gear
5   Roller on roller ring
6   Cam plate
7   Hydraulic timing device
8   Overflow restriction
9   Governor assembly (mechanical governor)
10 Solenoid-operated shutoff valve (ELAB)
11 Distributor plunger
12 Delivery valve

5   Component assemblies and their locations

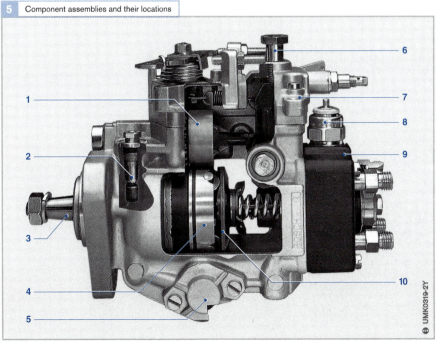

Fig. 5
1   Governor assembly
2   Pressure-control valve
3   Input shaft
4   Roller ring
5   Hydraulic timing device
6   Overflow restriction
7   Governor cover
8   Solenoid-operated shutoff valve (ELAB)
9   Distributor head with high-pressure pump
10 Cam plate

Low-pressure stage

The distributor injection pump's low-pressure stage comprises the following components (Fig. 1):
- The *vane-type supply pump* (4) supplies the fuel
- The *pressure-control valve* (3) maintains the specified fuel pressure in the system
- The *overflow restriction* (9) returns a defined amount of fuel to the pump to promote cooling

Vane-type supply pump

The vane-type supply pump extracts the fuel from the tank and conveys it through the supply lines and filters. As each rotation supplies an approximately constant amount of fuel to the inside of the fuel-injection pump, the supply volume increases as a function of engine speed. Thus the volume of fuel that the pump delivers reflects its own rotational speed, with progressively more fuel being supplied as pump speed increases. Pressur-

ized fuel for the high-pressure side is available in the fuel-injection pump.

Design

The vane-type supply pump is mounted on the main pump unit's input shaft (Fig. 2). The impeller (10) is mounted concentrically on the input shaft (8) and powers the former through a disk spring (7). An eccentric ring (2) installed in the pump housing (5) surrounds the impeller.

Operating concept

As the impeller rotates, centrifugal force presses the four floating blades (9) outward against the eccentric ring. The fuel in the gap between the bottom of the blade and the impeller body supports the blade's outward motion.

Fuel travels through the fuel-injection pump housing's inlet passage and supply channel (4) to a chamber formed by the impeller, blades and eccentric ring, called the cell (3). The rotation presses the fuel from between

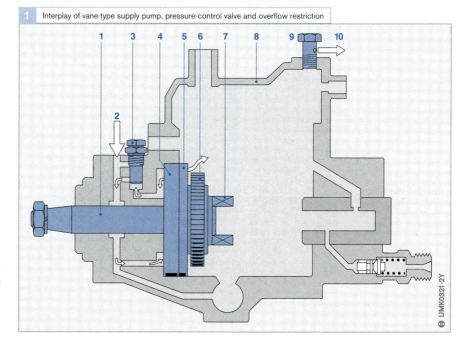

1   Interplay of vane-type supply pump, pressure-control valve and overflow restriction

Fig. 1
1 Pump drive
2 Fuel supply
3 Pressure-control valve
4 Eccentric ring on vane supply pump
5 Support ring
6 Governor drive gear
7 High-pressure pump drive claw
8 Pump housing
9 Overflow restriction
10 Fuel return

the blades toward the spill port (6), from where it proceeds through a bore to the pump's inner chamber. The eccentric shape of the ring's inner surface decreases the volume of the cell as the vane-type supply pump rotates to compress the fuel. A portion of the fuel proceeds through a second bore to the pressure-control valve (see Fig. 1).

The inlet and discharge sides operate using suction and pressure cells and have the shape of kidneys.

Pressure-control valve

As fuel delivery from the vane-type supply pump increases as a function of pump speed, the pump's internal chamber pressure is proportional to the engine's rotational speed. The hydraulic timing device relies on these higher pressurization levels to operate (see section on "Auxiliary control modules for distributor injection pumps"). The pressure-control valve is needed to govern pressurization and ensure that pressures correspond to the levels required for optimized operation of both the timing device and the

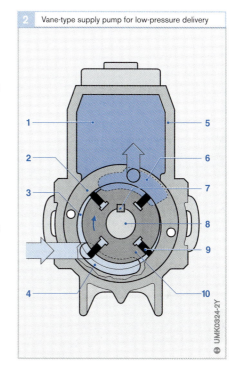

2 Vane-type supply pump for low-pressure delivery

Fig. 2
1 Pump inner chamber
2 Eccentric ring
3 Crescent-shaped cell
4 Fuel inlet (suction cells)
5 Pump housing
6 Fuel discharge (pressure cells)
7 Woodruff key
8 Input shaft
9 Blade
10 Impeller

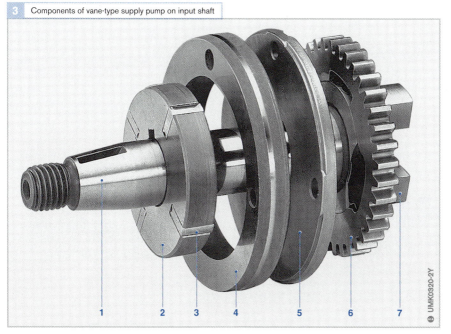

3 Components of vane-type supply pump on input shaft

Fig. 3
1 Input shaft
2 Impeller
3 Blade
4 Eccentric ring
5 Support ring
6 Governor drive gear
7 High-pressure pump drive claw

Fig. 4
1 Valve body
2 Compression spring
3 Valve plunger
4 Bore
5 Supply from vane-
   type supply pump
6 Return to vane-type
   supply pump

Fig. 5
1 Housing
2 Filter
3 Governor cover
4 Fuel supply
5 Throttle bore
6 Return line to fuel
   tank

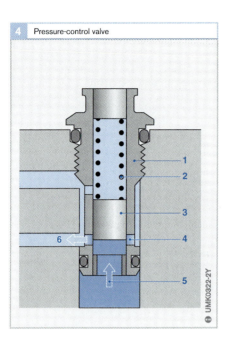

4   Pressure-control valve

UMK0322-2Y

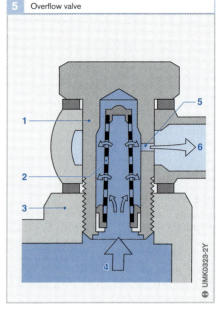

5   Overflow valve

UMK0323-2Y

engine itself. The regulator controls internal chamber pressures in the fuel-injection pump according to the quantity of fuel supplied by the vane-type supply pump. At a specific rotational speed, a specific internal pump pressure occurs, which then induces a defined shift in start of delivery.

A passage connects the pressure-control valve to the pressure cell (Fig. 2). It is directly adjacent to the vane-type supply pump.

The pressure-control valve is a spring-loaded slide valve (Fig. 4). When fuel pressure rises beyond a specified level, it pushes back the valve plunger (3), compressing the spring (2) and simultaneously exposing the return passage. The fuel can now proceed through the passage to the vane-type supply pump's suction side (6). When fuel pressure is low, the spring holds the return passage closed. The actual opening pressure is defined by the adjustable tension of the spring.

Overflow restriction

The distributor injection pump is cooled by fuel that flows back to the tank via an overflow restriction screwed to the governor cover (Fig. 5). The overflow restriction is located at the fuel-injection pump's highest point to bleed air automatically. In applications where higher internal pump pressures are required for low-speed operation, an overflow valve can be installed in place of the overflow restriction. This spring-loaded ball valve functions as a pressure-control valve.

The amount of fuel that the overflow restriction's throttle port (5) allows to return to the fuel tank varies as a function of pressure (6). The flow resistance furnished by the port maintains the pump's internal pressure. Because a precisely defined fuel pressure is required for each individual engine speed, the overflow restriction and the pressure-control valve must be matched.

High-pressure pump with fuel distributor

The fuel-injection pump's high-pressure stage pressurizes the fuel to the levels required for injection and then distributes this fuel to the cylinders at the specified delivery quantities. The fuel flows through the delivery valve and the high-pressure line to the nozzle-and-holder assembly, where the nozzle injects it into the engine's combustion chamber.

Distributor plunger drive

A power-transfer assembly transmits the rotational motion of the input shaft (Fig. 1, Pos. 1) to the cam plate (6), which is coupled to the distributor plunger (10). In this process, the claws from the input shaft and the cam plate engage in the intermediate yoke (3).

The cam plate transforms the input shaft's rotation into a motion pattern that combines rotation with reciprocation (total stroke of 2.2...3.5 mm, depending on pump version). The shaft's motion is translated by the motion of the cams on the cam plate (4) against the rollers on the roller ring (5). While the latter runs on bearing surfaces in the housing, there is no positive connection joining it to the input shaft. Because the profiles of the cams on the cam plate extend along the plane defined by the input shaft, they are sometimes referred to as "axial cams".

The distributor plunger's base (10) rests in the cam plate, where its position is maintained by a locating stud. Plunger diameters range from 8...12 mm, depending on the desired injected-fuel quantity.

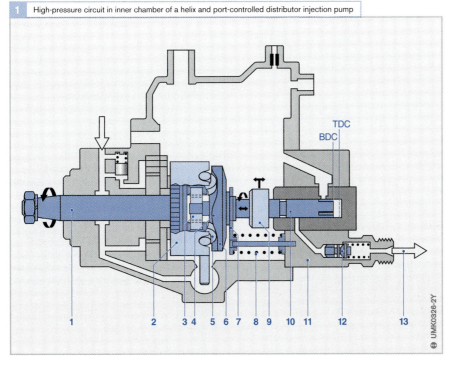

1   High-pressure circuit in inner chamber of a helix and port-controlled distributor injection pump

TDC
BDC

Fig. 1
1   Input shaft
2   Roller ring
3   Yoke
4   Cams
5   Roller
6   Cam plate
7   Spring-loaded
    cross brace
8   Plunger return
    spring
9   Control collar
10 Distributor plunger
11 Distributor head
12 Delivery valve
13 Discharge to high-
    pressure line

TDC Top **D**ead **C**enter
    for pump plunger
BDC Bottom **D**ead
    Center for pump
    plunger

1   2   3 4   5 6 7   8 9   10 11   12   13

UMK0326-2Y

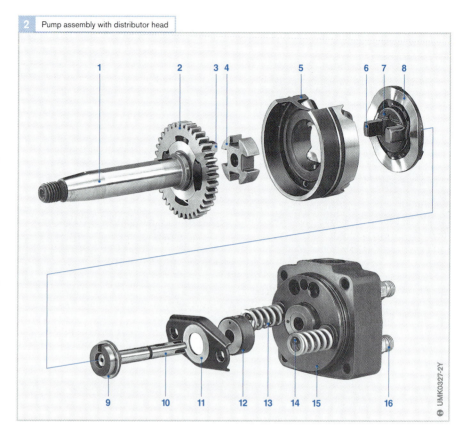

2   Pump assembly with distributor head

Fig. 2
1 Input shaft
2 Governor drive gear
3 Claw
4 Yoke
5 Roller ring
6 Claw
7 Cam plate
8 Cams
9 Distributor plunger
   base
10 Distributor plunger
11 Spring-loaded cross
   brace
12 Control collar
13 Plunger return
   spring
14 Guide pin
15 Distributor head
16 Delivery valve
   (discharge to
   high-pressure line)
9...16 Distributor head
   assembly

The cam plate's cams drive the plunger toward its **T**op **D**ead Center (TDC) position. The two symmetrically arranged plunger return springs (Fig. 2, Pos. 13) push the plunger back to **B**ottom **D**ead Center (BDC). At one end, these springs rest against the distributor body (15), while the force from the other end is transferred to the distributor plunger (10) through a spring coupling (11). The plunger return springs also prevent the cam plate (7) from slipping off the roller ring's rollers (5) in response to high rates of acceleration.

The heights of the plunger return springs are precisely matched to prevent the plunger from tilting in the distributor body.

Cam plates and cam profiles

The number of cams and rollers is determined by the number of cylinders in the engine and the required injection pressure (Fig. 3). The cam profile affects injection pressure as well as the maximum potential injection duration. Here the primary criteria are cam pitch and stroke velocity.

The conditions of injection must be matched to the combustion chamber's configuration and the engine's combustion process (DI or IDI). This is reflected in the cam profiles on the face of the cam plate, which are specially calculated for each engine type. The cam plate serves as a custom component in the specified pump type. This is why cam plates in different types of VE injection types are not mutually interchangeable.

Distributor body

The plunger (5) and the plunger barrel (2) are precisely matched (lapped assembly) in the distributor body (Fig. 4, Pos. 3) which is screwed to the pump housing. The control collar (1) is also part of a custom-matched assembly with the plunger. This allows the components to provide a reliable seal at extremely high pressures. At the same time, a slight pressure loss is not only unavoidable, it is also desirable as a source of lubrication for the plunger. Precise mutual tolerances in these assemblies mean that the entire distributor group must be replaced as a unit; no attempt should ever be made to replace the plunger, distributor body or control collar as individual components.

Also mounted in the distributor body are the solenoid-operated shutoff valve (ELAB) (not shown in this illustration) used to interrupt the fuel supply along with the screw cap (4) with vent screw (6) and delivery valves (7).

3 Various roller and cam combinations

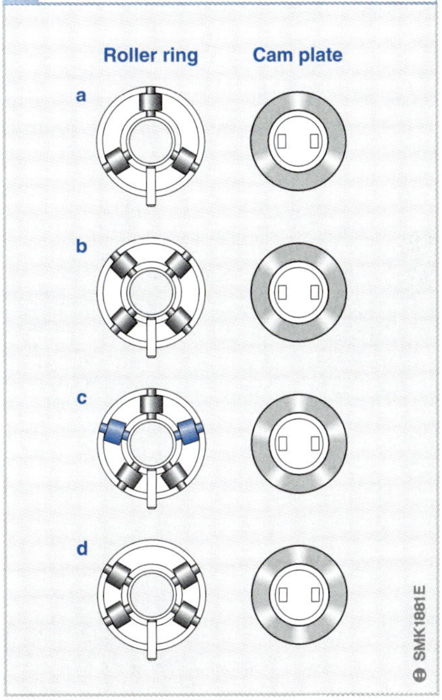

Roller ring   Cam plate

a

b

c

d

SMK1881E

Fig. 3
a **Three-cylinder engine**
   Six-cylinder version (d) is also available. In this version, every second spill port is rerouted to the inner chamber of the pump.
b **Four-cylinder**
   On two-cylinder engines, every second spill port is rerouted to the inner chamber of the pump.
c **Five-cylinder engine**
   The rollers in blue are not fitted to pumps for IDI engines as these operate at lower fuel-injection pressures and reduced physical loads.
d **Six-cylinder engine**
   Only four rollers used in this application.

4 Distributor head component assembly

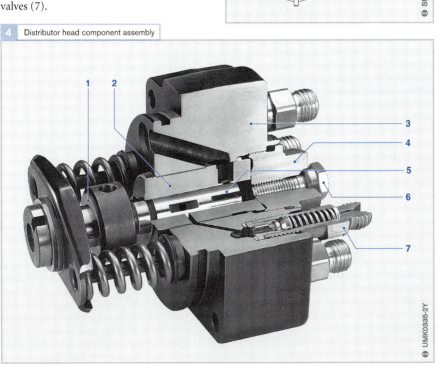

1 2

3

4

5

6

7

UMK0335-2Y

Fig. 4
1 Control collar
2 Plunger barrel
3 Distributor head
4 Screw cap
5 Distributor plunger
6 Vent screw
7 Delivery valve

Fuel metering

The distributor head assembly generates the pressure required for injection. It also distributes the fuel to the various engine cylinders. This dynamic process correlates with several different phases of the plunger stroke, called delivery phases.

The plunger's stroke phases as illustrated in Figure 6 show the process of metering fuel for a single cylinder. Although the plunger's motion is horizontal (as in the in-line fuel-injection pump), the extremes of its travel are still refereed to as top and bottom dead center (TDC and BDC). On four-cylinder engines the plunger rotates by one fourth of a turn during a delivery phase, while a sixth of a turn is available on six cylinder engines.

Suction (6a)

As the plunger travels from top to bottom dead center, fuel flows from the pump chamber and through the exposed inlet passage (2) into the plunger chamber (6) above the plunger. The plunger chamber is also called the element chamber.

Prestroke (option, 6b)

As the plunger rotates, it closes the inlet passage at the bottom dead center end of its travel range and opens the distributor slot (8) to provide a specific discharge.

After reaching bottom dead center, the plunger reverses direction and travels back toward TDC. The fuel flows back to the pump's inner chamber through a slot (7) at the front of the plunger.

This preliminary stroke delays fuel supply and delivery until the profile on the cam plate's cam reaches a point characterized by a more radical rise. The result is a more rapid rise in injection pressure for improved engine performance and lower emissions.

Fuel delivery (effective stroke, 6c)

The plunger continues moving toward TDC, closing the prestroke passage in the process. The enclosed fuel is now compressed. It travels through the distributor slot (8) to the delivery valve on one of the delivery ports (9). The port opens and the fuel is propelled through the high-pressure line to the nozzle-and-holder assembly.

Residual stroke (6d)

The effective stroke phase terminates when the lateral control passage (11) on the plunger reaches the control collar's helix (10). This allows the fuel to escape into the pump's inside chamber, collapsing the pressure in the element chamber. This terminates the delivery process. No further fuel is delivered to the nozzle (end of delivery). The delivery valve closes off the high-pressure line.

Fuel flows through the connection to the inside of the pump for as long as the plunger continues to travel toward TDC. The inlet passage is opened once again in this phase.

The governor or actuator mechanism can shift the control collar to vary the end of delivery and modulate the injected-fuel quantity.

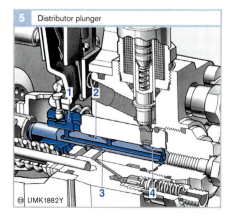

5   Distributor plunger

UMK1882Y

6   Stroke phases on helix and port-controlled distributor injection pump

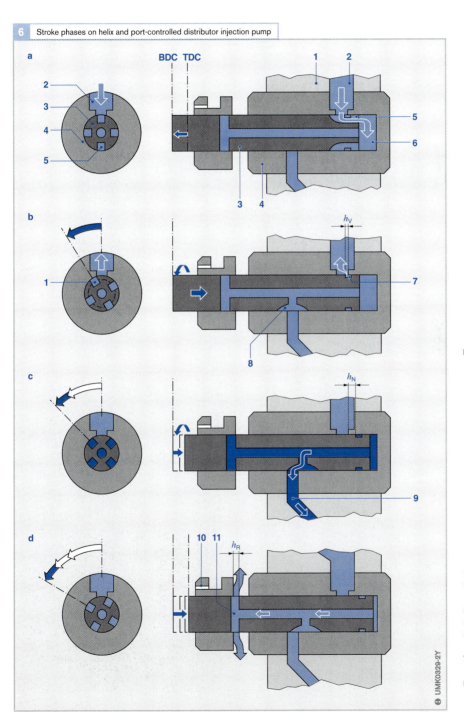

Fig. 6
a   Suction
b   Prestroke
c   Effective stroke
d   Residual stroke

1   Distributor head
2   Inlet passage
    (fuel supply)
3   Distributor plunger
4   Plunger barrel
5   Inlet metering slot
6   Plunger chamber
    (element chamber)
7   Prestroke groove
8   Distributor slot
9   Inlet passage to
    delivery valve
10 Control collar
11 Control bore

h_N Effective stroke
h_R Residual stroke
h_V Prestroke

TDC **T**op **D**ead **C**enter
      on pump plunger
BDC **B**ottom **D**ead
      Center on pump
      plunger

UMK0329-2Y

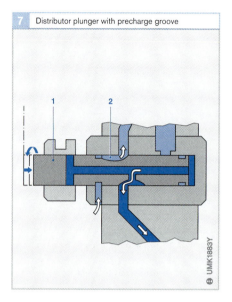

7 Distributor plunger with precharge groove

UMK1883Y

Precharge slot

During rapid depressurization at the end of delivery, the venturi effect extracts the remaining fuel from the area between the delivery valve and the plunger. At high rotational speeds with substantial delivery quantities, this area is closed off before enough fuel can flow back in. As a result, the pressure in this area is lower than that in the pump's inner chamber. The slot must thus be recharged before the next fuel injection. This reduces the delivery quantity.

The precharge slot (Fig. 7, Pos. 2) connects the pump's inner chamber to the area between the delivery valve and the plunger. The fuel always flows through the discharge opposite the discharge controlled for delivering fuel.

▶ Off-road applications

In addition to their use in on-road vehicles, helix and port-controlled distributor injection pumps are also used in a multitude of off-road applications. This sector includes stationary powerplants as well as construction and agricultural machines. Here, the primary requirements are robust construction and ease of maintenance.

The conditions that fuel-injection systems meet in off-road use can be exceptionally challenging (for instance, when exposed engines are washed down with high-pressure steam cleaners, poor fuel quality, frequent refuelings from canisters, etc.). Special filtration systems operate in conjunction with water separators to protect fuel-injection pumps against damage from fuel of poor quality).

PTOs of the kind employed to drive pumps and cranes rely on constant engine speeds with minimum fluctuation in response to load shifts (low speed droop). Heavy flyweights in the governor assembly provide this performance.

Excavator with 112 kW (152 bhp)

Combine harvester with 125 kW (170 bhp)

Harvester with 85 kW (116 bhp)
Tractor with 98 kW (133 bhp)

▶1   Off-road vehicles with VE pump

SAV0058Y

Delivery valve

The delivery valve closes the high-pressure line to the pump in the period between fuel supply phases. It insulates the high-pressure line from the distributor head.

The delivery valve (Fig. 8) is a slide valve. The high pressure generated during delivery lifts the valve plunger (2) from its seat. The vertical grooves (8) terminating in the ring groove (6) carry the fuel through the delivery-valve holder (4), the high-pressure line and the nozzle holder to the nozzle.

At the end of delivery, the pressure in the plunger chamber above the plunger and in the high-pressure lines falls to the level present in the pump's inner chamber. The valve spring (3) and the static pressure in the high-pressure line push the valve plunger back against its seat.

Yet another function of the delivery valve is to relieve injection pressure from the injection line after completion of the delivery phase by increasing a defined volume on the line side. This function is discharged by the retraction piston (7) that closes the valve before the valve plunger (2) reaches its seat. This pressure relief provides precisely calibrated termination of fuel discharge

through the nozzle at the end of the injection process. It also stabilizes pressure in the high-pressure lines between injection processes to compensate for shifts in injected-fuel quantities.

Delivery valve with torque control

The dynamic response patterns associated with high-pressure delivery processes in the fuel-injection pump cause flow to increase as a function of rotational speed. However, the engine needs less fuel at high speeds. A positive torque control capable of reducing flow rates as rotational speed increases is thus required in many applications. This function is usually executed by the governor. Another option available on units designed for low injection pressures (IDI powerplants) is to use the pressure-control valve for this function.

Pressure-control valves with torque-control functions feature a torque-control collar (2) adjacent to the retraction piston (Fig. 9, Pos. 1) with either one or two polished recesses (3), according to specific requirements. The resulting restricted opening reduces delivery quantity at high rpm.

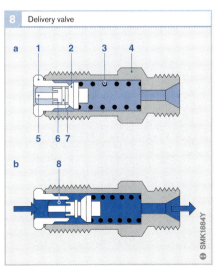

8 Delivery valve

a

1   2   3   4

5   6 7

b

8

SMK1884Y

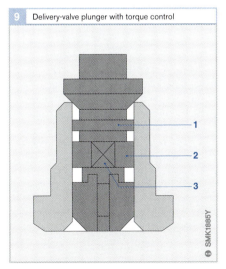

9 Delivery-valve plunger with torque control

1

2

3

SMK1885Y

Fig. 8
a   Closed
b   Open

1   Valve holder
2   Valve plunger
3   Valve spring
4   Delivery-valve holder
5   Stem
6   Ring groove
7   Retraction piston
8   Vertical groove

Fig. 9
1   Retraction piston
2   Control-torque collar
3   Specially ground recess

Cutoff bores with flattened surfaces in the control collar, specially designed for the individual engine application, can also provide a limited degree of torque control.

Delivery valve with return-flow restriction

The precise pressure relief required at the end of the injection event generates pressure waves. These are reflected by the delivery valve. At high injection pressures, these pulses have the potential to reopen the nozzle needle or induce phases of negative pressure in the high-pressure line. These processes cause post-injection dribble, with negative consequences on emission properties and/or cavitation and wear in the high-pressure line or at the nozzle.

Harmful reflections are inhibited by a calibrated restriction mounted upstream of the delivery valve, where it affects return flow only. This calibrated restriction attenuates pressure waves but is still small enough to allow maintenance of static pressure between injection events.

The return-flow restriction consists of a valve plate (Fig. 10, Pos. 4) with a throttle bore (3) and a spring (2). The valve plate lifts to prevent the throttle from exercising any effect on delivery quantity. During return flow, the valve plate closes to inhibit pulsation.

Constant-pressure valve

On high-speed diesel engines, the volumetric relief provided by the retraction piston and delivery valve is often not enough to prevent cavitation, dribble and blowback of combustion gases into the nozzle-and-holder assembly under all conditions.

Under these conditions, constant-pressure valves (Fig. 11) are installed. These valves use a single-action non-return valve with adjustable pressure (such as 60 bar) to relieve pressure on the high-pressure system (line and nozzle-and-holder assembly).

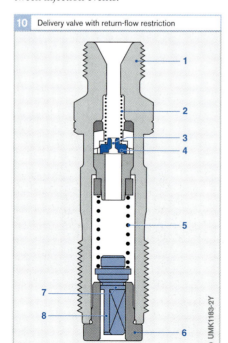

10 Delivery valve with return-flow restriction

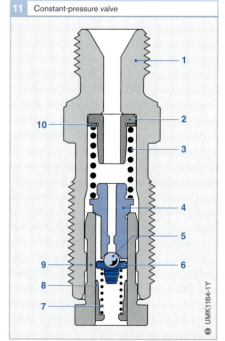

11 Constant-pressure valve

Fig. 10
1 Pressure valve holder
2 Valve spring (valve plate)
3 Return-flow restrictor
4 Valve plate
5 Valve spring (delivery valve)
6 Valve holder
7 Relief plunger
8 Plunger stem

Fig. 11
1 Delivery-valve holder
2 Spring-guided filler piece
3 Valve spring (delivery valve)
4 Delivery-valve plunger
5 Ball (constant-pressure valve)
6 Spring seat
7 Valve spring (constant-pressure valve)
8 Setting sleeve
9 Valve holder
10 Shims

Diesel records in 1972

In 1972 a modified Opel GT set a total of 20 international records for diesel-powered vehicles. The original roof structure was replaced to reduce aerodynamic resistance, while the 600-liter fuel tank was installed in place of the passenger seat.

This vehicle was propelled by a 2.1-liter 4-cylinder diesel engine fitted with combustion swirl chambers. A distinguishing factor relative to the standard factory diesel was the exhaust-gas turbocharger. This engine produced 95 DIN bhp (approx. 70 kW) at 4,400 rpm while consuming fuel at the rate of 13 liters per hundred kilometers. Optimized fuel injection was provided by an EP/VA CL 163 rotary axial-piston pump with mechanical injection from Bosch.

Unlike today's diesel engines, this unit was designed around an existing gasoline engine.

As a result, the distributor injection pump had to be installed in the location originally intended for the ignition distributor. This led to a virtually vertical installation in the engine compartment. Numerous issues awaited resolution: control variables based on horizontal installation had to be recalculated for vertical operation, problems with air in the fuel led to difficult starting, the pump tended to cavitate, etc.

Here is a sampling of the records posted for this vehicle:

- It covered 10,000 kilometers in 52 hours and 23 minutes. This translates into an average speed of 190.9 kph. The car was driven by six drivers relieving each other every three to four hours.
- Absolute world records for all vehicle classes were posted for 10 km (177.4 kph) and 10 miles (184.5 kph) from standing start.

NMM0600Y

▶ 1   2.1-*l*-Diesel engine from Opel GT record vehicle of 1972

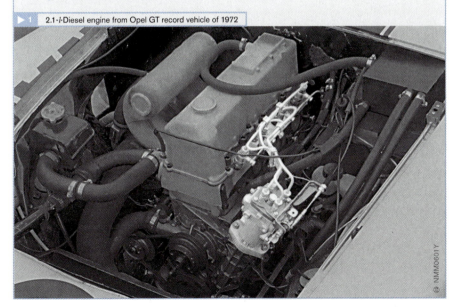

NMM0601Y

Auxiliary control modules for distributor injection pumps

The auxiliary control modules installed on distributor injection pumps with axial pistons govern start of delivery and regulate the volume of fuel discharged into the combustion chamber during injection. These control modules are composed of mechanical control elements, or actuators. They respond to variations in operating conditions (load factor, rotational speed, charge-air pressure, etc.) with precise adjustments for ideal performance. On distributor injection pumps with Electronic Diesel Control (EDC), an electric actuator mechanism replaces the mechanical actuators.

Overview

Auxiliary control modules adapt the start of delivery and delivery period to reflect both driver demand and the engine's instantaneous operating conditions (Fig. 1).

In the period since the distributor injection pump's introduction in 1962, an extensive range of controllers has evolved to suit a wide range of application environments. Numerous component configurations are also available to provide ideal designs for a variety of engine versions. Any attempt to list all possible versions of auxiliary control modules would break the bounds of this section. Instead, we will concentrate on the most important control modules. The individual units are the:

- Speed governor
- Timing device
- Adjustment and torque control devices
- Switches and sensors
- Shutoff devices
- Electric actuator mechanisms and
- Diesel immobilizers (component in the electronic vehicle immobilizer)

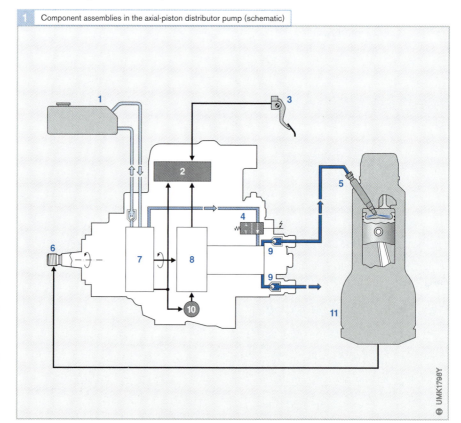

1 Component assemblies in the axial-piston distributor pump (schematic)

Fig. 1
1 Fuel supply (low-pressure)
2 Controlling system
3 Accelerator pedal
4 ELAB Electric shutoff device
5 Nozzle-and-holder assembly
6 Pump drive assembly
7 Low-pressure stage (vane-type supply pump with pressure-control valve and overflow throttle valve)
8 High-pressure pump with fuel rail
9 Delivery valve
10 Hydraulic timing device
11 Diesel engine

UMK1798Y

History of the mechanically controlled distributor injection pump from Bosch

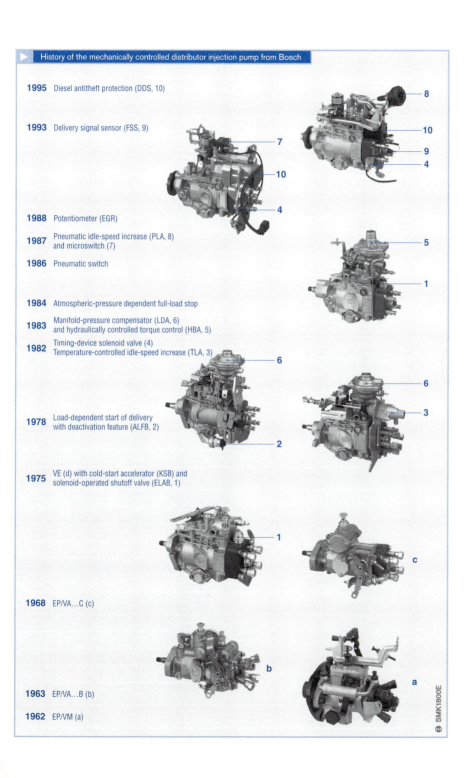

1995 Diesel antitheft protection (DDS, 10)

1993 Delivery signal sensor (FSS, 9)

1988 Potentiometer (EGR)

1987 Pneumatic idle-speed increase (PLA, 8) and microswitch (7)

1986 Pneumatic switch

1984 Atmospheric-pressure dependent full-load stop

1983 Manifold-pressure compensator (LDA, 6) and hydraulically controlled torque control (HBA, 5)

1982 Timing-device solenoid valve (4) Temperature-controlled idle-speed increase (TLA, 3)

1978 Load-dependent start of delivery with deactivation feature (ALFB, 2)

1975 VE (d) with cold-start accelerator (KSB) and solenoid-operated shutoff valve (ELAB, 1)

1968 EP/VA...C (c)

1963 EP/VA...B (b)

1962 EP/VM (a)

SMK1800E

Governors

Function

Engines should not stall when exposed to progressive increases in load during acceleration when starting off. Vehicles should accelerate or decelerate smoothly and without any surge in response to changes in accelerator pedal position. Speed should not vary on gradients of a constant angle when the accelerator pedal does not move. Releasing the pedal should result in engine braking.

The distributor injection pump governor offers active control to help cope with these operating conditions.

The basic function of every governor is to limit the engine's high idle speed. Depending on the individual unit's configuration, other functions can involve maintaining constant engine speeds, which may include specific selected rotational speeds or the entire rev band as well as idle. The different governor designs arise from the various assignments (Fig. 1):

Idle controller

The governor in the fuel-injection pump controls the idle speed of the diesel engine.

Maximum-speed governor

When the load is removed from a diesel engine operated at WOT, its rotational speed should not rise above the maximum permitted idle speed. The governor discharges this function by retracting the control collar toward "stop" at a certain speed. This reduces the flow of fuel into the engine.

Variable-speed governor

This type of governor regulates intermediate engine speeds, holding rotational speed constant within specified limits between idle and high idle. With this setup, fluctuations in the rotational speed n of an engine operating under load at any point in the power band are limited to a range between a speed on the full-load curve n_{VT} and an unloaded engine speed n_{LT}.

Other requirements

In addition to its primary functions, the governor also controls the following:
- Releases or blocks supplementary fuel flow for starting
- Modulates full-throttle delivery quantity as a function of engine speed (torque control)

Special torque-control mechanisms are required for some of these operations. They are described in the text below.

Control precision

The index defining the precision with which the governor regulates rotational speed when load is removed from the engine is the droop-speed control, or droop-speed control. This is the relative increase in engine

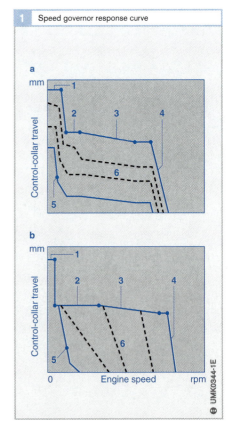

1   Speed governor response curve

speed that occurs when load is removed from the diesel engine while control lever position (accelerator-pedal travel) remains constant. The resulting rise in engine speed should not exceed a certain level in the controlled range. The maximum rpm specified for an unloaded engine represents the high-idle speed. This figure is encountered when the load factor of a diesel engine operating at wide-open throttle (WOT) decreases from 100 % to 0 %. The rise in rotational speed is proportional to the variation in load factor. Larger load shifts produce progressively larger increases in rotational speed.

$$\delta = \frac{n_{no} - n_{vo}}{n_{vo}}$$

or in %:

$$\delta = \frac{n_{no} - n_{vo}}{n_{vo}} \cdot 100\,\%$$

where: δ droop-speed control, n_{no} upper no-load speed, n_{vo} upper full-load speed (some sources refer to the upper no-load speed as the high-idle speed).

The desired droop-speed control is defined by the diesel engine's intended application environment. Thus low degrees of proportionality (on the order of 4 %) are preferred for electric power generators, as they respond to fluctuations in load factor by holding changes and engine speed and the resultant frequency shifts to minimal levels. Larger degrees of proportionality are better in motor vehicles because they furnish more consistent control for improved driveability under exposure to minor variations in load (vehicle acceleration and deceleration). In vehicular applications, limited droop-speed control would lead to excessively abrupt response when load factors change.

Design structure

The governor assembly (Fig. 2) consists of the mechanical governor (2) and the lever assembly (3). Operating with extreme precision, it controls the position of the control collar (4) to define the effective stroke and with it the injected-fuel quantity. Different versions of the control lever assembly can be employed to vary response patterns for various applications.

2   Axial-piston distributor pump with governor assembly (section)

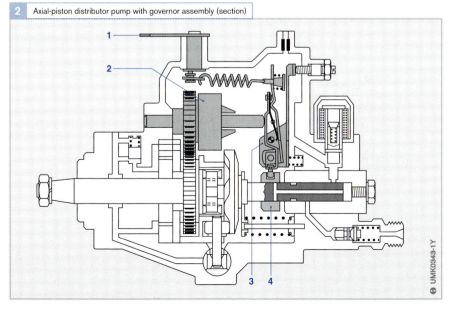

Fig. 2
1   Rotational-speed control lever (accelerator pedal)
2   Mechanical governor
3   Lever assembly
4   Control collar

Variable-speed governor

The variable-speed governor regulates all engine speeds from start or high idle. In addition to these two extremes, it also controls operation in the intermediate range. The speed selected at the rotational-speed control lever (accelerator pedal or supplementary lever) is held relatively constant, with actual consistency varying according to the droop-speed control (Fig. 4).

This sort of adjustment is required when ancillary equipment (winch, extinguisher water pump, crane, etc.) is used on commercial vehicles or in stand-alone operation. This function is also frequently used in commercial vehicles and with agricultural machinery (tractors, combine harvesters).

Design

The governor assembly, consisting of flyweight housing and flyweights, is powered by the input shaft (Fig. 2). A governor shaft mounted to allow rotation in the pump housing supports the assembly (Fig. 3, Pos. 12). The radial travel of the flyweights (11) is translated into axial movement at the

sliding sleeve (10). The force and travel of the sliding sleeve modify the position of the governor mechanism. This mechanism consists of the control lever (3), tensioning lever (2) and starting lever (4).

The control lever's mount allows it to rotate in the pump housing, where its position can be adjusted using the WOT adjusting screw (1). The starting and tensioning levers are also mounted in bearings allowing them to rotate in the control lever. At the lower end of the starting lever is a ball pin that engages with the control collar (7), while a starting spring (6) is attached to its top. Attached to a retaining stud (18) on the upper side of the tensioning lever is the idle-speed spring (19). The governor spring (17) is also mounted in the retaining stud. A lever (13) combines with the control-lever shaft (16) to form the link with the rotational-speed control lever (14).

The spring tension operates together with the sliding sleeve's force to define the position of the governor mechanism. The adjustment travel is transferred to the control collar to determine the delivery quantity h_1 and h_2, etc.).

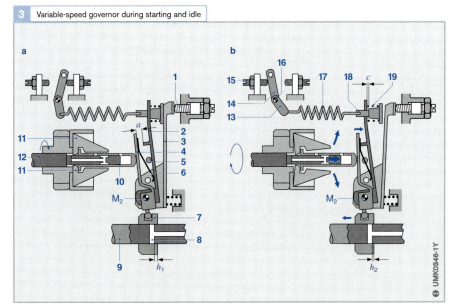

3 Variable-speed governor during starting and idle

Starting response

When the distributor injection pump is stationary, the flyweights and the sliding sleeve are at their base positions (Fig. 3a). The starting spring pushes the starting lever into the position for starting by rotating it about its axis M_2. The starting lever simultaneously transfers force through the ball pin onto the distributor plunger to shift the control collar to its starting position. This causes the distributor plunger (9) to execute substantial effective stroke. This provides maximum fuel start quantity.

Low rotational speeds are enough to overcome the compliant starting spring's tension and shift the sliding sleeve by the distance a. The starting lever again rotates about its pivot axis M_2, and the initial enhanced start quantity is automatically reduced to idle delivery quantity.

Idle-speed control

After the diesel engine starts and the accelerator pedal is released, the control lever returns to its idle position (Fig. 3b) as defined by the full-load stop on the idle adjusting screw (15). The selected idle speed ensures that the engine consistently continues to turn over smoothly when unloaded or under minimal loads. The control mechanism regulates the idle-speed spring mounted on the retaining stud. It maintains a state of equilibrium with the force generated by the flyweights.

By defining the position of the control collar relative to the control passage in the plunger, this equilibrium of forces determines the effective stroke. At rotational speeds above idle, it traverses the spring travel c to compress the idle-speed spring. At this point, the effective spring travel c is zero.

The housing-mounted idle-speed spring allows idle adjustments independent of accelerator pedal position while accommodating increases to compensate for temperature and load factor shifts (Fig. 5).

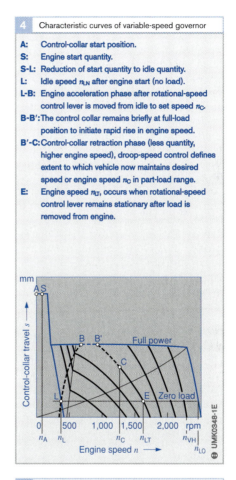

4   Characteristic curves of variable-speed governor

A: Control-collar start position.
S: Engine start quantity.
S-L: Reduction of start quantity to idle quantity.
L: Idle speed n_{LN} after engine start (no load).
L-B: Engine acceleration phase after rotational-speed control lever is moved from idle to set speed n_C.
B-B': The control collar remains briefly at full-load position to initiate rapid rise in engine speed.
B'-C: Control-collar retraction phase (less quantity, higher engine speed), droop-speed control defines extent to which vehicle now maintains desired speed or engine speed n_C in part-load range.
E: Engine speed n_{LT}, occurs when rotational-speed control lever remains stationary after load is removed from engine.

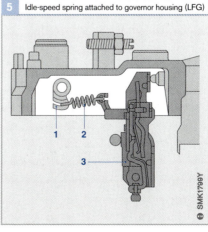

5   Idle-speed spring attached to governor housing (LFG)

Fig. 5
1   Rocker (fixed to pump housing)
2   Idle-speed spring
3   Lever assembly

Operation under load

During normal operation, the rotational-speed control lever assumes a position in its overall travel range that corresponds to the desired rotational or speed. The driver dictates this position by depressing the accelerator pedal to the desired angle. Because the starting and idle-speed springs are both fully compressed at rotational speeds above idle, they have no influence on control in this range. Control is exercised by the governor spring.

Example (Fig.6):
The driver uses the accelerator pedal or auxiliary control lever to move the flow-control lever (2) to a position corresponding to a specific (higher) speed. This adjustment motion compresses the governor spring (4) by a given increment. At this point, the governor spring's force exceeds the centrifugal force exerted by the flyweights (1). The spring's force rotates the starting lever (7) and tensioning lever (6) around their pivot axis M_2 to shift the control collar toward its WOT position by the increment defined by the design's ratio of conversion. This increases delivery quantity to raise the engine

speed. The flyweights generate additional force, which the sliding sleeve (11) then transfers to oppose the spring force.

The control collar remains in its WOT position until the opposed forces achieve equilibrium. From this point onward, any additional increase in engine speed will propel the flyweights further outwards and the force exerted by the sliding sleeve will predominate. The starting and tensioning levers turn about their shared pivot axis (M_2) to slide the control collar toward "stop" to expose the cutoff bore earlier. The mechanism ensures effective limitation of rotational speeds with its ability to reduce delivery capacity all the way to zero. Provided that the engine is not overloaded, each position of the flow-control lever corresponds to a specific rotational range between WOT and no-load. As a result, the governor maintains the selected engine speed with the degree of intervention defined by the system's droop-speed control (Fig. 4).

Once the imposed loads have reached such a high order of magnitude as to propel the control collar all the way to its WOT position (hill gradients, etc.), no additional increases in fuel quantity will be available,

6   Variable-speed governor under load (shown without control lever)

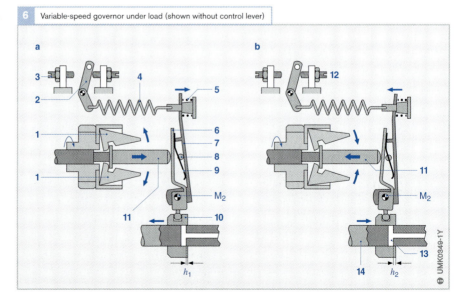

even in response to lower rotational speed. The engine is overloaded, and the driver should respond by downshifting to a lower gear.

Overrun

When the vehicle descends a steep gradient, or when the accelerator pedal is released at high speeds (overrun), the engine is driven by the vehicle's inertia. The sliding sleeve responds by pressing against the starting and tensioning levers. Both levers move to shift the control collar to decrease delivery quantity; this process continues until the fuel-delivery quantity reflects the requirements of the "new" load factor, or zero in extreme cases. The response pattern of the variable-speed governor described here is valid at all flow-control lever positions, and occurs whenever any factor causes load or rpm to vary so substantially as to shift the control collar all the way to its WOT or "stop" end position.

Minimum/maximum speed governors

This governor regulates the idle and high-idle engine speeds only. Response in the intermediate range is regulated exclusively by

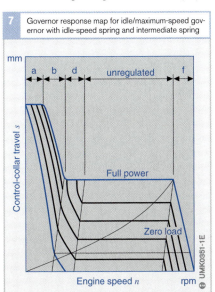

7 Governor response map for idle/maximum-speed governor with idle-speed spring and intermediate spring

mm

a   b   d   unregulated   f

Control-collar travel *s*

Full power

Zero load

Engine speed *n*   rpm

UMK0351-1E

the accelerator pedal (Fig. 6). This reduces surge, but is not suitable for use in light commercial vehicles equipped with ancillaries.

Design

The governor assembly with its flyweights and assembly of control levers is essentially comparable to the variable-speed governor described above. One difference in the idle and maximum-speed governor's design is the governor spring (Figure 7 on next page, Pos. 4) and its installation. The compression spring applies pressure from its position in a guide sleeve (5). A retaining stud (7) provides the link between the tensioning lever and the governor spring.

Starting response

Because the flyweights (1) are at rest, the sliding sleeve (15) is in its base position. This allows the starting spring (12) to transfer force through the starting lever (9) and the sliding sleeve to push the flyweights to the inside. The control collar (13) on the distributor plunger is positioned to provide the delivery quantity prescribed for starting.

Idle-speed control

Once the engine starts and the accelerator pedal is released, the rotational-speed control lever (2) responds to the force exerted by the return spring against the pump housing by returning to the idle position. As engine speed climbs, the centrifugal force exerted by the flyweights (Fig. 8a, next page) rises, and their slots press the sliding sleeve back against the starting lever. The control process relies on the idle-speed spring (8) mounted on the tensioning lever (10). The starting lever's rotation moves the control collar to reduce the delivery quantity. The control collar's position is regulated by the combined effects of centrifugal and spring force.

Operation under load

When the driver varies the position of the accelerator pedal the rotational-speed control lever rotates by a given angle. This action negates the effect of the starting and idle-

Fig. 7
a   Starting spring range
b   Starting and idle-speed spring range
d   Intermediate spring range
f   Governor spring range

speed spring and engages the intermediate spring (6), which smoothes the transition to the unregulated range on engines with idle/maximum-speed governors. Rotating the rotational-speed control lever further toward WOT allows the intermediate spring to expand until the retainer's shoulder is against the tensioning lever (Fig. 8b). The assembly moves outside the intermediate spring's effective control sector to enter the uncontrolled range. The uncontrolled range is defined by the tension on the governor spring, which can be considered as rigid in this speed range. The driver's adjustments to the rotational-speed control lever (accelerator pedal) can now be transferred directly through the control mechanism to the control collar. Delivery quantity now responds directly to movement at the accelerator pedal.

The driver must depress the pedal by an additional increment in order to raise the vehicle speed or ascend a gradient. If the object is to reduce engine output, the pressure on the accelerator pedal is reduced.

If the load on the engine decreases while the control lever position remains unchanged, the delivery quantity will remain constant, and rotational speed will increase.

This leads to greater centrifugal force, and the flyweights push the sliding sleeve against the starting and tensioning levers with increased force. The full-load speed governor assumes active control only once the sleeve's force overcomes the spring's tension.

When the load is completely removed from the engine, it accelerates to its high-idle speed, where the system protects it against further increases in speed.

Part-load governor

Passenger vehicles are usually equipped with a combination of variable-speed and idle/maximum-speed governors. This type of part-load governor features supplementary governor springs allowing it to function as a variable-speed governor at the low end of the rev band while also operating as an idle/maximum-speed governor when engine speed approaches the specified maximum. This arrangement provides a stable rotational speed up to roughly 2,000 rpm, after which no further constraints are imposed on further rises if the engine is not under load.

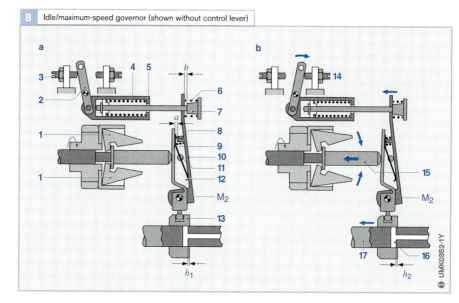

8 Idle/maximum-speed governor (shown without control lever)

Timing device

Function

The injection event must be initiated at a specific crankshaft angle (piston position) to ensure efficient combustion and optimal power generation. The pump's start of delivery and the resultant start of delivery must thus vary as rotational speed changes in order to compensate for two primary factors:

Injection lag

Fuel delivery to the nozzles starts when the plunger in the distributor head closes the inlet passage (start of delivery). The resulting pressure wave initiates the injection event when it reaches the injector (start of delivery). The pulse propagates through the injection line at the speed of sound. The travel duration remains essentially unaffected by rotational speed. The pressure wave's propagation period is defined by the length of the injection line and the speed of sound, which is approximately 1500 m/s in diesel fuel. The period that elapses between the start of delivery and the start of injection is the injection lag.

Thus the actual start of injection always occurs with a certain time offset relative to the start of delivery. As a result of this phenomenon, the nozzle opens later (relative to engine piston position) at high than at low rotational speeds. To compensate for this, start of delivery must be advanced, and this forward shift is dependent on pump and engine speed.

Ignition lag

Following injection, the diesel fuel requires a certain period to evaporate into a gaseous state and form a combustible mixture with the air. Engine speed does not affect the time required for mixture formation. The time required between the start of injection and the start of combustion in diesel engines is the ignition lag.

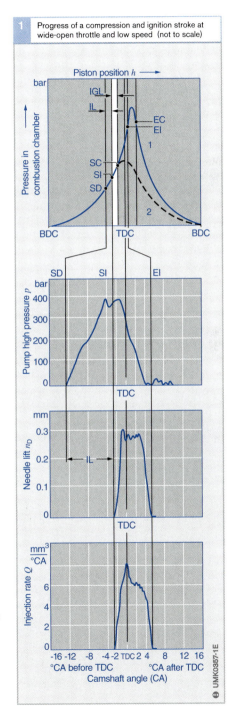

1   Progress of a compression and ignition stroke at wide-open throttle and low speed (not to scale)

Fig. 1
1   Combustion pressure
2   Compression

SD   **S**tart of **d**elivery
SI   **S**tart of **i**njection
IL   **I**njection **l**ag
SC   **S**tart of **c**ombustion
IGL   **I**gnition **l**ag
EI   **E**nd of **i**njection
EC   **E**nd of **c**ombustion
BDC   Engine piston at **B**ottom **D**ead **C**enter
TDC   Engine piston at **T**op **D**ead **C**enter

Factors influencing ignition lag include
- The ignitability of the diesel fuel
  (as defined by the cetane number)
- The compression ratio
- The air temperature and
- The fuel discharge process

The usual ignition lag is in the order of one millisecond.

If the start of injection remains constant, the crankshaft angle traversed between the start of injection and the start of combustion will increase as rotational speed rises. This, in turn, prevents the start of combustion from occurring at the optimal moment relative to piston position.

The hydraulic timing device compensates for lag in the start of injection and ignition by advancing the distributor injection pump's start of delivery by varying numbers of degrees on the diesel engine's crankshaft. This promotes the best-possible combustion and power generation from the diesel engine at all rotational speeds.

Spill and end of combustion

Opening the cutoff bore initiates a pressure drop in the pump's high-pressure system (spill), causing the nozzle to close (end of injection). This is followed by the end of combustion. Because spill depends on the start of delivery and the position of the control collar, it is indirectly adjusted by the timing device.

2   Axial-piston distributor injection pump with hydraulic timing control

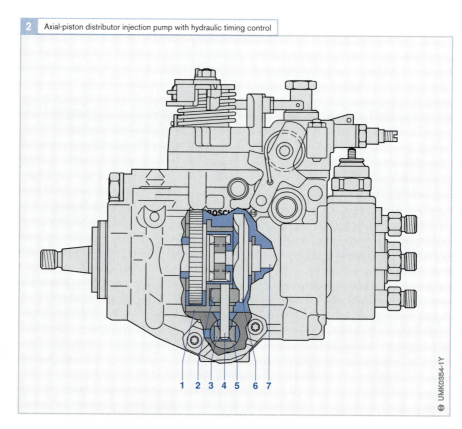

Fig. 2
1   Roller ring
2   Rollers on roller ring
3   Sliding block
4   Pin
5   Timing plunger
6   Cam plate
7   Distributor plunger

UMK0354-1Y

Design

The hydraulic timing device is located in the distributor injection pump's underside at a right angle to the pump's longitudinal axis (Fig. 2 and 3).

The timing plunger (Fig.3, Pos.7) is guided in the pump housing (1). Both ends of this housing are sealed by covers (6). A bore (5) in the timing plunger allows fuel inlet, and a spring (9) is installed on the opposite side.

A sliding block (8) and pin (4) connect the timing plunger to the roller ring (2).

Operating concept

The timing plunger in the distributor injection pump is held in its base position by the pre-loaded tension of the compressed timing spring (Fig. 3a). As the pump operates, the pressure-control valve regulates the pressure of the fuel in its inner chamber in proportion to rotational speed. The fuel thus acts on the side of the plunger opposite the timing spring at a pressure proportional to the pump's rotational speed.

The pump must reach a specific rotational speed, such as 300 rpm, before the fuel pressure (pressure in inner chamber) overcomes the spring's tension to compress it and shift the timing plunger (to the left in Fig, 3b). The sliding block and pin transfer the plunger's axial motion to the rotating roller ring. This modifies the relative orientation of the cam disk and roller ring, and the rollers in the roller ring lift the turning cam disk earlier. Thus the rollers and roller ring rotate relative to the cam disk and plunger by a specific angular increment (a) determined by rotational speed. (α). The maximum potential angle is usually twelve camshaft degrees (24 degrees crankshaft).

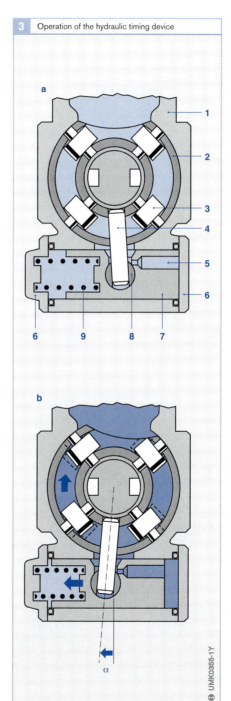

3   Operation of the hydraulic timing device

Fig. 3
a   Passive state
b   In operation

1   Pump housing
2   Roller ring
3   Rollers on roller ring
4   Pin
5   Bore in timing plunger
6   Cover plate
7   Timing plunger
8   Sliding block
9   Timing spring

α Roller ring pivot angle

Mechanical torque-control modules

Application

Distributor injection pumps are designed around modular building blocks, with various supplementary control devices available to satisfy individual engine requirements (as in Fig. 1). This concept affords an extended range of torque-control options allowing pumps to provide maximum torque, power and fuel economy along with low emissions. This overview serves as a compilation of the various torque-control modules and their effects on the diesel engine. The schematic diagram illustrates the interrelationships between the basic distributor injection pump and the different torque-control modules (Fig. 2).

Torque control

The torque-control process is the operation in which delivery quantity is adjusted to the engine's full-load requirement curve in response to changes in rotational speed.

This adaptive strategy may be indicated when special demands on full-load characteristics (improved exhaust-gas composition, torque generation and fuel economy) are encountered.

1 Axial-piston distributor injection pump (VE..F) with mechanical adjusters, potentiometer and timing-device solenoid valve (illustration)

5 cm

Fig. 1
1   Potentiometer
2   Hydraulic cold-start accelerator KSB
3   Load-dependent start of delivery with deactivation feature-ALFB
4   Connector
5   Pneumatic idle-speed increase PLA
6   Hydraulically controlled torque-control HBA

SMK1801Y

2   Block diagram of VE distributor injection pump with mechanical/hydraulic full-load torque control

Charge-air pressure-dependent manifold-pressure compensator LDA
Delivery-quantity control relative to charge-air pressure (engine with turbocharger).

Hydraulically controlled torque control device HBA
Control of delivery quantity based on engine speed (not on turbocharged engines with LDA).

Load-dependent start of delivery LFB
Adjust start of delivery to reflect load to reduce noise emissions and, as primary aim, to reduce exhaust-gas emissions.

Atmospheric-pressure dependent full-load stop ADA
Control of delivery quantity based on atmospheric pressure.

Cold-start accelerator KSB
Adjusts start of delivery for improved performance during cold starts.

Graduated (or adjustable) start quantity GST
Avoid start enrichment during hot starts.

Temperature-controlled idle-speed increase TLA
Improved post-start warm-up and smoother operation with idle-speed increase on cold engine.

Solenoid-operated shutoff valve ELAB
Makes it possible to shut down engine with "ignition key".

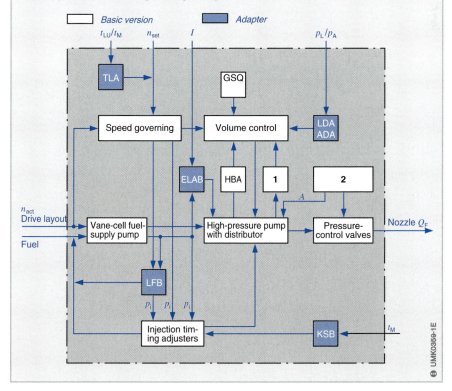

Fig. 2
1 Full-load torque control with control-lever assembly
2 Hydraulic full-load torque control

A Cutoff bore
n_{act} Actual speed (control parameter)
n_{set} Set speed (reference parameter)
Q_F Delivery quantity
t_M Engine temperature
t_{LU} Ambient air temperature
p_L Charge-air pressure
p_A Atmospheric pressure
p_i Pump inner chamber pressure
I Current at ELAB (PWM signal)

The object is to inject precisely the amount of fuel required by the engine. Following an initial rise, the engine's fuel requirement falls slightly as its speed increases. Fig. 3 shows the curve for fuel delivery on an fuel-injection pump without torque control (1).

As the curve indicates, the distributor injection pump delivers somewhat more fuel at high rotational speeds, while the control collar remains stationary relative to the plunger. This increase in the pump's delivery quantity is traceable to the venturi effect at the cutoff bore on the plunger.

Permanently defining the fuel-injection pump's delivery quantity for maximum torque generation at low rotational speeds results in an excessive rate of high-speed fuel injection, and the engine is unable to burn the fuel without producing smoke. The results of excessive fuel injection include engine overheating, particulate emissions and higher fuel consumption.

The contrasting case occurs when maximum fuel delivery is defined to reflect the engine's requirements in full-load operation at maximum rotational speed; this strategy fails to exploit the engine's low-speed power-pro-

duction potential. Again, delivery quantity decreases as engine speed increases. In this case, power generation is the factor that is less than optimal. Conclusion: A means is required to adjust injected fuel quantities to reflect the engine's actual instantaneous requirements.

On distributor injection pumps, this torque-control can be executed by the delivery valve, the cutoff bore, an extended control lever assembly or hydraulically controlled torque control (HBA). The control lever assembly is employed when negative full-load torque control is required.

Positive torque control

Positive full-load torque control is required on fuel-injection pumps that would otherwise deliver too much fuel at the top of the speed range. This type of system is employed to avoid this issue by reducing the fuel-injection pump's high-speed delivery quantity.

Positive torque control with the pressure-control valve
in certain limits, pressure-control valves can provide positive torque control, for instance, when equipped with more compliant springs. This strategy limits the magnitude of the high-speed rise in the pump's internal pressure.

Positive torque control using cutoff bore
Selected shapes and dimensions for the plunger's cutoff bore can be selected to reduce delivery quantities delivered at high speed.

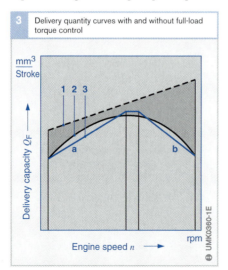

3 Delivery quantity curves with and without full-load torque control

Negative torque control

Negative full-load torque control may be indicated with engines that suffer from a low-speed tendency to generate black smoke or need special torque-rise characteristics. Turbocharged engines often require negative torque control to take over from the manifold-pressure compensator (LDA). The response to these scenarios is to increase delivery quantity as engine speed rises (Fig. 3, Sector a).

Negative torque control with control lever assembly (Fig. 4)

The starting spring (4) compresses and the torque-control lever (9) presses against the tensioning lever (2) via the stop pin (3). The torque-control shaft (11) is also pressed against the tensioning lever. When speed rises, increasing the sleeve force F_M, the torque-control lever presses against the torque-control spring. If the sleeve force exceeds the torque-control spring's force, the torque-control lever (9) is pressed toward the pin shoulder (5). This shifts the pivot axis M_4 shared by the starting lever and the torque-control lever. The starting lever simultaneously rotates about M_2 to slide the control collar (8) for increased delivery quantity. The torque-control process terminates with the torque-control lever coming to rest against the pin shoulder.

Negative torque control with hydraulically controlled torque control (HBA)

A torque control mechanism similar to the manifold-pressure compensator (LDA) can be used to define the full-load delivery characteristics as a function of engine speed (Fig. 5). As engine speed increases, rising pressure in the pump's internal chamber p_i is transferred to the control plunger (6). This system differs from spring-based adjustment by allowing use of a cam profile on the control pin to define full-throttle curves (to a limited degree).

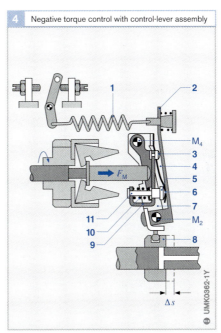

4   Negative torque control with control-lever assembly

Fig. 4
1 Governor spring
2 Tensioning lever
3 Full-load stop pin
4 Starting spring
5 Pin shoulder
6 Impact point
7 Starting lever
8 Control collar
9 Torque-control lever
10 Torque-control spring
11 Torque-control shaft

M_2 Pivot point for 2 and 7
M_4 Pivot axis for 7 and 9

F_M Sleeve force
Δs Control-collar travel

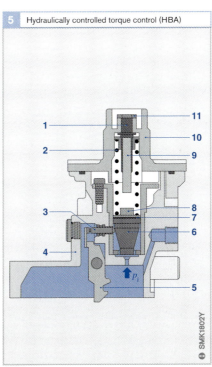

5   Hydraulically controlled torque control (HBA)

Fig. 5
1 Adjusting screw
2 Spring
3 Pin
4 Pump cover
5 Reverse-transfer lever with full-load stop
6 Adjustment piston
7 Shim
8 Base disk
9 Stop pin
10 Cover plate
11 Locknut

p_i Pump inner chamber pressure

Charge-air pressure torque control

During artificial induction a (turbo)super-charger forces pressurized fresh air into the intake tract. This charge-air pressure allows a diesel engine of any given displacement to generate more power and torque than its atmospheric-induction counterpart in any given speed band. The rise in effective power corresponds to the increase in air mass (Fig. 6). In many cases, it proves possible to reduce specific fuel consumption at the same time. A standard means of generating charge-air pressure for diesel engines is the exhaust-gas turbocharger.

Manifold-pressure compensator (LDA)

Function

Fuel delivery in the turbocharged diesel engine is adapted to suit the increased density of the air charges produced in charging mode. When the turbocharged diesel operates with cylinder charges of relatively low density (low induction pressure), fuel delivery must be adjusted to reflect the lower air mass. The manifold-pressure compensator performs this function by reducing fuel-throttle fuel flow below a specific (selected) charge-air pressure (Fig. 7).

Design

The manifold-pressure compensator is mounted on top of the distributor injection pump (Fig. 8 and 9). On top of this mechanism are the charge-air-pressure connection (7) and the vent port (10). A diaphragm (8) separates the inside chamber into two air-tight and mutually isolated sections. A compression spring (9) acts against the diaphragm, while the adjusting screw (5) holds the other side. The adjusting screw is used to adjust the spring's tension. This process adapts the manifold-pressure compensator to the charge-air pressure generated by the turbocharger. The diaphragm is connected to the sliding bolt (11). The sliding bolt features a control cone (12) whose position is monitored by a probe (4). The probe transfers the sliding bolt's motion through the reverse-transfer lever (3) to vary the full-load stop. The adjusting screw (6) on top of the LDA defines the initial positions of the diaphragm and sliding bolt.

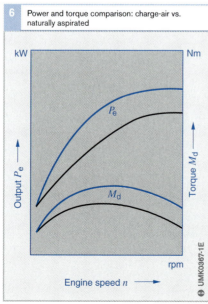

6 Power and torque comparison: charge-air vs. naturally aspirated

kW ... Nm
Output P_e
Torque M_d
P_e
M_d
Engine speed n
rpm
UMK0367-1E

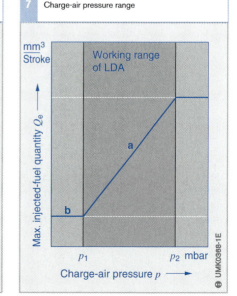

7 Charge-air pressure range

mm³ / Stroke
Working range of LDA
Max. injected-fuel quantity Q_e
a
b
p_1 p_2 mbar
Charge-air pressure p
UMK0368-1E

Operating concept

At the lower end of the speed range, the charge-air pressure generated by the turbocharger is not powerful enough to compress the spring. The diaphragm remains in its initial position. Once the rising charge-air pressure p_L starts to deflect it, the diaphragm pushes the sliding bolt and control cone down against the pressure of the spring.

This vertical movement in the sliding bolt shifts the position of the probe, causing the reverse-transfer lever to rotate about its pivotal point M_1. The tensile force exerted by the governor spring creates a positive connection between tensioning lever, reverse-transfer lever, probe and control cone.

The tensioning lever thus mimics the reverse-transfer lever's rotation, and the starting and tensioning levers execute a turning motion around their shared pivot axis to shift the control lever and raise delivery quantity. This process adapts fuel flow to meet the demands of the greater air mass with the engine's combustion chambers.

| 8 | Axial-piston distributor injection pump with manifold-pressure compensator |

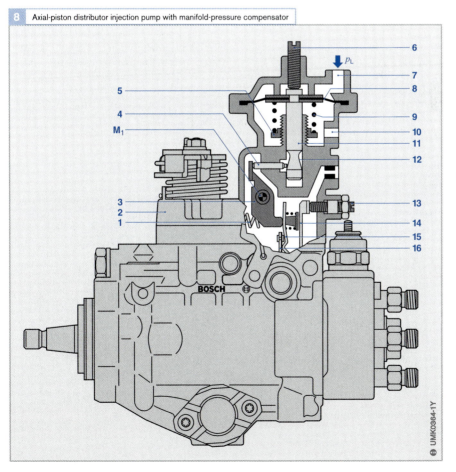

Fig. 8

1 Governor spring
2 Governor cover
3 Reverse-transfer lever
4 Sensor pin
5 Adjustment nut
6 Adjustment pin
7 Charge-air pressure connection
8 Diaphragm
9 Spring
10 Vent
11 Sliding bolt
12 Control cone
13 Adjusting screw for full-load delivery
14 Control lever
15 Tensioning lever
16 Starting lever

p_L Charge-air pressure

M_1 Pivot axis for 3

As charge-air pressure falls, the spring be-
neath the diaphragm presses the sliding bolt
back up. The control mechanism now acts in
the opposite direction, and fuel quantity is
reduced to reflect the needs of the lower
charge-air pressure.

The manifold-pressure compensator responds to turbocharger failure by reverting
to its initial position, and limits full-throttle
fuel delivery to ensure smoke-free combustion. Full-load fuel quantity with charge-air
pressure is adjusted by the full-throttle stop
screw in the governor cover.

Atmospheric pressure-sensitive torque control

Owing to the lower air density, the mass
of inducted air decreases at high altitudes.
If the standard fuel quantity prescribed for
full-load operation is injected, there will not
be enough air to support full combustion.
The immediate results are smoke generation
and rising engine temperatures. The atmospheric pressure-sensitive full-load stop can
help prevent this condition. It varies full-
load fuel delivery in response to changes
in barometric pressure.

9   Axial-piston distributor injection pump with manifold-pressure compensator (section)

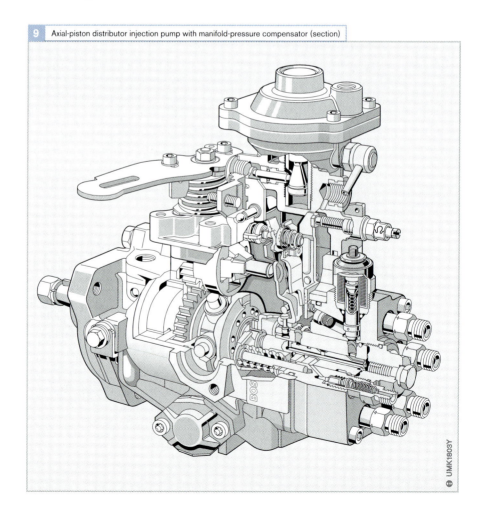

UMK1803Y

Atmospheric-pressure sensitive full-load stop (ADA)

Design

The basic structure of the atmospheric pressure-sensitive full-load stop is identical to that of the charge-air pressure-sensitive full-load stop. In this application, it is supplemented by a vacuum unit connected to a vacuum-operated pneumatic device (such as the power brake system). The vacuum unit provides a constant reference pressure of 700 mbar (absolute).

Operating concept

Atmospheric pressure acts on the upper side of the ADA's internal diaphragm. On the other side is the constant reference pressure supplied by the vacuum unit.

Reductions in atmospheric pressure (as encountered in high-altitude operation) cause the adjustment piston to rise away from the lower full-load stop. As with the LDA, a reverse-transfer lever then reduces the injected fuel quantity.

Load-sensitive torque control

Load-sensitive start of delivery (LFB)

Function

As the diesel engine's load factor changes, the start of injection – and thus the start of delivery – must be advanced or retarded accordingly.

The load-sensitive start of delivery is designed to react to declining loads (from full-load to part throttle, etc.) at constant control lever positions by retarding start of delivery. It responds to rising load factors by shifting the start of delivery forward. This adaptive process provides smoother engine operation along with cleaner emissions at part throttle and idle.

Fuel-injection pumps with load-sensitive start of delivery can be recognized by the press-fit ball plug inserted during manufacture (Fig. 10, Pos. 10).

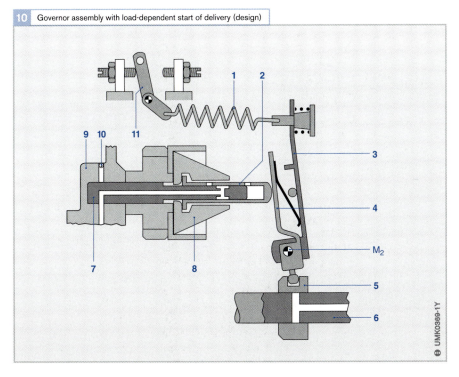

10   Governor assembly with load-dependent start of delivery (design)

Fig. 10
1 Governor spring
2 Sliding sleeve
3 Tensioning lever
4 Starting lever
5 Control collar
6 Distributor plunger
7 Controller base
8 Flyweight
9 Pump housing
10 Cone
11 Rotational-speed control lever

M_2 Pivot for 3 and 4

Design
Load-sensitive start of delivery torque control (Fig. 10) relies on modifications to the sliding sleeve (2), governor base (7) and pump housing (9). In this configuration, the sliding sleeve is equipped with a supplementary control port while the base includes a ring groove, one longitudinal and two transverse passages. The pump housing contains an additional bore allowing this assembly to link the pump's inner chamber with the suction-side of the vane-type supply pump.

Operating concept
Rises in engine speed and the accompanying increases in the pump's internal pressure advance the start of delivery. The reduction in the pump's inner chamber supplied by the LFB provide a (relative) retardation of the timing. The control function is regulated by the ring groove on the base and the control port on the sliding sleeve. Individual full-load speeds can be specified by the rotational-speed control lever (11).

If the engine attains this speed without reaching wide-open throttle, the rotational speed can continue to rise. The flyweights (8) respond by spinning outward to shift the position of the sliding sleeve. Under normal control conditions, this reduces delivery quantity, while the sliding sleeve's control port is exposed by the ring groove in the base (open, Fig. 11). At this point, a portion of the fuel flows through the base's longitudinal and transverse passages to the suction side, reducing the pressure in the pump's inner chamber.

This pressure reduction moves the timing plunger to a new position. This forces the roller ring to rotate in the pump's rotational direction, retarding the start of delivery.

The engine speed will drop if load factor continues to rise while the control lever's position remains constant. This retracts the flyweights to shift the sliding sleeve and reclose its control port. The flow of fuel to the pump's inside chamber is blocked, and pressure in the chamber starts to rise. The timing plunger acts against the spring pressure to turn the roller ring against the pump's direction of rotation and the start of delivery is advanced.

Load-dependent start of delivery with deactivation feature (LFB)
The LFB can be deactivated to reduce HC emissions generated by the diesel engine when it is cold ($< 60°\,C$). This process employs a solenoid valve (8) to block the fuel flow. This solenoid open when de-energized.

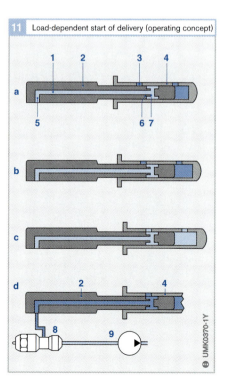

Fig. 11
a　Start (base) position
b　Full-load position just before activation
c　Activation (pressure reduction in inner chamber)
d　Load-dependent start of delivery with deactivation feature ALFB

1　Longitudinal passage in plunger
2　Controller axis
3　Sliding sleeve control port
4　Sliding sleeve
5　Transverse passage in controller axis
6　Control helix on ring groove on controller axis
7　Transverse passage in controller axis
8　Solenoid valve
9　Vane-type supply pump

11　Load-dependent start of delivery (operating concept)

Cold-start compensation

The cold-start compensation device improves the diesel engine's cold-start response by advancing the start of delivery. This feature is controlled by a driver-operated cable or by an automatic temperature-sensitive control device (Fig. 13).

Mechanical cold-start accelerator (KSB) on roller ring

Design

The KSB is mounted on the pump's housing. In this assembly, a shaft (Fig. 12, Pos. 12) connects the stop lever (Fig. 13, Pos. 3) with an inner lever featuring an eccentrically mounted ball head (3). This lever engages with the roller ring. The stop lever's initial position is defined by the full-load stop and the leg spring (13). The control cable attached to the upper end of the stop lever serves as the link to the adjustment mechanism which is either manual or automatic. The automatic adjuster is installed in a bracket on the distributor injection pump (Fig. 13), while the manual adjustment cable terminates in the passenger compartment.

There also exists a version in which the adjuster intervenes through the timing plunger.

Operating concept

The only difference between the manual and automatic versions of the cold-start accelerator is the external control mechanisms. The key process is always the same. When the control cable is not tensioned, the leg spring presses the stop lever against its full-load stop. Both ball head and roller ring (6) remain in their initial positions. Tension on the control cable causes the stop lever, shaft and inside lever to turn with the ball head.

This rotation changes the position of the roller ring to advance the start of delivery. The ball head engages a vertical groove in the roller ring. This prevents the timing plunger from advancing the start of delivery further before a specified engine speed is reached.

When the driver activates the cold-start acceleration device (KSB timing device), a position shift of approximately 2.5° camshaft (b) remains, regardless of the adjustment called for by the timing device (Fig. 14a).

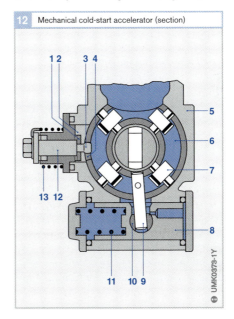

12 Mechanical cold-start accelerator (section)

Fig. 12
1 Lever
2 Adjustment window
3 Ball head
4 Vertical groove
5 Pump housing
6 Roller ring
7 Rollers in roller ring
8 Timing plunger
9 Pin
10 Sliding block
11 Timing spring
12 Shaft
13 Leg spring

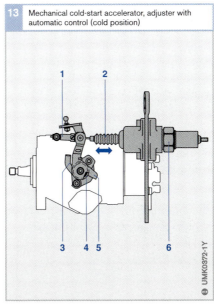

13 Mechanical cold-start accelerator, adjuster with automatic control (cold position)

Fig. 13
1 Retainer
2 Bowden cable
3 Stop lever
4 Leg spring
5 Control lever KSB
6 Adjuster dependent on coolant and ambient temperatures

In cold-start systems with automatic control, the actual increment depends on engine temperature and/or ambient temperature.

Automatic adjustment relies on a control mechanism in which a temperature-sensitive expansion element translates variations in engine temperature into linear motion.

This arrangement's special asset is that it provides the optimum start of delivery and start of injection for each individual temperature.

Various lever layouts and actuation mechanisms are available according to the mounting side and rotational direction encountered in individual installation environments.

Temperature-controlled idle-speed increase (TLA)

The temperature-controlled idle-speed increase, which is combined with the automatic KSB, is also operated by the control mechanism (Fig. 15). The ball pin in the extended KSB control lever presses against the rotational-speed control lever to lift it from the idle-speed stop screw when the engine is cold. This raises the idle speed to promote smoother engine operation. The KSB control lever rests against its full-load stop when the engine is warm. This allows the rota-

tional-speed control lever to return to its own full-load stop, at which point the temperature-controlled idle-speed increase system is no longer active.

Hydraulic cold-start accelerator

There are inherent limits on the use of strategies that shift the timing plunger to advance start of injection. Hydraulic start of injection advance applies speed-controlled pressure in the pump's inside chamber to the timing plunger. The system employs a bypass valve in the pressure-control valve to modify the inner chamber's automatic pressure control, automatically increasing internal pressure to obtain additional advance extending beyond the standard advance curve.

Design

The hydraulic cold-start accelerator comprises a modified pressure-control valve (Fig. 17, Pos. 1), a KSB ball valve (7), an electrically heated expansion element (6) and a KSB control valve (9).

14 Effects of mechanical cold-start accelerator KSB

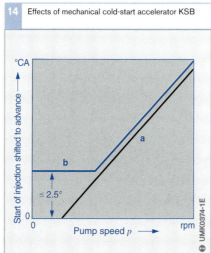

15 Mechanical cold-start accelerator (automatic) with temperature-controlled idle-speed increase

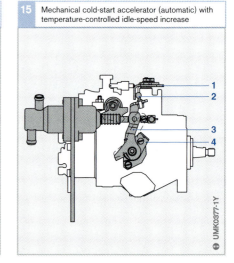

Operating concept
The fuel supplied by the supply pump (5) flows through the distributor injection pump's inner chamber and to the end of the timing plunger (11), which compresses the return spring (12) and shifts position in response to the inner chamber pressure to vary the start of injection. The pressure-control valve controls the pressure in the inner chamber, raising pressure for greater delivery quantity as engine speed increases (Fig. 16).

The throttle port in the pressure-control valve's plunger (Fig. 17, Pos. 3) supplies the added pressure that allows the KSB to provide the forward offset in start of injection advance (Fig. 16, blue curve). This conveys an equal pressure to the spring side of the pressure-control valve. The KSB ball valve, with its higher pressure setting, regulates activation and deactivation (with the thermal element) while also serving as a safety release. An adjusting screw on the integrated KSB control valve is available for adjusting KSB operation to a specific engine speed. Pressure from the supply pump presses the KSB control-valve plunger (10) against a spring. A damping throttle inhibits pulsation against the control plunger. The KSB

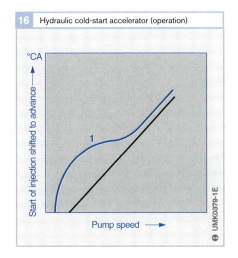

16 Hydraulic cold-start accelerator (operation)

°CA

Start of injection shifted to advance →

1

Pump speed →

UMK0379-1E

Fig. 16
1 Start of delivery
   advanced

pressure curve is controlled by the timing edge on the control plunger and the opening on the valve holder. The spring rate on the control valve and the control port configuration can be modified to match KSB functionality to the individual application. The ambient temperature will act on the expansion element to open the cold-start accelerator ball valve before starting the hot engine.

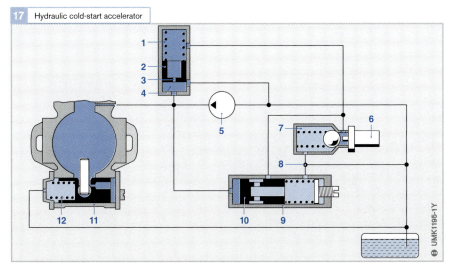

17 Hydraulic cold-start accelerator

UMK1195-1Y

Fig. 17
1 Pressure-control
   valve
2 Valve plunger
3 Throttled bore
4 Inner chamber
   pressure
5 Vane-type supply
   pump
6 Electrically heated
   expansion element
7 Ball valve KSB
8 Fuel drains without
   pressure
9 Adjustable KSB
   control valve
10 Control plunger
11 Timing device
12 Return spring

Soft-running device

To achieve the desired emission properties, the system injects the fuel charge into the engine's combustion chamber in the briefest possible time span; it operates at high fuel-delivery rates.

Depending on system configuration, these high fuel-delivery rates can have major consequences, especially in the form of diesel knock at idle. Remedial action is available through the use of extended injection periods to achieve smoother combustion processes at and near idle.

Design and operating concept

In distributor injection pumps featuring an integral soft-running device, the plunger is equipped with two longitudinal passages (Fig. 19, Pos. 3 and 5) connected by a ring groove (6). The longitudinal passage 3 is connected to a cutoff bore (7) with a restrictor in the area adjacent to the control collar (1).

The plunger travels through the stroke h_1 on its path toward TDC. The cutoff bore (7) connected to passage 3 emerges from the control collar earlier than the cutoff bore (2) on passage 5.

Because the ring groove (6) links passages 3 and 5, this causes a portion of the fuel to seep from the plunger chamber back to the pump's inner chamber. This reduces the fuel-delivery rate (less fuel discharged for

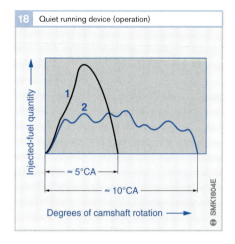

18 Quiet running device (operation)

each degree of camshaft travel). The travel that the camshaft executes for any given quantity of fuel injected is roughly doubled (Fig. 18).

Under extreme loads, the control collar is closer to the distributor head. This makes the distance h_2 smaller than h_1. When the plunger now moves toward TDC, the ring groove (6) is covered before the cutoff bore (7) emerges from the control collar. This cancels the link joining passages 3 and 5, deactivating the soft-running mechanism in the high-load range.

Fig. 18
1 without quiet running
2 with quiet running

Fig. 19
1 Control collar
2 Cutoff bore
3 Port 3
4 1-way non-return valve for nozzle
5 Port 5
6 Ring groove
7 Cutoff bore

h_1 Lift 1
h_2 Lift 2

TDC **T**op **D**ead **C**enter on distributor plunger

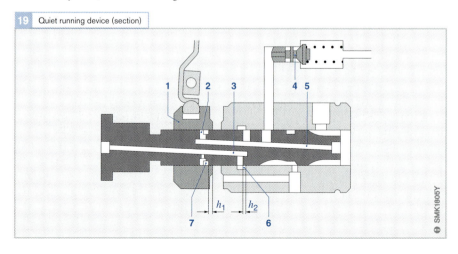

19 Quiet running device (section)

Load switch

The load switches are installed at the distributor injection pump's main rotational-speed control lever. These switches control operation of assemblies outside of the fuel-injection pump. They rely on microswitches for electronic control, or valves for pneumatic switching of these external units. Their primary application is in triggering the exhaust-gas recirculation valve. In this application, the switch opens and closes the valve in response to variations in the main control lever's position.

Two different elements are available to control microswitches and pneumatic valves:
● Angular stop plate with control recess and
● Cast aluminum control cams

The control plate or cam attached to the main control lever initiates activation or deactivation based on the lever's travel. The switching point corresponds to a specific coordinate on the pump's response curve (rotational speed vs. delivery quantity).
   Either two microswitches or one pneumatic valve can be installed on the fuel-injection pump due to differences in the respective layouts.

Microswitch
The electric microswitch is a bipolar make-and-break switch. It consists of leaf spring elements with a rocker arm. A control pin presses on the springs. A supplementary control element (Fig. 1, Pos. 6) limits mechanical wear and provides a unified control stroke for all applications on distributor injection pumps.

Pneumatic valve
The pneumatic valve (Fig. 2) blocks air flow in a vacuum line.

Potentiometer

The potentiometer is an option available for cases in which the object is to control several different points on the curve for rotational speed vs. load factor. It is mounted on top of the rotational-speed control lever, to which it is attached by clamps (Fig. 1). When energized, the potentiometer transmits a continuous electrical voltage signal that reflects the control lever's position with a linear response curve. A suitable governor design then provides useful information in the defined range.

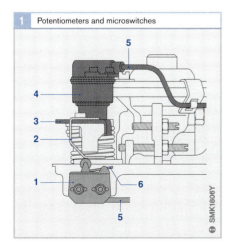

1   Potentiometers and microswitches

© SMK1806Y

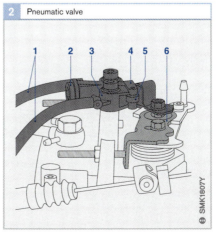

2   Pneumatic valve

© SMK1807Y

Fig. 1
1   Microswitch
2   Angular stop bracket
3   Rotational-speed
     control lever
4   Potentiometer
5   Electrical
     connection
6   Auxiliary controller

Fig. 2
1   Pneumatic
     connections
2   Vacuum adjusting
     screw
3   Pneumatic valve
4   Control lever
5   Control roller
6   Rotational-speed
     control lever

Delivery-signal sensor

Application
The FSS delivery-signal sensor is a dynamic pressure sensor installed in the threaded center plug of the diesel fuel-injection pump in place of the vent screw (Fig. 2). It monitors pressure in the element chamber. The sensor signal can be employed to register the start of delivery, the delivery period and the pump speed n.

The pressure-signal measurement range extends from 0...40 MPa, or 0...400 bar, at which it is suitable for use with IDI fuel-injection pumps. This device is a "dynamic sensor". It registers pressure variations instead of monitoring static pressure levels.

Design and operating concept
The delivery-signal sensor relies on piezo-electric principles for operation. The pressure in the element chamber acts on a sensor measurement cell (Fig. 2, Pos. 13). This contains the piezoelectric ceramic layer. Pressure variations modify electrical charge states in this layer. These charge shifts generate minute electric voltages, which the integrated circuit (11) in the sensor converts to a square-wave signal (Fig. 1).

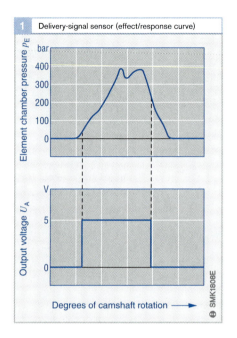

1 Delivery-signal sensor (effect/response curve)

Element chamber pressure p_E (bar): 400, 300, 200, 100, 0

Output voltage U_A (V): 5, 0

Degrees of camshaft rotation ⟶

SMK1808E

Fig. 1
a Pressure curve for element chamber
b Signal from delivery-signal sensor

Fig. 2
1 Disk
2 Spring
3 Seal
4 Contact pin
5 Housing
6 Contact spring
7 Cable housing
8 Power supply connection (yellow/white)
9 Signal connection (gray/white)
10 O-ring
11 Circuit board with integrated circuit
12 Insulator
13 Sensor measurement cell (piezoelectric ceramic material)
14 Fuel-injection pump element chamber

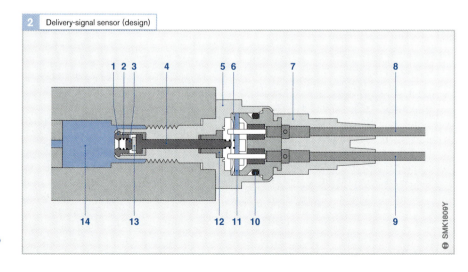

2 Delivery-signal sensor (design)

1 2 3   4   5 6   7   8

14   13   12 11 10   9

SMK1809Y

Shutoff devices

Shutoff

One inherent property of the diesel engine's auto-ignition concept is that interrupting the fuel supply represents the only way to switch off the engine.

The mechanically controlled distributor injection pump is usually deactivated by an electric shutoff valve (ELAB). It is only in rare special application environments that these pumps are equipped with mechanical shutoff devices.

Solenoid-operated shutoff valve (ELAB)

The electric shutoff device is a solenoid valve. Operation is controlled by the "ignition" key to offer drivers a high level of convenience.

The solenoid valve employed to interrupt fuel delivery in the distributor injection pump is mounted on top of the distributor head (Fig. 1). When the diesel engine is running, the solenoid (4) is energized, and the armature retracts the sealing cone (6) to hold the inlet passage to the plunger chamber open. Switching off the engine with the key interrupts the current to the solenoid coil. The magnetic field collapses and the spring (5) presses the armature and sealing cone back against the seat. By blocking fuel flow through the plunger chamber's inlet passage, the seal prevents delivery from the plunger. The solenoid can be designed as either pulling or pushing electromagnet.

The ELABs used on marine engines open when de-energized. This allows the vessel to continue in the event of electrical system failure. Minimizing the number of electrical consumers is a priority on ships, as the electromagnetic fields generated by continuous current flow promote salt-water corrosion.

Mechanical shutoff device

The mechanical shutoff devices installed on distributor injection pumps consist of lever assemblies (Fig. 2). This type of assembly consists of an inside and outside stop lever mounted on the governor cover (1, 5). The driver operates the outer stop lever from the

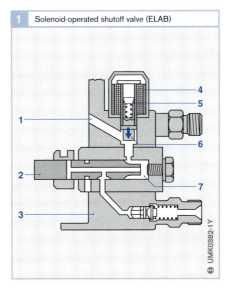

1 Solenoid-operated shutoff valve (ELAB)

Fig. 1
1   Inlet passage
2   Distributor plunger
3   Distributor head
4   Solenoid (pulling electromagnet in this instance)
5   Spring
6   Armature with sealing cone
7   Plunger chamber

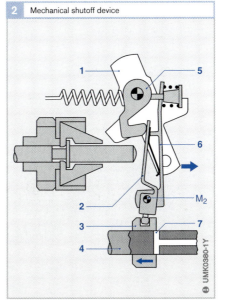

2 Mechanical shutoff device

Fig. 2
1   External stop lever
2   Starting lever
3   Control collar
4   Distributor plunger
5   Inner stop lever
6   Tensioning lever
7   Control port

M_2 Pivot axis for 2 and 6

vehicle's interior via Bowden cable, etc. Cable operation causes the two stop levers to rotate about their pivot axis, with the inside stop lever pressing against the control mechanism's (2) starting lever. The starting lever pivots about its axis M_2 to push the control collar (3) into its stop position. This acts on the control passage (7) in the plunger to prevent fuel delivery.

Electric actuator mechanism

Systems with electronic diesel control employ the fuel-delivery actuator to switch off the engine (control parameter from electronic control unit (ECU): injected-fuel quantity = zero; refer to following section). The separate solenoid-operated shutoff valve (ELAB) serves only as a backup system in the event of a defect in the actuator mechanism.

Electronic Diesel Control

Systems with mechanical rotational-speed control respond to variations in operating conditions to reliably guarantee high-quality mixture formation.

EDC (Electronic Diesel Control) simultaneously satisfies an additional range of performance demands. Electric monitoring, flexible electronic data processing and closed-loop control circuits featuring electric actuators extend performance potential to include parameters where mechanical systems have no possibility of supply regulation. At the core of EDC is the electronic control unit that governs distributor injection pump operation.

As Electronic Diesel Control supports data communications with other electronic systems (traction control, electronic transmission-shift control, etc.) it can be integrated in an all-encompassing vehicle system environment.

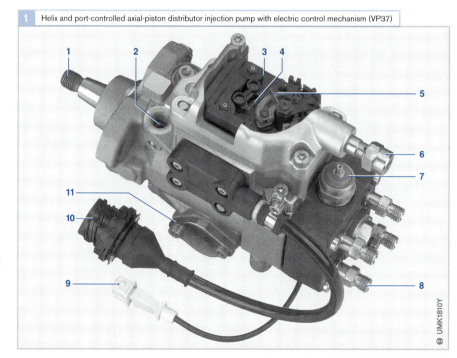

1   Helix and port-controlled axial-piston distributor injection pump with electric control mechanism (VP37)

Fig. 1
1 Pump drive
2 Fuel inlet
3 Fuel-injection pump delivery positioner
4 Fuel temperature sensor
5 Angle sensor
6 Fuel return
7 Type ELAB electric shutoff valve
8 Delivery valve (discharge to nozzle)
9 Connection for start of injection solenoid valve
10 Connection for fuel-injection pump delivery positioner
11 Timing device

Helix and port-controlled electronic distributor injection pumps use a fuel-delivery actuator mechanism to regulate injected-fuel quantities and a solenoid valve to control start of injection.

Solenoid actuator for fuel-delivery control

The solenoid actuator (rotary actuator, Fig. 2, Pos. 2) acts on the control collar via shaft. On mechanically controlled fuel-injection pumps, the position determines the point at which the cutoff bores are exposed.

Injected fuel quantity can be progressively varied continuously from zero to the system's maximum capacity (for instance, during cold starts). This valve is triggered by a pulse-width modulated (PWM) signal. When current flow is interrupted, the rotary-actuator return springs reduce the fuel-delivery quantity to zero.

A sensor with a semidifferential short-circuit ring (1) relays the actuator-mechanism rotation angle, and thus the control collar position, back to the control unit[1]. This action determines the injected-fuel quantities for the instantaneous engine speed.

Solenoid valve for start of injection timing

As with mechanical timing devices, the timing plunger reacts to pressure in the pump's inner chamber, which rises proportionally to rotational speed. This force against the timing device's pressure-side is controlled by the timing-device solenoid valve (5). This valve is triggered by a PWM signal.

When the solenoid valve remains constantly open, pressure falls to retard start of injection, while closing the valve advances the timing. The electronic control unit can continuously vary the PWM signal's pulse-duty factor (ratio of open to closed periods at the valve) in the intermediate range.

[1] On 1st generation pumps, a potentiometer was employed to register rotation angle.

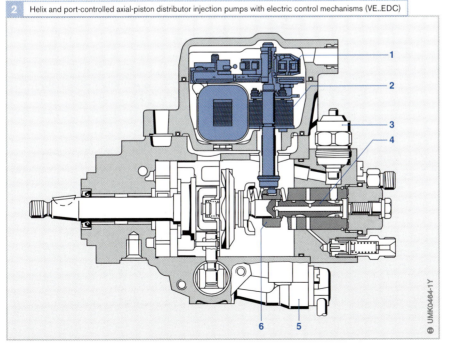

2 Helix and port-controlled axial-piston distributor injection pumps with electric control mechanisms (VE..EDC)

UMK0464-1Y

Fig. 2
1 Semi-differential short-circuiting ring sensor
2 Solenoid control mechanism
3 Electric shutoff valve ELAB
4 Distributor plunger
5 Timing-device solenoid valve
6 Control collar

Sensor with semidifferential short-circuiting ring

Applications

Semidifferential short-circuit ring sensors are position sensors for monitoring travel and angles. These sensors are extremely precise and robust. They are used as:

- Control-rack travel sensors for monitoring control-rack position on in-line diesel fuel-injection pumps and
- Angle sensors for fuel-delivery actuator assemblies in diesel distributor injection pumps

Design and operating concept

The sensors (Figures 3 and 4) are built around a laminated soft-iron core. Attached to each leg of the soft-iron core are a sensor coil and a reference coil.

As the AC current transmitted by the control unit permeates the coils, it creates alternating magnetic fields. The copper short-circuiting rings surrounding the legs of the soft-iron core serve as a shield against these alternating magnetic fields. While the short-circuiting ring for reference remains stationary, the measuring ring is attached to the control rack or control-collar shaft (control-rack travel s or advance angle φ).

Changes in the measuring ring's position change the magnetic flux to vary the coil's voltage, while the control unit maintains voltage at a constant level (impressed current).

Evaluation circuitry calculates the ratio of output voltage U_A to reference voltage U_{Ref} (Fig. 5), which is proportional to the deflection angle of the measurement short-circuiting ring. This curve's gradient can be adjusted by bending the reference short-circuiting ring while the base point is repositioned by shifting the position of the measuring ring.

Fig. 3
1   Sensor coil
2   Sensor short-circuiting ring
3   Soft iron core
4   Control collar shaft
5   Reference coil
6   Reference short-circuiting ring

φ_{max} Adjustment range of control-collar shaft
φ Registered angle

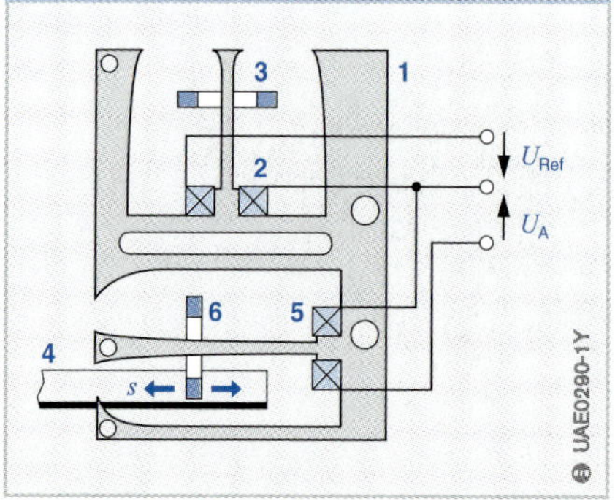

Fig. 4
1   Soft-iron core
2   Reference coil
3   Reference short-circuiting ring
4   Control rack
5   Sensor coil
6   Sensor short-circuiting ring
s Control-rack travel

Fig. 5
U_A Voltage output
U_{Ref} Reference voltage

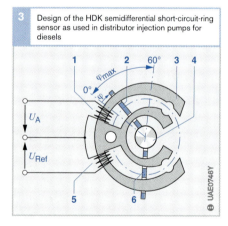

3   Design of the HDK semidifferential short-circuit-ring sensor as used in distributor injection pumps for diesels

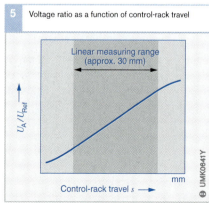

4   Design of RWG Control-rack travel sensor for in-line diesel fuel-injection pumps

5   Voltage ratio as a function of control-rack travel

Linear measuring range (approx. 30 mm)

U_A/U_{Ref}

Control-rack travel s mm

Diesel-engine immobilizers

Diesel immobilizers is a component in the electronic vehicle immobilizer installed on vehicles with a mechanically-controlled distributor injection pump. The diesel immobilizer is mounted above the solenoid-operated shutoff valve (ELAB) of the distributor injection pump. It responds to signals from the immobilizer control unit by switching current flow to the ELAB off and on to allow or block the flow of diesel fuel to the engine. The diesel immobilizer unit and ELAB are mounted behind a metal plate which must be destroyed for removal.

The diesel-engine immobilizer's input wiring usually comprises three terminals, consisting of power supply, data line and ground wire. The single exit wire carries the current to the ELAB.

Control units

The DDS 1.1. and DDS 3.1 control units are available for diesel-engine immobilizers.

DDS1.1

DDS 1.1 is a PCB-equipped ECU featuring an integrated protective housing made of plastic with injected sheet metal. On its own, DDS 1.1 provides only the lowest degree of protection. The security rating can be raised by adding various external protection features to the pump.

DDS3.1

DDS 3.1 is a hybrid control unit with a supplementary protective housing made of manganese steel. DDS 3.1 can provide optimum security without any supplements.

The DDS system must be released on the pump for operation on the test bench.

1 Helix and port-controlled axial-piston distributor injection pump with diesel antitheft protection

5 cm

SMK1811Y

Fig. 1
1 DDS 1
   (via ELAB)
2 Plug connection

Solenoid-valve controlled distributor injection pumps

Ever stricter emission limits for diesel engines and the demand for further reductions in fuel consumption have resulted in the continuing refinement of electronically controlled distributor injection pump. High-pressure control using a solenoid valve permits greater flexibility in the variation of start and end of delivery and even greater accuracy in the metering of the injected-fuel quantity than with port-controlled fuel-injection pumps. In addition, it permits pre-injection and correction of injected-fuel quantity for each cylinder.

The essential differences from port-controlled distributor injection pumps are the following:
- A control unit mounted on the pump
- Control of fuel injection by a high-pressure solenoid valve and
- Timing of the high-pressure solenoid valve by an angle-of-rotation sensor integrated in the pump

As well as the traditional benefits of the distributor injection pump such as light weight and compact dimensions, these characteristics provide the following additional benefits:
- A high degree of fuel-metering accuracy within the program map
- Start of injection and injection duration can be varied independently of factors such as engine speed or pump delivery quantity
- Injected-fuel quantity can be corrected for each cylinder, even at high engine speeds
- A high dynamic volume capability
- Independence of the injection timing device range from the engine speed and
- The capability of pre-injection

Areas of application

Solenoid-valve controlled distributor injection pumps are used on small and medium-sized diesel engines in cars, commercial vehicles and agricultural tractors. They are fitted both on direct-injection (DI) and indirect-injection (IDI) engines.

The nominal speed, power output and design of the diesel engine determine the type and size of the fuel-injection pump. Distributor injection pumps are used on car, commercial-vehicle, agricultural tractor and fixed-installation engines with power outputs of up to 45 kW per cylinder. Depending on the type of control unit and solenoid valve, they can be run off either a 12-volt or a 24-volt electrical system.

Solenoid-valve controlled distributor injection pumps are available with high-pressure outlets for either four or six cylinders. The maximum injected-fuel capacity per stroke is in the range of 70...175 mm^3. The required maximum injection pressures depend on the requirements of the engine (DI or IDI). They range from 800...1,950 bar.

All distributor injection pumps are lubricated by the fuel. Consequently, they are maintenance-free.

Designs

There are two basic designs:
- Axial-piston distributor pumps (Type VE..MV or VP29/VP30) and
- Radial-piston distributor pumps (Type VR or VP44)

There are a number of design variations (e.g. number of outlets, pump drive by ring gear) according to the various types of application and engine.

The hydraulic performance capacity of the axial-piston pump with injection pressures up to 1,400 bar at the nozzle will remain sufficient for many direct-injection and indirect-injection engines in the future.
    Where higher injection pressures are required for direct-injection engines, the radial-piston distributor pump introduced in 1996 is the more suitable choice. It has the capability of delivering injection pressures of up to 1950 bar at the nozzle.

▶ Family tree of Bosch electronically controlled distributor injection pumps

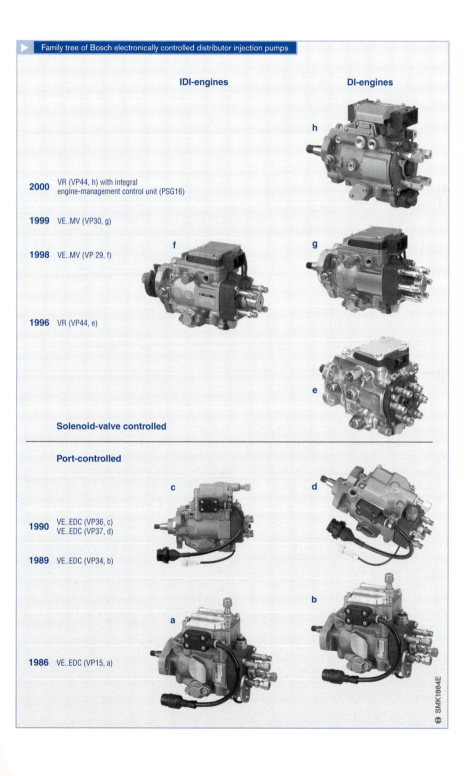

IDI-engines   **DI-engines**

h

2000 VR (VP44, h) with integral
          engine-management control unit (PSG16)

1999 VE..MV (VP30, g)

1998 VE..MV (VP 29, f)   f   g

1996 VR (VP44, e)

                                                e

Solenoid-valve controlled

Port-controlled

                              c   d

1990 VE..EDC (VP36, c)
          VE..EDC (VP37, d)

1989 VE..EDC (VP34, b)

                                                b

                              a

1986 VE..EDC (VP15, a)

SMK1864E

Fitting and drive system

Distributor injection pumps are flange-mounted directly on the diesel engine. The engine's crankshaft drives the fuel-injection pump's heavy-duty drive shaft by means of a toothed-belt drive, a pinion, a gear wheel or a chain. In axial-piston pumps, the drive shaft runs on two plain bearings in the pump housing. In radial-piston pumps, there is a plain bearing at the flange end and a deep-groove ball bearing at the opposite end. The distributor injection pump is fully synchronous with the engine crankshaft and pistons (positive mechanical link).

On four-stroke engines, the pump speed is half that of the diesel-engine crankshaft speed. In other words, it is the same as the camshaft speed.

There are distributor injection pumps for clockwise and counterclockwise rotation[1]. Consequently, the injection sequence differs according to the direction of rotation – but is always consecutive in terms of the geometrical arrangement of the outlets (e.g. A-B-C or C-B-A).

In order to prevent confusion with the numbering of the engine cylinders (cylinder no. 1, 2, 3, etc.), the distributor-pump outlets are identified by letters (A, B, C, etc., Fig. 1). For example, the assignment of pump outlets to engine cylinders on a four-cylinder engine with the firing sequence 1-3-4-2 is A-1, B-3, C-4 and D-2.

In order to achieve good hydraulic characteristics on the part of the fuel-injection system, the high-pressure fuel lines between the fuel-injection pump and the nozzle-and-holder assemblies must be as short as possible. For this reason, the distributor injection pump is fitted as close as possible to the diesel-engine cylinder head.

The interaction of all components has to be optimized in order to insure that the diesel engine and the fuel-injection system function correctly. The fuel lines and the nozzle-and-holder assemblies must therefore not be interchanged when carrying out servicing work.

In the high-pressure system between the fuel-injection pump and the nozzles, there are degrees of free play manufactured to an accuracy of a few thousandths of a millimeter. That means that dirt particles in the fuel can impair the function of the system. Consequently, high fuel quality and a fuel filter are required specifically designed to meet the requirements of the fuel-injection system. They prevent damage to pump components, delivery valves and nozzles and guarantee trouble-free operation and long service life.

Diesel fuel can absorb water in solution in quantities ranging between 50...200 ppm (by weight), depending on temperature. This dissolved water does not damage the fuel-injection system. However, if greater quantities of water find their way into the fuel (e.g. condensate resulting from changes in temperature), they remain present in undissolved form. If undissolved water accesses the fuel-injection pump, it may result in corrosion damage. Distributor injection pumps therefore need fuel filters with water chambers (see chapter "Fuel supply system"). The collected water has to be drained off at appropriate intervals. The increased use of diesel engines in cars has resulted in the demand for an automatic water warning device. It indicates whenever water needs to be drained off by means of a warning light.

[1] Direction of rotation as viewed from drive-shaft end of the pump

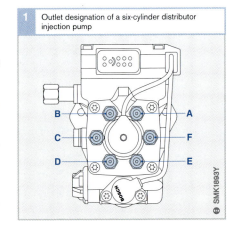

1   Outlet designation of a six-cylinder distributor injection pump

Fig. 1
Outlets are always numbered counterclockwise. Numbering starts at top right.

1998 Diesel Records

The 24-hour race on Germany's famous "Nürburgring" race track is not just about speed but about the durability of automotive technology. On 14th June 1998 the race was won for the first time by a diesel-engined car. The BMW 320 d left its gasoline-engined rivals trailing.

The average speed achieved during practice was approximately 160 kph. The maximum speed was around 250 kph. The 1,040-kg vehicle accelerated from 0 to 100 kph in 4.5 seconds.

The racing car was powered by a four-cylinder direct-injection diesel engine with a capacity of 1950 cc and a maximum power output of more than 180 kW (245 bhp) at 4,200...4,600 rpm. The engine produced its maximum torque of 430 N·m at 2,500...3,500 rpm.

The car's performance was due in no small part to its high-performance fuel-injection system designed by Bosch. The central component of the system is the Type VP44 radial-piston distributor injection pump that has been in volume production since 1996.

Compared with normal production engines, the racing diesel is an "enhanced-performance" model achieved by larger injected-fuel quantities and even higher injection pressures and engine speeds. For this purpose, a new high-pressure stage with 4 delivery plungers instead of 2 was fitted in combination with new nozzles. The software for the control units was also modified.

These design modifications and the fact that the engines are run continuously at full power obviously reduce the life of the components (designed for 24...48 hours of service), and increase the concentration levels of the noxious constituents and fuel consumption. The car used around 23 l/100 km (a comparable gasoline racing engine uses almost twice as much).

That impressive victory demonstrates once again that the diesel engine is no longer the "lame duck" that it used to be.

BMW 320d: Winner of the 1998 24-hour race

SMM0607Y

Design and method of operation

Assemblies

Solenoid-valve controlled distributor injection pumps are modular in design. This section explains the interaction between the individual modular components. They are described in more detail later on. The modular components referred to are the following (Figs. 1 and 2):

- *The low-pressure stage* (7), consisting of the vane-type supply pump, pressure-control valve and overflow throttle valve
- *The high-pressure stage* (8)
- *The delivery valves* (11)
- *The high-pressure solenoid valve* (10)
- *The timing device* (9) with timing-device solenoid valve and angle-of-rotation sensor and
- *The pump ECU* (4)

The combination of these modular components in a compact unit allows the interaction of individual functional units to be very precisely coordinated. This allows compli-

ance with strict tolerances and fully meets demands regarding performance characteristics.

The high pressures and the associated physical stresses require a wide variety of fine-tuning adjustments in the design of the components. Lubrication is carried out by the fuel itself. In some cases, lubricant-film thickness is less than 0.1 μm and is much less than the minimum achievable surface roughness. For this reason, the use of special materials and manufacturing methods is necessary in addition to special design features.

Fuel supply and delivery

Low-pressure stage

The vane-type supply pump in the low-pressure stage (7) pumps fuel from the fuel tank and produces a pressure of 8...22 bar inside the fuel-injection pump depending on pump type and speed.

The vane-type supply pump delivers more fuel than is required for fuel injection. The excess fuel flows back to the fuel tank.

Fig. 1

1　Fuel supply system (low-pressure stage)
2　Type MSG engine ECU
3　Accelerator pedal
4　Type PSG pump ECU
5　Nozzle-holder assembly
6　Pump drive shaft
7　Low-pressure stage (vane-type supply pump with pressure-control valve and overflow throttle valve)
8　High-pressure stage (axial or radial-piston high-pressure pump with fuel rail)
9　Timing device with timing-device solenoid valve and angle-of-rotation sensor
10　High-pressure solenoid valve
11　Delivery valve
12　Diesel engine

1　Assemblies and their functions (schematic diagram)

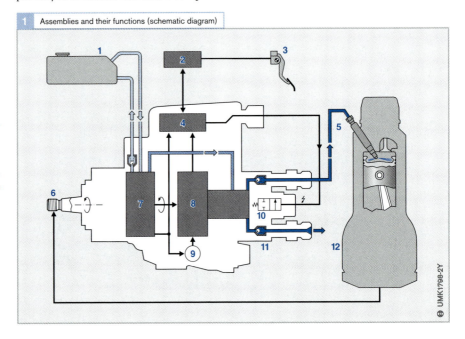

High-pressure stage

The high-pressure pump (8) generates the high pressure required for fuel injection and controls the injected-fuel quantity as required. At the same time the fuel rail opens the outlet for the appropriate engine cylinder so that the fuel is delivered via the delivery valve (11) to the nozzle-and-holder assembly (5). The assembly then injects the fuel into the combustion chamber (12) of the engine. It is in the design of the high-pressure pump and fuel distributor that the axial-piston and radial-piston distributor pumps differ the most.

The high-pressure pump is able to deliver fuel as long the high-pressure solenoid valve (10) keeps the pump's plunger chamber sealed. In other words, it is the high-pressure solenoid valve that determines the delivery period and therefore, in conjunction with pump speed, the injected-fuel quantity and injection duration.

The injection pressure increases during the period of fuel injection. The maximum pressure depends on the pump speed and the injection duration.

Timing

As with port-controlled distributor injection pumps, the timing device (9) varies the start of injection. It alters the cam position in the high-pressure pump.

An injection-timing solenoid valve controls the supply-pump pressure acting on the spring-loaded timing-device piston. This controls the start of injection independently of engine speed.

The integrated angle-of-rotation sensor detect and, as necessary, correct the pump speed and, together with the speed sensor on the engine crankshaft, the position of the timing device.

Electronic control unit

The pump ECU (4) calculates the triggering signals for the high-pressure solenoid valve and the timing-device solenoid valve on the basis of a stored map.

Fig. 2
For the sake of clarity, various components are shown end-on rather than side-on.
The index figures are the same as for Fig. 1.

2 Engine ECU
4 Pump ECU
5 Nozzle-holder assembly
6 Pump drive shaft
7 Low-pressure stage (vane-type supply pump with pressure-control valve and overflow throttle valve)
8 High-pressure pump with fuel rail
9 Timing device with timing-device solenoid valve and angle-of-rotation sensor
10 High-pressure solenoid valve
11 Delivery valve

2 Assemblies and their functions (schematic section diagram of a radial-piston distributor injection pump)

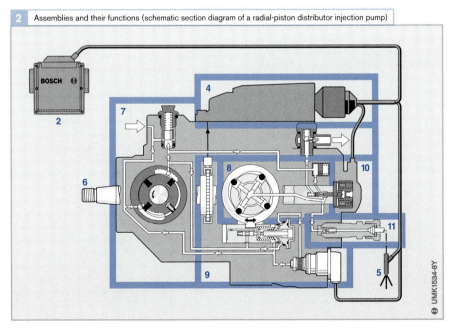

UMK1534-8Y

Low-pressure stage

The low-pressure stage delivers sufficient fuel for the high-pressure stage and generates the pressure for the high-pressure pump and the timing device (8...25 bar depending on pump type). Its basic components include the vane-type supply pump, the pressure-control valve and the overflow valve.

Vane-type supply pump

The purpose of the vane-type supply pump (Fig. 1) is to draw in a sufficient quantity of fuel and to generate the required internal pressure.

The vane-type supply pump is positioned around the drive shaft (4) in the distributor injection pump. Between the inner surface of the pump housing and a support ring acting as the end plate is the retaining ring (2) which forms the inner surface of the vane-pump stator. On the inner surface of the pump housing, there are two machined recesses which form the pump inlet (5) and outlet (6). Inside the retaining ring is the impeller (3) which is driven by an interlocking gear (Type VP44) or a Woodruff key (VP29/30) on the drive shaft. Guide slots

in the impeller hold the vanes (8) which are forced outwards against the inside of the retaining ring by centrifugal force. Due to the higher delivery pressures involved in radial-piston distributor injection pumps, the vanes have integrated springs which also help to force the vanes outwards. The compression-chamber "cells" (7) are formed by the following components:
- The inner surface of the pump housing ("base")
- The support ring ("cap")
- The shaped inner surface of the retaining ring
- The outer surface of the impeller and
- Two adjacent vanes

The fuel that enters through the inlet passage in the pump housing and the internal passages in the compression-chamber cell is conveyed by the rotation of the impeller to the compression-chamber outlet. Due to the eccentricity (VP29/30) or profile (VP44) of the inner surface of the retaining ring, the cell volume reduces as the impeller rotates. This reduction in volume causes the fuel pressure to rise sharply until the fuel escapes through the compression-chamber outlet – in other words, the fuel is compressed. From

Fig. 1
1   Pressure-control
    valve
2   Eccentric retaining
    ring
3   Impeller
4   Pump drive shaft
5   Fuel inlet
6   Fuel outlet to pump
    intake chamber
7   Compression-
    chamber cell
8   Vane

Fig. 2
1   Valve body
2   Compression spring
3   Valve plunger
4   Outlet to pump
    intake
5   Inlet from pump
    outlet
6   Bore

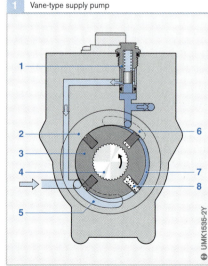

1   Vane-type supply pump

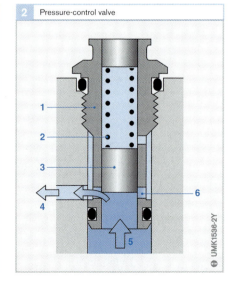

2   Pressure-control valve

the compression-chamber outlet, the various components are supplied with pressurized fuel via internal passages in the pump housing.

The pressure level required in a radial-piston distributor injection pump is relatively high compared with other types of distributor injection pump. Due to high pressure, the vanes have a bore in the center of the end face so that only one of the end-face edges is in contact with the inner surface of the retaining ring at one time. This prevents the entire end face of the vane from being subjected to pressure, which would result in an undesirable radial movement. At the point of changeover from one edge to the other (e.g. when changing over from inlet to outlet), the pressure acting on the end face of the vane can transfer to the other side of the vane through the bore. The opposing pressures balance each other out to a large extent and the vanes are pressed against the inner surface of the retaining ring by centrifugal force and the action of the springs as described above.

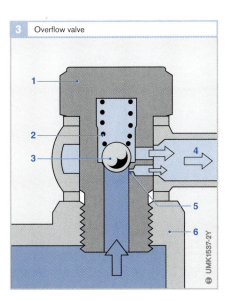

Pressure-control valve

The fuel pressure created at the pressure outlet of the vane-type supply pump depends on the pump speed. To prevent the pressure from reaching undesirably high levels at high pump speeds, there is a pressure-control valve in the immediate vicinity of the vane-type supply pump which is connected to the pressure outlet by a bore (Fig. 1, Pos. 1). This spring-loaded slide valve varies the delivery pressure of the vane-type supply pump according to the fuel quantity delivered. If the fuel pressure rises beyond a certain level, the valve plunger (Fig. 2, Pos. 3) opens radially positioned bores (6) through which the fuel can flow via an outlet (4) back to the intake port of the vane-type supply pump. If the fuel pressure is too low, the pressure-control valve remains closed and the entire fuel quantity is pumped into the distributor-pump intake chamber. The adjustable tension on the compression spring (2) determines the valve opening pressure.

Overflow valve

In order to vent and, in particular, cool the distributor injection pump, excess fuel flows back to the fuel tank through the overflow valve (Fig. 3) screwed to the pump housing.

The overflow valve is connected to the overflow valve (4). Inside the valve body (1), there is a spring-loaded ball valve (3) which allows fuel to escape when the pressure exceeds a preset opening pressure.

In the overflow channel to the ball valve, there is a bore that is connected to the pump overflow via a very small throttle bore (5). Since the overflow valve is mounted on top of the pump housing, the throttle bore facilitates automatic venting of the fuel-injection pump.

The entire low-pressure stage of the fuel-injection pump is precisely coordinated to allow a defined quantity of fuel to escape through the overflow valve and return to the fuel tank.

Fig. 3
1   Valve body
2   Compression spring
3   Valve ball
4   Fuel overflow
5   Throttle bore
6   Pump housing

High-pressure stage of the axial-piston distributor injection pump

Solenoid-valve controlled (Fig. 1) and port-controlled distributor injection pumps have essentially the same dimensions, fitting requirements and drive system including cam drive.

Design and method of operation

A clutch unit transmits the rotation of the drive shaft (Fig. 2 overleaf, Pos. 1) to the cam plate (5). The claws on the drive shaft and the cam plate engage in the yoke (3) positioned between them.

The cam plate converts the purely rotational movement of the drive shaft into a combined rotating-reciprocating movement. This is achieved by the fact that the cams on the cam plate rotate over rollers held in the roller ring (2). The roller ring is mounted inside the pump housing but has no connection to the drive shaft.

The cam plate is rigidly connected to the distributor plunger (8). Consequently, the distributor plunger also performs the rotating-reciprocating movement described by the cam plate. The reciprocating movement of the distributor plunger is aligned axially with the drive shaft (hence the name axial-piston pump).

The movement of the plunger back to the roller ring takes place the symmetrically arranged plunger return springs (7). They are braced against the distributor body (9) and act against the distributor plunger by means of a thrust plate (6). The piston return springs also prevent the cam plate from jumping away from the rollers in the roller ring when subjected to high acceleration forces.

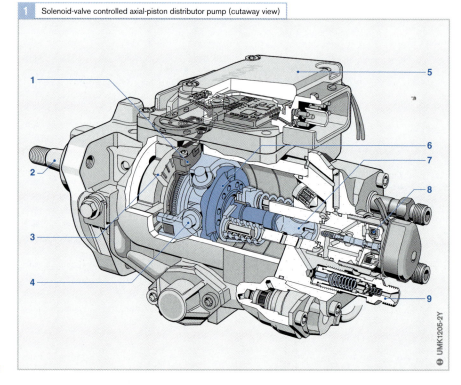

1　Solenoid-valve controlled axial-piston distributor pump (cutaway view)

UMK1205-2Y

Fig. 1
1　Angle-of-rotation sensor
2　Pump drive shaft
3　Support ring of vane-type supply pump
4　Roller ring
5　Pump control unit
6　Cam plate
7　Distributor plunger
8　High-pressure solenoid valve
9　High-pressure outlet

The lengths of the plunger return springs are precisely matched to one another in order to prevent lateral forces from acting on the distributor plunger.

Although the distributor plunger moves horizontally, the limits of its travel are still referred to as top dead center (TDC) and bottom dead center (BDC). The length of the plunger stroke between bottom and top dead center is application-specific. It can be up to 3.5 mm.

The number of cams and rollers is determined by the number of cylinders in the engine. The cam shape affects injection pressure (injection pattern and maximum injection pressure) and the maximum possible injection duration. The factors determining this connection are cam lift and the speed of movement.

When designing the fuel-injection pump, the injection parameters must be individually adapted to suit the design of the combustion chamber and the nature of the combustion process (DI or IDI) employed by the engine on which the pump is to be used. For this reason, a specific cam profile is calculated for each type of engine and is then machined on the end face of the cam plate. The cam plate produced in this way is an application-specific component of the distributor injection pump. Cam plates are not interchangeable between different types of pump.

2   High-pressure stage inside a solenoid-valve controlled axial-piston distributor injection pump

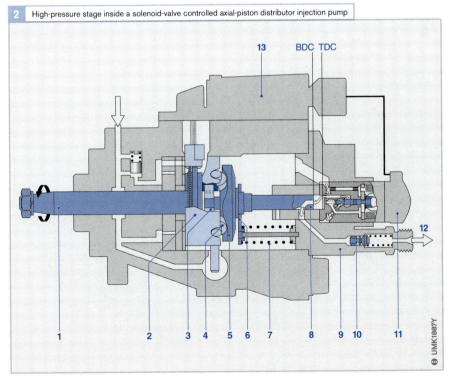

Fig. 2
1 Drive shaft
2 Roller ring
3 Yoke
4 Roller
5 Cam plate
6 Spring plate
7 Piston return spring
   (only one shown)
8 Distributor plunger
9 Distributor body
   (also called
   distributor head
   or distributor-head
   flange)
10 Delivery valve
11 High-pressure
   solenoid valve
12 Outlet to high-
   pressure delivery
   line
13 Pump ECU

TDC Distributor-plunger
    top dead center
BDC Distributor-plunger
    bottom dead
    center

Delivery phases

Induction (Fig. 4a)

As the distributor plunger (4) moves towards bottom dead center (BDC), it draws fuel into the plunger chamber (6) through the fuel inlet (1) in the distributor body (3) and the open high-pressure solenoid valve (7).

Effective stroke (Fig. 4b)

At bottom dead center, before the cam lobes are in contact with the rollers, the pump ECU sends a control signal to the high-pressure solenoid valve. The valve needle (9) is pressed against the valve seat (7). The high-pressure solenoid valve is then closed.

When the distributor plunger then starts to move towards top dead center (TDC), the fuel cannot escape. It passes through channels and passages in the distributor plunger to the high-pressure outlet (10) for the appropriate cylinder. The rapidly developed high pressure opens the orifice check valve (DI) or delivery valve (IDI), as the case may be, and forces fuel along the high-pressure fuel line to the nozzle integrated in the nozzle holder (start of delivery). The maximum injection pressure at the nozzle is around 1,400 bar.

The rotation of the distributor plunger directs the fuel to the next outlet on the next effective stroke.

The rapid release of pressure when the high-pressure solenoid valve opens can cause the space between the delivery valve and the distributor plunger to be over-depressurized. The filler channel (5) simultaneously fills the space for the outlet opposite to the outlet which is currently being supplied by the delivery stroke.

Residual stroke (Figure 4c).

Once the desired injected-fuel quantity has been delivered, the ECU cuts off the power supply to the solenoid coil. The high-pressure solenoid valve opens again and pressure in the high-pressure stage collapses (end of delivery). As the pressure drops, the nozzle and the valve in the high-pressure outlet close again and the injection sequence comes to an end. The point of closure and the open duration of the high-pressure solenoid valve, the cam pitch during the delivery stroke and the pump speed determine the injected-fuel quantity.

The remaining travel of the pump plunger to top dead center forces the fuel out of the plunger chamber and back into the pump intake chamber.

As there are no other inlets, failure of the high-pressure solenoid valve prevents fuel injection altogether. If the valve remains open, the required high pressure cannot be generated. If it remains closed, no fuel can enter the plunger chamber. This prevents uncontrolled over-revving of the engine and no other shutoff devices are required.

Fig. 3
The index figures are the same as for Fig. 4.

2 Filter
4 Distributor plunger
6 Plunger chamber
7 Valve seat
9 Solenoid-valve needle
10 Outlet to high-pressure delivery line

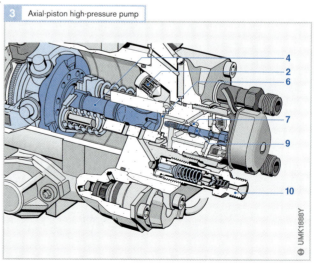

3   Axial-piston high-pressure pump

UMK1888Y

4 Delivery phases of solenoid-valve controlled axial-piston distributor injection pumps

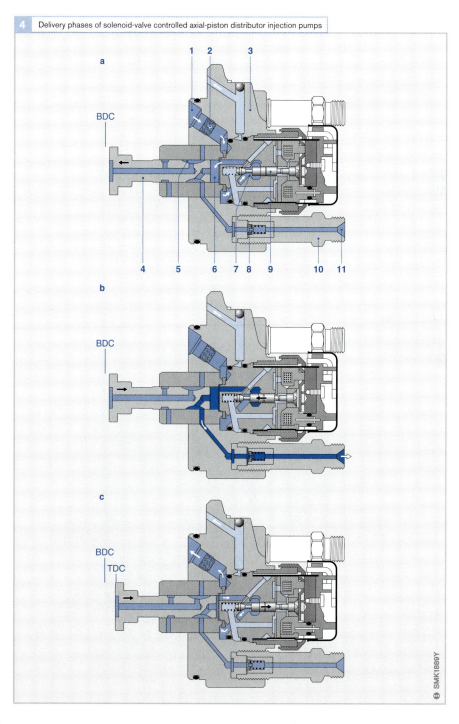

Fig. 4
a Induction
b Effective stroke
c Residual stroke

1 Inlet passage
   (fuel inlet)
2 Filter
3 Distributor body
4 Distributor plunger
5 Filler groove
6 Plunger chamber
7 Valve seat
8 Delivery valve
9 Solenoid-valve
   needle
10 High-pressure outlet
11 Outlet to high-
   pressure delivery
   line

TDC Pump-plunger
    top dead center
BDC Pump-plunger
    bottom dead
    center

High-pressure stage of the radial-piston distributor injection pump

Radial-piston high-pressure pumps (Fig. 1) produce higher injection pressures than axial-piston high-pressure pumps. Consequently, they also require more power to drive them (as much as 3.5...4.5 kW compared with 3 kW for axial-piston pumps).

Design

The radial-piston high-pressure pump (Fig. 2 overleaf) is driven directly by the distributor-pump drive shaft. The main pump components are
- The cam ring (1)
- The roller supports (4) and rollers (2)
- The delivery plungers (5)
- The drive plate and
- The front section (head) of the distributor shaft (6)

The drive shaft drives the drive plate by means of radially positioned guide slots. The guide slots simultaneously act as the locating slots for the roller supports. The roller supports and the rollers held by them run around the inner cam profile of the cam ring that surrounds the drive shaft. The number of cams corresponds to the number of cylinders in the engine.

The drive plate drives the distributor shaft. The head of the distributor shaft holds the delivery plungers which are aligned radially to the drive-shaft axis (hence the name "radial-piston high-pressure pump").

The delivery plungers rest against the roller supports. As the roller supports are forced outwards by centrifugal force, the delivery plungers follow the profile of the cam ring and describe a cyclical-reciprocating motion (plunger lift 3.5...4.15 mm).

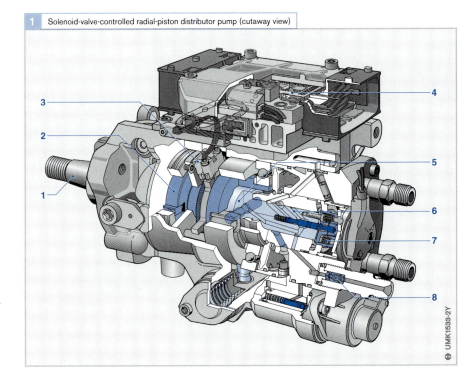

1 Solenoid-valve-controlled radial-piston distributor pump (cutaway view)

3
2
1
4
5
6
7
8

UMK1533-2Y

Fig. 1
1 Pump drive shaft
2 Vane-type supply pump
3 Angle-of-rotation sensor
4 Pump ECU
5 Radial-piston high-pressure pump
6 Distributor shaft
7 High-pressure solenoid valve
8 Delivery valve

When the delivery plungers are pushed in-wards by the cams, the volume in the central plunger chamber between the delivery plungers is reduced. This compresses and pumps the fuel. Pressures of up to 1,200 bar are achievable at the pump.

Through passages in the distributor shaft, the fuel is directed at defined times to the appropriate outlet delivery valves (Fig. 1, Pos. 8 and Fig. 3, Pos. 5).

There may be 2, 3 or 4 delivery plungers de-pending on the number of cylinders in the engine and the type of application (Fig. 2). Sharing the delivery work between at least two plungers reduces the forces acting on the mechanical components and permits the use of steep cam profiles with good delivery rates. As a result, the radial-piston pump achieves a high level of hydraulic efficiency.

The direct transmission of force within the cam-ring drive gear minimizes the amount of "give", which also improves the hydraulic performance of the pump.

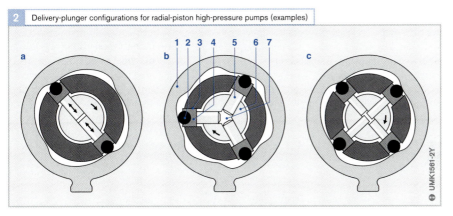

2 Delivery-plunger configurations for radial-piston high-pressure pumps (examples)

a   b   1 2 3 4 5 6 7   c

UMK1561-2Y

Fig. 2
a   For 4- or 6-cylinder engines
b   For 6-cylinder engines
c   For 4-cylinder engines

1   Cam ring
2   Roller
3   Guide slot in drive shaft
4   Roller support
5   Delivery plunger
6   Distributor shaft
7   Plunger chamber

3 High-pressure stage within a solenoid-valve controlled radial-piston distributor injection pump

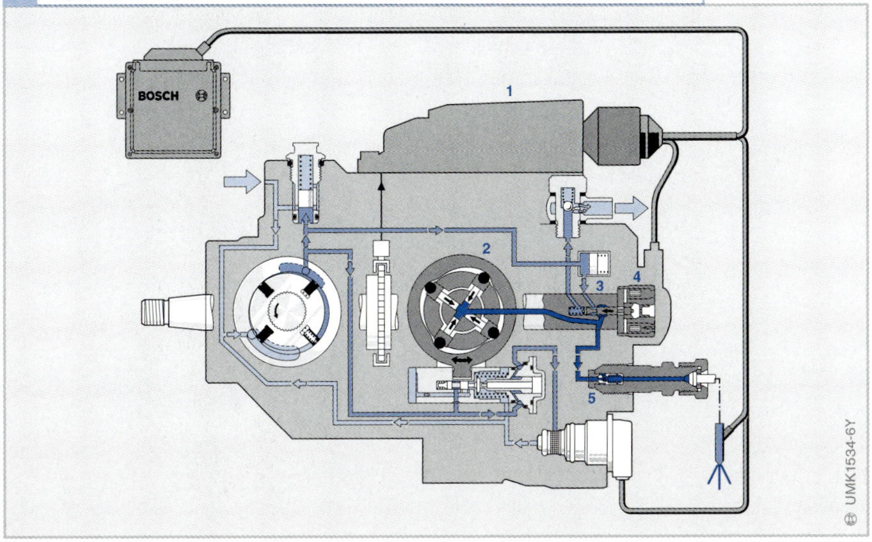

BOSCH

1

2

3

4

5

UMK1534-6Y

Fig. 3
For the sake of clarity, various components are shown end-on rather than side-on.

1   Pump ECU
2   Radial-piston high-pressure pump (end-on view)
3   Distributor shaft
4   High-pressure solenoid valve
5   Delivery valve

Distributor-body assembly

The distributor-body assembly (Fig. 4) consists of the following:
- The distributor body (2)
- The control sleeve (5) which is shrink-fitted in the distributor body
- The rear section of the distributor shaft (4) which runs in the control sleeve
- The valve needle (6) of the high-pressure solenoid valve
- The accumulator diaphragm (1) and
- The delivery valve (7) with orifice check valve

In contrast with the axial-piston distributor pump, the intake chamber that is pressurized with the delivery pressure of the vane-type supply pump consists only of the diaphragm chamber enclosed by an accumulator diaphragm (1). This produces higher pressures for supplying the high-pressure pump.

Delivery phases (method of operation)

Induction

During the induction phase (Fig. 5a), the delivery plungers (1) are forced outwards by the supply-pump pressure and centrifugal force. The high-pressure solenoid valve is open. Fuel flows from the diaphragm chamber (12) past the solenoid-valve needle (4) and through the low-pressure inlet (13) and the annular groove (10) to the plunger chamber (8). Excess fuel escapes via the return passage (5).

Effective stroke

The high-pressure solenoid valve (Fig. 5b, Pos. 7) is closed by a control pulse from the pump ECU when the cam profile is at bottom dead center. The plunger chamber is now sealed and fuel delivery starts as soon as the cams start to move the pistons inwards (start of delivery).

Residual stroke

Once the desired injected-fuel quantity has been delivered, the ECU cuts off power supply to the solenoid coil. The high-pressure

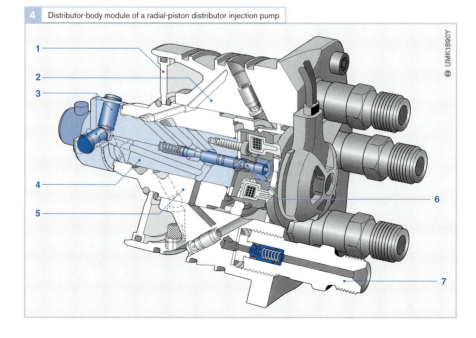

4   Distributor-body module of a radial-piston distributor injection pump

UMK1890Y

Fig. 4

1   Accumulator diaphragm
2   Distributor body
3   Delivery plunger
4   Distributor shaft
5   Control sleeve
6   Valve needle
7   Delivery valve

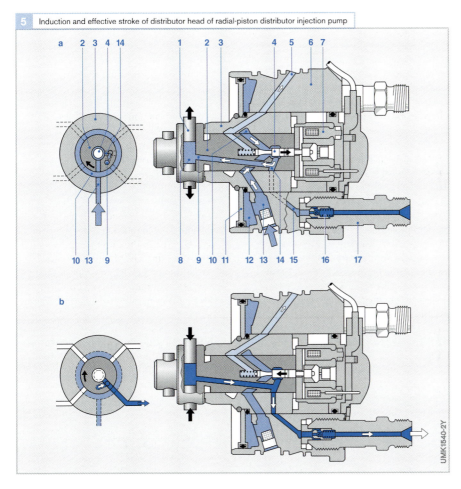

5 Induction and effective stroke of distributor head of radial-piston distributor injection pump

Fig. 5
For the sake of clarity, the pump plungers are shown end-on rather than side-on.

a Induction
b Effective stroke

1 Delivery plunger
2 Distributor shaft
3 Control sleeve
4 Valve needle
5 Fuel return
6 Distributor body
7 Solenoid coil
8 Plunger chamber
9 High-pressure fuel
10 Annular groove
11 Accumulator diaphragm
12 Diaphragm chamber
13 Low-pressure inlet
14 Distributor slot
15 High-pressure outlet
16 Orifice check valve
17 Delivery-valve body

UMK1540-2Y

solenoid valve opens again and pressure in the high-pressure stage collapses (end of delivery). The pressure drop closes the nozzle and the delivery valve again and the injection sequence comes to an end.

The excess fuel that is delivered by the pump while the pistons continue to move toward the cam top dead center is diverted back to the diaphragm chamber (12). The high pressure peaks that are thus produced in the low-pressure stage are damped by the accumulator diaphragm (11). In addition, the fuel stored in the diaphragm chamber helps to fill the plunger chamber for the next injection cycle.

Fuel metering takes place between the start of the cam lift and opening of the high-pressure solenoid valve. This phase is referred to as the delivery period. It determines the injected-fuel quantity in conjunction with the pump speed.

The high-pressure solenoid valve can completely shut off high-pressure fuel delivery in order to stop the engine. For this reason, an additional shutoff valve as used with port-controlled distributor injection pumps is not necessary.

Delivery valves

Between injection cycles, the delivery valve shuts off the high-pressure delivery line from the pump. This isolates the high-pressure delivery line from the outlet port in the distributor head. The residual pressure retained in the high-pressure delivery line ensures rapid and precise opening of the nozzle during the next injection cycle.

Integrated orifice check valve

The delivery valve with integral Type RSD orifice check valve (Figure 1) is a piston valve. At the start of the delivery sequence, the fuel pressure lifts the valve cone (3) away from the valve seat. The fuel then passes through the delivery-valve holder (5) to the high-pressure delivery line to the nozzle-and-holder assembly. At the end of the delivery sequence, the fuel pressure drops abruptly. The valve spring (4) and the pressure in the delivery line forces the valve cone back against the valve seat (1).

With the high pressures used for direct-injection engines, reflected pressure waves can occur at the end of the delivery lines. They can cause the nozzle to reopen, resulting in undesired dribble which adversely affects pollutant levels in the exhaust gas. In addition, areas of low pressure can be created and this can lead to cavitation and component damage.

A throttle bore (2) in the valve cone dampens the reflected pressure waves to a level at which they are no longer harmful. The throttle bore is designed so that the static pressure in the high-pressure delivery line is retained between injection cycles. Since the throttle bore means that the space between the fuel-injection pump and the nozzle is no longer hermetically sealed, this type of arrangement is referred to as an open system.

Separate Type RDV orifice check valve

On axial-piston pumps and some types of radial-piston pump, a delivery valve with a separate orifice check valve is used (Fig. 2). This valve is also referred to as a Type GDV constant-pressure valve. It creates dynamic pressure in the delivery line. During the injection sequence, the valve plate (5) opens so that the throttle bore (6) has no effect. When fuel flows in the opposite direction, the valve plate closes and the throttle comes into action.

Fig. 1
a   Valve closed
b   Valve open

1   Valve seat
2   Throttle bore
3   Valve cone
4   Valve spring
5   Delivery-valve holder

Fig. 2
1   Valve holder
2   Pressure-valve stem
3   Retraction piston
4   Valve spring
    (delivery valve)
5   Valve plate
6   Throttle bore
7   Valve spring
    (valve plate)
8   Delivery-valve holder

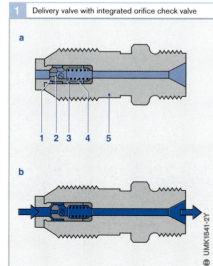

1   Delivery valve with integrated orifice check valve

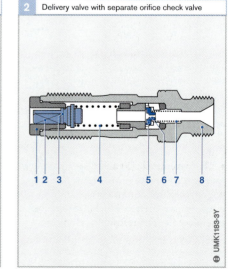

2   Delivery valve with separate orifice check valve

High-pressure solenoid valve

The high-pressure solenoid valve (Fig. 3) is a 2/2-way valve, i.e. it has two hydraulic ports and two possible settings. It is built into the distributor-body assembly. The valve needle (3) protrudes into the distributor shaft (5) and rotates in synchronization with it.

Arranging the solenoid coil (7) concentrically with the valve needle creates a compact combination of high-pressure solenoid valve and distributor body.

The high-pressure solenoid valve opens and closes in response to control signals from the pump ECU (valve-needle stroke 0.3...0.4 mm). The length of time it remains closed determines the delivery period of the high-pressure pump. This means that the fuel quantity can be very precisely metered for each individual cylinder.

The high-pressure solenoid valve has to meet the following criteria:
- Large valve cross-section for complete filling the plunger chamber, even at high speeds
- Light weight (small mass movements) in order to minimize component stress
- Rapid switching times for precise fuel metering and
- Generation of high magnetic forces in keeping with the loads encountered at high pressures

The high-pressure solenoid valve consists of:
- The valve body consisting of the housing (4), and attachments
- The valve needle (3) and solenoid armature (6)
- The solenoid plate and
- The electromagnet (7) with its electrical connection to the pump ECU (8)

The high-pressure solenoid valve is controlled by regulating the current. Steep current-signal edges must be used to achieve a high level of reproducibility on the part of the injected-fuel quantity. In addition, control must be designed in such a way as to minimize power loss in the ECU and the solenoid valve. This is achieved by keeping the control currents as low as possible, for example. Therefore, the solenoid valve reduces the current to the holding current (approx. 10 A) after the pickup current (approx. 18 A) has been applied.

The pump control unit can detect when the solenoid-valve needle meets the valve seat by means of the current pattern (BIP signal; Beginning of Injection Period). This allows the exact point at which fuel delivery starts to be calculated and the start of injection to be controlled very precisely.

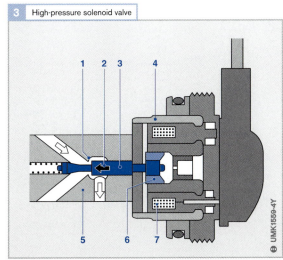

3 High-pressure solenoid valve

Fig. 3
1 Valve seat
2 Closing direction
3 Valve needle
4 Housing
5 Distributor shaft
6 Solenoid armature
7 Solenoid coil
8 Electrical connection

UMK1559-4Y

Injection timing adjustment

Purpose

The point at which combustion commences relative to the position of the piston/crankshaft has a decisive impact on engine performance, exhaust emission levels and noise output. If the start of delivery remains constant without adjusting the injection timing as engine speed increases, the amount of crankshaft rotation between the start of delivery and the start of combustion would increase to such an extent that combustion would no longer take place at the correct time.

As with port-controlled distributor injection pumps, the timing device rotates the roller ring/cam ring so that the start of delivery occurs earlier or later relative to the position of the engine crankshaft. The interaction between the high-pressure solenoid valve and the timing device thus varies the start of injection and the injection pattern to suit the operating status of the engine.

Explanation of terms

For a proper understanding of injection timing adjustment, a number of basic terms require explanation.

Injection lag

The start of delivery (SD, Fig. 1) occurs after the point at which the high-pressure solenoid valve closes. High pressure is generated inside the fuel line. The point at which that pressure reaches the nozzle opening pressure and opens the nozzle is the start of injection (SI). The time between the start of delivery and the start of injection is called the injection lag (IL).

The injection lag is largely independent of pump/engine speed. It is essentially determined by the propagation of the pressure wave along the high-pressure delivery line. The propagation time of the pressure wave is determined by the length of the delivery line and the speed of sound. In diesel fuel, the speed of sound is approx. 1,500 m/s.

If the engine speed increases, the amount of degrees of crankshaft rotation during the injection lag also increases. As a consequence, the nozzle opens later (relative to the position of the engine piston). This is undesirable. For this reason, the start of delivery must be advanced as engine speed increases.

Ignition lag

The diesel fuel requires a certain amount of time after the start of injection to form a combustible mixture with the air and to ignite. The length of time required from the start of injection to the combustion start (SC) is the ignition lag (IGL). This, too, is independent of engine speed and is affected by the following variables:

- The ignition quality of the diesel fuel (indicated by the cetane number)
- The compression ratio of the engine
- The temperature in the combustion chamber
- The degree of fuel atomization and
- The exhaust-gas recirculation rate

The ignition lag is in the range of 2...9° of crankshaft rotation.

Fig. 1

1 Combustion
   pressure
2 Compression
   pressure

SD **S**tart of **d**elivery
TDC Engine-piston
    top **d**ead **c**enter
SI **S**tart of **i**njection
EI **E**nd of **i**njection
IL **I**njection **l**ag
BDC Engine-piston
    bottom **d**ead
    center
SC **S**tart of **c**ombustion
EC **E**nd of **c**ombustion
IGL **I**gnition **l**ag

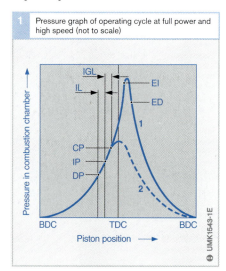

1 Pressure graph of operating cycle at full power and high speed (not to scale)

End of injection

When the high-pressure solenoid valve opens again, the high fuel pressure is released (end of injection, EI) and the nozzle closes. This is followed by the end of combustion (EC).

Other impacts

In order to limit pollutant emissions, the start of injection also has to be varied in response to engine load and temperature. The engine ECU calculates the required start of injection for each set of circumstances.

Design and method of operation

The timing device "advances" the position of the cams in the high-pressure pump relative to the diesel-engine crankshaft position as engine speed increases. This advances the start of injection. This compensates for the timing shift resulting from the injection lag and ignition lag. The impacts of engine load and temperature are also taken into account.

The injection timing adjustment is made up of the timing device itself, a timing-device solenoid valve and an angle-of-rotation sensor. Two types are used:

- The hydraulic timing device for axial-piston pumps and
- The hydraulically assisted timing device for radial-piston pumps

Hydraulic timing device

The hydraulic timing device (Fig. 2) is used on Type VP29 and VP30 axial-piston distributor injection pumps. Its design is the same as the version used on the electronically modulated Type VE..EDC port-controlled distributor injection pump.

The hydraulic timing device with its timing-device solenoid valve (5) and timing-device piston (3), which is located transversely to the pump axis, is positioned on the underside of the fuel-injection pump. The timing-device piston rotates the roller ring (1) according to load conditions and speed, so as to adjust the position of the rollers according to the required start of delivery.

As with the mechanical timing device, the pump intake-chamber pressure, which is proportional to pump speed, acts on the timing-device piston. That pressure on the

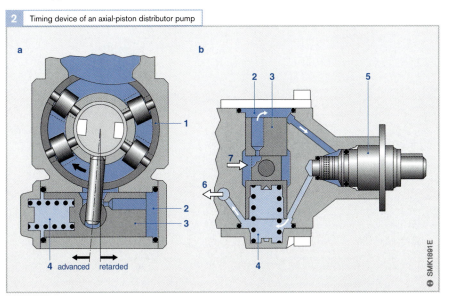

2   Timing device of an axial-piston distributor pump

a   Front view
b   Top view

SMK1891E

Fig. 2
a   Front view
b   Top view

1   Roller ring
2   Pump intake-chamber pressure
3   Timing-device piston
4   Pressure controlled by solenoid valve
5   Timing-device solenoid valve
6   Fuel return
7   Fuel inlet from pump intake chamber

4   advanced   retarded

pressure side of the timing device is regulated by the timing-device solenoid valve.

When the solenoid valve is open (reducing the pressure), the roller ring moves in the "retard" direction; when the valve is fully closed (increasing the pressure), it moves in the "advance" direction.

In between those two extremes, the solenoid valve can be "cycled", i.e. opened and closed in rapid succession by a pulse-width modulation signal (PWM signal) from the pump control unit. This is a signal with a constant voltage and frequency in which the ratio of "on" time to "off" time is varied. The ratio between the "on" time and the "off"

time determines the pressure acting on the adjuster piston so that it can be held in any position.

Hydraulically assisted timing device

The hydraulically assisted timing device is used for radial-piston distributor injection pumps. It can produce greater adjustment forces. This is necessary to securely brace the cam ring with the greater drive power of the radial-piston pump. This type of timing device responds very quickly and regardless of the friction acting on the cam ring and the adjuster piston.

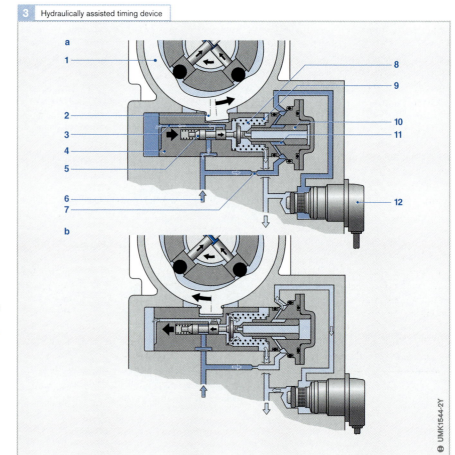

3   Hydraulically assisted timing device

Fig. 3
a Advance setting
b Retard setting

1 Cam ring
2 Ball pivot
3 Inlet channel/
   outlet channel
4 Timing-device piston
5 Control collar
6 Inlet from vane-type
   supply pump
7 Throttle
8 Control-plunger
   spring
9 Return spring
10 Control plunger
11 Annular chamber
   of hydraulic stop
12 Timing-device
   solenoid valve

UMK1544-2Y

The hydraulically assisted timing device, like the hydraulic timing device, is positioned on the underside of the injection pump (Fig. 3). It is also referred to by the Bosch type designation NLK.

The cam ring (1) engages in a cross-slot in the adjuster piston (4) by means of an adjuster lug (2) so that the axial movement of the adjuster piston causes the cam ring to rotate. In the center of the adjuster piston is a control sleeve (5) which opens and closes the control ports in the adjuster piston. In axial alignment with the adjuster piston is a spring-loaded hydraulic control piston (10) which defines the required position for the control sleeve.

At right-angles to the adjuster piston is the injection-timing solenoid valve (Pos. 12, shown schematically in Fig. 3 in the same plane as the timing device). Under the control of the pump control unit, the solenoid valve modulates the pressure acting on the control piston.

Advancing injection
When at rest, the adjuster piston (4) is held in the "retarded" position by a return spring (9). When in operation, the fuel supply pump pressure is regulated according to pump speed by means of the pressure control valve. That fuel pressure acts as the control pressure on the annular chamber (11) of the hydraulic stop via a restrictor bore (7) and when the solenoid valve (12) is closed moves the control piston (10) against the force of the control-piston spring (8) towards an "advanced" position (to the right in Fig. 3). As a result, the control sleeve (5) also moves in the "advance" direction so that the inlet channel (3) opens the way to the space behind the adjuster piston. Fuel can then flow through that channel and force the adjuster piston to the right in the "advance" direction.

The rotation of the cam ring relative to the resulting pump drive shaft causes the rollers to meet the cams sooner in the advanced position, thus advancing the start of injection. The degree of advance possible can be as much as 20° of camshaft rotation. On four-stroke engines, this corresponds to 40° of crankshaft rotation.

Retarding injection
The timing-device solenoid valve (12) opens when it receives the relevant PWM signal from the pump ECU. As a result, the control pressure in the annular chamber of the hydraulic stop (11) drops. The control plunger (10) moves in the "retard" direction (to the left in Figure 3) by the action of the control-plunger spring (8).

The timing-device piston (4) remains stationary to begin with. Only when the control collar (5) opens the control bore to the outlet channel can the fuel escape from the space behind the timing-device piston. The force of the return spring (9) and the reactive torque on the cam ring then force the timing-device piston back in the "retard" direction and to its initial position.

Regulating the control pressure
The timing-device solenoid valve acts as a variable throttle. It can vary continuously the control pressure so that the control plunger can assume any position between the fully advanced and fully retarded positions. The hydraulically assisted timing device is more precise in this regard than the straightforward hydraulic timing device.

If, for example, the control plunger is to move more in the "advance" direction, the on/off ratio of the PWM signal from the pump ECU is altered so that the valve closes more (low ratio of "on" time to "off" time). Less fuel escapes through the timing-device solenoid valve and the control plunger moves to a more "advanced" position.

Timing-device solenoid valve

The timing-device solenoid valve (Fig. 4) is the same in terms of fitting and method of control as the one used on the port-controlled type VE..EDC electronically regulated distributor injection pump. It is also referred to as the Type DMV10 diesel solenoid valve.

The pump ECU controls the solenoid coil (6) by means of a PWM signal. The solenoid armature (5) is drawn back against the force of the valve spring (7) while the solenoid valve is switched on. The valve needle (3) connected to the solenoid armature opens the valve. The longer the "on" times of the PWM signal are, the longer the solenoid valve remains open. The on/off ratio of the signal thus determines the flow rate through the valve.

In order to avoid problems caused by resonance effects, the otherwise fixed timing frequency of the PWM signal (Fig. 5) does not remain constant over the entire speed range. It changes over to a different frequency (30...70 Hz) in specific speed bands.

Incremental angle-time system with angle-of-rotation sensor

The closed-loop position control of the timing device uses as its input variables the signal from the crankshaft speed sensor and the pump's internal incremental angle-time system signal from the angle-of-rotation sensor.

The incremental angle-time system is located in the fuel-injection pump between the vane-type supply pump and the roller ring/cam ring (Fig. 6). Its purpose is to measure the angular position of the engine camshaft and the roller ring/cam ring relative to one another. This information is used to calculate the current timing device setting. In addition, the angle-of-rotation sensor (2) supplies an accurate speed signal.

Design and method of operation

Attached to the pump drive shaft is the increment ring (4). This has a fixed number of 120 teeth (i.e. one tooth for every 3°). In addition, there are reference tooth spaces (3) according to the number of cylinders in the engine. The increment ring is also called the angle-sensor ring or the sensor ring.

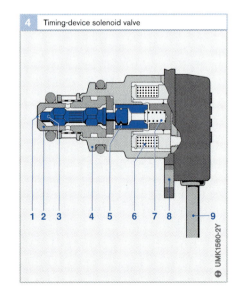

4   Timing-device solenoid valve

1 2 3   4   5   6   7 8   9

UMK1560-2Y

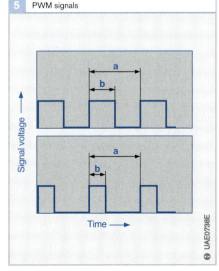

5   PWM signals

Signal voltage ⟶

Time ⟶

UAE0738E

Fig. 4
1   Throttle bore
2   Valve body
3   Valve needle
4   Valve housing
5   Solenoid armature
6   Solenoid coil
7   Valve spring
8   Mounting flange
9   Electrical
    connection

Fig. 5
a   Fixed timing
    frequency
b   Variable switch-on
    time

The speed sensor is connected to the bearing ring (5) of the timing device. As the increment ring rotates, the sensor produces an electrical signal by means of magnetically controllable semiconductor resistors and relative to the number of teeth that pass by the pickup. If the position of the timing device changes, the sensor moves along with the roller ring/cam ring. Consequently, the positions of the reference tooth spaces in the increment ring alter relative to the TDC signal from the crankshaft speed sensor.

The angular separation between the reference tooth spaces (or the synchronization signal produced by the tooth spaces) and the TDC signal is continuously detected by the pump ECU and compared with the stored reference figure. The difference between the two signals represents the actual position of the timing device.

If even greater accuracy in determining the start of injection is required, the start-of-delivery control system can be supplemented by a start-of-injection control system using a needle-motion sensor.

Start-of-injection control system

The start of delivery and start of injection are directly related to one another. This relationship is stored in the "wave-propagation time map" in the engine ECU. The engine ECU uses this data to calculate a start-of-injection setpoint according to the operating status of the engine (load, speed, temperature). It then sends the information to the pump ECU. The pump ECU calculates the necessary control signals for the high-pressure solenoid valve and the setpoint position for the timing-device piston.

The timing-device controller in the pump ECU continuously compares the actual position of the timing-device piston with the setpoint specified by the engine ECU and, if there is a difference, alters the on/off ratio of the signal which controls the timing-device solenoid valve. Information as to the actual start-of-injection setting is provided by the signal from an angle-of-rotation sensor or

alternatively, from a needle-motion sensor in the nozzle. This variable control method is referred to as "electronic" injection timing adjustment.

The benefits of a start-of-delivery control system are its rapid response characteristics, since all cylinders are taken into account. Another benefit is that it also functions when the engine is overrunning, i.e. when no fuel is injected. It means that the timing device can be preset for the next injection sequence.

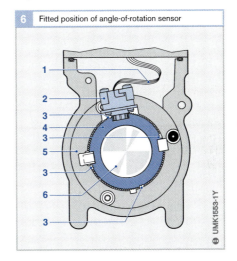

6   Fitted position of angle-of-rotation sensor

Fig. 6
1   Flexible conductive foil to ECU
2   Type DWS angle-of-rotation sensor
3   Tooth space
4   Increment ring
5   Bearing ring (connected to roller ring/cam ring)
6   Pump drive shaft

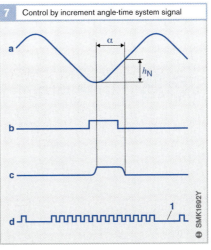

7   Control by increment angle-time system signal

Fig. 7
a   Cam pitch
b   Control pulse for high-pressure solenoid valve
c   Valve lift of high-pressure solenoid valve
d   Signal of angle-of-rotation sensor

1   Tooth space

h_N Effective stroke
α Delivery angle

Electronic control unit

Requirements

Screwed to the top of the fuel-injection pump and identifiable by its cooling fins is the pump ECU or combined pump/engine ECU. It is an LTCC (Low Temperature Cofired Ceramic) hybrid design. This gives it the capability to withstand temperatures of up to 125°C and vibration levels up to 100 g (acceleration due to gravity). These ECUs are available in 12-volt or 24-volt versions.

In addition to withstanding the external conditions in the engine compartment, the pump ECU has to perform the following tasks:
- Exchange data with the separate engine ECU via a serial bus system
- Analyze the increment angle-time system signals
- Control the high-pressure solenoid valve
- Control the timing-device solenoid valve and
- Detect fuel temperature with the aid of an integrated temperature sensor in order to take fuel density into account when calculating injected-fuel quantity

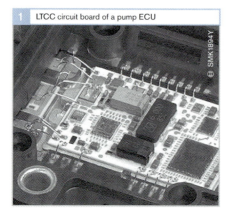

1   LTCC circuit board of a pump ECU

The control of fuel injection must operate very precisely so that the fuel-injection pump delivers precisely and consistently the required quantity of fuel at the required time in all engine operating conditions. Even minute discrepancies in start of injection and injection duration have a negative effect on the smooth running and noise levels in the diesel engine as well as its pollutant emissions.

The fuel-injection system also has to respond very quickly to changes. For this reason, the calculations taking place in the microcontroller and conversion of the control signals in the output stages are performed in real time (approx. 50 μs). In the case of a fuel-injection pump for a six-cylinder engine, the fuel-injection data is calculated up to 13,000 times a minute.

A pre-injection (PI) phase controlled by the high-pressure solenoid valve is also a viable option. This involves the injection of 1...2 mm³ of fuel before the main injection (MI) phase. This produces a more gradual increase in combustion pressure and therefore reduces combustion noise.

During the pre-injection phase, the fuel-quantity solenoid valve is operated ballistically, i.e. it is only partially opened. Consequently, it can be closed again more quickly. This keeps the injection gap as short as possible so that even at high speeds, there is sufficient cam lift remaining for the main injection phase. The entire injection sequence lasts approx. 1...2 ms.

The timing of the individual control phases is calculated by the microcontroller in the pump ECU. It also makes use of stored data maps. The maps contain settings for the specific vehicle application and certain engine characteristics along with data for checking the plausibility of the signals received. They also form the basis for determining various calculated variables.

In order to achieve specific and rapid opening of the solenoid valve, "fast extinction" of the energy stored in the solenoid valve takes place combined with a high extinction potential.

Two-ECU concept

Separate ECUs are used in diesel fuel-injection systems with solenoid-valve controlled axial-piston distributor injection pumps and first-generation radial-piston distributor injection pumps. These systems have a Type MSG engine ECU in the engine compartment and a Type PSG pump ECU mounted directly on the fuel-injection pump. There are two reasons for this division of functions: Firstly, it prevents the overheating of certain electronic components by removing them from the immediate vicinity of pump and engine. Secondly, it allows the use of short control leads for the high-pressure solenoid valve, so eliminating interference that may occur as a result of the very high currents (up to 20 A).

The pump ECU detects and analyzes the pump's internal sensor signals for angle of rotation and fuel temperature in order to adjust the start of injection. On the other hand, the engine ECU processes all engine and ambient data signals from external sensors and interfaces and uses them to calculate actuator adjustments on the fuel-injection pump.

The two control units communicate via a CAN interface.

Integrated engine and pump ECU on the fuel-injection pump

Increasing levels of integration using hybrid technology have made it possible to combine the engine-management control unit with the pump control unit on second-generation solenoid-valve controlled distributor injection pumps. The use of integrated ECUs allows a space-saving configuration. Other advantages are simpler installation and lower system costs due to fewer electrical interfaces.

The integrated engine/pump ECU is only used with radial-piston distributor injection pumps.

Summary

The overall system and the many modular assemblies are very similar on different types of solenoid-valve controlled distributor injection pump. Nevertheless, there are a number of differences. The main distinguishing features are detailed in Table 1.

1 Essential distinguishing features of solenoid-valve controlled distributor injection pumps			
Type	**VP29** (VE..MV)	**VP30** (VE..MV)	**VP44** (VRV)
Application	IDI engines	DI engines	DI engines
Maximum injection pressure at nozzle	800 bar	1,400 bar	1,950 bar
High-pressure pump	Axial-piston	Axial-piston	Radial-piston
Delivery valve	Separate orifice check valve	Integrated orifice check valve	Integrated orifice check valve
Timing device	Hydraulic	Hydraulic	Hydraulically assisted
Vane-type delivery pump	Circular retaining ring Vanes without springs	Circular retaining ring Vanes without springs	Profiled retaining ring Vanes with springs
Engine ECU and pump ECU	Separate	Separate	Integrated or separate

Table 1

Overview of discrete cylinder systems

Diesel engines with discrete cylinder systems have a separate fuel-injection pump for each cylinder of the engine. Such individual fuel-injection pumps are easily adaptable to particular engines. The short high-pressure fuel lines enable the achievement of particularly good injection characteristics and extremely high injection pressures.

Continually increasing demands have led to the development of a variety of diesel fuel-injection systems, each of which is suited to different requirements. Modern diesel engines must offer low emissions, good fuel economy, high torque and power output while also being quiet-running.

There are basically three types of discrete cylinder system: the Type PF port-controlled discrete fuel-injection pump, and the solenoid-valve controlled unit injector and unit pump systems. Those systems differ not only in their design but also in their performance data and areas of application (Figure 1).

Single-plunger fuel-injection pumps PF

Application
Type PF discrete injection pumps are particularly easy to maintain. They are used in the "off-highway" sector as
- Fuel-injection pumps for diesel engines with outputs of 4...75 kW/cylinder in small construction-industry machines, pumps, tractors and power generators and
- Fuel-injection pumps for large-scale engines with outputs of between 75 kW and 1,000 kW per cylinder. These versions are capable of working with high-viscosity diesel fuel and heavy oil

Design and method of operation
Type PF discrete fuel-injection pumps operate in the same way as Type PE in-line fuel-injection pumps. They have a single pump unit on which the injection quantity can be varied by means of a helix.

Each discrete fuel-injection pump is separately flanged-mounted to the engine and driven by the camshaft that controls the en-

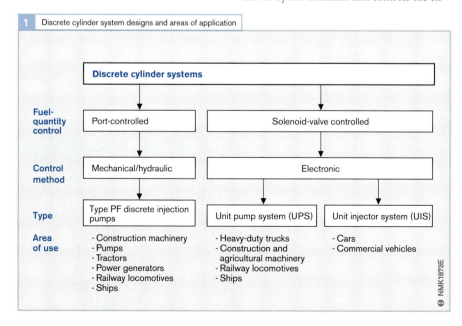

1 Discrete cylinder system designs and areas of application

	Discrete cylinder systems		
Fuel-quantity control	Port-controlled	Solenoid-valve controlled	
Control method	Mechanical/hydraulic	Electronic	
Type	Type PF discrete injection pumps	Unit pump system (UPS)	Unit injector system (UIS)
Area of use	- Construction machinery - Pumps - Tractors - Power generators - Railway locomotives - Ships	- Heavy-duty trucks - Construction and agricultural machinery - Railway locomotives - Ships	- Cars - Commercial vehicles

NMK1873E

gine valve timing. They can therefore be described as externally driven pumps. They may also be referred to as plug-in pumps.

Some of the smaller Type PF pumps come in 2, 3 and 4-cylinder versions. However, the majority of designs supply only a single cylinder and are therefore known as discrete or cylinder fuel-injection pumps.

Many discrete fuel-injection pumps have an integral roller tappet. In such cases they have the type designation PFR. With some designs for smaller engines, the roller tappet is mounted on the engine. Those versions have the type designation PFE.

Control

As with in-line fuel-injection pumps, a control rod incorporated in the engine acts on the fuel-injection pump units. A governor or control system moves the control rack, thereby varying the fuel delivery and injected-fuel quantity.

On large-scale engines, the governor is mounted directly on the engine block. Hydro-mechanical governors or electronic control systems may be used, or more rarely, purely mechanical governors.

Between the control rack for the discrete fuel-injection pumps and the actuating linkage from the governor, there is a sprung compensating link so that, in the event that the adjusting mechanism on one of the pumps jams, control of the other pumps is not compromised.

Fuel supply

Supply and filtering of the fuel and removal of air from the fuel-injection system is performed in the same way with Type PF dis-

crete fuel-injection pumps as for in-line fuel-injection pumps.

The fuel is fed to the individual fuel-injection pumps by a gear-type presupply pump. It delivers around 3...5 times as much fuel as the maximum full-load delivery of all individual fuel-injection pumps. The fuel pressure in this part of the system is around 3...10 bar.

The fuel is filtered by fine-pore filters with a pore size of 5...30 μm in order to keep suspended particles out of the fuel-injection system. Such particles would otherwise cause premature wear on the part of the high-precision fuel- injection components.

Heavy oil operation

Discrete fuel-injection pumps for engines with outputs of over 100 kW/cylinder are not only used to pump diesel fuel. They are also suitable for use with high-viscosity heavy oils with viscosities up to 700 mm²/s at 50 °C. In order to do so, the heavy oil has to be pre-heated to temperatures as high as 150 °C. This ensures that the required fuel-injection viscosity of 10...20 mm²/s is obtained.

| 2 | Examples of Type PF discrete fuel-injection pumps |

Fig. 2
a Type PFE 1 for small engines
b Type PFR 1 for small engines
c Type PFR 1 W for large-scale engines
d Type PF 1 D for large-scale engines

Unit injector system (UIS) and unit pump system (UPS)

The unit injector and unit pump fuel-injection systems achieve the highest injection pressures of all diesel fuel-injection systems currently available. They are capable of high-precision fuel injection that is infinitely variable in response to engine operating status. Diesel engines equipped with these systems produce low emission levels, are economical and quiet to run, and offer high performance and torque characteristics.

Areas of application

Unit injector system (UIS)

The unit injector system (UIS) went into volume production for commercial vehicles in 1994 and for cars in 1998. It is a fuel-injection system with timer-controlled discrete fuel-injection pumps for diesel engines with direct injection (DI). This system offers a significantly greater degree of adaptability to individual engine designs than conventional port-controlled systems. It can be used on a wide range of modern diesel engines for cars and commercial vehicles extending to

- *Cars* and *light commercials* with engines ranging from three-cylinder 1.2 *l* units producing 45 kW (61 bhp) of power and 195 Nm of torque to 10-cylinder, 5 *l* engines with power outputs of 230 kW (bhp) and torque levels of 750 Nm.
- *Heavy-duty trucks* developing up to 80 kW/cylinder.

As it requires no high-pressure fuel lines, the unit injector system has excellent hydraulic characteristics. That is the reason why this system is capable of producing the highest injection pressures (up to 2,050 bar). The unit injector system for cars also offers the option of pre-injection.

Unit pump system (UPS)

The unit pump system (UPS) is also referred to by the type designation PF..MV for large-scale engines.

Like the unit injector system, the unit pump system is a fuel-injection system with timer-controlled discrete fuel-injection pumps for direct injection (DI) diesel engines. There are three versions:

- The UPS12 for commercial-vehicle engines with up to 8 cylinders and power outputs of up to 35 kW/cylinder
- The UPS20 for heavy commercial-vehicle engines with up to 8 cylinders and power outputs of up to 80 kW/cylinder
- UPS for engines in construction and agricultural machinery, railway locomotives and ships with power outputs of up to 500 kW/cylinder and up to 20 cylinders

Design

System structure

The unit injector and unit pump systems are made up of four subsystems (Figure 1):

- The *fuel supply system* (low-pressure system) provides suitably filtered fuel at the correct pressure.
- The *high-pressure system* generates the necessary injection pressure and injects the fuel into the combustion chamber.
- The *EDC electronic control system* consisting of the sensors, control unit and actuators performs all diesel engine management and control functions as well as providing all electrical and electronic interfaces.
- The *air-intake and exhaust-gas-systems* handle the supply of air for combustion, exhaust-gas recirculation and exhaust-gas treatment.

The modular design of the individual subsystems allows the entire fuel-injection system to be easily adapted to individual engine designs.

Differences

The essential difference between the unit injector system and the unit pump system lies in the way in which high pressure is generated (Figure 2).

In the *unit injector system*, the high-pressure pump and the nozzle form a single unit – the "unit injector". There is a unit injector fitted in each cylinder of the engine. As there are no high-pressure fuel lines, extremely high injection pressures can be generated

and precisely controlled injection patterns can be produced.

With the *unit pump system*, the high-pressure pump – the "unit pump" – and the nozzle-and-holder assembly are separate units that are connected by a short length of high-pressure pipe. This arrangement has advantages in terms of use of space, pump-drive system, and servicing and maintenance.

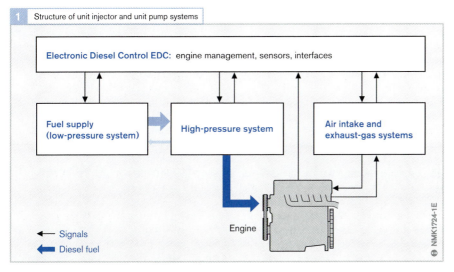

1 Structure of unit injector and unit pump systems

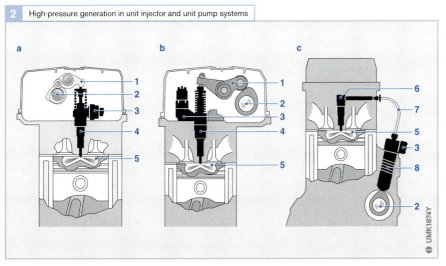

2 High-pressure generation in unit injector and unit pump systems

Fig. 2
a Unit injector system for cars
b Unit injector system for commercial vehicles
c Unit pump system for commercial vehicles

1 Rocker arm
2 Camshaft
3 High-pressure solenoid valve
4 Unit injector
5 Engine combustion chamber
6 Nozzle-and-holder assembly
7 Short high-pressure line
8 Unit pump

Method of operation

UIS and UPS are both diesel fuel-injection systems that use timer-controlled integral solenoid valves. The point at which the solenoid valve is actuated – and consequently at which the valve closes – determines the start of delivery. The length of time the valve remains closed is a measure of the injected-fuel quantity. The valve actuation point and closed period are determined by the electronic control unit on the basis of the programmed engine data maps. The calculation process also takes account of the current engine operating status and the ambient conditions. The input data includes the following:

- Accelerator pedal position
- Crankshaft angle of rotation
- Camshaft speed
- Charge-air pressure
- Temperature of intake air, engine coolant and fuel
- Vehicle road speed, etc.

These parameters are recorded by sensors and processed by the control unit. With this information, the control unit is able to control the vehicle, and in particular the operation of the engine, in such a way as to ensure optimum efficiency.

Generation of high pressure

The high-pressure pumps are driven directly by one of the engine's camshafts or by rocker arms running off the camshaft. While the solenoid valve is closed, the piston in the high-pressure pump generates pressure and the nozzle opens. When the solenoid valve opens, the pressure is dissipated and the nozzle closes again.

Control

Basic functions

The basic functions involve the precise control of injection timing, quantity and pressure. In this way, they ensure that the diesel engine has low-emission, low-consumption and smooth-running characteristics.

Additional functions

Additional control functions perform the tasks of reducing exhaust-gas emissions and fuel consumption or providing added safety and convenience. Some examples are:

- Exhaust-gas recirculation
- Charge-air pressure control
- Cruise control
- Electronic immobilizer

A diagnosis interface enables analysis of stored system data when the vehicle is serviced.

Control unit configuration

The Type MSG engine control unit is fitted inside the engine compartment (partially engine-mounted). For some commercial-vehicle applications, the heat given off by the electronic components has to be dissipated through an integral heat sink to the fuel (control unit cooler). In addition to the input circuitry and the microcontroller, the control unit also incorporates all output stages for controlling the solenoid valves.

Master-and-slave configuration

Present-day control units contain six output stages for the injectors. For engines with more than six cylinders, two engine control units are used. They are linked via a dedicated high-speed CAN interface in a master-and-slave configuration. As a result, there is also a higher microcontroller processing capacity available. Some functions are permanently allocated to a specific control unit (e.g. volume balancing). Other functions can be dynamically allocated to one or other of the control units as situations demand (e.g. recording of sensor signals).

Air-intake and exhaust-gas systems

Exhaust-gas recirculation for cars

Exhaust-gas recirculation is an effective method of reducing NO_X components in the exhaust gas. It involves the use of a valve which returns some of the exhaust gas to the intake manifold. If the recirculated exhaust gas is also cooled, further advantages can be gained. This method has been the state of the art for diesel cars for a number of years. The exhaust gas is recirculated at low engine loads and speeds.

Exhaust-gas recirculation for commercial vehicles

The vast majority of modern diesel engines are fitted with exhaust-gas turbochargers. Such engines do not generally have a negative pressure differential between the exhaust manifold upstream of the turbine and the inlet manifold downstream of the compressor at high engine loads. Since exhaust-gas recirculation and cooling cannot be dispensed with even at the higher end of the load curve on commercial-vehicle engines, additional features such as turbochargers with variable turbine geometry (VTG), wastegates or flutter valves are necessary.

Exhaust-gas treatment

In order to be able to comply with stricter emission-control legislation, exhaust-gas treatment will become increasingly important for diesel engines in the future despite advances in internal engine design.

This is particularly true for larger cars and commercial vehicles. There are many systems currently in the process of development. Which of them will eventually become established remains an unanswered question. The possibilities include:

- Diesel-oxidation catalytic converters
- Various particulate filters (PF)
- NO_X accumulator-type catalytic converters
- SCR (selective catalytic reduction) catalytic converters

In *combination systems* (also called four-way systems), several individual systems are combined. They can then reduce not only NO_X but also HC, CO and particulate emissions. Such systems demand very powerful engine management systems.

The most important emission control systems are dealt with in a separate chapter.

| 3 | Example of a unit injector for cars |

UMK1875Y

Fig. 3
1 Nozzle
2 High-pressure solenoid valve
3 Ball pin for driving pump plunger

System diagram of UIS for cars

Figure 1 shows all the components of a fully equipped unit injector system for an eight-cylinder diesel car engine. Depending on the type of vehicle and application, some of the components may not be used.

For the sake of clarity of the diagram, the sensors and desired-value generators (A) are not shown in their fitted positions. Exceptions to this are the components of the exhaust-gas treatment systems (F) as their proper fitted positions are necessary in order to understand the system.

The CAN bus in the interfaces section (B) enables exchange of data between a wide variety of systems and components including:

- The starter motor
- The alternator
- The electronic immobilizer
- The transmission control system
- The traction control system, TCS and
- The electronic stability program ESP

Even the instrument cluster (12) and the air-conditioning system (13) can be connected to the CAN bus.

For emission control, three alternative combination systems are shown (a, b and c).

Fig. 1

Engine, engine control unit and high-pressure fuel-injection components

24 Fuel rail
25 Camshaft
26 Unit injector
27 Glow plug
28 Diesel engine (DI)
29 Engine control unit (master)
30 Engine control unit (slave)
M Torque

A   Sensors and desired-value generators

1 Accelerator-pedal sensor
2 Clutch switch
3 Brake switches (2)
4 Operator unit for cruise control
5 Glow plug/starter switch ("ignition switch")
6 Vehicle-speed sensor
7 Crankshaft speed sensor (inductive)
8 Engine-temperature sensor (in coolant system)
9 Intake-air temperature sensor
10 Charge-air pressure sensor
11 Hot-film air-mass flow sensor (intake air)

B   Interfaces

12 Instrument cluster with signal output for fuel consumption, engine speed, etc.
13 Air-conditioning compressor with control
14 Diagnosis interface
15 Glow plug control unit
CAN **C**ontroller **A**rea **N**etwork (vehicle's serial data bus)

C   Fuel supply system (low-pressure system)

16 Fuel filter with overflow valve
17 Fuel tank with filter and electric presupply pump
18 Fuel level sensor
19 Fuel cooler
20 Pressure limiting valve

D   Additive system

21 Additive metering unit
22 Additive control unit
23 Additive tank

E   Air-intake system

31 Exhaust-gas recirculation cooler
32 Charge-air pressure actuator
33 Charge-air (in this case with variable turbine geometry)
34 Intake manifold flap
35 Exhaust-gas recirculation actuator
36 Vacuum pump

F   Emission control systems

37 Exhaust temperature sensor
38 Oxidation catalytic converter
39 Particulate filter
40 Differential-pressure sensor
41 Exhaust heater
42 NO_x sensor
43 Broadband oxygen sensor Type LSU
44 NO_x accumulator-type catalytic converter
45 Two-point oxygen sensor Type LSF
46 Catalyzed soot filter Type CSF

1   Diesel fuel-injection system for cars using unit injector system

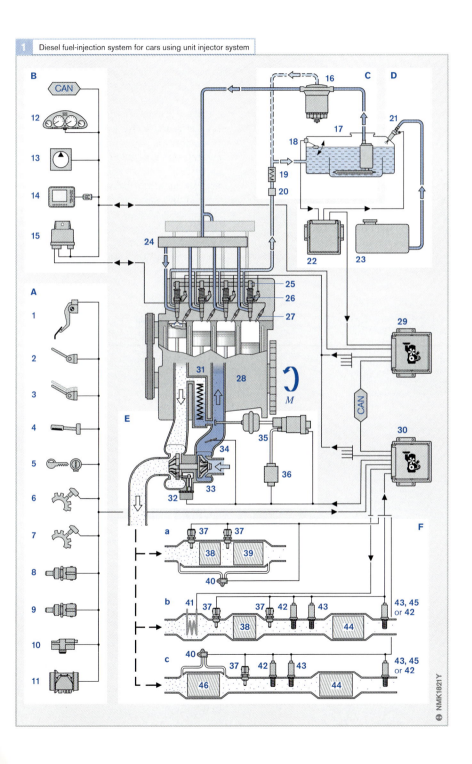

System diagram of UIS/UPS for commercial vehicles

Figure 2 shows all the components of a unit injector system for a six-cylinder diesel commercial-vehicle engine. Depending on the type of vehicle and application, some of the components may not be used.

The components of the electronic diesel control system EDC (sensors, interfaces and engine control unit), the fuel-supply system, air-intake system and exhaust-gas treatment are very similar in the unit injector and unit pump systems. They differ only in the high-pressure section of the overall system.

For the sake of clarity of the diagram, only those sensors and desired-value generators

whose true position is necessary in order to understand the system are shown in their fitted locations.

Data exchange with a wide range of other systems (e.g. transmission control system, traction control system TCS, electronic stability program ESP, oil quality sensor, tachograph, radar ranging sensor, vehicle management system, brake co-ordinator, fleet management system) involving up to 30 control units is possible via the CAN bus in the "Interfaces" section. Even the alternator (18) and the air-conditioning system (17) can be connected to the CAN bus.

For exhaust-gas treatment, three alternative combination systems are shown (a, b and c).

Fig. 2
Engine, engine control unit and high-pressure injection components
22 Unit pump and nozzle-and-holder assembly
23 Unit injector
24 Camshaft
25 Rocker arm
26 Engine control unit
27 Relay
28 Auxiliary equipment (e.g. retarder, exhaust flap for engine brake, starter motor, fan)
29 Diesel engine (DI)
30 Flame glow plug (alternatively grid heater)
M Torque

A Sensors and setpoint generators
1 Accelerator-pedal sensor
2 Clutch switch
3 Brake switches (2)
4 Engine brake switch
5 Parking brake switch
6 Control switch (e.g. cruise control, intermediate speed control, engine speed and torque reduction)
7 Starter switch ("ignition switch")
8 Charge-air speed sensor
9 Crankshaft speed sensor (inductive)
10 Camshaft speed sensor
11 Fuel temperature sensor
12 Engine-temperature sensor (in coolant system)
13 Charge-air temperature sensor
14 Charge-air pressure sensor
15 Fan speed sensor
16 Air-filter differential-pressure sensor

B Interfaces
17 Air-conditioning compressor with control
18 Alternator
19 Diagnosis interface
20 SCR control unit

21 Air compressor
CAN **C**ontroller **A**rea **N**etwork (vehicle's serial data bus) (up to three data busses)

C Fuel supply system (low-pressure system)
31 Fuel pump
32 Fuel filter with water-level and pressure sensors
33 Control unit cooler
34 Fuel tank with filter
35 Fuel level sensor
36 Pressure limiting valve

D Air intake system
37 Exhaust-gas recirculation cooler
38 Control flap
39 Exhaust-gas recirculation actuator with exhaust-gas recirculation valve and position sensor
40 Intercooler with bypass for cold starting
41 Turbocharger (in this case with VTG) with position sensor
42 Charge-air pressure actuator

E Emission control systems
43 Exhaust-gas temperature sensor
44 Oxidation-type catalytic converter
45 Differential-pressure sensor
46 Particulate filter
47 Soot sensor
48 Fluid level sensor
49 Reducing agent tank
50 Reducing agent pump
51 Reducing agent injector
52 NO_x sensor
53 SCR catalytic converter
54 NH_3 sensor
55 Blocking catalytic converter
56 Catalyzed soot filter Type CSF
57 Hydrolyzing catalytic converter

2   Diesel fuel-injection system for commercial vehicles using unit injector or unit pump system

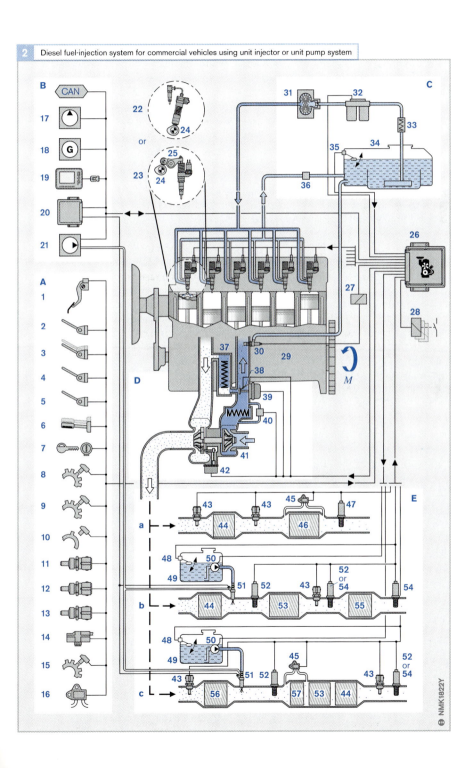

Single-plunger fuel-injection pumps PF

Single-plunger fuel-injection pumps PF are suitable for use on small, medium- and large-size engines. They are used for "off-highway" applications (agricultural and construction machinery, fixed installations such as pumps and power generators, railway locomotives, and small and large marine vessels). They are fitted to direct-injection and indirect-injection engines. On large-bore engines, they can also run on heavy oil. A single-plunger fuel-injection pump is fitted to each engine cylinder. Single-plunger fuel-injection pumps feature durability and ease of maintenance.

Design and method of operation

Single-plunger fuel-injection pumps PF operate in the same way as Type PE inline fuel-injection pumps. The layout of the timing edges on the pump plungers is identical on both types of pump. The delivery quantity is varied by a control rack and control sleeve which act as the pump plunger rotates.

In contrast with in-line fuel-injection pumps PE, however, single-plunger fuel-injection pumps PF are driven by the diesel engine's camshaft and not by a camshaft integrated in the pump housing. The designation code "PF" indicates that they are driven by an external camshaft.

This fuel-injection system has a particularly wide range of maximum injected-fuel quantities due to the wide variety of applications. Depending on version, it can range from 13,000 to 18,000 mm³ per stroke.

Single-plunger fuel-injection pumps are manufactured with aluminum or diecast housings. They are flanged to the engine. The standard design is a single-plunger variant, i.e. each engine cylinder is fitted with a fuel-injection pump. On multicylinder engines, this allows the use of very short fuel-injection tubing as each cylinder has a fuel-injection pump mounted close to it. The shorter the fuel-injection tubing, the better the fuel-injection system performance. Each type of diesel engine requires only one type of pump and fuel-injection tubing. This makes spare-parts inventory management much easier.

1 Technical data of Bosch single-plunger fuel-injection pumps PF

Type designation	Area of application	Max. output per engine cylinder (kW)	Max. injection pressure at nozzle (bar)	Max. plunger lift (mm)	Piston diameter (mm)
Up to 75 kW per cylinder (light, medium and heavy duty)					
PFE 1Q..	Light duty	10	500	7	5...7
PFE 1A..	Medium duty	20	800	9	5...9
PFR 1K..	Medium duty	20	600	8	5...9
PFM [1] 1P	Heavy duty	50	1,150	12	9...10
75 kW per cylinder and above (large-bore engines)					
PF..Z	Large-bore engines	150	1,200	12	10...14
PF(R)..C	Large-bore engines	300	1,500	24	15...23
PF(R)..W	Large-bore engines	400	1,500	26	20...24
PF(R)..D	Large-bore engines	600	1,500	34	22...34
PF..E	Large-bore engines	700	1,200	45	25...36
PF(R)..H	Large-bore engines	1,000	1,500	48	32...46

Table 1
[1] M stands for monoblock

In exceptional cases, some of the smaller Type PF pumps come in 2-, 3- and 4-cylinder versions.

Single-plunger fuel-injection pumps are fitted a fixing device which maintains the injected-fuel quantity setting at maximum while the pump is being shipped. This prevents any accidental change in the setting. It is simplifies adjustment work when the pump is fitted to the engine.

There are basically two different types of single-plunger fuel-injection pump:
- Type PFR pumps have an integrated roller tappet. The roller is in direct contact with the cam of the engine camshaft.
- On Type PF and PFE pumps, the roller tappet is an integral component of the engine, and not part of the pump.

Injection timing

The drive cams for single-plunger fuel-injection pumps are located on the same camshaft that drives the engine valve timing. Therefore, it is not possible to change the injection timing by rotating the common camshaft relative to its drive gear.

If an intermediate device is adjusted – for example, an eccentric oscillating crank between the cam and the roller tappet – it is possible to obtain an advance angle of a few degrees of rotation (Fig. 1). This helps to optimize fuel consumption or exhaust-gas emissions, or adapt to the different ignition qualities of various types of fuel.

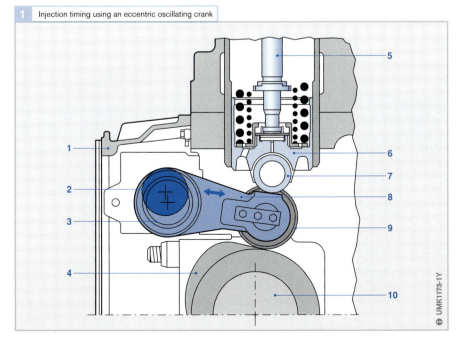

1   Injection timing using an eccentric oscillating crank

UMK1173-1Y

Fig. 1
1 Diesel engine
2 Profile of setting shaft
3 Oscillating crank bearing
4 Injection cam
5 Pump plunger
6 Roller tappet
7 Tappet roller
8 Oscillating crank
9 Camshaft roller
10 Engine camshaft

Sizes

Small fuel-injection pumps for power outputs of up to 50 kW/cylinder

These fuel-injection pumps are used for diesel engines on small construction machines, pumps, tractors and power generators.

Type *PFE 1A..* and *PFE 1Q..* pumps are single-plunger designs with no integrated roller tappet (Fig. 1). In these types of pump, the roller tappet runs in a guide bore inside the motor housing. With these type ranges, a control rack held in the motor housing engages in a control-sleeve lever of the control sleeve (6), which then rotates the pump plunger (5). This adjusts the delivery quantity.

The Type *PFR..K* pumps with integrated roller tappet (Fig. 2) are produced in 1, 2, 3, and 4-cylinder versions. They incorporate several plunger-and-barrel assemblies in a common pump housing, depending on the version. The plunger-and-barrel assemblies are driven by suitably positioned cams on the engine camshaft. With this type of pump, the delivery quantity is varied by means of a toothed control sleeve which engages in a control rack (5) guided inside the pump housing.

The maximum camshaft speed for driving small fuel-injection pumps is around 1,800 rpm. Depending on the plunger diameter, which can vary from 5 mm to 9 mm, the maximum WOT injected-fuel quantity may be as much as 95 mm^3 per stroke. The maximum peak injection pressure may be up to 600 bar on the pump side of the high-pressure fuel-injection tubing.

These types of pump are fitted as standard with constant-pressure valves with or without a return-flow restriction. Constant-pressure valves may be used if higher injection pressures or specific injected-fuel quantity stability are required.

Large fuel-injection pumps for power outputs greater than 50 kW/cylinder

These types of single-plunger fuel-injection pump are used for diesel engines with outputs of up to 1,000 kW per cylinder. They are capable of delivering high-viscosity diesel fu-

Fig. 1
1 Delivery valve
2 Delivery-valve holder
3 Pump housing
4 Pump barrel
5 Pump plunger
6 Control sleeve
7 Plunger return
   spring

Fig. 2
1 Delivery-valve holder
2 Delivery valve
3 Pump barrel
4 Pump plunger
5 Control rack
6 Control sleeve
7 Piston control arm
8 Roller tappet

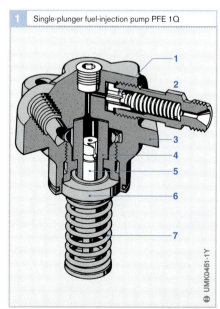

1   Single-plunger fuel-injection pump PFE 1Q

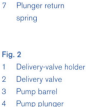

UMK0451-1Y

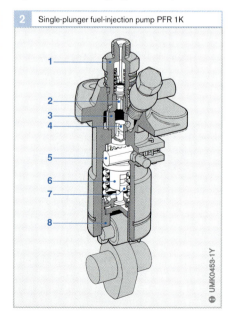

2   Single-plunger fuel-injection pump PFR 1K

UMK0453-1Y

els and heavy oil (Fig. 3). Pump barrels are designed with a plunger through-bore to achieve peak injection pressures at the pump of up to approx. 1,200 bar. For applications involving higher pressures, barrels are provided with a blind pocket to prevent the high fuel pressure from deforming the top of the plunger-and-barrel assembly (Fig. 4).

Pump plungers are symmetrical in design so that they can be guided centrally inside the pump barrels. Anti-erosion screws close to the spill ports in the pump barrel protect the pump housing from damage caused by the high-energy cutoff jet at the end of the spill.

The delivery valve is high-pressure-sealed against the pump barrel and the flange by lapped flat sealing faces. There is a rack-travel indicator attached to the control rack.

There may be several annular grooves machined into the pump barrel. They perform the following functions: The top groove closest to the pump interior acts as a leakage-return duct (5). It returns the leak fuel that escapes through the gap in the plunger-

and-barrel assembly back to the intake chamber via a hole in the pump barrel. Below that, there may be an oil-block groove. Blocking oil from the engine lube-oil circuit first passes through a fine filter and is then forced through a hole in this oil-block groove at a pressure of 3 to 5 bar. This pressure is higher than the pressure in the pump's fuel gallery at normal operating speeds. This prevents the engine lube oil from becoming diluted by the fuel. Between those two grooves, there may be another groove (13) to drain off mixed fuel and engine oil (emulsion). Emulsion goes into a separate collection tank.

For heavy-oil applications, the roller tappet on single-plunger fuel-injection pump PFR 1CY or, as the case may be, the guide bushing on Type PF pumps and the control sleeve, are lubricated by engine lube oil via a special connection.

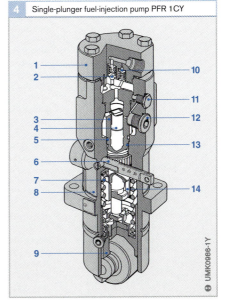

| 3 | Single-plunger fuel-injection pump PF 1D |

1
2
3

4

5

6

7

⊕ UMK0452-1Y

| 4 | Single-plunger fuel-injection pump PFR 1CY |

1
2

3
4
5

6

7
8

9

10

11

12

13

14

⊕ UMK0986-1Y

Fig. 3
1 Delivery valve
2 Vent screw
3 Pump barrel
4 Pump plunger
5 Control rack
6 Control sleeve
7 Guide bushing

Fig. 4
1 Flange
2 Forward-delivery valve
3 Pump barrel
4 Pump plunger
5 Leakage-return channel
6 Control rack
7 Pump spring
8 Pump housing
9 Roller tappet
10 Pressure-holding valve
11 Vent screw
12 Anti-erosion screw
13 Emulsion drain
14 Control sleeve

Unit Injector (UI)

The Unit Injector (UI) injects into the engine cylinders the exact amount of fuel at the correct pressure and precise instant in time as calculated by the ECU. This accuracy must be maintained in all operating ranges and throughout the engine's useful life. The UIS supersedes the nozzle-and-holder assembly of the conventional fuel-injection system. With the UIS though, high-pressure delivery lines have become redundant, a fact which has a positive effect upon the fuel-injection characteristics.

Installation and drive

Each cylinder has its own Unit Injector (UI) which is installed directly in the cylinder head (Fig. 1). The nozzle assembly (4) is integrated in the UI and projects into the combustion chamber (8). The engine camshaft (2) has an individual cam for each UI,

the particular cam pitch being transferred to the pump plunger (6) by a rocker (1) so that the plunger moves up and down under the combined action of rocker and plunger follower spring.

In addition to the electrical triggering, start of injection and injected fuel quantity are a function of instantaneous plunger velocity which itself is defined by the cam shape. This is one of the reasons for such high precision being required in camshaft manufacture. Torsional vibration is induced in the camshaft by the forces applied to it during operation, and adversely affects injected-fuel-quantity tolerance and injection characteristics. It is therefore imperative that in order to reduce these vibrations the individual-pump drives are designed to be as rigid as possible (this applies to the camshaft drive, the camshaft itself, the rocker, and the rocker bearings).

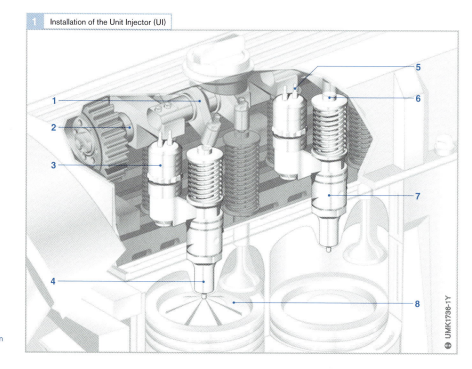

1   Installation of the Unit Injector (UI)

Fig. 1
1   Rocker
2   Engine camshaft
3   Solenoid valve
4   Nozzle assembly
5   Electrical
    connection
6   Pump plunger
7   Unit injector (UI)
8   Engine combustion
    chamber

UMK1736-1Y

Design and construction

The UI body assembly (Fig. 2, Pos. 4) serves as the pump barrel. It has an extension arm in which the high-pressure solenoid valve (1) is integrated. Passages in the injector's body provide the connections between high-pressure chamber (5) and solenoid valve/low-pressure stage, and between high-pressure chamber and nozzle assembly (6). The unit injector's shape is such that the UI can be fastened in the engine's cylinder head (3) by means of a special clamp (9). The follower spring (2) forces the pump plunger against the rocker (7) and the rocker against the actuating cam (8). This ensures that plunger, rocker, and actuating cam are always in mechanical contact during actual operation. As soon as injection has finished, the follower spring forces the plunger back to its initial position.

Figs. 3 and 4 on the following pages show details of the design and construction of unit injectors for passenger cars and commercial vehicles.

The unit injector is sub-divided into the following function units:

High-pressure generation
The major components involved in high-pressure generation are the pump body assembly, the pump plunger, and the follower spring (Figs. 3 and 4 on the next pages, Positions (4), (3), and (2)).

High-pressure solenoid valve
The high-pressure solenoid valve controls the start (instant) of injection and the duration of injection. Its major components are: Coil (10), solenoid-valve needle (8), armature (9), magnet core and solenoid-valve spring (26).

Nozzle assembly
The nozzle assembly (20) atomises the fuel and distributes it in the combustion chamber in precisely metered quantities. It shapes the rate-of-discharge curve. The nozzle assembly is attached to the unit-injector body assembly by the nozzle nut (19).

---

2 Installation of the Unit Injector in the engine's cylinder head

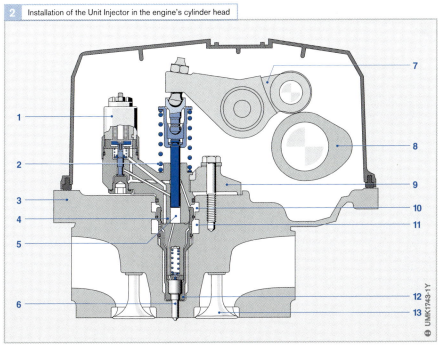

UMK1743-1Y

Fig. 2
1 High-pressure solenoid valve
2 Follower spring
3 Engine cylinder head
4 Unit-injector body assembly
5 High-pressure chamber
6 Nozzle assembly
7 Rocker
8 Actuating cam
9 Clamp
10 Fuel return
11 Fuel inlet
12 Nozzle nut
13 Engine valve

2 cm

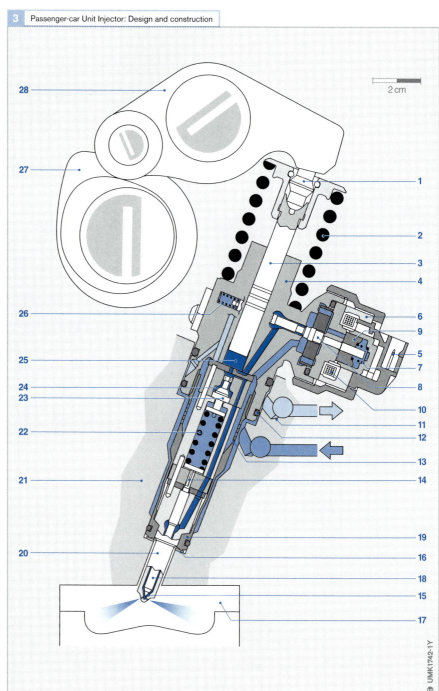

Fig. 3

1   Ball pin
2   Follower spring
3   Pump plunger
4   Pump-body
    assembly
5   Plug-in connection
6   Magnet core
7   Compensating
    spring
8   Solenoid-valve
    needle
9   Armature
10 Solenoid-valve coil
11 Fuel return
    (low-pressure stage)
12 Seal
13 Inlet passages
    (approx. 350 laser-
    drilled holes acting
    as a filter)
14 Hydraulic stop
    (damping unit)
15 Needle seat
16 Sealing disc
17 Engine combustion
    chamber
18 Nozzle needle
19 Retaining nut
20 Integral nozzle
    assembly
21 Engine cylinder
    head
22 Needle-valve spring
23 Accumulator plunger
24 Accumulator
    chamber
25 High-pressure
    chamber
26 Solenoid-valve
    spring
27 Drive camshaft
28 Roller rocker

UMK1742-1Y

4   Commercial-vehicle Unit Injector: Design and construction

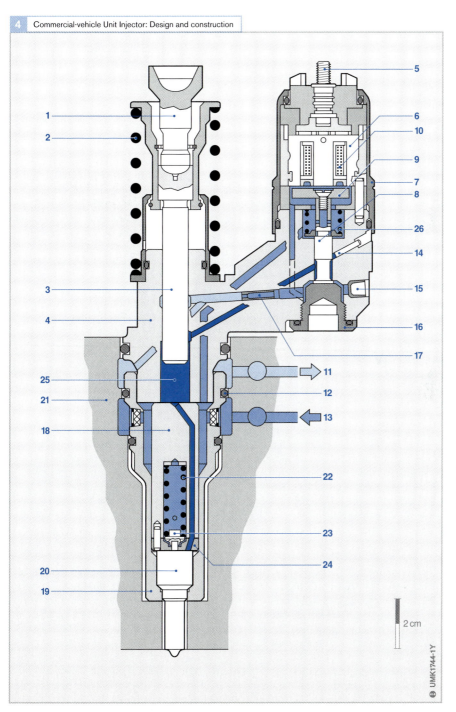

2 cm

UMK1744-1Y

Fig. 4
1 Ball pin
2 Return spring
3 Pump plunger
4 Pump-body
   assembly
5 Plug-in connection
6 Magnet core
7 Solenoid-valve
   retaining nut
8 Solenoid-valve
   needle
9 Armature plate
10 Solenoid-valve coil
11 Fuel return
   (low-pressure stage)
12 Seal
13 Fuel inlet
14 High-pressure plugs
15 Low-pressure plugs
16 Solenoid-valve
   stroke stop
17 Throttling orifice
18 Spring retainer
19 Retaining nut
20 Integral nozzle
   assembly
21 Engine cylinder
   head
22 Needle-valve spring
23 Pressure pin
24 Shim
25 High-pressure
   chamber
26 Solenoid-valve
   spring

Operating concept

Main injection
The function of these single-cylinder injection-pump systems can be subdivided into four operating states (Fig. 1):

Suction stroke (a)
The follower spring (3) forces the pump plunger (2) upwards. The fuel in the fuel supply's low-pressure stage is permanently under pressure and flows from the low-pressure stage into the solenoid-valve chamber (6) via the bores in the engine block and the inlet passage (7).

Initial stroke (b)
The actuating cam (1) continues to rotate and forces the pump plunger downwards. The solenoid valve is open so that the pump plunger can force the fuel through the fuel-return passage (8) into the fuel supply's low-pressure stage.

Delivery stroke and injection of fuel (c)
At a given instant in time, the ECU outputs the signal to energise the solenoid-valve coil (9) so that the solenoid-valve needle is pulled into the seat (10) and the connection between the high-pressure chamber and the low-pressure stage is closed. This instant in time is designated the "electrical start of injection" or "Beginning of the Injection Period", BIP, (also known as the "Begin of injection period"). The closing of the solenoid-valve needle causes a change of coil current. This is recognized by the ECU (BIP detection) as the actual start of delivery and is taken into account for the next injection process.

Further movement of the pump plunger causes the fuel pressure in the high-pressure chamber to increase, so that the fuel pressure in the injection nozzle also increases.

Upon reaching the nozzle-needle opening pressure of approx. 300 bar, the nozzle needle (11) is lifted from its seat and fuel is sprayed into the engine's combustion chamber (this is the so-called "actual start of injection" or start of delivery). Due to the pump plunger's high delivery rate, the pressure continues to increase throughout the whole of the injection process.

Residual stroke (d)
As soon as the solenoid-valve coil is switched off, the solenoid valve opens after a brief delay and opens the connection between the high-pressure chamber and the low-pressure stage.

The peak injection pressure is reached during the transitional phase between delivery stroke and residual stroke. Depending upon pump type, it varies between max. 1,800 and 2,050 bar. As soon as the solenoid valve opens, the pressure collapses abruptly, and when the nozzle-closing pressure is dropped below, the nozzle closes and terminates the injection process.

The remaining fuel which is delivered by the pumping element until the cam's crown point is reached is forced into the low-pressure stage via the fuel-return passage.

These single-cylinder injection systems are intrinsically safe. In other words, in the unlikely event of a malfunction, one uncontrolled injection of fuel is the most that can happen. For instance:

If the *solenoid valve remains open*, no injection can take place since the fuel flows back into the low-pressure stage and it is impossible for pressure to be built up. And since the high-pressure chamber can only be filled via the solenoid valve, when this *remains closed* no fuel can enter the high-pressure chamber. In this case, at the most only a single injection can take place.

The unit injector is installed in the engine's cylinder head and is therefore subject to very high temperatures. In order to keep its temperatures as low as possible, it is cooled by the fuel flowing back to the low-pressure stage.

Special measures applied in the fuel inlet to the unit injector ensure that differences in fuel temperature from cylinder to cylinder are kept to a minimum.

1   Unit Injector (UI) and Unit Pump (UP): Functional principle

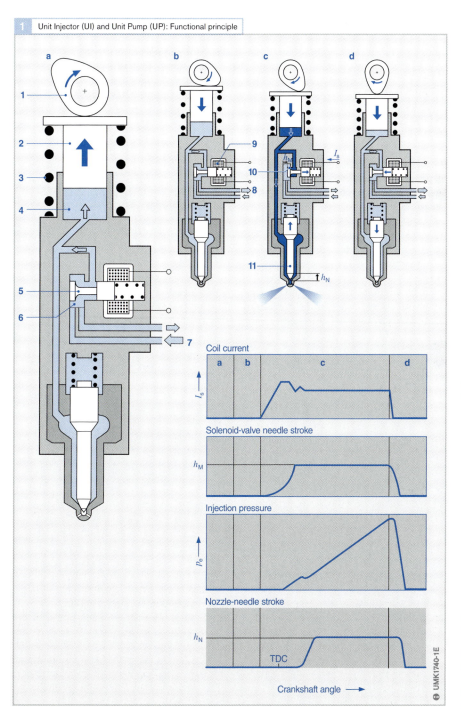

Coil current

Solenoid-valve needle stroke

Injection pressure

Nozzle-needle stroke

TDC

Crankshaft angle ⟶

Fig. 1
Operating states:
a   Suction stroke
b   Initial stroke
c   Prestroke
d   Residual stroke

1   Actuating cam
2   Pump plunger
3   Follower spring
4   High-pressure
     chamber
5   Solenoid-valve
     needle
6   Solenoid-valve
     chamber
7   Feed passage
8   Fuel-return passage
9   Coil
10   Solenoid-valve seat
11   Nozzle assembly

I_S Coil current
h_M Solenoid-valve
        needle stroke
p_e Injection pressure
h_N Nozzle-needle stroke

UMK1740-1E

Pilot injection (passenger cars)

Pilot injection with mechanical-hydraulic control is incorporated in the passenger-car unit injector. This serves to reduce both noise and pollutant emissions (refer to the Chapter "Diesel combustion"). This facility can be subdivided into four operating states (Fig. 2):

Initial position

Nozzle needle (7) and accumulator plunger (3) are up against their seats. The solenoid valve is open which means that no pressure can build up.

Start of pilot injection

Pressure buildup starts as soon as the solenoid valve closes. When the nozzle opening pressure is reached, the needle lifts from its seat and pilot injection commences. During this phase, the nozzle needle's stroke is limited hydraulically by a damping unit.

End of pilot injection

Further pressure increase leads to the accumulator plunger lifting from its seat so that a connection is set up between the high-pressure (2) and the low-pressure chambers (4). The resulting pressure drop, and the accompanying increase in the spring's (5) initial tension, lead to the nozzle needle closing. This marks the end of pilot injection.

For the most part, the pilot-injection quantity of approx. 1.5 mm3 is defined by the accumulator-plunger opening pressure. The interval between the main and pilot injection phases is essentially a function of the accumulator-plunger stroke.

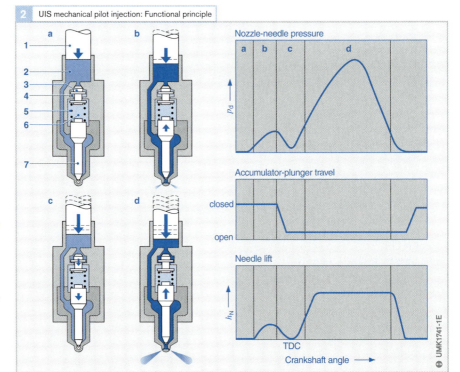

2   UIS mechanical pilot injection: Functional principle

Nozzle-needle pressure

Accumulator-plunger travel

Needle lift

TDC

Crankshaft angle ⟶

UMK1741-1E

Fig. 2
a   Initial position
b   Start of pilot
    injection
c   End of pilot injection
d   Main injection

1   Pump plunger
2   High-pressure
    chamber
3   Accumulator plunger
4   Accumulator
    chamber
5   Spring
6   Spring-retainer
    chamber
7   Nozzle needle

Start of main injection

The continuing movement of the pump plunger leads to the pressure in the high-pressure chamber continuing to increase. The main injection phase starts once the now higher nozzle opening pressure is reached.

During the actual main injection phase, the injection pressure increases to about 2,050 bar.

The opening of the solenoid valve marks the termination of the main injection phase. Nozzle needle and accumulator plunger return to their initial positions.

▶ The history and the future of the unit injector (UI)

1905

The idea behind the unit injector is practically as old as the diesel engine itself, and originated from Rudolf Diesel.

1999

It took the ideas, ingenuity, and hard work of countless engineers and technicians to turn Diesel's idea into a modern fuel-injection system.

But first of all, numerous difficulties and problems in the areas of materials technology, production engineering, control engineering, electronics, and fluid mechanics had to be overcome.

The Electronic Diesel Control (EDC) today places the unit injector system (UIS) in the position of being able to optimally control the various diesel-engine functions in a wide variety of operating states.

This system is capable of generating the highest injection pressures on today's market.

2000 onwards

Even considering today's high level of UIS development, there are still perspectives for the future. Further refinement of the electronic control, and even higher injection pressures, are only two of the possibilities on which the development departments are expending great effort.

In other words, the unit injector systems (UIS) are well prepared for the future.

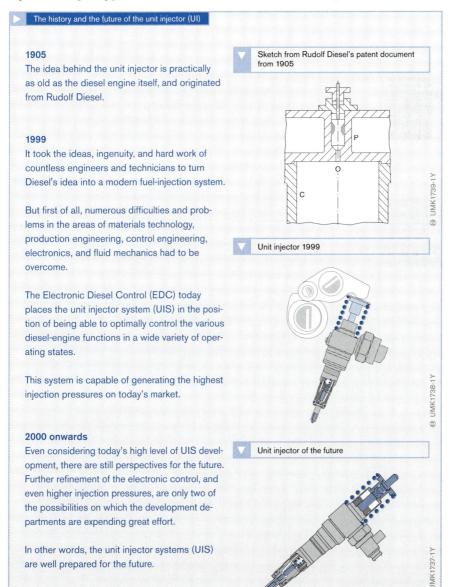

▼ Sketch from Rudolf Diesel's patent document from 1905

▼ Unit injector 1999

▼ Unit injector of the future

High-pressure solenoid valve

It is the job of the high-pressure solenoid valve to initiate injection at the correct instant in time, and to ensure precision metering to the engine cylinder of the correct amount of fuel for a precisely measured length of time (injection duration).

Design and construction

The high-pressure solenoid valve is subdivided into two major subassemblies:

Valve

The valve itself is comprised of the valve needle, the valve body as an integral part of the pump body, and the valve spring (Figs. 1 and 2, Pos. 2, 12, and 1 respectively).

The valve body's sealing surface is conically ground (10), and the valve needle is also provided with a conical sealing surface (11). The angle of the needle's ground surface is slightly larger than that of the valve body. With the valve closed, when the needle is forced up against the valve body, valve body and needle are only in contact along a line (and not a surface) which represents the valve seat. As a result, sealing is very efficient (dual-conical sealing). High-precision processing must be applied to perfectly match the valve needle and valve body to each other.

Magnet

The magnet is comprised of the fixed stator and the movable armature (16).

The *stator* itself is composed of the magnet core (15), a coil (6), and the corresponding electrical contacting with plug (8).

The *armature* is fastened to the valve needle.

In the non-energized position, there is an initial air gap between the stator and the armature.

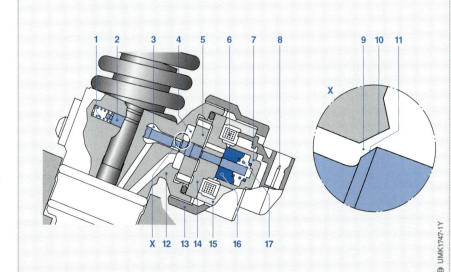

1   Passenger-car Unit Injector: High-pressure solenoid valve

Fig. 1
1 Valve spring
2 Valve needle
3 High-pressure area
4 Low-pressure area
5 Shim
6 Coil
7 Retainer
8 Plug
9 Valve flow cross-
   section
10 Valve-body sealing
    surface
11 Valve-needle sealing
    surface
12 Integral valve body
13 Union nut
14 Magnetic disc
15 Magnet core
16 Armature
17 Compensating
    spring

Operating concept

The solenoid valve has two switched positions: Open and closed. It is open when no voltage is applied across the coil (non-energized). It closes when energized by the ECU driver stage.

Valve open

The force exerted by the valve spring pushes the valve needle up against the stop so that the valve flow cross-section (9) between the valve needle and the valve body is opened in the vicinity of the valve seat. The pump's high-pressure (3) and low-pressure (4) areas are now connected with each other. In this initial position, it is possible for fuel to flow into and out of the high-pressure chamber.

Valve closed

When fuel injection is required, the coil is energized by the ECU driver stage (refer also to the "ECU" Chapter). The pickup current causes a magnetic flux in the magnetic-circuit components (magnet core and armature) which generates a magnetic force to shift the armature towards the stator. Armature movement is halted by the needle and the valve body coming into contact at the seal seat. A residual air gap remains between the armature and the stator. The valve is now closed, and injection takes place when the pump plunger is forced downwards.

The magnetic force is not only used to pull in the armature, but must at the same time overcome the force exerted by the valve spring and hold the armature against the spring force. Apart from this, the magnetic force must apply a certain force to keep the sealing surfaces in contact with each other. The force at the armature is maintained as long as the coil is energized.

The nearer the armature is to the magnet stator, the greater is the magnetic flux. When the valve is closed, therefore, it is thus possible to reduce the current to the holding-current level. The valve remains closed nevertheless, and the power loss (heat) due to current flow is kept to a minimum.

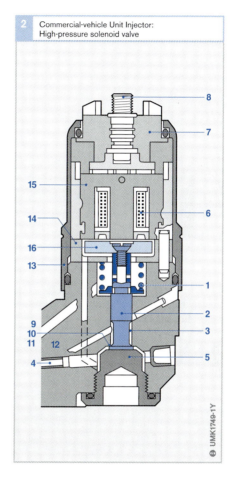

2   Commercial-vehicle Unit Injector: High-pressure solenoid valve

Fig. 2
1 Valve spring
2 Valve needle
3 High-pressure area
4 Low-pressure area
5 Stop
6 Coil
7 Cover
8 Plug
9 Valve flow cross-section
10 Valve-body sealing surface
11 Valve-needle sealing surface
12 Integral valve body
13 Solenoid-valve clamping nut
14 Adjusting element for residual air gap
15 Magnet core
16 Armature plate

UMK1749-1Y

To stop the fuel-injection process, the current through the solenoid coil is switched off. As a result, the magnetic flux and the magnetic force collapse, and the spring forces the valve needle to its normal position against the stop. The valve seat is open.

Solenoid-valve switching must be extremely rapid and very precise in order to comply with the tolerances demanded from the injection system regarding start of injection and injected fuel quantity. Irrespective of operating conditions, this high-level precision is maintained from stroke to stroke, and from pump to pump.

Unit pump (UP)

The assignment and operating concept of the unit pump (UP) correspond to those of the unit injector (UI). The fact that the UP separates the function units "High-pressure generation" and "High-pressure solenoid valve" from the "Injection nozzle" by a short high-pressure delivery line is the only difference between it and the UI system.

The unit pump is of modular design, and the fact that the pumps are integrated in the side of the engine block (Fig. 1) has the following advantages:
- New cylinder-head designs are unnecessary
- Rigid drive, since no rockers needed
- Simple handling for the workshop since the pumps are easy to remove

The UPS nozzles are installed in nozzle holders (refer to the Chapter "Nozzles and nozzle holders").

Design and construction

High-pressure delivery lines
All pumps have a very short delivery line (6). These are all of the same length, and must be able to permanently withstand the maximum pump pressure and the to some extent high-frequency pressure fluctuations which occur during the injection pauses. High-tensile, seamless steel tubing is therefore used for these delivery lines which normally have an OD of 6 mm and an ID of 1.8 mm.

Unit Pump
Each pump is driven directly by an injection cam on the engine camshaft (4). Connection to the pump plunger is through the return spring (8) and the roller tappet (9). A flange on the pump body is used to fasten the pump in the engine block.

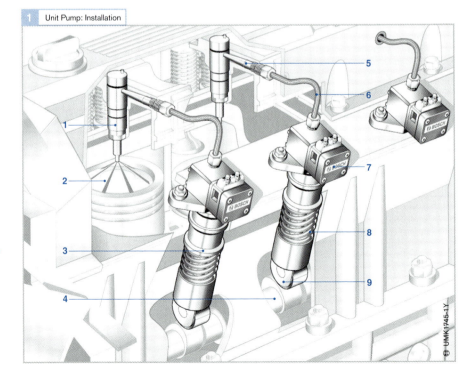

1   Unit Pump: Installation

Fig. 1
1   Stepped nozzle holder
2   Engine combustion chamber
3   Unit pump
4   Engine camshaft
5   Pressure fittings
6   High-pressure delivery line
7   Solenoid valve
8   Return spring
9   Roller tappet

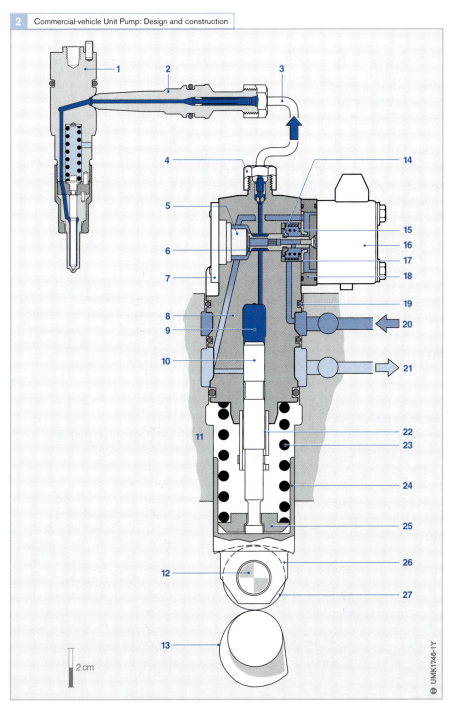

2   Commercial-vehicle Unit Pump: Design and construction

Fig. 2
1 Nozzle holder
2 Pressure fitting
3 High-pressure
  delivery line
4 Connection
5 Stroke stop
6 Solenoid-valve
  needle
7 Plate
8 Pump housing
9 High-pressure
  chamber
10 Pump plunger
11 Engine block
12 Roller-tappet pin
13 Cam
14 Spring seat
15 Solenoid-valve
  spring
16 Valve housing with
  coil and magnet
  core
17 Armature plate
18 Intermediate plate
19 Seal
20 Fuel inlet
  (low pressure)
21 Fuel return
22 Pump-plunger
  retention device
23 Tappet spring
24 Tappet body
25 Spring seat
26 Roller tappet
27 Tappet roller

2 cm

UMK1748-1Y

Unit pump for large-size engines

Applications and requirements

Solenoid-valve controlled single-plunger fuel-pumps known as unit pumps (UP) (Bosch Type PF..MV pumps) are used on large-size diesel engines for railway locomotives and ships. The maximum permissible engine speed is 1,000 rpm with power outputs of 150...450 kW per cylinder.

Statutory requirements for exhaust-gas emission limits have already been implemented for certain applications and are in the course of preparation for others. Such regulations, combined with the continuing efforts to optimize fuel consumption and reduce noise, have a decisive effect on engine development and, therefore, are a fundamental factor in the design of fuel-injection systems.

Fuel-injection systems that are required not only to offer high performance but also satisfy demands for low exhaust-gas and noise emissions and economic fuel consumption must be capable of an infinitely variable injection timing. This demand is met by the use of unit pumps where the start and end of injection are determined by an electronically controlled solenoid valve.

In a fully electronically controlled engine management system, there is no longer a need for the control linkage/control shaft with the links to the control rods of the single-plunger fuel-pumps (as required for the Type PF pumps) and the governor control lever. Instead, these components are replaced by control leads to the power output stages and from there to the solenoid valves of the unit pumps. The advantages of this arrangement are as follows:

- Smaller space requirements
- Greater versatility in terms of installed configuration and suitability for conventional cylinder-head designs
- Very rigid and compact drive system
- Low material and adaptation costs
- Rapid response times and precise control capability
- Ease of replacement for servicing or repair
- Precise fuel metering with the facility for cylinder-specific control of injected-fuel quantity and start of delivery for each cylinder and
- Capability of shutting down individual cylinders in order to be able to operate the remaining cylinders within a range suitable for optimum fuel consumption or exhaust-gas emissions in part-load range

Furthermore, it is also possible to implement different control data for every fuel-injection cycle for a particular cylinder. This takes account of feedback from the solenoid valves in different operating conditions such as fuel temperature and/or degree of component wear can be directly taken into account.

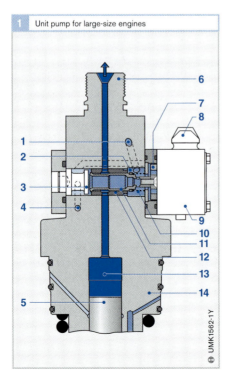

1   Unit pump for large-size engines

Fig. 1
1 Fuel inlet
2 Fuel gallery
3 Stop plate
4 Fuel return
5 Pump plunger
6 High-pressure delivery line
7 Armature plate
8 Connection to engine ECU
9 Solenoid-valve housing
10 Valve spring
11 Control plunger
12 Sleeve
13 Plunger chamber
14 Pump housing

UMK1562-1Y

Design

In comparison with the port-controlled type PF single-plunger fuel-pump, the pump plunger of a unit pump is a substantially simpler design. It has neither helical grooves nor vertical control grooves. In addition, there are no spill ports in the pump cylinder. Instead of the delivery valve on the pump barrel there is a solenoid valve with a control plunger.

A sleeve (Fig. 1, Item 12) in the upper part of the pump barrel acts as a guide for a cylindrical control plunger (11). On the end of the plunger that faces away from the solenoid-valve body (9), it has a tapered face that operates in conjunction with a tapered valve seat on the sleeve. Attached to the other end of the control plunger is an armature plate (7) which is drawn inwards against the action of the valve spring (10) by the solenoid when an electric current is applied. This moves the control plunger to its closed position against the valve seat. In closed position, there remains a residual air gap between the armature plate and the solenoid body. As soon as the power supply to the solenoid is switched off, the valve spring moves the armature plate and the control plunger away from the solenoid, thereby opening the valve, and presses the control plunger against the stop plate.

Ring grooves and holes in the pump barrel allow the non-pressurized removal of leakage fuel and/or pressurized lubrication of the pump plunger.

The fuel galleries are arranged in such a way that the fresh diesel fuel entering the pump continuously cools the solenoid; the cooling flow rate is around 3...5 times as great as the maximum delivery rate of the fuel-injection pump.

When fuel injection is required, the output stage of the engine ECU energizes the solenoid coil when the pump is on its upstroke. As long as the electric current is flowing through the solenoid coil, the valve remains closed and maintains delivery of fuel to the nozzle. While the valve is closed, the holding current flowing through the solenoid coil can be less than the pickup current because the distance between the armature plate and the solenoid is then substantially smaller. In this way, the power loss and heat generation by the electric current can be kept to a minimum.

To terminate the fuel-injection cycle the control unit switches off the power supply to the solenoid coil. The valve spring then returns the control plunger to the open position. The remaining fuel delivered by the pump plunger flows through the open valve to the fuel return outlet. Thus the pressure in the high-pressure delivery line is reduced to fuel-inlet pressure.

The rate-of-discharge curve of a solenoid-valve controlled unit injection pump can be controlled by the shape of the injection cam in the same way as a mechanically controlled fuel-injection pump.

Charging the pump chamber through the valve aperture at the tapered valve seat has the advantage that, if the control plunger sticks – either in the open or closed position – it will not result in an undesirable engine operating status (e.g. the engine overrevs due to an excessive injected-fuel quantity). The same applies to a fault in the control of the solenoid valve.

The sleeve that guides the control plunger is press-fitted in a hole in the pump located in the upper part of the pump barrel running at right angles to its axis. If, after a long period of service, the tapered seat is so worn that it requires replacement, the sleeve and the control plunger can be cost-effectively replaced in the repair shop. After the specified gap and clearances have been adjusted, the repaired unit pump is then virtually "as new".

The pumps currently available have plunger diameters of 18...22 mm and plunger strokes of 20...28 mm.

Overview of common-rail system

The demands placed on diesel-engine fuel-injection systems are continually increasing. Higher pressures, faster switching times and greater adaptability of the injection pattern to engine operating conditions make diesel engines more economical, cleaner and more powerful. In addition, the fuel-injection system is becoming more and more integrated in the overall network of vehicle systems. As a result, diesel engines have even broken into the luxury-car market.

One of the most advanced of these fuel-injection systems is the pressure-accumulator system known as common-rail fuel injection. The main advantage of the common-rail system is its ability to vary injection pressure and timing over a broad scale. This is made possible by separating the functions of pressure generation and fuel injection.

Areas of application

The pressure-accumulator common-rail fuel-injection system for diesel engines with direct injection (DI) is used in the following types of vehicle:
- *Cars* ranging from economy models with three-cylinder, 0.8 *l* engines producing 30 kW (41 bhp) of power and 100 Nm of torque, and with fuel consumption of 3.5 *l*/100 km (NETC) to luxury sedans with eight-cylinder, 3.9 *l* engines developing 180 kW (245 bhp) of power and 560 Nm of torque
- *Light commercial vehicles* with power outputs of up to 30 kW/cylinder and
- *Heavy-duty trucks*, *railway locomotives* and *ships* with engines producing up to 200 kW/cylinder

The common-rail system offers a significantly higher level of adaptability to engine design on the part of the fuel-injection system than cam-operated systems, as evidenced by its:
- Wide range of applications (see above)
- High injection pressures (up to approx. 1,600 bar)

- Variable injection timing
- Capability of multiple pre- and post-injection phases (even extremely retarded post-injection is possible)
- Variation of injection pressure (230...1,600 bar) according to engine operating conditions

The common-rail system thus plays a major role in increasing specific power output, lowering fuel consumption and decreasing noise and exhaust emissions from diesel engines.

Design

The engine control unit using the common-rail fuel-injection system is made up of four subsystems (Figure 1):
- The low-pressure system comprising the components of the *fuel-supply system*
- The *high-pressure system* consisting of the high-pressure pump, high-pressure accumulator (fuel rail), the nozzles and the high-pressure fuel lines
- The electronic control system EDC made up of the sensors, control unit and actuators and
- The *air-intake and exhaust-gas systems* (air intake, emission control and exhaust-gas recirculation)

Among the most important components of the common-rail system are the injectors. They incorporate a fast-switching solenoid valve by means of which the nozzle is opened and closed. This enables the injection cycle to be individually controlled for each cylinder. In contrast with other solenoid-valve controlled fuel-injection systems, the common-rail injector injects fuel whenever the solenoid valve is open.

All the injectors are fed by a common fuel rail, hence the name "common-rail fuel injection".

The modular design of the common-rail system simplifies adaptation to individual engine designs.

Method of operation

In the common-rail pressure-accumulator fuel-injection system, the functions of pressure generation and fuel injection are separate. The EDC electronic-control system controls the individual fuel-injection components.

Pressure generation

A continuously operating high-pressure pump driven by the engine produces the desired injection pressure. As that pressure is stored in the pressure accumulator, it is largely independent of engine speed and injected-fuel quantity. The speed of the high-pressure pump is directly proportional to the engine speed as it is driven by a system with a fixed transmission ratio. Because of the almost uniform injection pattern, the high-pressure pump can be significantly smaller and designed for a lower peak drive-system torque than conventional fuel-injection systems.

Pressure is controlled by means of a pressure control valve and/or a controlled inlet on the high-pressure pump. The pressurized fuel is held in the fuel rail ready for injection.

Fuel injection

The nozzles inject the fuel directly into the engine's combustion chambers. They are supplied by short high-pressure fuel lines connected to the fuel rail. A nozzle consists essentially of an injector nozzle and a fast-switching solenoid valve that controls the injector nozzle by means of mechanical actuators. The solenoid valve is controlled by the electronic engine control unit.

At a constant system pressure, the fuel quantity injected is proportional to the length of time that the solenoid valve is open and thus entirely independent of the engine or pump speed (time-based fuel-injection system).

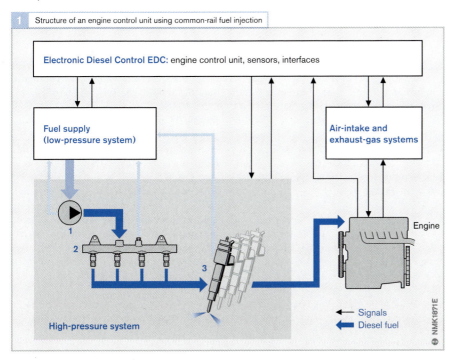

1 Structure of an engine control unit using common-rail fuel injection

Electronic Diesel Control EDC: engine control unit, sensors, interfaces

Fuel supply (low-pressure system)

Air-intake and exhaust-gas systems

High-pressure system

Engine

1

2

3

Signals

Diesel fuel

NMK1871E

Fig. 1
1 High-pressure pump
2 Fuel rail
3 Nozzle

Control

With the aid of a range of sensors, the engine control unit records the accelerator-pedal position and the current status of the engine and the vehicle (see also the chapter "Electronic diesel control EDC"). The data collected includes:

- The crankshaft angle of rotation
- The camshaft speed
- The fuel rail pressure
- The charge-air pressure
- The temperature of intake air, engine coolant and fuel
- The mass of the air charge
- The road speed of the vehicle, etc.

The control unit analyzes the input signals and calculates within a split second the control signals required for the high-pressure pump, the nozzles and the other actuators. The latter may include the exhaust-gas recirculation valve or the charge-air actuator, for example.

The extremely fast switching times demanded of the nozzles are achieved with the aid of optimized high-pressure solenoid valves and a special control method.

The position-time system matches the start of injection to the rotation of the engine using the data from the crankshaft and camshaft sensors (time-based control). The electronic diesel control EDC makes it possible to precisely meter fuel injection. In addition, EDC offers the potential for additional functions that can improve engine response and convenience.

Basic functions

The basic functions involve the precise control of injection timing and quantity at the set pressure. In this way, they ensure that the diesel engine has low consumption and smooth running characteristics.

Additional functions

Additional control functions perform the tasks of reducing exhaust-gas emissions and fuel consumption or providing added safety and convenience. Some examples are:

- Control of exhaust-gas recirculation
- Charge-air pressure control
- Cruise control
- Electronic immobilizer, etc.

Integration of EDC in an overall network of vehicle systems also opens up a range of new possibilities (e.g. data exchange with the climate-control system or the transmission-control system).

A diagnosis interface enables analysis of stored system data when the vehicle is serviced.

Control unit configuration

As the engine control unit has only six output stages for the nozzles, engines with more than six cylinders require two engine control units. They are linked via an internal high-speed CAN interface in a master-and-slave configuration. As a result, there is also a higher microcontroller processing capacity available. Some functions are permanently allocated to a specific control unit (e.g. volume balancing). Others can be dynamically allocated to one or other of the control units as situations demand (e.g. recording of sensor signals).

Air-intake and exhaust-gas systems

Exhaust-gas recirculation

Cars

Exhaust-gas recirculation is an effective method of reducing NO_X components in exhaust gas. It involves the use of a valve which returns some of the exhaust gas to the intake manifold. If the recirculated exhaust gas is also cooled, further advantages can be gained. This method has been the state of the art for diesel cars for a number of years. The exhaust gas is recirculated at low engine loads and speeds.

Commercial vehicles

The vast majority of modern diesel engines are fitted with an exhaust-gas turbocharger. Such engines do not generally have a negative pressure differential between the exhaust manifold upstream of the turbine and the inlet manifold downstream of the compressor at high engine loads. Since exhaust-gas recirculation and cooling is essential even at the higher end of the power curve on commercial vehicle engines, additional features such as turbochargers with variable turbine geometry (VTG), wastegate or flutter valve are necessary.

Exhaust-gas treatment systems

In order to be able to comply with stricter emission-control legislation, emission control will become increasingly important for diesel engines in the future despite advances in internal engine design. This is particularly true for larger cars and commercial vehicles. There are many systems under development. Which of them will eventually become established remains an

unanswered question. As a rule, however, greater adaptability on the part of the fuel-injection system will be necessary. The common-rail system offers a broad range of possibilities:

- A *diesel-oxidation catalytic converter* (DOC) primarily reduces hydrocarbon (HC) and carbon monoxide (CO) emissions as well as a proportion of the volatile particulate components.
- Various types of particulate filter (PF) filter the soot particles from the exhaust gas (e.g. *CRT* (**C**ontinuous **R**egeneration **T**rap) system or an additive system).
- An NO_X *accumulator-type catalytic converter* reduces the nitrogen oxides NO and NO_2. A version for use in car diesel engines is currently in the process of development.
- The *SCR* (**S**elective **C**atalytic **R**eduction) catalytic converter reduces NO_X emissions with the help of ammonia. Ammonia is obtained from the reducing agent urea by passing it through a hydrolyzing catalytic converter. In more recent SCR systems, the hydrolyzing catalytic converter is integrated in the SCR catalytic converter.

In *combination systems* (also called four-way systems), several individual systems are combined. They can then reduce not only NO_X but also HC, CO and particulate emissions. Such systems demand very powerful engine control units.

The most important emission control systems are dealt with in more detail in a separate chapter.

UMK1872Y

System diagram for cars

Figure 3 shows all the components of a fully equipped common-rail system for an eight-cylinder diesel car engine. Depending on the type of vehicle and application, some of the components may not be used.

For the sake of clarity of the diagram, the sensors and setpoint generators (A) are not shown in their fitted positions. Exceptions to this are the sensors of the exhaust-gas treatment systems (F) and the fuel-rail pressure sensors, as their proper fitted positions are necessary in order to understand the system.

The CAN bus in the interfaces section (B) enables exchange of data between a wide variety of systems and components including
● The starter motor
● The alternator
● The electronic immobilizer
● The transmission control system
● The traction control system TCS and
● The electronic stability program ESP

Even the instrument cluster (12) and the air-conditioning system (13) can be connected to the CAN bus.

For emission control, three alternative combination systems are shown (a, b and c).

Fig. 3

Engine, engine control unit and high-pressure fuel-injection components
16 High-pressure pump
23 Engine control unit (master)
24 Engine control unit (slave)
25 Fuel rail
26 Fuel-rail pressure sensor
27 Injector
28 Glow plug
29 Diesel engine (DI)
M Torque

A   Sensors and setpoint generators
1   Accelerator-pedal sensor
2   Clutch switch
3   Brake switches (2)
4   Operator unit for cruise control
5   Glow plug/starter switch ("ignition switch")
6   Vehicle-speed sensor
7   Crankshaft speed sensor (inductive)
8   Engine-temperature sensor (in coolant system)
9   Intake-air temperature sensor
10 Charge-air pressure sensor
11 Hot-film air-mass flow sensor (intake air)

B   Interfaces
12 Instrument cluster with signal output for fuel consumption, engine speed, etc.
13 Air-conditioning compressor with control
14 Diagnosis interface
15 Glow plug control unit
CAN **C**ontroller **A**rea **N**etwork (vehicle's serial data bus)

C   Fuel supply system (low-pressure system)
17 Fuel filter with overflow valve
18 Fuel tank with filter and electric fuel pump
19 Fuel level sensor

D   Additive system
20 Additive metering unit
21 Additive control unit
22 Additive tank

E   Air-intake system
30 Exhaust-gas recirculation cooler
31 Charge-air pressure actuator
32 Turbocharger (in this case with variable turbine geometry)
33 Control flap
34 Exhaust-gas recirculation actuator
35 Vacuum pump

F   Emission control systems
36 Exhaust temperature sensor
37 Oxidation catalytic converter
38 Particulate filter
39 Differential-pressure sensor
40 Exhaust heater
41 NO_X sensor
42 Broadband oxygen sensor Type LSU
43 NO_X accumulator-type catalytic converter
44 Two-point oxygen sensor Type LSF
45 Catalyzed soot filter Type CSF

3   Common-rail diesel fuel-injection system for cars

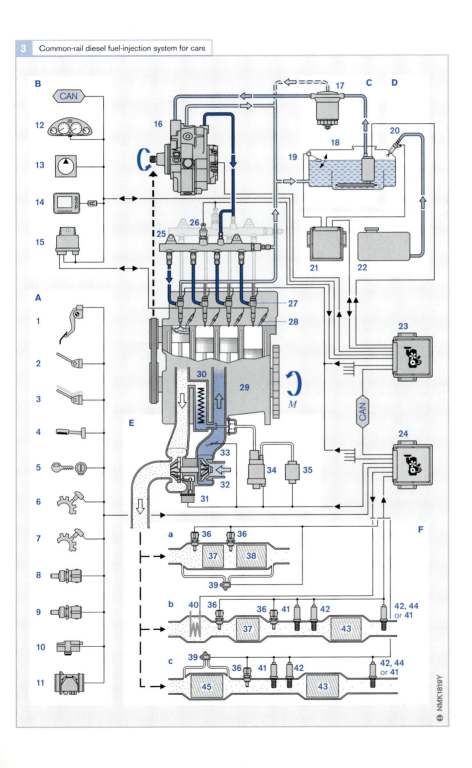

NMK1819Y

System diagram for commercial vehicles

Figure 4 shows all the components of a fully equipped common-rail system for a six-cylinder diesel commercial-vehicle engine. Depending on the type of vehicle and application, some of the components may not be used.

For the sake of clarity of the diagram, only the sensors and desired-value generators whose true position is necessary to the understanding of the system are shown in their fitted locations.

Data exchange with a wide range of other systems (e.g. transmission control system, traction control system TCS, electronic stability program ESP, oil quality sensor, tachograph, radar sensor ACC, vehicle management system, brake co-ordinator, fleet management system) involving up to 30 control units is possible via the CAN bus in the "Interfaces" section (B). Even the alternator (18) and the air-conditioning system (17) can be connected to the CAN bus.

For exhaust-gas treatment, three alternative combination systems are shown (a, b and c).

Fig. 4

Engine, engine control unit and high-pressure injection components

22 High-pressure pump
29 Electronic engine control unit
30 Fuel rail
31 Fuel-rail pressure sensor
32 Fuel injector
33 Relay
34 Auxiliary equipment (e.g. retarder, exhaust flap for engine brake, starter motor, fan)
35 Diesel engine (DI)
36 Flame glow plug (alternatively grid heater)
M Torque

A   Sensors and desired-value generators

1   Accelerator-pedal sensor
2   Clutch switch
3   Brake switches (2)
4   Engine brake switch
5   Parking brake switch
6   Control switch (e.g. cruise control, intermediate speed control, engine speed and torque reduction)
7   Starter switch ("ignition switch")
8   Charge-air speed sensor
9   Crankshaft speed sensor (inductive)
10 Camshaft speed sensor
11 Fuel temperature sensor
12 Engine-temperature sensor (in coolant system)
13 Charge-air temperature sensor
14 Charge-air pressure sensor
15 Fan speed sensor
16 Air-filter differential-pressure sensor

B   Interfaces

17 Air-conditioning compressor with control
18 Alternator
19 Diagnosis interface
20 SCR control unit
21 Air compressor

CAN **C**ontroller **A**rea **N**etwork (vehicle's serial data bus) (up to three data busses)

C   Fuel supply system (low-pressure system)

23 Fuel pump
24 Fuel filter with water-level and pressure sensors
25 Control unit cooler
26 Fuel tank with filter
27 Pressure limiting valve
28 Fuel level sensor

D   Air intake system

37 Exhaust-gas recirculation cooler
38 Control flap
39 Exhaust-gas recirculation actuator with exhaust recirculation valve and position sensor
40 Intercooler with bypass for cold starting
41 Turbocharger (in this case with variable turbine geometry) with position sensor
42 Charge-air pressure actuator

E   Exhaust-gas treatment systems

43 Exhaust-gas temperature sensor
44 Oxidation-type catalytic converter
45 Differential-pressure sensor
46 Particulate filter
47 Soot sensor
48 Fluid level sensor
49 Reducing agent tank
50 Reducing agent pump
51 Reducing agent injector
52 NO_x sensor
53 SCR catalytic converter
54 NH3 sensor
55 Blocking catalytic converter
56 Catalyzed soot filter Type CSF
57 Hydrolyzing catalytic converter

4 Common-rail diesel fuel-injection system for commercial vehicles

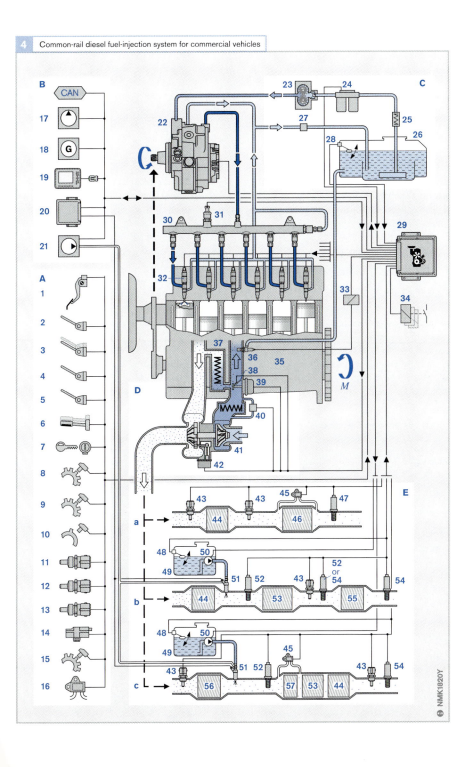

High-pressure components of common-rail system

The high-pressure section of the common-rail system is subdivided in pressure-generation, pressure-accumulation and fuel-metering stages (Figs. 1 and 2). The high-pressure pump assumes the role of pressure generation in conjunction with the pressure-control valve and the optional element switchoff valve. Pressure is accumulated in the fuel rail by means of the rail-pressure sensor, pressure limiter and flow limiter. The function of the nozzles is correct timing and metering the quantity of fuel injected. High-pressure delivery lines interconnect the three sections.

able in all engine operating conditions. At the same time it must operate for the entire service life of the vehicle. This includes providing a fuel reserve that is required for quick engine starting and rapid pressurization in the fuel rail.

The high-pressure pump constantly maintains a system pressure of up to 1,600 bar in the high-pressure accumulator (fuel rail). As a result, the fuel does not have to be pressurized during the fuel-injection cycle (as opposed to conventional fuel-injection systems).

High-pressure pump (pressure generation)

Functions

The high-pressure pump is the interface between the low-pressure and high-pressure stages. Its function is to make sure there is always sufficient fuel under pressure avail-

Design

On cars, the pump used to generate the required pressure is a radial-piston pump. It generates that pressure independently of the fuel-injection cycle.

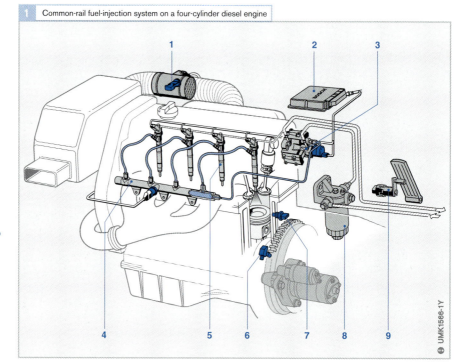

1   Common-rail fuel-injection system on a four-cylinder diesel engine

UMK1566-1Y

Preferably, the high-pressure pump is fitted to the diesel engine in the same position as a conventional distributor injection pump. It is driven by the engine via a coupling, gear, chain or toothed belt and rotates at a maximum speed of 3,000 rpm. The speed of the high-pressure pump is directly proportional to the engine speed as it is driven by a system with a fixed transmission ratio. It is lubricated by the fuel.

Depending on the available fitting space, a pressure-control valve is either mounted directly on the high-pressure pump or separately from it.

Three radially aligned pump plungers inside the high-pressure pump pressurize the fuel. The three pistons are arranged at an angle of 120° to each other. With three delivery strokes per revolution, the peak drive torque levels are kept low and the pump drive system is subjected to a constant load. At

16 Nm, the torque is only about a ninth of the drive torque required for a comparable distributor injection pump. As a result, the common-rail system places less demands on the pump-drive system than conventional fuel-injection systems. The power required to drive the pump increases in proportion to the pressure in the fuel rail and the rotational speed of the pump (delivery rate). In the case of a two-liter engine running at rated speed and a fuel-rail pressure of 1,350 bar, the high-pressure pump uses 3.8 kW of power (assuming a pump efficiency of approx. 90 %). The higher power requirement originates from the leakage and control volumes in the nozzle and the return of fuel via the pressure-control valve.

Method of operation
The presupply pump delivers fuel via a filter and a water separator to the high-pressure pump inlet (Fig. 3, Item 13). It forces fuel through the restrictor bore of the safety

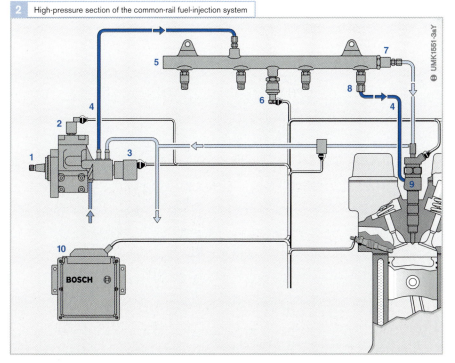

2 High-pressure section of the common-rail fuel-injection system

⊕ UMK1551-3eY

Fig. 2

1 High-pressure pump
2 Element shutoff valve
3 Pressure-control valve
4 High-pressure delivery line
5 High-pressure accumulator (fuel rail)
6 Fuel-rail pressure sensor
7 Pressure-limiting valve
8 Flow limiter
9 Nozzle
10 Engine ECU

valve (14) to the lube and cooling circuits of the high-pressure pump. The drive shaft (1) with its drive cam (2) moves the three pump plungers (3) up and down in accordance with the shape of the cam.

If the delivery pressure exceeds the opening pressure of the safety valve (0.5 to 1.5 bar), the presupply pump can force fuel through the inlet valve of the high-pressure pump to the pump-plunger chamber when the plunger is on its downstroke (intake stroke). When the pump plunger passes bottom-dead center position, the inlet valve closes and the fuel in the pumping-element chamber (4) can no longer escape. It can then be pressurized beyond the delivery pressure of the presupply pump. The rising pressure opens the outlet valve (7) as soon as pressure reaches the level in the fuel rail. The pressurized fuel then passes to the high-pressure stage.

The pump plunger continues to deliver fuel until it reaches its top-dead center position (delivery stroke). The pressure then drops so that the outlet valve closes. The remaining fuel is depressurized and the pump plunger moves downwards.

If the pressure in the pumping-element chamber falls below the fuel-pump delivery pressure, the inlet valve opens again and the process starts over from the beginning.

Delivery rate

As the high-pressure pump is designed for high delivery rates, there is a surplus of pressurized fuel when the engine is idling or running in part-load range. Excess fuel is returned to the fuel tank via the pressure-control valve. However, since the pressurized fuel depressurizes in the fuel tank, the energy introduced to pressurize the fuel is lost. In addition to heating up the fuel, this decreases overall efficiency.

3 High-pressure pump (schematic, axial cross-section)

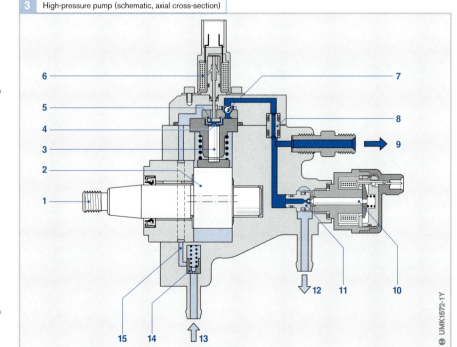

Fig. 3

1 Drive shaft
2 Eccenter cam
3 Plunger-and-barrel assembly with pump plunger
4 Pumping-element chamber
5 Intake valve
6 Element shutoff valve
7 Outlet valve
8 Sealing cap
9 High-pressure connection to fuel rail
10 Pressure-control valve
11 Ball valve
12 Fuel return
13 Fuel inlet
14 Safety valve with calibrated restriction
15 Low-pressure duct to plunger-and-barrel assembly

4 High-pressure pump (schematic, transverse cross-section)

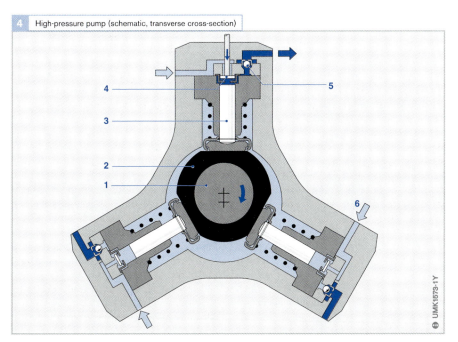

Fig. 4
1 Drive shaft
2 Drive cam
3 Pump element
   with pump plunger
4 Intake valve
5 Outlet valve
6 Fuel inlet

This can be remedied to a certain extent by adjusting the delivery rate to the fuel demand by shutting down one of the pumping elements.

Pumping-element shutdown

If one of the pumping elements (3) is shut down, the quantity of fuel pumped to the high-pressure fuel rail is reduced. This is achieved by keeping the intake valve (5) permanently open. When the pumping-element shutdown solenoid valve is activated, a pin attached to its armature holds the intake valve continuously open. Consequently, the fuel drawn in cannot be compressed during the delivery stroke. As a result, there is no pressure rise in the pumping-element chamber because the fuel can escape back to the low-pressure duct.

As one of the pumping elements is shut down in response to low fuel demand, the high-pressure pump no longer delivers fuel continuously; instead, there is a pause in fuel delivery.

Transmission ratio

The delivery rate of a high-pressure pump is proportional to its rotational speed. In turn, the pump speed is dependent on the engine speed. The transmission ratio between the engine and the pump is determined in the process of adapting the fuel-injection system to the engine so as to limit the volume of excess fuel delivered. At the same time it makes sure that the engine's fuel demand at WOT is covered to the full extent. Possible transmission ratios are 1:2 or 2:3 relative to the crankshaft.

Pressure-control valve

Function

The purpose of the pressure-control valve is to adjust and maintain the pressure in the fuel rail as a factor of engine load, i.e.:
- When the pressure in the fuel rail is too high, the pressure-control valve opens so that some of the fuel can escape from the fuel rail and return to the fuel tank via a return line.

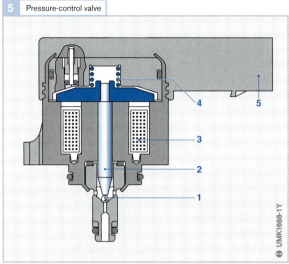

5   Pressure-control valve

UMK1668-1Y

Fig. 5
1   Valve ball
2   Armature
3   Electromagnet
4   Valve spring
5   Electrical
    connection

Pressure-control valve not activated:
The high pressure present in the fuel rail or at the outlet of the high-pressure pump acts on the pressure-control valve via the high-pressure inlet. As the electromagnet is not applying any force, the high-pressure exceeds the force of the spring so that the pressure-control valve opens and remains open to a greater or lesser extent, depending on the pump delivery rate. The spring is dimensioned to maintain a pressure of approx. 100 bar.

- When the pressure in the fuel rail is too low, the pressure-control valve closes and seals the high-pressure stage from the low-pressure stage.

Design
The pressure-control valve (Fig. 5) has a mounting flange which attaches it to the high-pressure pump or the fuel rail. The armature (2) forces a valve ball (1) against the valve seat in order to isolate the high-pressure stage from the low-pressure stage; this is achieved by the combined action of a valve spring (4) and an electromagnet (3) which force the armature downwards.

Fuel flows around the whole of the armature for lubrication and cooling purposes.

Method of operation
The pressure-control valve has two closed control loops:
- A slower, closed electrical control loop for setting a variable average pressure level in the fuel rail and
- A faster hydromechanical control loop for balancing out high-frequency pressure pulses

Pressure-control valve activated:
When the pressure in the high-pressure stage needs to be increased, the force of the electromagnet is added to that of the spring. The pressure-control valve is activated and closes until a state of equilibrium is reached between the high pressure and the combined force of the electromagnet and the spring. At this point, it remains in partly open position and maintains a constant pressure. Variations in the delivery rate of the high-pressure pump and the withdrawal of fuel from the fuel rail by the nozzles are compensated by varying the valve aperture. The magnetic force of the electromagnet is proportional to the control current. The control current is varied by pulse-width modulation. A pulse frequency of 1 kHz is sufficiently high to prevent adverse armature movement or pressure fluctuations in the fuel rail.

On more recent systems, the pressure is regulated by controlling the quantity of fuel drawn in by the high-pressure pump. This reduces energy losses.

Fuel rail (high-pressure accumulator)

Function

The function of the high-pressure accumulator (fuel rail) is to maintain the fuel at high pressure. In so doing, the accumulator volume has to dampen pressure fluctuations caused by the fuel pulses delivered by the pump, and the fuel-injection cycles. This ensures that, when the nozzles open, the injection pressure remains constant.
The fuel rail also acts as a fuel distributor.

Design

The design of the tubular fuel rail (Fig. 1, Item 1) can vary depending on the various constraints which determine the way it is fitted to the engine. It has connections for fitting the rail-pressure sensor (3), the pressure limiter (4), the flow limiter (6, optional) and a pressure-control valve (if it is not fitted to the high-pressure pump).

Method of operation

The pressurized fuel delivered by the high-pressure pump passes via a high-pressure delivery line to the fuel-rail inlet (2). From there, it enters the fuel rail and is distributed to the individual nozzles (this explains the term "common rail" as the fuel rail is shared by all the nozzles).

Fuel pressure is measured by the rail-pressure sensor (see the chapter on "Sensors") and is adjusted to the required level by the pressure-control valve. The pressure limiter has the job of preventing fuel pressure in the fuel rail from exceeding the maximum level. The high-pressure fuel passes from the fuel rail to the nozzles via a flow limiter (optional), which prevents any fuel from flowing into the engine combustion chamber.

The cavity inside the fuel rail is permanently filled with pressurized fuel. The compressibility of the fuel under high pressure is utilized to achieve an accumulator effect. When fuel is released from the fuel rail for injection, the pressure in the high-pressure accumulator remains virtually constant, even when large quantities of fuel are released.

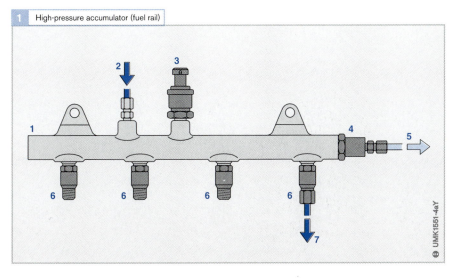

1 High-pressure accumulator (fuel rail)

Fig. 1
1 Fuel rail
2 Inlet from high-pressure pump
3 Fuel-rail pressure sensor
4 Pressure-limiting valve
5 Return pipe from fuel rail to fuel tank
6 Flow limiter
7 Fuel line to nozzle

Pressure-limiting valve
Function
The function of the pressure-limiting valve is equivalent to that of a pressure-relief valve. The pressure-limiting valve limits the pressure in the fuel rail by opening a pressure outlet when the pressure exceeds a specified limit.

Design and method of operation
The pressure-limiting valve (Fig. 2) is a mechanical component. It consists of the following parts:
- A valve housing (7) with an external thread for screwing to the fuel rail
- A connection to the fuel-return line to the fuel tank (9)
- A movable piston (4) and
- A plunger return spring (5)

At the end which is screwed to the fuel rail, there is a hole in the valve housing which is sealed by the tapered end of the plunger resting against the valve seat inside the valve housing. At normal operating pressure, a spring presses the plunger against the valve seat so that the fuel rail remains sealed. Only if the pressure rises above the maximum system pressure is the plunger forced back against the action of the spring by the pressure in the fuel rail so that the high-pressure fuel can escape. As it does so, the fuel is ducted through passages (3) to a central hole in the plunger. It then passes into the fuel return line and back to the fuel tank. As the valve opens, fuel can escape from the fuel rail to produce a reduction in fuel-rail pressure.

Flow limiter
Function
The flow limiter is used on heavy-duty trucks in particular. Its function is to prevent any unlikely event of continuous injection by a nozzle. To perform these tasks, the flow limiter shuts off the fuel line to the nozzle concerned if a specific maximum flow rate to the nozzle is exceeded.

Design
The flow limiter (Fig. 3) consists of a metal housing (5) with an external thread for screwing to the fuel rail (high pressure) and another external thread for screwing to the nozzle fuel line. The body has a hole at either end to provide the hydraulic connection to the fuel rail (1) and the nozzle fuel line (6).

Fig. 2
1 High-pressure connection
2 Valve
3 Duct
4 Plunger
5 Spring
6 Stop
7 Housing
8 Outlet hole
9 Return outlet

Fig. 3
1 Connection to fuel rail
2 Sealing washer
3 Plunger
4 Spring
5 Housing
6 Connection to nozzle
7 Valve seat
8 Restrictor

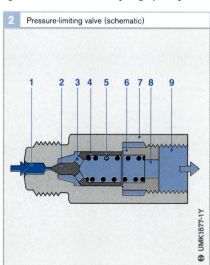

2 Pressure-limiting valve (schematic)

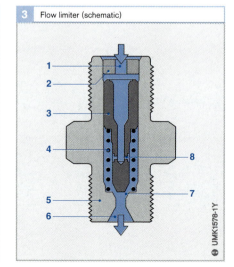

3 Flow limiter (schematic)

Inside the flow limiter, there is a plunger (3) which is forced towards the fuel rail by a spring (4). This plunger is sealed against the wall of the flow-limiter housing. The longitudinal passage through the plunger forms the hydraulic connection between the flow-limiter inlet and outlet. The diameter of the passage is restricted at its end. This constriction acts as a flow restrictor (8) with a definite throughflow rate.

Method of operation
Normal mode (Figs. 3 and 4)
The plunger (3) is at its resting position, i.e. resting against the stop at the fuel-rail end of the limiter. During a fuel-injection cycle, the pressure on the nozzle side drops slightly so that the piston moves towards the nozzle. The flow of fuel to the nozzle is compensated by the flow limiter by means of the volume displaced by the plunger and not by the flow restrictor (8), as the latter is too small to do this. At the end of the fuel-injection cycle, the plunger stops without sealing against the valve seat (7). The spring (4) then forces it back to its resting position and there is a residual flow of fuel through the flow restrictor.

The spring and the calibrated restriction are dimensioned so that, at maximum injected-fuel quantity (including a safety margin), the plunger can be returned to the stop at the fuel-rail end. The resting position is maintained until the next fuel-injection cycle.

Fault mode in case of major leakage
The large flow volume forces the plunger away from its resting position and against the valve seat at the outlet end. It then remains in this position at the nozzle end of the flow limiter until the engine is switched off. Then it closes off the fuel line to the nozzle.

Fault mode in case of minor leakage (Fig. 4)
Due to the leakage flow, the plunger does not come to rest in its resting position. After several fuel-injection cycles, the plunger moves far enough to seal against the valve seat at the limiter outlet.

As with a major leak, it remains in this position at the nozzle end of the flow limiter until the engine is switched off. It then closes off the fuel line to the nozzle.

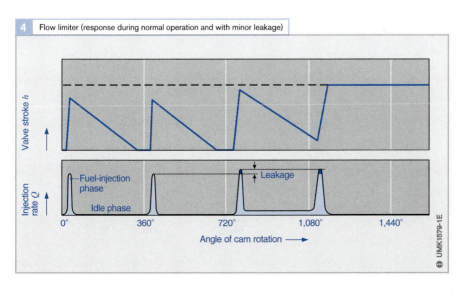

4 Flow limiter (response during normal operation and with minor leakage)

Valve stroke *h*

Injection rate *Q*

Fuel-injection phase

Leakage

Idle phase

0°   360°   720°   1,080°   1,440°

Angle of cam rotation ⟶

UMK1579-1E

Nozzle (fuel injection)

The nozzles are connected to the fuel rail by short high-pressure fuel lines. As with the nozzle sockets on diesel engines with direct injection (DI), the nozzles are ideally attached to the cylinder head by clamps. Consequently, common-rail fuel nozzles can be fitted to DI diesel engines without major modifications.

Functions

Injection timing and injected-fuel quantity are adjusted by means of an electrically operated nozzle. Injection timing is controlled by the crankshaft-position/time-based system of the electronic diesel control (EDC) system. This requires the use of sensors to detect the crankshaft position and the camshaft position (phase detection).

The one-piece nozzle replaces the nozzle-and-holder assembly used on conventional diesel fuel-injection systems.

In future there will also be nozzles with a piezoelectric actuator instead of a solenoid valve.

Design

The nozzle can be subdivided into a number of functional blocks:
- The hole-type nozzle (see the chapter on "Nozzles")
- The hydraulic servo system and
- The solenoid valve

Fuel passes from the high-pressure connection (Fig. 1a, Item 9) via an inlet channel to the nozzle and via the inlet restrictor (10) to the valve-control chamber (5). The valve control chamber is connected to the fuel return outlet (1) via the outlet restrictor (8) which can be opened by a solenoid valve.

---

1   Nozzle (principle of operation)

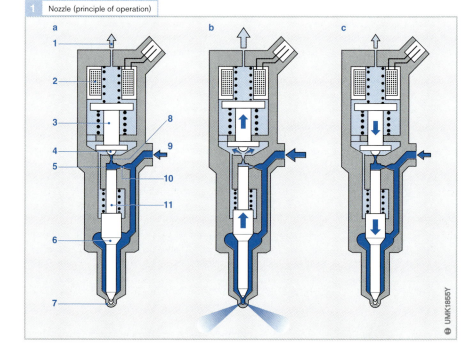

Fig. 1

a   Resting position
b   Nozzle opens
c   Nozzle closes

1   Fuel-return outlet
2   Solenoid coil
3   Solenoid armature
4   Valve ball
5   Valve-control
    chamber
6   Pressure shoulder
    of nozzle needle
7   Nozzle jet
8   Outlet restrictor
9   High-pressure
    connection
10   Inlet restrictor
11   Valve plunger

UMK1855Y

When the outlet restrictor is closed (Fig. 1a), the hydraulic force acting on the valve plunger (11) is greater than that acting on the pressure shoulder of the nozzle needle (6). As a result, the nozzle needle is forced against its seat and seals off the high-pressure duct from the combustion chamber. No fuel can therefore enter the combustion chamber.

When the solenoid valve is actuated, the outlet restrictor opens (Fig. 1b). As a result, pressure in the valve-control chamber drops and the force acting on the valve-control plunger is then reduced. As soon as the hydraulic force drops below the force acting on the pressure shoulder of the nozzle needle, the nozzle needle moves away from its seat so that the fuel can escape through the nozzle jets (7) into the engine combustion chamber. This indirect method is used to operate the nozzle needle by means of a hydraulic servo system because the forces required to open the nozzle needle rapidly cannot be generated directly by the solenoid valve. The extra fuel quantity (known as the control volume) required in addition to the fuel quantity passes through the control-chamber restrictors to the fuel return outlet.

In addition to the control volume, there are also leakage volumes through the nozzle and valve-plunger guides. The control and leakage volumes are ducted back to the fuel tank via the fuel return to which the overflow valve, high-pressure pump and pressure-control valve are also connected.

Method of operation

The function of the nozzle can be subdivided into four operating statuses when the engine and the high-pressure pump are operating:

- Nozzle closed (with high-pressure applied)
- Nozzle opens (start of injection)
- Nozzle fully open and
- Nozzle closes (end of injection)

Those operating statuses are brought about by the balance of forces acting on the nozzle components. When the engine is not running and the fuel rail is not pressurized, the nozzle spring closes the nozzle.

Nozzle closed (resting position)

The solenoid valve is in its resting position and not energized, and therefore closed (Figs. 1a and 2a overleaf).

When the outlet restrictor is closed, the armature ball is held against the outlet-restrictor seat by the force of the spring. Inside the valve-control chamber, the pressure rises to the pressure in the fuel rail. The same pressure is also present in the nozzle chamber. The forces acting on the end faces of the control plunger resulting from the fuel-rail pressure and the force of the nozzle spring hold the nozzle needle in the closed position against the opening force acting on the pressure shoulder of the needle.

Nozzle opens (start of injection)

To begin with, the nozzle is in its resting position. The pickup current is applied to the solenoid valve (see the chapter on "Electronic open- and closed-loop control"), which makes the solenoid valve to open rapidly (Figs. 1b and 2b overleaf). The required rapid switching times are achieved by configuring solenoid-valve activation in the ECU with high voltages and currents. The force of the now energized electromagnet overcomes the force of the valve spring and the armature opens the outlet restrictor. Within a very short space of time, the higher pickup current applied to the electromagnet is reduced to a lower holding current. This is possible because the air gap of the magnetic circuit is now smaller. Once the outlet restrictor is open, the fuel can flow out of the valve-control chamber into the chamber above it and through the fuel return outlet back to the fuel tank. The inlet restrictor prevents complete pressure compensation and the pressure in the valve-control chamber drops. As a result, pressure in the valve-control chamber falls below the pressure in the

2   Nozzle (schematic)

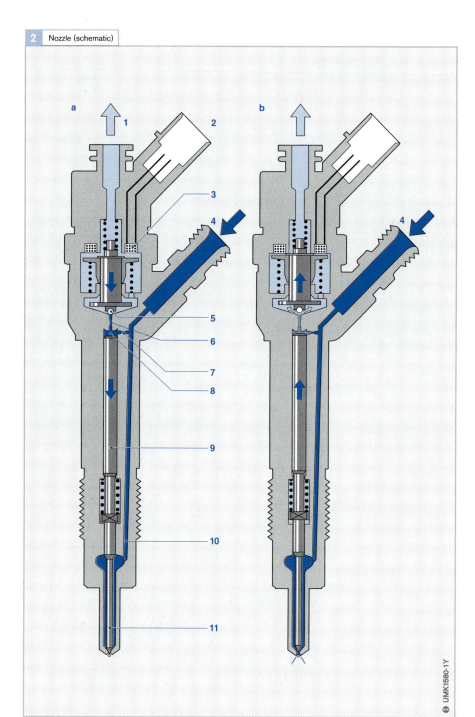

Fig. 2
a   Nozzle closed
    (resting position)
b   Nozzle open
    (injection)

1   Fuel-return outlet
2   Electrical
    connection
3   Actuator unit
    (solenoid valve)
4   Fuel inlet (high-
    pressure) from
    fuel rail
5   Valve ball
6   Outlet restrictor
7   Inlet restrictor
8   Valve-control
    chamber
9   Valve-control
    plunger
10 Inlet duct to nozzle
11 Nozzle needle

nozzle chamber, which is still the same as the pressure in the fuel rail. The reduced pressure in the valve-control chamber means that the force acting on the control plunger is also reduced and allows the nozzle needle to open. Fuel injection commences.

The rate of movement of the nozzle needle is determined by the difference in the flow rates through the inlet and outlet restrictors. The control plunger reaches its upper stop and remains in that position, resting against a cushion of fuel. The cushion is created by the flow of fuel between the inlet and outlet restrictors. The nozzle jet is now fully open and the fuel is injected into the combustion chamber at a pressure virtually equivalent to the fuel-rail pressure. The balance of forces in the nozzle is similar to that during the opening phase.

At a given system pressure, the fuel quantity injected is proportional to the length of time that the solenoid valve is open. This is entirely independent of the engine or pump speed (time-based injection system).

Nozzle closes (end of injection)

Once the solenoid valve is de-energized, the armature is pressed down by the force of the spring and the ball seals off the outlet restrictor (Fig. 1c). The armature is constructed in two parts. Although the armature plate is pressed down by a lug, it is also independently cushioned by the return spring so that it does not exert any downward force on the armature or the ball.

When the outlet restrictor closes, the pressure in the control chamber again rises to that in the fuel rail via the inlet restrictor. The higher pressure exerts a greater force on the control plunger. The force from the control chamber and the force of the nozzle spring now exceed the force acting on the pressure shoulder of the nozzle needle. The needle is pushed back to its closed position.

The closing rate of the nozzle needle is determined by the rate of flow through the inlet restrictor. The fuel-injection cycle comes to an end when the nozzle needle is resting against its seat, thus closing off the nozzle jets.

---

3   Diesel engine with common-rail fuel injection on engine test bench

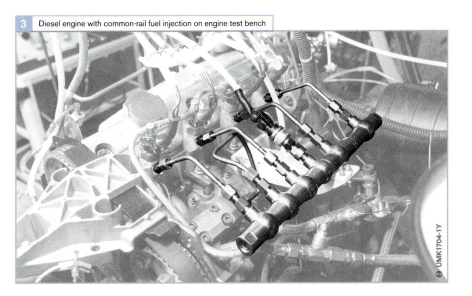

UMK1704-1Y

Nozzles

The nozzle injects the fuel into the combustion chamber of the diesel engine. It is a determining factor in the efficiency of mixture formation and combustion and therefore has a fundamental effect on engine performance, exhaust-gas behavior and noise. In order that nozzles can perform their function as effectively as possible, they have to be designed to match the fuel-injection system and engine in which they are used.

The nozzle is a central component of any fuel-injection system. It requires highly specialized technical knowledge on the part of its designers. The nozzle plays a major role in

- Shaping the rate-of-discharge curve (precise progression of pressure and fuel distribution relative to crankshaft rotation)
- Optimum atomization and distribution of fuel in the combustion chamber and
- Sealing off the fuel-injection system from the combustion chamber

Because of its exposed position in the combustion chamber, the nozzle is subjected to constant pulsating mechanical and thermal stresses from the engine and the fuel-injection system. The fuel flowing through the nozzle must also cool it. When the engine is overrunning, when no fuel is being injected, the nozzle temperature increases steeply. Therefore, it must have sufficient high-temperature resistance to cope with these conditions.

In fuel-injection systems based on in-line injection pumps (Type PE) and distributor injection pumps (Type VE/VR), and in unit pump systems (UPS), the nozzle is combined with the nozzle holder to form the nozzle-and-holder assembly (Figure 1) and installed in the engine. In high-pressure fuel-injection systems such as the common rail (CR) and unit injector systems (UIS) the nozzle is a single integrated unit so that the nozzle holder is not required.

Indirect-injection (IDI) engines use pintle nozzles, while direct-injection engines have hole-type nozzles.

The nozzles are opened by the fuel pressure. The nozzle opening, injection duration and rate-of-discharge curve (injection pattern) are the essential determinants of injected fuel quantity. The nozzles must close rapidly and reliably when the fuel pressure drops. The closing pressure is at least 40 bar above the maximum combustion pressure in order to prevent unwanted post-injection or intrusion of combustion gases into the nozzle. The nozzle must be designed specifically for the type of engine in which it is used as determined by

- The injection method (direct or indirect)
- The geometry of the combustion chamber
- The required injection-jet shape and direction
- The required penetration and atomization of the fuel jet
- The required injection duration and
- The required injected fuel quantity relative to crankshaft rotation

Standardized dimensions and combinations provide the required degree of adaptability combined with the minimum of component diversity. Because of the superior performance combined with lower fuel consumption that it offers, all new engine designs use direct injection (and therefore hole-type nozzles).

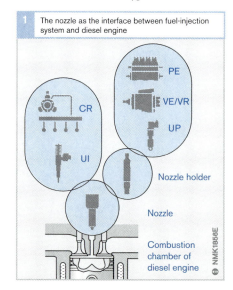

1   The nozzle as the interface between fuel-injection system and diesel engine

PE
VE/VR
CR
UP
UI
Nozzle holder
Nozzle
Combustion chamber of diesel engine

NMK1856E

▷ Dimensions of diesel fuel-injection technology

The world of diesel fuel injection is a world of superlatives.

The valve needle of a commercial-vehicle nozzle will open and close the nozzle more than a billion times in the course of its service life. It provides a reliable seal at pressures as high as 2,050 bar as well as having to withstand many other stresses such as

- The shocks caused by rapid opening and closing (on cars this can take place as frequently as 10,000 times a minute if there are pre- and post-injection phases)
- The high flow-related stresses during fuel injection and
- The pressure and temperature of the combustion chamber

The facts and figures below illustrate what modern nozzles are capable of.

- The pressure in the fuel-injection chamber can be as high as 2,050 bar. That is equivalent to the pressure produced by the weight of a large executive car acting on an area the size of a fingernail.

- The injection duration is 1...2 milliseconds (ms). In one millisecond, the sound wave from a loudspeaker only travels about 33 cm.
- The injection durations on a car engine vary between 1 mm^3 (pre-injection) and 50 mm^3 (full-load delivery); on a commercial vehicle between 3 mm^3 (pre-injection) and 350 mm^3 (full-load delivery). 1 mm^3 is equivalent to half the size of a pinhead. 350 mm^3 is about the same as 12 large raindrops (30 mm^3 per raindrop). That amount of fuel is forced at a velocity of 2,000 km/h through an opening of less than 0.25 mm^2 in the space of only 2 ms.
- The valve-needle clearance is 0.002 mm (2 μm). A human hair is 30 times as thick (0.06 mm).

Such high-precision technology demands an enormous amount of expertise in development, materials, production and measurement techniques.

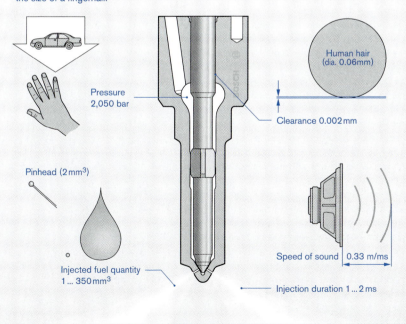

Pressure 2,050 bar

Human hair (dia. 0.06 mm)

Clearance 0.002 mm

Pinhead (2 mm^3)

Injected fuel quantity 1 ... 350 mm^3

Speed of sound | 0.33 m/ms

Injection duration 1 ... 2 ms

NMK1708-2E

Pintle nozzles

Usage

Pintle nozzles are used on indirect injection (IDI) engines, i.e. engines that have prechambers or whirl chambers. In this type of engine, the mixing of fuel and air is achieved primarily by the whirl effects created inside the cylinder. The shape of the injection jet can also assist the process. Pintle nozzles are not suitable for direct-injection engines as the peak pressures inside the combustion chamber would open the nozzle. The following types of pintle nozzle are available:

- Standard pintle nozzles
- Throttling pintle nozzles and
- Flatted-pintle nozzles

Design and method of operation

The fundamental design of all pintle nozzles is virtually identical. The differences between them are to be found in the geometry of the pintle t(Figure 1, Item 7). Inside the nozzle body is the nozzle needle (3) It is pressed downwards by the force F_F exerted by the spring and the pressure pin in the nozzle holder so that it seals off the nozzle from the combustion chamber. As the pressure of the fuel in the pressure chamber (5) increases, it acts on the pressure shoulder (6) and forces the nozzle needle upwards (force F_D). The pintle lifts away from the injector orifice (8) and opens the way for fuel to pass through into the combustion chamber (the nozzle "opens"; opening pressure 110...170 bar). When the pressure drops, the nozzle closes again. Opening and closing of the nozzle is thus controlled by the pressure inside the nozzle.

Design variations
Standard pintle nozzle

The nozzle needle of (Figure 1, Item 3) of a standard pintle nozzle has a pintle (7) that fits into the injector orifice (8) of the nozzle with a small degree of play. By varying the dimensions and geometry of the of the pintle, the characteristics of the injection jet produced can be modified to suit the requirements of different engines.

Throttling pintle nozzle

One of the variations of the pintle nozzle is the throttling pintle nozzle. The profile of the pintle allows a specific rate-of-discharge curve to be produced. As the nozzle needle opens, at first only a very narrow annular orifice is provided which allows only a small amount of fuel to pass through (throttling effect).

As the pintle draws further back with increasing fuel pressure, the size of the gap through which fuel can flow increases. The greater proportion of the injected fuel quantity is only injected as the pintle approaches the limit of its upward travel. By modifying the rate-of-discharge curve in this way, "softer" combustion is produced because the pressure in the combustion chamber does not rise so quickly. As a result, combustion noise is reduced in the part-load range. This means that the shape of the pintle in combination with the throttling gap and the characteristic of the compression spring in the nozzle holder produces the desired rate-of-discharge curve.

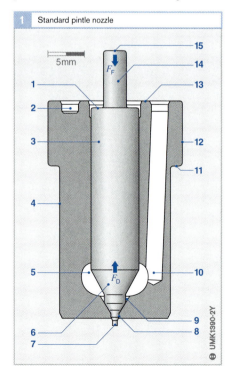

1   Standard pintle nozzle

5mm

F_F

F_D

UMK1390-2Y

Flatted-pintle nozzle

The flatted-pintle nozzle (Figure 3) has a pintle with a flatted face on its tip which, as the nozzle opens (at the beginning of needle lift travel) produces a wider passage within the annular orifice. This helps to prevent deposits at that point by increasing the volumetric flow rate. As a result, flatted-pintle nozzles "coke" to a lesser degree and more evenly. The annular orifice between the jet orifice and the pintle is very narrow (< 10 μm). The flatted face is frequently parallel to the axis of the nozzle needle. By setting the flatted face at an angle, the volumetric flow rate, Q, can be increased in the flatter section of the rate-of-discharge curve (Figure 4). In this way, a smoother transition between the initial phase and the fully-open phase of the rate-of-discharge curve can be obtained. Specially designed variations in pintle geometry allow the flow-rate pattern to be modified to suit particular engine requirements. As a result, engine noise in the part-load range is reduced and engine smoothness improved.

Heat shielding

Temperatures above 220 °C also promote nozzle coking. Thermal-protection plates or sleeves (Figure 2) help to overcome this problem by conducting heat from the combustion chamber into the cylinder head.

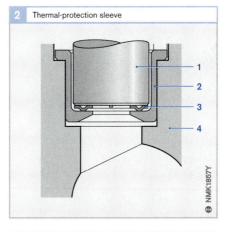

2   Thermal-protection sleeve

Fig. 2
1 Pintle nozzle
2 Thermal-protection
   sleeve
3 Protective disc
4 Cylinder head

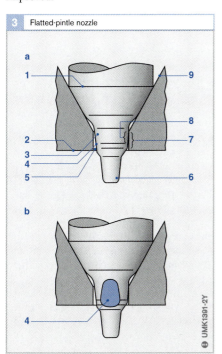

3   Flatted-pintle nozzle

a
b

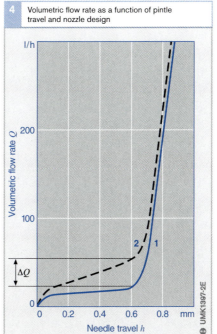

4   Volumetric flow rate as a function of pintle travel and nozzle design

l/h

Volumetric flow rate Q

200

100

ΔQ

0
   0   0.2   0.4   0.6   0.8   mm
   Needle travel h

Fig. 3
a Side view
b Front view
   (rotation of 90°
   relative to side view)

1 Pintle seat face
2 Nozzle-body base
3 Throttling pintle
4 Flatted face
5 Injection orifice
6 Profiled pintle
7 Total contact ratio
8 Cylindrical overlap
9 Nozzle-body seat face

Fig. 4
1 Throttling pintle
   nozzle
2 Flatted-pintle nozzle
   (throttling pintle nozzle
   with flatted face)

ΔQ Difference in
   volumetric flow rate
   due to flatted face

Hole-type nozzles

Usage

Hole-type nozzles are used on direct-injection (DI) engines. The position in which the nozzles are fitted is generally determined by the engine design. The injector orifices are set at a variety of angles according to the requirements of the combustion chamber (Figure 1). Hole-type nozzles are subdivided into
● Blind-hole nozzles and
● Sac-less (vco) nozzles

Hole-type nozzles are also divided according to size into
● *Type P* which have a needle diameter of 4 mm (blind-hole and sac-less (vco) nozzles) and
● *Type S* which have a needle diameter of 5 or 6 mm (blind-hole nozzles for large engines)

In the common rail (CR) and unit injector (UI) fuel-injection systems, the hole-type nozzle is a single integrated unit. It therefore combines the functions of nozzle and nozzle holder.

The opening pressure of hole-type nozzles is in the range 150...350 bar.

Design

The injection orifices (Figure 2, Item 6) are positioned around the cladding of the nozzle cone (7). The number and size are dependent on
● The required injected fuel quantity
● The shape of the combustion chamber and
● The air vortex (whirl) inside the combustion chamber

The bore of the injection orifices is slightly larger at the inner end than at the outer end. This difference is defined by the port taper factor. The leading edges of the injection orifices may be rounded by using the hydro-erosion (HE) process. This involves the use of an HE fluid that contains abrasive particles which smooth off the edges at points where high flow velocities occur (leading edges of injection orifices). Hydro-erosion can be used both on blind-hole and sac-less (vco) nozzles. Its purpose is to
● optimize the flow resistance coefficient,
● pre-empt crosion of edges caused by particles in the fuel, and/or
● tighten flow-rate tolerances.

Nozzles have to be carefully designed to match the engine in which they are used. Nozzle design plays a decisive role in
● Precise metering of injected fuel (injection duration and injected fuel quantity relative to degrees of crankshaft rotation)
● Fuel conditioning (number of jets, spray shape and atomization of fuel)
● Fuel dispersal inside the combustion chamber and
● Sealing the fuel-injection system against the combustion chamber

The pressure chamber (10) is created by electrochemical machining (ECM). An electrode through which an electrolyte solution is passed is introduced into the pre-bored nozzle body. Material is then removed from the positively charged nozzle body (anodic dissolution).

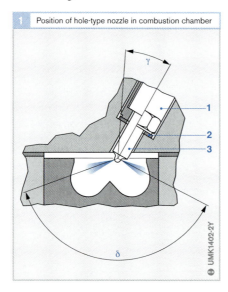

Fig. 1
1 Nozzle
2 Sealing washer
3 Hole-type nozzle

γ Inclination
δ Jet cone angle

1 Position of hole-type nozzle in combustion chamber

UMK1402-2Y

Design variations

The fuel in the space below the seat of the nozzle needle evaporates after combustion and, therefore, contributes significantly to the hydrocarbon (HC) emissions produced by the engine. For this reason, it is important to keep that dead volume or "detrimental" volume as small as possible.

In addition, the geometry of the needle seat and the shape of the nozzle cone have a decisive influence on the opening and closing characteristics of the nozzle. This in turn affects the soot and NO_X emissions produced by the engine.

The consideration of these various factors in combination with the demands of the engine and the fuel-injection system has led to a variety of nozzle designs.

There are two basic types of injector:
● Blind-hole nozzles and
● Sac-less (vco) nozzles

Among the blind-hole nozzles, there are a number of variations.

Blind-hole nozzles

On a blind-hole nozzle (Figure 2, Item 6) the injection orifices exit from a blind hole in the tip of the nozzle.

If the nozzle has a *cone*, the injection orifices are drilled either mechanically or by electro-erosion depending on design.

In blind-hole nozzles with a *conical tip*, the injection orifices are generally created by electro-erosion.

Blind-hole nozzles may have a cylindrical or conical blind hole of varying dimensions.

Blind-hole nozzles with a cylindrical blind hole and conical tip (Figure 3), which consists of a cylindrical and a hemispherical section, offer a large amount of scope with regard to the number of holes, length of injection orifices and orifice taper angle. The nozzle cone is hemispherical in shape, which – in combination with the shape of the blind hole – ensures that all the spray holes are of equal length.

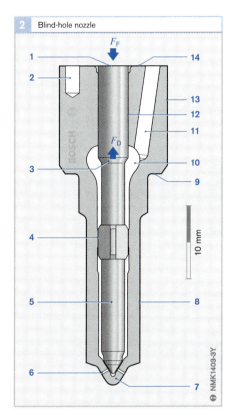

Fig. 2 Blind-hole nozzle

Fig. 2
1 Stroke-limiting shoulder
2 Fixing hole
3 Pressure shoulder
4 Secondary needle guide
5 Needle shaft
6 Injection orifice
7 Nozzle cone
8 Nozzle body
9 Nozzle-body shoulder
10 Pressure chamber
11 Inlet passage
12 Needle guide
13 Nozzle-body collar
14 Sealing face

F_F Spring force
F_D Force acting on pressure shoulder due to fuel pressure

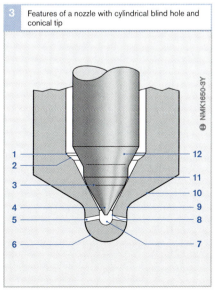

Fig. 3 Features of a nozzle with cylindrical blind hole and conical tip

Fig. 3
1 Shoulder
2 Seat lead-in
3 Needle-seat face
4 Needle tip
5 Injection orifice
6 Conical nozzle tip
7 Cylindrical blind hole (dead volume)
8 Injection orifice leading edge
9 Neck radius
10 Nozzle-cone taper
11 Nozzle-body seat face
12 Damping taper

Blind-hole nozzles with cylindrical blind holes and conical tip (Figure 4a) are produced only with a spray-hole length of 0.6 mm. The conical-shaped tip increases the strength of the cone by virtue of the greater wall thickness between the neck radius (3) and the nozzle-body seat (4).

Blind-hole nozzles with conical blind holes and conical tip (Figure 4b) have a smaller dead volume than nozzles with a cylindrical blind hole. The volume of the blind hole is between that of a sac-less (vco) nozzle and a blind-hole nozzle with a cylindrical blind hole. In order to obtain an even wall thickness throughout the cone, it is shaped conically to match the shape of the blind hole.

A further refinement of the blind-hole nozzle is the *micro-blind-hole nozzle* (Figure 4c). Its blind-hole volume is around 30 % smaller than that of a conventional blind-hole nozzle. This type of nozzle is particularly suited to use in common-rail fuel-injection systems, which operate with a relatively slow needle lift and consequently a comparatively long nozzle-seat restriction. The micro-blind-hole nozzle currently represents the best compromise between minimizing dead volume and even spray dispersal when the nozzle opens for common-rail systems.

Sac-less (vco) nozzles

In order to minimize the dead volume – and therefore the HC emissions – the injection orifice exits from the nozzle-body seat face. When the nozzle is closed, the nozzle needle more or less covers the injection orifice so that there is no direct connection between the blind hole and the combustion chamber (Figure 4d). The blind-hole volume is considerably smaller than that of a blind-hole nozzle. Sac-less (vco) nozzles have a significantly lower stress capacity than blind-hole nozzles and can therefore only be produced with a spray-hole length of 1 mm. The nozzle tip has a conical shape. The injection orifices are generally produced by electro-erosion.

Special spray-hole geometries, secondary needle guides and complex needle-tip geometries are used to further improve spray dispersal, and consequently mixture formation, on both blind-hole and sac-less (vco) nozzles.

4   Nozzle cones

a

4
3
1
2

b

5
2

c

d

⊕ NMK1858Y

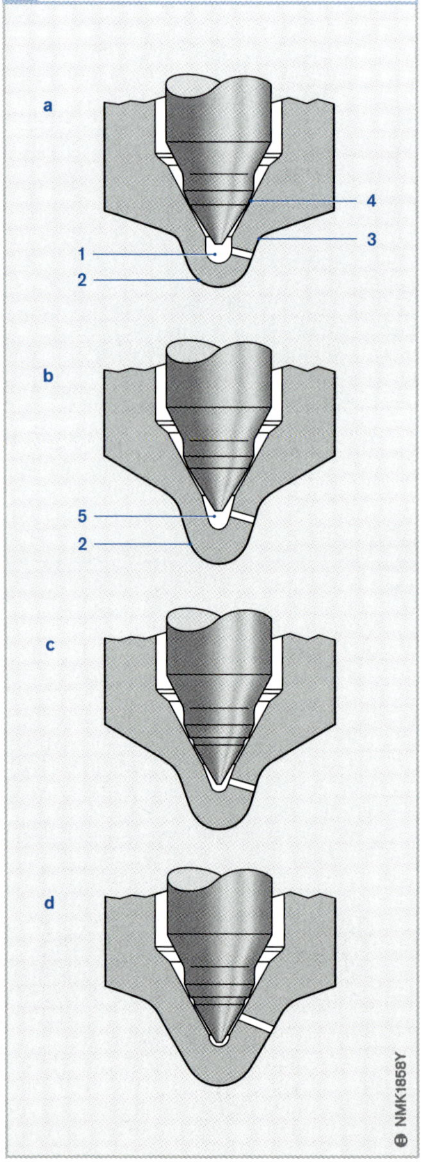

Fig. 4
a Cylindrical blind hole and conical tip
a Conical blind hole and conical tip
c Micro-blind-hole
d Sac-less (vco) nozzle

1 Cylindrical blind hole
2 Conical nozzle tip
3 Neck radius
4 Nozzle-body seat face
5 Conical blind hole

Heat shielding

The maximum temperature capacity of hole-type nozzles is around 300 °C (heat resistance of material). Thermal-protection sleeves are available for operation in especially difficult conditions, and there are even cooled nozzles for large-scale engines.

Effect on emissions

Nozzle geometry has a direct effect on the engine's exhaust-gas emission characteristics.
- The *spray-hole geometry* (Figure 5, Item 1) affects particulate and NO_X emissions.
- The *needle-seat geometry* (2) affects engine noise due to its effect on the pilot volume, i.e. the volume injected at the beginning of the injection process. The aim of optimizing spray-hole and seat geometry is to produce a durable nozzle capable of mass production to very tight dimensional tolerances.
- *Blind-hole geometry* (3) affects HC emissions, as previously mentioned. The designer can select and combine the various nozzle characteristics to obtain the optimum design for a particular engine and vehicle concerned.

For this reason, it is important that the nozzles are designed specifically for the vehicle, engine and fuel-injection system in which they are to be used. When servicing is required, it is equally important that genuine OEM parts are used in order to ensure that engine performance is not impaired and exhaust-gas emissions are not increased.

Spray shapes

Basically, the shape of the injection jet for car engines is long and narrow because these engines produce a large degree of swirl inside the combustion chamber. There is no swirl effect in commercial-vehicle engines. Therefore, the injection jet tends to be wider and shorter. Even where there is a large amount of swirl, the individual injection jets must not intermingle otherwise fuel would be injected into areas where combustion has already taken place and therefore where there is a lack of air. This would result in the production of large amounts of soot.

Hole-type nozzles have up to six injection orifices in cars and up to ten in commercials. The aim of future development will be to further increase the number of injection orifices and to reduce their bore size (<0.12 mm) in order to obtain even finer dispersal of fuel.

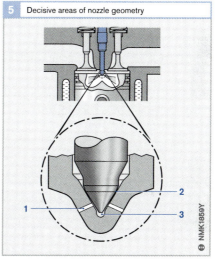

5   Decisive areas of nozzle geometry

Fig. 5
1   Injection-orifice geometry
2   Seat geometry
3   Blind-hole geometry

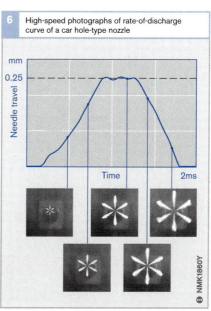

6   High-speed photographs of rate-of-discharge curve of a car hole-type nozzle

Future development of the nozzle

In view of the rapid development of new, high-performance engines and fuel-injection systems with sophisticated functionality (e.g. multiple injection phases), continuous development of the nozzle is a necessity. In addition, there are number of aspects of nozzle design which offer scope for innovation and further improvement of diesel engine performance in the future. The most important aims are:

- Minimizing untreated emissions in order to reduce or even eliminate the expense of costly exhaust-gas treatment equipment that also presents difficulties with regard to waste disposal (e.g. soot filters)
- Minimizing fuel consumption
- Optimizing engine noise

There various different areas on which attention can be focused in the future development of the nozzle (Figure 1) and a corresponding variety of development tools (Figure 2). New materials are also constantly being developed which offer improvements in durability. The use of multiple injection phases also has consequences for the design of the nozzle.

If different types of fuel (e.g. designer fuels) are used, this also affects nozzle design because of the differences in viscosity or flow characteristics. Such changes will in some cases also demand new production processes such as laser drilling for the injection orifices.

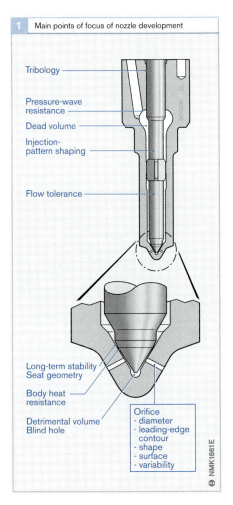

1   Main points of focus of nozzle development

Tribology

Pressure-wave resistance

Dead volume

Injection-pattern shaping

Flow tolerance

Long-term stability
Seat geometry

Body heat resistance

Detrimental volume
Blind hole

Orifice
- diameter
- leading-edge contour
- shape
- surface
- variability

NMK1861E

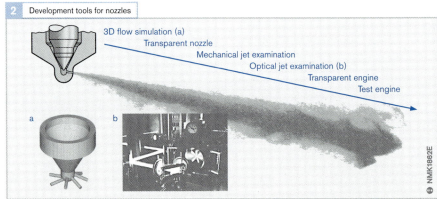

2   Development tools for nozzles

3D flow simulation (a)
Transparent nozzle
Mechanical jet examination
Optical jet examination (b)
Transparent engine
Test engine

a   b

NMK1862E

▶ High-precision technology

The image associated with diesel engines in many people's minds is more one of heavy-duty machinery than high-precision engineering. But modern diesel fuel-injection systems are made up of components that are manufactured to the highest degrees of accuracy and required to withstand enormous stresses.

The nozzle is the interface between the fuel-injection system and the engine. It has to open and close precisely and reliably for the entire life of the engine. When it is closed, it must not leak. This would increase fuel consumption, adversely affect exhaust-gas emissions and might even cause engine damage.

To ensure that the nozzles seal reliably at the high pressures generated in modern fuel-injection systems such as the VR (VP44), CR, UPS and UIS designs (up to 2,050 bar), they have to be specially designed and very precisely manufactured. By way of illustration, here are some examples:

- To ensure that the sealing face of the nozzle body (1) provides a reliable seal, its has a dimensional tolerance of 0.001 mm (1 μm). That means it must be accurate to within approximately 4,000 metal atom layers!
- The nozzle-needle guide clearance (2) is 0.002...0.004 mm (2...4 μm). The dimensional tolerances are similarly less than 0.001 mm (1 μm).

The injection orifices (3) in the nozzles are created by an electro-erosion machining process. This process erodes the metal by vaporization caused by the high temperature generated by the spark discharge between an electrode and the workpiece. Using high-precision electrodes and accurately configured parameters, extremely precise injection orifices with diameters of 0.12 mm can be produced. This means that the smallest injection orifice diameter is only twice the thickness of a human hair (0.06 mm). In order to obtain better injection characteristics, the leading edges of the nozzle injection orifices are rounded off by special abrasive fluids (hydro-erosion machining).

The minute tolerances demand the use of highly specialized and ultra-accurate measuring equipment such as
- optical 3-D coordinate measuring machine for measuring the injection orifices, or
- laser interferometers for checking the smoothness of the nozzle sealing faces.

The manufacture of diesel fuel-injection components is thus "high-volume, high-technology".

▼ A matter of high-precision

NMK1709-2Y

1 Nozzle body sealing face
2 Nozzle-needle guide clearance
3 Injection orifice

Nozzle holders

A nozzle holder combines with the matching nozzle to form the nozzle-and-holder assembly. There is a nozzle-and-holder assembly fitted in the cylinder head for each engine cylinder (Figure 1). These components form an important part of the fuel-injection system and help to shape engine performance, exhaust emissions and noise characteristics. In order that they are able to perform their function properly, they must be designed to suit the engine in which they are used.

The nozzle (4) in the nozzle holder sprays fuel into the diesel-engine combustion chamber (6). The nozzle holder contains the following essential components:

● *Valve spring(s)* (9)
  which act(s) against the nozzle needle so as to close the nozzle
● *Nozzle-retaining nut* (8)
  which retains and centers the nozzle
● *Filter* (11)
  for keeping dirt out of the nozzle
● *Connections* for the fuel supply and return lines which are linked via the *pressure channel* (10)

Depending on design, the nozzle holder may also contain seals and spacers. Standardized dimensions and combinations provide the required degree of adaptability combined with the minimum of component diversity.

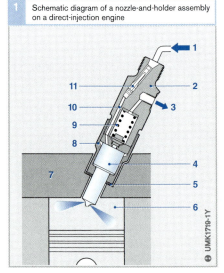

1 Schematic diagram of a nozzle-and-holder assembly on a direct-injection engine

UMK1719-1Y

Fig. 1
1 Fuel supply
2 Holder body
3 Fuel return
4 Nozzle
5 Sealing gasket
6 Combustion chamber of diesel engine
7 Cylinder head
8 Nozzle-retaining nut
9 Valve spring
10 Pressure channel
11 Filter

2 Bosch type designation codes for nozzle holders

$$K \; B \; A \; L \; Z \; 105 \; S \; V \; XX...$$

K Nozzle holder

B Attached by flange or clamp
C External thread on nozzle-retaining nut
D Sleeve nut

A Spring at bottom
Nozzle-holder dia. 17 mm (Type P nozzle), dia. 25 mm (Type S nozzle)
E Spring at bottom
Nozzle-holder dia. 21 mm (Type P and S nozzle)
N Spring at bottom
Nozzle-holder dia. 17/21 mm (Type P nozzle)

L Long nozzle collar
No letter = Short nozzle collar

Z Two inlet passages
No letter = One inlet passage

Ser. no.
Specimen: last 7 digits of the drawing number

V Test holder
No letter = Standard nozzle holder

P Nozzle (collar dia. 14.3 mm)
S Nozzle (collar dia. 17 mm)

Length (mm)

SMK1831E

The design of the nozzle holder for direct-injection (DI) and indirect-injection (IDI) engines is basically the same. But since modern diesel engines are almost exclusively direct-injection, the nozzle-and-holder assemblies illustrated here are mainly for DI engines. The descriptions, however, can be applied to IDI nozzles as well, but bearing in mind that the latter use pintle nozzles rather than the hole-type nozzles found in DI engines.

Nozzle holders can be combined with a range of nozzles. In addition, depending on the required injection pattern, there is a choice of
● *standard nozzle holder* (single-spring nozzle holder) or
● *two-spring nozzle holder* (not for unit pump systems).

A variation of those designs is the *stepped holder* which is particularly suited to situations where space is limited.

Depending on the fuel-injection system in which they are used, nozzle holders may or may not be fitted with *needle-motion sensors*.

The needle-motion sensor signals the precise start of injection to the engine control unit.

Nozzle holders may be attached to the cylinder block by flanges, clamps, sleeve nuts or external threads. The fuel-line connection is in the center or at the side.

The fuel that leaks past the nozzle needle acts as lubrication. In many nozzle-holder designs, it is returned to the fuel tank by a fuel-return line.

Some nozzle holders function without fuel leakage – i.e. without a fuel-return line. The fuel in the spring chamber has a damping effect on the needle stroke at high injection volumes and engine speeds so that a similar injection pattern to that of a two-spring nozzle holder is generated.

In the common-rail and unit-injector high-pressure fuel-injection systems, the nozzle is integral with the injector, so that a nozzle-and-holder assembly is unnecessary.

For large-scale engines with a per-cylinder output of more than 75 kW, there are application-specific fuel-injector assemblies which may also be cooled.

Fig. 3
a Stepped nozzle holder for commercial vehicles
b Standard nozzle holder for various engine types
c Two-spring nozzle holder for cars
d Standard nozzle holder for various engine types
e Stepped nozzle holder without fuel-leakage connection for commercial vehicles
f Stepped nozzle holder for commercial vehicles
g Stepped nozzle holder for various engine types
h Two-spring nozzle holder for cars
i Stepped nozzle holder for various engine types
j Standard nozzle holder with pintle nozzle for various types of IDI engine

3 Examples of nozzle-and-holder assemblies

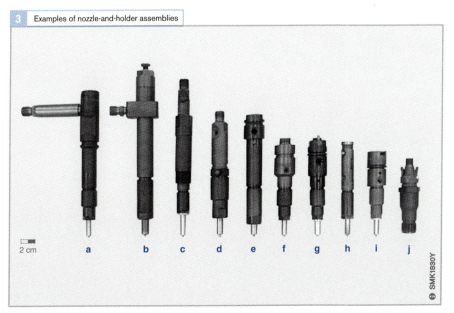

2 cm   a   b   c   d   e   f   g   h   i   j

SMK1830Y

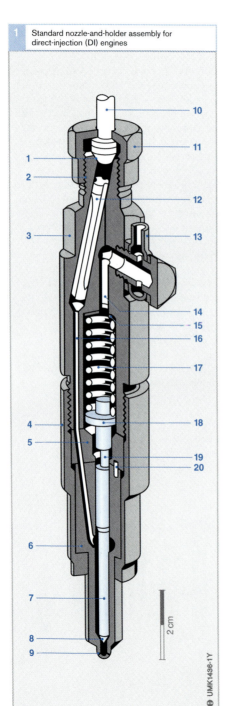

1 Standard nozzle-and-holder assembly for direct-injection (DI) engines

Fig. 1
1 Sealing cone
2 Screw thread for
   central pressure
   connection
3 Holder body
4 Nozzle-retaining nut
5 Intermediate disk
6 Nozzle body
7 Nozzle needle
8 Nozzle-body seat
   face
9 Injection orifice
10 Fuel inlet
11 Sleeve nut
12 Edge-type filter
13 Leak fuel connection
14 Leak fuel port
15 Shim
16 Pressure passage
17 Compression spring
18 Pressure pin
19 Pressure pin
20 Locating pin

Standard nozzle holders

Design and usage

The key features of standard nozzle holders are as follows:
- Cylindrical exterior with diameters of 17, 21, 25 and 26 mm
- Non-twist hole-type nozzles for engines with direct injection and
- Standardized individual components (springs, pressure pins, nozzle retaining nuts) that permit different combinations

The nozzle-and-holder assembly is made up of nozzle holder and nozzle (Figure 1, with hole-type nozzle). The nozzle holder consists of the following components:
- Holder body (3)
- Intermediate disk (5)
- Nozzle-retaining nut (4)
- Pressure pin (18)
- Compression spring (17)
- Shim (15) and
- Locating pin (20)

The nozzle is attached centrally to the holder by the nozzle-retaining nut. When the retaining nut and holder body are screwed together, the intermediate disk is pressed against the sealing faces of the holder and nozzle body. The intermediate disk acts as a limiting stop for the needle lift and also centers the nozzle relative to the nozzle holder by means of the locating pins.

The pressure pin centers the compression spring and is guided by the nozzle-needle pressure pin (19).

The pressure passage (16) inside the nozzle holder body connects through the channel in the intermediate disk to the inlet passage of the nozzle, thus connecting the nozzle to the high-pressure line of the fuel-injection pump. If required, an edge-type filter (12) may be fitted inside the nozzle holder. This keeps out any dirt that may be contained in the fuel.

Method of operation

The compression spring inside the nozzle holder acts on the nozzle needle via the pressure pin. The spring tension is set by means of a shim. The force of the spring thus determines the opening pressure of the nozzle.

The fuel passes through the edge-type filter (12) to the pressure passage (16) in the holder body (3), through the intermediate disk (5) and finally through the nozzle body (6) to the space (8) surrounding the nozzle needle. During the injection process, the nozzle needle (7) is lifted upwards by the pressure of the fuel (110...170 bar for pintle nozzles and 150...350 bar for hole-type nozzles). The fuel passes through the injection orifices (9) into the combustion chamber. The injection process comes to an end when the fuel pressure drops to a point where the compression spring (17) is able to push the nozzle needle back against its seat. Start of injection is thus controlled by fuel pressure. The injected fuel quantity depends essentially on how long the nozzle remains open.

In order to limit needle lift for pre-injection, some designs have a nozzle-needle damper (Figure 2).

Stepped nozzle holders

Design and usage

On multi-valve commercial-vehicle engines in particular, where the nozzle-and-holder assembly has to be fitted vertically because of space constraints, stepped nozzle-and-holder assemblies are used (Figure 3). The reason for the name can be found in the graduated dimensions (1).

The design and method of operation are the same as for standard nozzle holders. The essential difference lies in the way in which the fuel line is connected. Whereas on a standard nozzle holder it is screwed centrally to the top end of the nozzle holder, on a stepped holder it is connected to the holder body (11) by means of a delivery connection (10). This type of arrangement is normally used to achieve very short injection fuel lines, and has a beneficial effect on the injection pressure because of the smaller dead volume in the fuel lines.

Stepped nozzle holders are produced with or without a leak fuel connection (9).

Fig. 2
a Closed nozzle
b Damped lift

1 Compression spring
2 Holder body
3 Leak gap
4 Hydraulic cushion
5 Damper piston
6 Nozzle needle

h_u Undamped lift
     (approx. 1/3 of full lift)

Fig. 3
1 Step
2 Pressure passage
3 Pressure pin
4 Intermediate disk
5 Nozzle-retaining nut
6 Nozzle body
7 Locating pin
8 Compression spring
9 Leak fuel port
10 Delivery connection
11 Holder body
12 Thread for extractor
     bolt

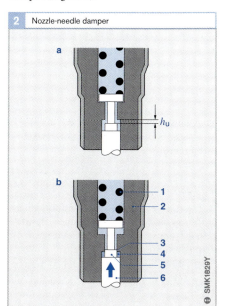

2 Nozzle-needle damper

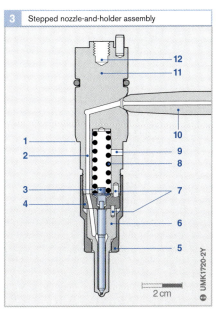

3 Stepped nozzle-and-holder assembly

2 cm

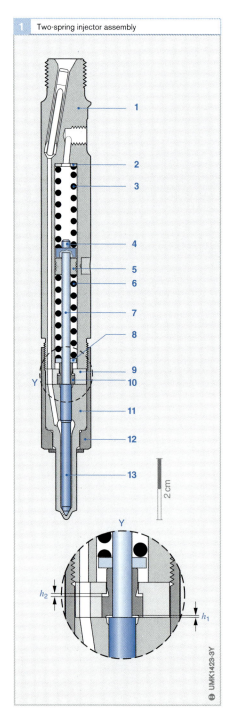

1 Two-spring injector assembly

Fig. 1
1 Holder body
2 Shim
3 Compression
   spring 1
4 Pressure pin
5 Guide washer
6 Compression
   spring 2
7 Pressure pin
8 Spring seat
9 Intermediate disk
10 Stop sleeve
11 Nozzle body
12 Nozzle-retaining nut
13 Nozzle needle

h_1 Plunger lift to port
    closing
h_2 Main lift

Fig. 2
a Standard nozzle
   holder (single-spring)
b Two-spring nozzle
   holder

h_1 Plunger lift to port
    closing
h_2 Main lift

Two-spring nozzle holders

Usage
The two-spring nozzle holder is a refinement of the standard nozzle holder. It has the same external dimensions. Its graduated rate-of-discharge curve (Figure 2) produces "softer" combustion and therefore a quieter engine, particularly at idle speed and part load. It is used primarily on direct-injection (DI) engines.

Design and method of operation
The two-spring nozzle holder (Figure 1) has two compression springs positioned one behind the other. Initially, only one of the compression springs (3) is acting on the nozzle needle (13) and thus determines the opening pressure. The second compression spring (6) rests against a stop sleeve (10) which limits the plunger lift to port closing. During the injection process, the nozzle needle initially moves towards the plunger lift to port closing, h_1 (0.03...0.06 mm for DI engines, 0.1 mm for IDI engines). This allows only a small amount of fuel into the combustion chamber.

As the pressure inside the nozzle holder continues to increase, the stop sleeve overcomes the force of both compression springs (3 and 6). The nozzle needle then completes the main lift ($h_1 + h_2$, 0.2...0.4 mm) so that the main injected fuel quantity is injected.

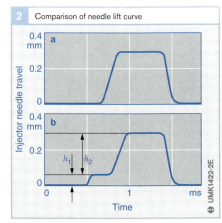

2 Comparison of needle lift curve

Nozzle holders with needle-motion sensors

Usage

Start of delivery is a key variable for optimizing diesel-engine performance. Detection of this variable allows the adjustment of start of delivery according to engine load and speed within a closed control loop. In systems with distributor and in-line fuel-injection pumps, this is achieved by means of a nozzle with a needle-motion sensor (Figure 2) which transmits a signal when the nozzle needle starts to move upwards. It is sometimes also called a needle-motion sensor.

Design and method of operation

A current of approximately 30 mA is passed through the detector coil (Figure 2, Item 11). This produces a magnetic field. The extended pressure pin (12) slides inside the guide pin (9). The penetration depth X determines the magnetic flux in the detector coil. By virtue of the change in magnetic flux in the coil, movement of the nozzle needle induces a velocity-dependent voltage signal (Figure 1) in the coil which is processed by an analyzer circuit in the electronic control unit. When the signal level exceeds a threshold voltage, it is interpreted by the analyzer circuit to indicate the start of injection.

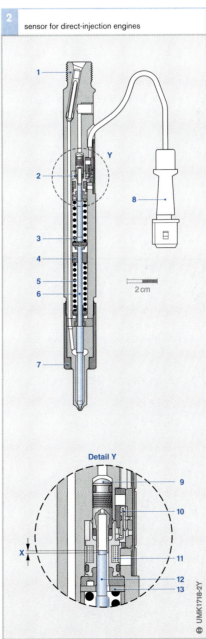

2 sensor for direct-injection engines

Detail Y

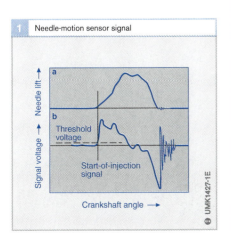

1 Needle-motion sensor signal

Fig. 1
a   Needle-lift curve
b   Corresponding coil
    signal voltage curve

Fig. 2
1   Holder body
2   Needle-motion
    sensor
3   Compression spring
4   Guide washer
    Compression spring
    Pressure pin
7   Nozzle-retaining nut
8   Connection to
    analyzer circuit
9   Guide pin
10 Contact tab
11 Detector coil
12 Pressure pin
13 Spring seat

X   Penetration depth

High-pressure lines

Regardless of the basic system concept – in-line fuel-injection pump, distributor injection pump or unit pump systems – it is the high-pressure delivery lines and their connection fittings that furnish the links between the fuel-injection pump(s) and the nozzle-and-holder assemblies at the individual cylinders. In common-rail systems, they serve as the connection between the high-pressure pump and the rail as well as between rail and nozzles. No high-pressure delivery lines are required in the unit-injector system.

High-pressure connection fittings

The high-pressure connection fittings must supply secure sealing against leakage from fuel under the maximum primary pressure. The following types of fittings are used:
- Sealing cone and union nut
- Heavy-duty insert fittings and
- Perpendicular connection fittings

Sealing cone with union nut
All of the fuel-injection systems described above use sealing cones with union nuts (Fig. 1). The advantages of this connection layout are:
- Easy adaptation to individual fuel-injection systems
- Fitting can be disconnected and reconnected numerous times
- The sealing cone can be shaped from the base material

At the end of the high-pressure line is the compressed pipe-sealing cone (3). The union nut (2) presses the cone into the high-pressure connection fitting (4) to form a seal. Some versions are equipped with a supplementary thrust washer (1). This provides a more consistent distribution of forces from the union nut to the sealing cone. The cone's open diameter should not be restricted, as this would obstruct fuel flow. Compressed sealing cones are generally manufactured in conformity with DIN 73 365 (Fig. 2).

Heavy-duty insert fittings
Heavy-duty insert fittings (Fig. 3) are used in unit-pump and common-rail systems as installed in heavy-duty commercial vehicles. With the insert fitting, it is not necessary to route the fuel line around the cylinder head to bring it to the nozzle holder or nozzle. This allows shorter fuel lines with associated benefits when it comes to space savings and ease of assembly.

The screw connection (8) presses the line insert (3) directly into the nozzle holder (1) or nozzle. The assembly also includes a mainte-

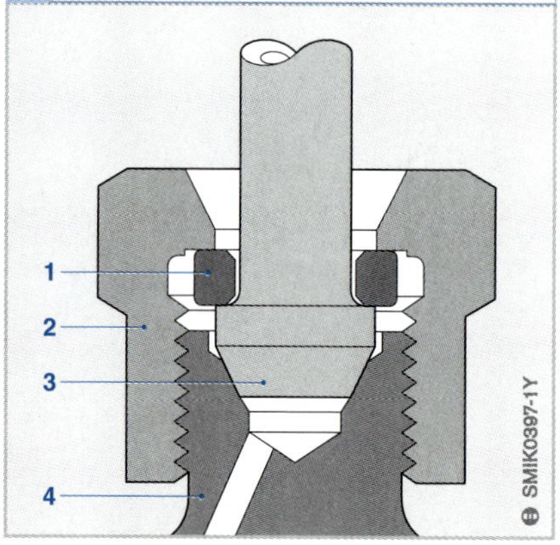

Fig. 1
1   Thrust washer
2   Union nut
3   Pipe sealing cone on high-pressure delivery line
4   Pressure connection on fuel-injection pump or nozzle holder

Fig. 2
1   Sealing surface

d Outer line diameter
d_1 Inner line diameter
d_2 Inner cone diameter
d_3 Outer cone diameter
k Length of cone
R_1, R_2 Radii

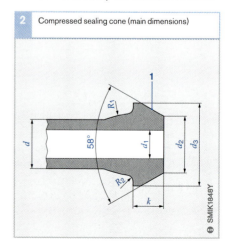

1   High-pressure connection with sealing cone and union nut

SMIK0397-1Y

2   Compressed sealing cone (main dimensions)

SMIK1848Y

nance-free edge-type filter (5) to remove coarse contamination from the fuel. At its other end, the line is attached to the high-pressure delivery line (7) with a sealing cone and union nut (6).

Perpendicular connection fittings

Perpendicular connection fittings (Fig. 4) are used in some passenger-car applications. They are suitable for installations in which there are severe space constraints. The fitting contains passages for fuel inlet and return (7, 9). A bolt (1) presses the perpendicular fitting onto the nozzle holder (5) to form a sealed connection.

High-pressure delivery lines

The high-pressure fuel lines must withstand the system's maximum pressure as well as pressure variations that can attain very high fluctuations. The lines are seamless precision-made steel tubing in killed cast steel which has a particularly consistent microstructure. Dimensions vary according to pump size (Table 1, next page).

All high-pressure delivery lines are routed to avoid sharp bends. The bend radius should not be less than 50 mm.

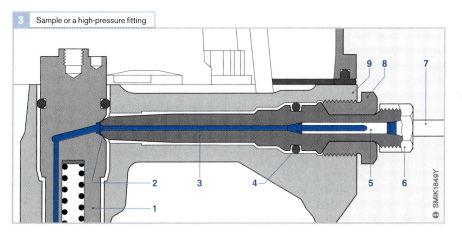

3 Sample or a high-pressure fitting

Fig. 3
1 Nozzle holder
2 Sealing cone
3 High-pressure fitting
4 Seal
5 Edge-type filter
6 Union nut
7 High-pressure delivery line
8 Screw connections
9 Cylinder head

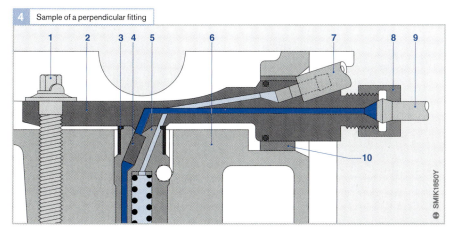

4 Sample of a perpendicular fitting

Fig. 4
1 Expansion bolt
2 Perpendicular fitting
3 Molded seal
4 Edge-type filter
5 Nozzle holder
6 Cylinder head
7 Fuel return line (leakage-fuel line)
8 Union nut
9 High-pressure delivery line
10 Clamp

Length, diameter and wall depth of the high-pressure lines all affect the injection process. To cite some examples: Line length influences speed-sensitive the rate of discharge, while internal diameter is related to throttling loss and compression effects, which will be reflected in the injected-fuel quantity. These considerations lead to prescribed line dimensions that must be strictly observed. Tubing of other dimensions should never be installed during service and repairs. Defective high-pressure tubing should always be replaced by OEM lines. During servicing or maintenance, it is also important to observe precautions against fouling entering the system. This applies in any case to all service work on fuel-injection systems.

A general priority in the development of fuel-injection systems is to minimize the length of high-pressure lines. Shorter lines produce better injection-system performance.

Injection is accompanied by the formation of pressure waves. These are pulses that propagate at the speed of sound before finally being reflected on impact at the ends. This phenomenon increases in intensity as engine speed rises. Engineers exploit it to raise injection pressure. The engineering process entails defining line lengths that are precisely matched to the engine and the fuel-injection system.

All cylinders are fed by high-pressure delivery lines of a single, uniform length. More or less angled bends in the lines compensate for the different distances between the outlets from the fuel-injection pump or rail, and the individual engine cylinders.

The primary factor determining the high-pressure line's compression-pulsating fatigue strength is the surface quality of the inner walls of the lines, as defined by material and peak-to-valley height. Especially demanding performance requirements are satisfied by prestressed high-pressure delivery lines (for applications of 1,400 bar and over). Before installation on the engine, these customized lines are subjected to extremely high pressures (up to 3,800 bar). Then pressure is suddenly relieved. The process compresses the material on the inner walls of the lines to provide increased internal strength.

The high-pressure delivery lines for vehicle engines are normally mounted with clamp brackets located at specific intervals. This means that transfer of external vibration to the lines is either minimal or nonexistent.

The dimensions of high-pressure lines for test benches are subject to more precise tolerance specifications.

Table 1

d Outer line diameter

d_1 Inner line diameter

Wall thicknesses indicated in **bold** should be selected when possible.

Dimensions for high-pressure lines are usually indicated as follows:

$d \times s \times l$

l Line length

1 Main dimensions of major high-pressure delivery lines in mm

Wall thickness s

d \ d_1	1.4	1.5	1.6	1.8	2.0	2.2	2.5	2.8	3.0	3.6	4.0	4.5	5.0	6.0	7.0	8.0	9.0
4	1.3	1.25	**1.2**														
5	1.8	1.75	**1.7**	1.6													
6		**2.25**	**2.2**	2.1	2	1.9	1.75	**1.6**	1.5								
8					3	2.9	**2.75**	**2.6**	2.5	2.2	2						
10							**3.75**	**3.6**	3.5	3.2	3	2.75	2.5				
12									4.5	4.2	4	3.75	3.5				
14											5	4.75	4.5	4		3	
17												6	5.5	5		4.5	
19																	5
22																7	

Cavitation in the high-pressure system

Cavitation can damage fuel-injection systems (Fig. 1). The process takes place as follows:

Local pressure variations occur at restrictions and in bends when a fluid enter an enclosed area at extremely high speeds (for instance, in a pump housing or in a high-pressure line). If the flow characteristics are less than optimum, low-pressure sectors can form at these locations for limited periods of time, in turn promoting the formation of vapor bubbles.

These gas bubbles implode in the subsequent high-pressure phase. If a wall is located immediately adjacent to the affected sector, the concentrated high energy can create a cavity in the surface over time (erosion effect). This is called cavitation damage.

As the vapor bubbles are transported by the fluid's flow, cavitation damage will not necessarily occur at the location where the bubble forms. Indeed, cavitation damage is frequently found in eddy zones.

The causes behind these temporary localized low-pressure areas are numerous and varied. Typical factors include:

- Discharge processes
- Closing valves
- Pumping between moving gaps and
- Vacuum waves in passages and lines

Attempts to deal with cavitation problems by improving material quality and surface-hardening processes cannot produce anything other than very modest gains. The ultimate objective is and remains to prevent the vapor bubbles from forming, and, should complete prevention prove impossible, to improve flow behavior to limit the negative impacts of the bubbles.

1 Cavitation damage in the distributor head of a VE pump

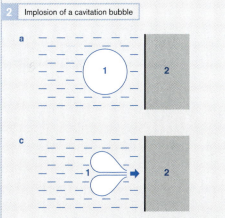

2 Implosion of a cavitation bubble

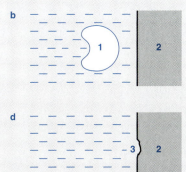

Fig. 1
1   Cavitation

Fig. 2
a   A vapor bubble
     is formed
b   The vapor bubble
     collapses
c   The collapsed
     sections form a
     sharp edge with
     extremely high
     energy
d   The imploding vapor
     bubble leaves a
     recess on the
     surface

1   Vapor bubble
2   Wall
3   Recess

Electronic Diesel Control (EDC)

Electronic control of a diesel engine enables precise and differentiated modulation of fuel-injection parameters. This is the only means by which a modern diesel engine is able to satisfy the many demands placed upon it. The EDC (Electronic Diesel Control) system is subdivided into three areas, "Sensors and desired-value generators", "Control unit" and "Actuators".

Requirements

The lowering of fuel consumption and harmful exhaust-gas emissions (NO_X, CO, HC, particulates) combined with simultaneous improvement of engine power output and torque are the guiding principles of current development work on diesel engine design.

In recent years, this has led to an increase in the popularity of the direct-injection (DI) diesel engine which uses much higher fuel-injection pressures than indirect-injection (IDI) engines with whirl or prechamber systems. Because of the more efficient mixture formation and the absence of flow-related losses between the whirl chamber/prechamber and the main combustion chamber, the fuel consumption of direct-injection engines is 10 … 15 % lower than that achieved by indirect-injection designs.

In addition, diesel engine development has been influenced by the high levels of comfort and convenience demanded in modern cars. Noise levels, too, are subject to more and more demanding demands.

As a result, the performance demanded of the fuel-injection and engine management systems has also increased, specifically with regard to

- High fuel-injection pressures
- Rate-of-discharge curve variability
- Pre-injection and, where applicable, post-injection
- Variation of injected fuel quantity, charge-air pressure and start of delivery to suit operating conditions

- Temperature-dependent excess fuel quantity for starting
- Control of idle speed independently of engine load
- Controlled exhaust-gas recirculation (cars)
- Cruise control and
- Tight tolerances for start of delivery and quantity, and maintenance of high precision over the service life of the system (long-term performance)

Conventional mechanical governing of engine speed uses a number of adjusting mechanisms to adapt to different engine operating conditions and ensures a high mixture formation quality. Nevertheless, it is restricted to a simple engine-based control loop and there are a number of important influencing variables that it cannot take account of or cannot respond quickly enough to.

As demands have increased, what was originally a straightforward system using electric actuator shafts has developed into the present-day EDC, a complex electronic control system capable of processing large amounts of data in real time. It can form part of an overall electronic vehicle control system ("drive-by-wire"). And as a result of the increasing integration of electronic components, the control-system circuitry can be accommodated in a very small space.

System overview

Due to the rapid improvement in microcontroller performance over recent years, the Electronic Diesel Control EDC system is capable of meeting all the demands outlined above.

In contrast with diesel-engine vehicles with conventional mechanically controlled fuel-injection pumps, the driver of a vehicle equipped with EDC has no direct control over the injected fuel quantity through the accelerator pedal and cable. Instead, the injected fuel quantity is determined by a number of variable factors. Those include:

- The vehicle response desired by the driver (accelerator pedal position)
- The engine operating status
- The engine temperature
- Intervention by other systems (e.g. traction control)
- The effect on exhaust-gas emission levels, etc.

The control unit calculates the injected fuel quantity on the basis of all those factors. Start of delivery can also be varied. This demands a comprehensive monitoring concept that detects inconsistencies and initiates appropriate actions in accordance with the effects (e.g. torque limitation or limp-home mode in the idle-speed range). EDC therefore incorporates a number of control loops.

The Electronic Diesel Control system is also capable of data exchange with other electronic systems such as traction control, transmission control or dynamic handling systems such as ESP (Electronic Stability Program). As a result, the engine management system can be integrated in the vehicle's overall control system network, thereby enabling functions such as reduction of engine torque when the automatic transmission changes gear, regulation of engine torque to compensate for wheel spin, disabling of fuel injection by the engine immobilizer, etc.

The EDC system is fully integrated in the vehicle's diagnostic system. It meets all OBD (On-Board Diagnosis) and EOBD (European OBD) requirements.

System structure

The Electronic Diesel Control EDC is subdivided into three sections (Figure 1):

1. The *sensors and desired-value generators* (1) detect the engine operating conditions (e.g. engine speed) and the driver's control commands (e.g. switch positions). They convert physical variables into electrical signals.

2. The *control unit* (2) processes the information received from the sensors and desired-value generators using specific mathematical calculation sequences (control algorithms). It controls the actuators by means of electrical output signals. It also provides interfaces with other systems (4) and with the vehicle's diagnostic system (5).

3. The *actuators* (3) convert the electrical output signals from the control unit into physical variables (e.g. the solenoid valve for fuel injection or the solenoid pump actuator).

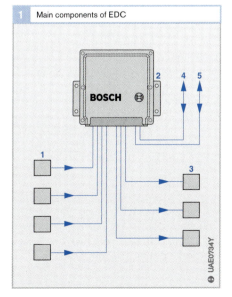

1   Main components of EDC

Fig. 1
1   Sensors and
    desired-value
    generators
    (input signals)
2   ECU
3   Actuators
4   Interface with other
    systems
5   Diagnosis interface

In-line fuel-injection pumps

1   Overview of the EDC components for in-line injection pumps

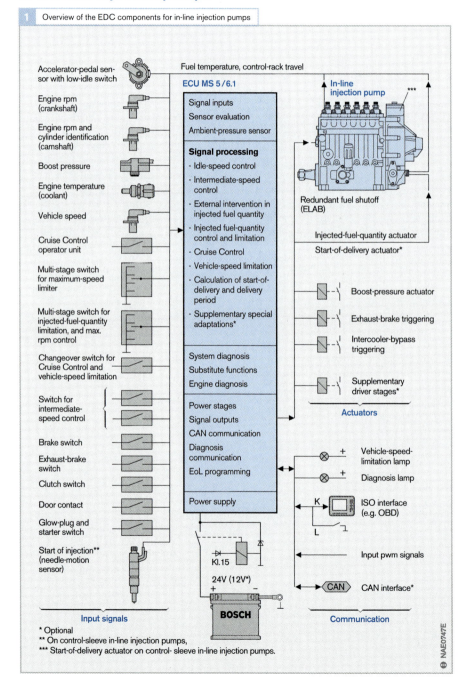

Input signals:
- Accelerator-pedal sensor with low-idle switch
- Engine rpm (crankshaft)
- Engine rpm and cylinder identification (camshaft)
- Boost pressure
- Engine temperature (coolant)
- Vehicle speed
- Cruise Control operator unit
- Multi-stage switch for maximum-speed limiter
- Multi-stage switch for injected-fuel-quantity limitation, and max. rpm control
- Changeover switch for Cruise Control and vehicle-speed limitation
- Switch for intermediate-speed control
- Brake switch
- Exhaust-brake switch
- Clutch switch
- Door contact
- Glow-plug and starter switch
- Start of injection** (needle-motion sensor)

Input signals

Fuel temperature, control-rack travel

ECU MS 5/6.1

Signal inputs
Sensor evaluation
Ambient-pressure sensor

Signal processing
- Idle-speed control
- Intermediate-speed control
- External intervention in injected fuel quantity
- Injected fuel-quantity control and limitation
- Cruise Control
- Vehicle-speed limitation
- Calculation of start-of-delivery and delivery period
- Supplementary special adaptations*

System diagnosis
Substitute functions
Engine diagnosis

Power stages
Signal outputs
CAN communication
Diagnosis communication
EoL programming

Power supply

Kl. 15

24V (12V*)
+ −

BOSCH

In-line injection pump ***

Redundant fuel shutoff (ELAB)

Injected-fuel-quantity actuator
Start-of-delivery actuator*

Boost-pressure actuator
Exhaust-brake triggering
Intercooler-bypass triggering
Supplementary driver stages*

Actuators

Vehicle-speed-limitation lamp
Diagnosis lamp

K   ISO interface (e.g. OBD)
L

Input pwm signals

CAN   CAN interface*

Communication

* Optional
** On control-sleeve in-line injection pumps,
*** Start-of-delivery actuator on control-sleeve in-line injection pumps.

NAE0747E

Helix and-Port-controlled axial-piston distributor pumps

2 | Overview of the EDC components for VE-EDC port-and-helix-controlled distributor pumps

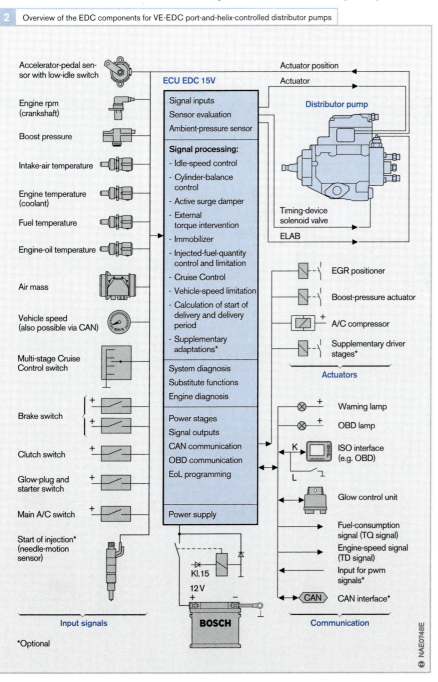

Accelerator-pedal sensor with low-idle switch

Engine rpm (crankshaft)

Boost pressure

Intake-air temperature

Engine temperature (coolant)

Fuel temperature

Engine-oil temperature

Air mass

Vehicle speed (also possible via CAN)

Multi-stage Cruise Control switch

Brake switch

Clutch switch

Glow-plug and starter switch

Main A/C switch

Start of injection* (needle-motion sensor)

Input signals

*Optional

ECU EDC 15V

Signal inputs
Sensor evaluation
Ambient-pressure sensor

Signal processing:
- Idle-speed control
- Cylinder-balance control
- Active surge damper
- External torque intervention
- Immobilizer
- Injected-fuel-quantity control and limitation
- Cruise Control
- Vehicle-speed limitation
- Calculation of start of delivery and delivery period
- Supplementary adaptations*

System diagnosis
Substitute functions
Engine diagnosis

Power stages
Signal outputs
CAN communication
OBD communication
EoL programming

Power supply

Kl.15
12 V

BOSCH

Actuator position
Actuator

Distributor pump

Timing-device solenoid valve

ELAB

EGR positioner

Boost-pressure actuator

A/C compressor

Supplementary driver stages*

Actuators

Warning lamp

OBD lamp

ISO interface (e.g. OBD)

Glow control unit

Fuel-consumption signal (TQ signal)

Engine-speed signal (TD signal)

Input for pwm signals*

CAN interface*

Communication

NAE0748E

Solenoid-valve-controlled axial-piston and radial-piston distributor pumps

3 | Overview of the EDC components for solenoid-valve-controlled distributor pumps VE-MV, VR

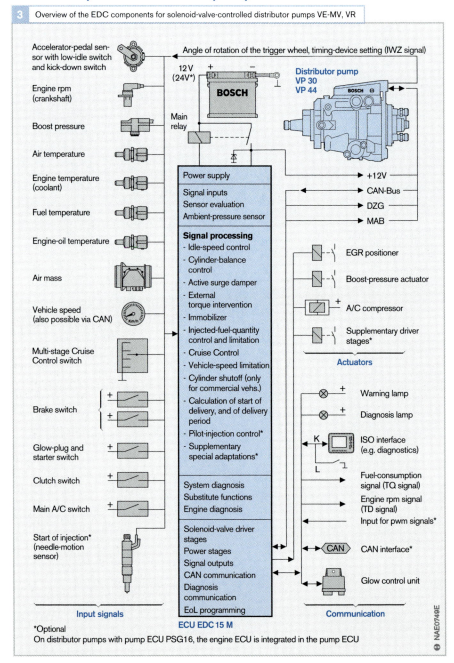

*Optional
On distributor pumps with pump ECU PSG16, the engine ECU is integrated in the pump ECU

Common Rail System (CRS)

2 Overview of the EDC components for the Common Rail System (CRS)

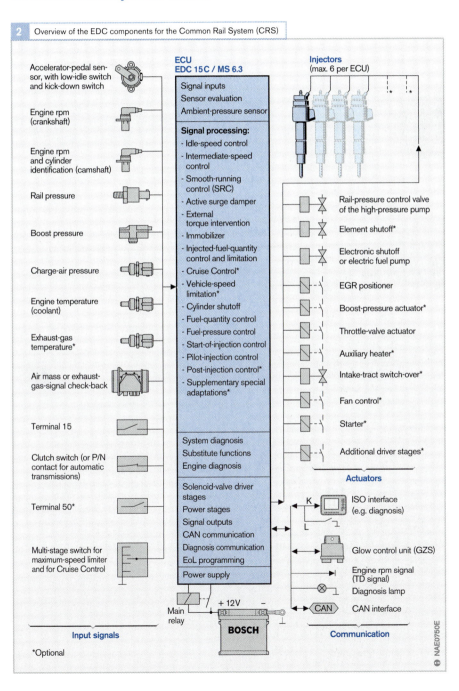

Input signals

Accelerator-pedal sensor, with low-idle switch and kick-down switch

Engine rpm (crankshaft)

Engine rpm and cylinder identification (camshaft)

Rail pressure

Boost pressure

Charge-air pressure

Engine temperature (coolant)

Exhaust-gas temperature*

Air mass or exhaust-gas-signal check-back

Terminal 15

Clutch switch (or P/N contact for automatic transmissions)

Terminal 50*

Multi-stage switch for maximum-speed limiter and for Cruise Control

*Optional

ECU EDC 15C / MS 6.3

Signal inputs
Sensor evaluation
Ambient-pressure sensor

Signal processing:
- Idle-speed control
- Intermediate-speed control
- Smooth-running control (SRC)
- Active surge damper
- External torque intervention
- Immobilizer
- Injected-fuel-quantity control and limitation
- Cruise Control*
- Vehicle-speed limitation*
- Cylinder shutoff
- Fuel-quantity control
- Fuel-pressure control
- Start-of-injection control
- Pilot-injection control
- Post-injection control*
- Supplementary special adaptations*

System diagnosis
Substitute functions
Engine diagnosis

Solenoid-valve driver stages
Power stages
Signal outputs
CAN communication
Diagnosis communication
EoL programming
Power supply

Main relay   +12V   −

BOSCH

Injectors (max. 6 per ECU)

Rail-pressure control valve of the high-pressure pump

Element shutoff*

Electronic shutoff or electric fuel pump

EGR positioner

Boost-pressure actuator*

Throttle-valve actuator

Auxiliary heater*

Intake-tract switch-over*

Fan control*

Starter*

Additional driver stages*

Actuators

K   ISO interface (e.g. diagnosis)
L

Glow control unit (GZS)

Engine rpm signal (TD signal)

Diagnosis lamp

CAN   CAN interface

Communication

NAE0750E

Unit Injector System (UIS) for passenger cars

1 Overview of the EDC components for a passenger-car unit injector system (UIS)

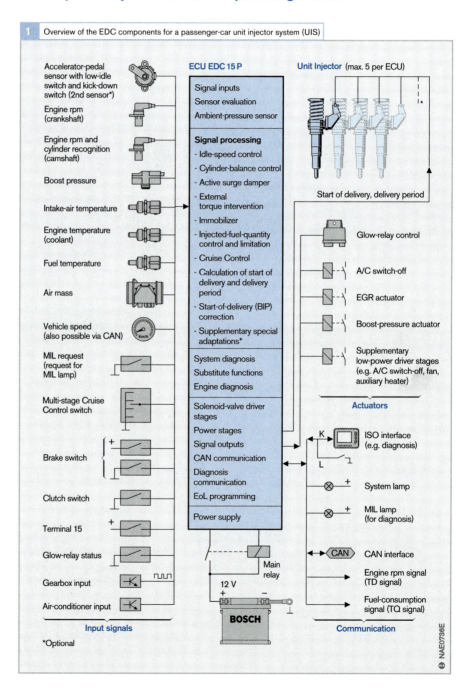

Unit Injector System (UIS) and Unit Pump System (UPS) for commercial vehicles

2 Overview of the EDC components for Unit Injector System (UIS) and Unit Pump Systems (UPS) for commercial vehicles

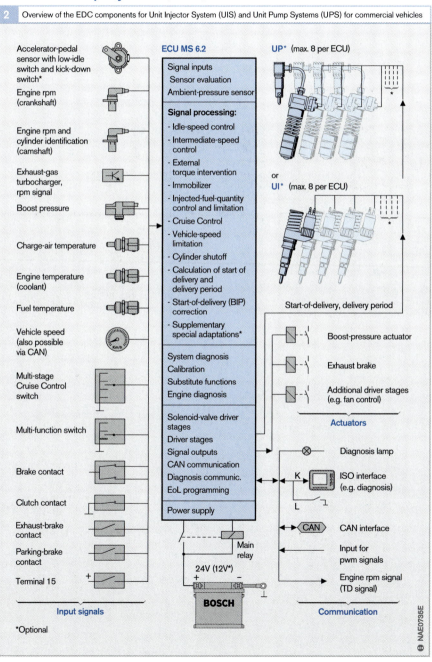

Accelerator-pedal sensor with low-idle switch and kick-down switch*

Engine rpm (crankshaft)

Engine rpm and cylinder identification (camshaft)

Exhaust-gas turbocharger, rpm signal

Boost pressure

Charge-air temperature

Engine temperature (coolant)

Fuel temperature

Vehicle speed (also possible via CAN)

Multi-stage Cruise Control switch

Multi-function switch

Brake contact

Clutch contact

Exhaust-brake contact

Parking-brake contact

Terminal 15

Input signals

*Optional

ECU MS 6.2

Signal inputs
  Sensor evaluation
Ambient-pressure sensor

Signal processing:
- Idle-speed control
- Intermediate-speed control
- External torque intervention
- Immobilizer
- Injected-fuel-quantity control and limitation
- Cruise Control
- Vehicle-speed limitation
- Cylinder shutoff
- Calculation of start of delivery and delivery period
- Start-of-delivery (BIP) correction
- Supplementary special adaptations*

System diagnosis
Calibration
Substitute functions
Engine diagnosis

Solenoid-valve driver stages
Driver stages
Signal outputs
CAN communication
Diagnosis communic.
EoL programming

Power supply

Main relay

24V (12V*)

BOSCH

UP* (max. 8 per ECU)

or

UI* (max. 8 per ECU)

Start-of-delivery, delivery period

Boost-pressure actuator

Exhaust brake

Additional driver stages (e.g. fan control)

Actuators

Diagnosis lamp

ISO interface (e.g. diagnosis)

CAN interface

Input for pwm signals

Engine rpm signal (TD signal)

Communication

NAE0735E

1) Some parts of the
adaptation process
are also referred to
as calibration.

Application-related adaptation[1]) of car engines

Application-related adaptation means modification of an engine to suit a particular type of vehicle intended for a specific type of use. Adaptation of the fuel-injection system – and specifically of electronic diesel control EDC – is a major part of that process.

All new diesel engines for cars are now direct-injection (DI) engines. And they all have to comply with the Euro III emission control standards that have been in force since 2000, or other comparable standards. These emission standards – combined with the higher expectations in the area of vehicle user-friendliness – can only be met by the use of sophisticated electronic control systems. Such systems have the capability – and reflect the necessity – of controlling thousands of parameters (approx. 6,000 in the case of the present EDC generation). Those parameters are subdivided into:
- Individual parameter values
  (e.g. temperature thresholds at which specific functions are activated) and
- Ranges of parameter values in the form of two-dimensional or multi-dimensional data maps (e.g. injection point t_E as a function of engine speed n, injected-fuel quantity m_e and start of delivery FB)

The optimization potential of EDC systems has become so great that it is now limited only by the constraints of time available and the cost of the personnel and the work involved in adapting and testing the various functions and their interaction.

Adaptation phases
Application-related adaptation of car engines is subdivided into the three stages described below.

Hardware adaptation
In the context of application-related adaptation of car engines, items such as the combustion chamber, the injection pump and the injectors are referred to as hardware. That hardware is primarily adapted in such a way that the performance and emission figures demanded are obtained. Hardware adaptation is performed initially on an engine test bench under static conditions. If dynamic tests are possible on the test bench, they are used to further optimize the engine and the fuel-injection system.

1 Vehicle-specific calibration using PC tools has become the standard

SAE 0922Y

Software adaptation

Once the hardware adaptation is complete, the control-unit software is accordingly configured and adapted for optimum mixture preparation and combustion control. For example, this includes calculating and programming the engine data maps for start of injection, exhaust-gas recirculation and charge-air pressure. As with hardware adaptation, this work is carried out on the test bench.

Vehicle-related adaptation

When the basis for the initial vehicle trials has been established, adaptation of all parameters that affect engine response and dynamic characteristics takes place. This third stage involves the essential adaptation to the particular vehicle concerned. The work is for the greater part performed with the engine in situ (Figure 1).

Interaction between the three phases

As there are reciprocal effects between the adaptation phases, recursions (repeated procedures) are required. As soon as possible, it is also necessary to run all three phases simultaneously with the engine in the vehicle and on the test bench.

For example, at low engine loads a very high exhaust-gas recirculation rate is aimed at in order to reduce the NO_X emissions. Under dynamic conditions, this can lead to poor "accelerator response" on the part of the engine. In order to obtain good acceleration characteristics, the static emissions settings programmed in the software adaptation phase must be re-adjusted. In turn, this may result in negative effects on emissions under certain engine operating conditions which have to be compensated for under other conditions.

In the example outlined, there is a fundamental conflict between the various objectives: on the one hand, strict requirements have to be met (e.g. statutory limits for exhaust emission levels), while on the other hand there are "optional" demands that are more attributable to the desire for comfort and performance (engine response, noise, etc.). The latter can result in opposing conclusions. A compromise between the different objectives offers the vehicle manufacturer the opportunity to imbue the vehicle with some of the features that make up its characteristic brand identity.

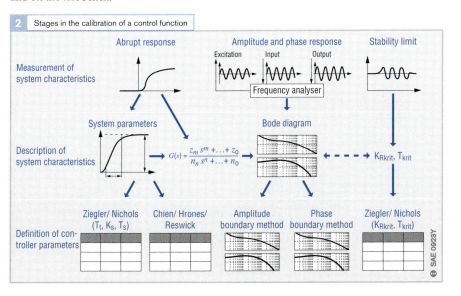

2   Stages in the calibration of a control function

SAE 0923Y

Adaptation to differing ambient conditions

The various controllers and adjustment parameters must be configured for a wide variety of different ambient conditions. To control idle speed, for example, there are several parameter sets for each individual gear which are further differentiated according to whether

- The vehicle is stationary or moving
- The engine is warm or cold
- The clutch is engaged or disengaged

That means that for this function alone, there are as many as 50 parameter sets.

The EDC also provides adaptation functions for extreme ambient conditions. These generally have to be verified by specifically targeted special trials involving

- Cold-weather testing in temperatures down to –25 °C (e.g. winter trials in Sweden)
- Hot-weather testing in temperatures over 40 °C (e.g. summer trials in Arizona)
- High-altitude/low atmospheric pressure testing (e.g. in the Alps) and
- Combined hot-weather and altitude or cold-weather and altitude testing, e.g. towing a heavy trailer over mountain passes (e.g. in Spain's Sierra Nevada or in the Alps)

3 Screen of an engine test-bench monitor (example)

For cold starting, very specific adjustments have to be made to the injected fuel quantity and the start of delivery based on engine coolant temperature. In addition, the glow plugs have to be switched on. At high altitudes with a cold engine, the effectively available pull-away torque is very low. For some applications, EDC suspends turbocharger operation for that short period because it would otherwise "use up" a large proportion of the engine's torque output. Particularly in the case of vehicles with automatic transmission, this would prevent the vehicle from pulling away at all, as the torque available at the driving wheels would be insufficient.

Altitude compensation for turbocharged engines demands limitation of the required turbocharger pressure in response to atmospheric pressure, as otherwise the turbocharger would be destroyed by over-revving.

Other adjustments

Safety functions

As well as the functions that determine emission levels, power output and user-friendliness, there are also numerous safety functions that require adaptation (e.g. response to failure of a sensor or actuator).

Such safety functions are primarily intended to restore the vehicle to a safe operating condition for the driver and/or to ensure the safe operation of the engine (e.g. to prevent engine damage).

Communication

There are also numerous functions which require communication between the engine control unit and other control units on the vehicle (e.g. traction control, ESP, transmission control for automatic transmission and electronic immobilizer). For this reason, a special communication code is employed (input and output variables). Where necessary, additional measured data has to be calculated and encoded in the appropriate form.

Examples of adaptation

Since the arrival of the EDC system in 1986, the possibilities for optimization, especially with regard to the convenience features, have considerably expanded. A wide variety of software functions (e.g. control functions) are used, all of which have to be specifically adapted to each individual vehicle. Some examples are outlined below.

Idle-speed control

This function controls the speed at which the engine runs when the accelerator pedal is not depressed. Idle-speed control must operate with absolute reliability under all possible engine operating conditions. Therefore, extensive adaptation work is required. Adjustment of the coasting response in all gears, for example, is highly involved, especially with regard to the interplay with the twin-mass flywheels generally used. This type of flywheel produces highly complex rotational vibration effects throughout the drivetrain.

The first stage of the process is an analytical definition (i.e. recording of the controlled system response, description of the controlled system by algorithms and definition of the control parameters).

This is followed by a comprehensive road test. A circular track (test track) provides the possibility for virtually unlimited flat-road driving. Particularly with active surge damping, conflict between objectives can arise as this function may prevent rapid compensation in response to abrupt changes in engine speed or load.

Apart from the drivetrain, the engine mountings also play an important part. In order to diminish the various conflicts in objectives, therefore, some applications employ variable-characteristic engine mountings which are controlled by the EDC. These can be set to a softer setting when the engine is idling and to a harder response when the engine is under load.

Smooth-running control

The engine smooth-running function ensures that the injection volumes are the same for all cylinders and in so doing improves engine smoothness and emission levels. Under certain circumstances, a malfunction can occur at very high or very low ambient temperatures if the vibration damping characteristics of the belt drive systems for auxiliary units (e.g. alternator, power-steering pump, air-conditioning compressor) significantly alter. Depending on the frequencies generated as a result of periodic speed fluctuations, the engine smooth-running function may attempt to even them out by alteration of the injected-fuel quantity volume for individual cylinders. Under unfavourable conditions, this may then result in higher exhaust-emission levels or make the engine run even more unevenly. For that reason, this function must be thoroughly tested under all operating conditions.

Pressure-charging controller

Almost all existing DI car diesel engines are fitted with turbochargers. On most of those engines, the charge-air pressure is controlled by the EDC system. The aim is to obtain optimum response characteristics (rapid generation of charge-air pressure) while ensuring reliable protection of the engine against excessive charge-air pressure and consequent excessively high cylinder pressure.

Exhaust-gas recirculation EGR

Exhaust-gas recirculation EGR is now a standard feature of DI car diesel engines. As previously indicated, together with the control of turbocharger pressure it is a determining factor in the amount of air that enters the engine. In order to ensure smokeless and low-NO_X combustion, the air-fuel mixture must conform to precisely defined parameters for all engine operating conditions. Those parameters are initially optimized under static conditions on the engine test bench. The control function then has the task of maintaining those parameters under dynamic operating conditions without adversely affecting the response characteristics of the engine.

[1]) Some parts of the adaptation process are also referred to as calibration.

Application-related adaptation[1]) of commercial-vehicle engines

Particularly because of its economy and durability, the diesel has established itself as the engine of choice for commercial vehicles. Today all new engines are direct-injection (DI) designs.

Optimization objectives

For commercial-vehicle engines, the following attributes are optimized.

Torque

The aim is to obtain the maximum possible torque under all operating conditions in order to be able to move heavy loads in even the most difficult situations (e.g. when negotiating steep gradients or using PTO drives). When pursuing that objective, the engine's limits (e.g. maximum permissible cylinder pressure and exhaust temperature) as well as the smoke emission limit have to be taken into account.

Fuel consumption

For commercial vehicles, economy is a decisive factor. For that reason fuel consumption occupies a position of greater importance for commercial vehicles than is the case with cars. Minimizing fuel consumption (or CO_2 emissions) is therefore of prime significance in engine adaptation.

Durability

Modern commercial-vehicle engines are expected to be able to complete over a million kilometers of service.

Pollutant emissions

Since October 2000, new commercial vehicles registered in the European Union have been required to conform to the Euro III emission-control standard. Engine adaptation must ensure that the limits for NO_X, particulate, HC and CO emission and exhaust opacity are reliably complied with.

Comfort/convenience

The demands relating to such aspects as engine response, quietness, smoothness and starting characteristics must also be taken into account.

Adaptation phases

The aim of adaptation is to ensure that the objectives outlined above are achieved as fully as possible, i.e. that the best possible compromise is reached between competing demands. This involves adaptation of engine and fuel-injection hardware components as

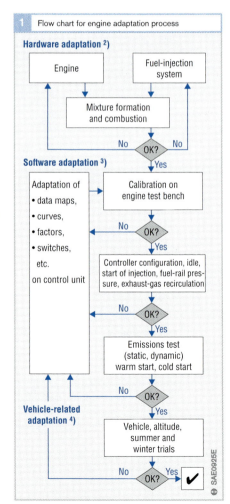

1   Flow chart for engine adaptation process

Fig. 1
2)   Criteria:
• Full-load response
• Emissions
• Fuel consumption

3)   Additional criterion:
• Dynamic adaptation

4)   Other criteria:
• Starting characteristics
• Smoothness, etc.

well as software functions performed by the engine-management module.

As with car engines, the phases of hardware, software and vehicle-related adaptation can be distinguished (Figure 1).

Hardware adaptation

Hardware adaptation involves making modifications to all significant "components" of the engine and fuel-injection system. Significant engine-hardware components include the combustion chamber, the turbocharger, the air-intake system (e.g. swirl-imparting

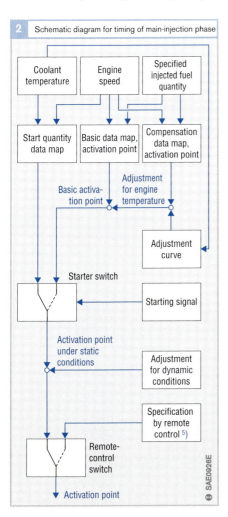

2   Schematic diagram for timing of main-injection phase

SAE0926E

components) and, if necessary, the exhaust-recirculation system. Significant components of the fuel-injection system are the injection pump, the high-pressure fuel lines if applicable, and the injectors. Hardware adaptation is carried out on the engine test bench.

Software adaptation

Once the hardware adaptation is complete, the control-unit software is configured accordingly. Stored in the software are the relationships between a vast number of engine and fuel-injection parameters (for examples, see Figure 2). This work too is carried out on the engine test bench. An application control unit, which – as with the adaptation of car engines – is linked to a PC with operator software, provides access to the software to be adapted.

The following tasks are performed in the course of software adaptation:
- Calibration of the basic engine-data maps under static operating conditions
- Control function configuration
- Calibration of compensation data maps
- Optimisation of engine-data maps under dynamic conditions

First of all, adjustments to the system-specific parameters – such as start of injection, injection pressure, exhaust-gas recirculation, charge-air pressure and, if applicable, pre- and post-injection – are carried out under static operating conditions on the engine test bench. The test results are assessed with reference to the target criteria (emission levels, fuel consumption, etc.). Based on those results, the appropriate parameter values, data curves and data maps are then calculated and programmed (Figure 3 overleaf). Because of the ever increasing number of such parameters, automation of parameter configuration is a continuing aim.

Following adaptation of the basic data maps, the effect of such variables as ambient temperature, atmospheric pressure, engine-coolant temperature and fuel temperature

Fig. 2
5) Specification of set values in order to bypass data maps during calibration

on the major parameters is factored into so-called compensation data maps. In addition, existing control functions are adapted (e.g. fuel-rail pressure control for common-rail injection systems, charge-air pressure control). The data established under static operating conditions is finally optimized under dynamic conditions.

Vehicle-related adaptation

The process of vehicle-related adaptation involves modifying the basic design of the engine arrived at on the test bench to the specifics of the vehicle in which it is to be used, and testing conformity with requirements under as wide a range as possible of real operating and ambient conditions.

The adaptation/testing of the basic functions such as idle-speed control, engine response and starting characteristics is essentially performed in the same way as for cars, though the assessment criteria may differ according to the particular type of application. When adapting an engine for use in a bus, for example, more emphasis is placed on comfort aspects or low noise output, whereas a truck engine for long-distance operation would be designed more for reliable and economical transportation of heavy loads.

Examples of adaptation

Idle-speed control

When adapting the idle-speed control function for a commercial-vehicle engine, major emphasis is generally placed on good load response and minimal undershoot. This ensures good pulling away and manoeuvring capabilities even when carrying heavy loads.

The behavior of the drivetrain as a controlled system depends heavily on temperature and transmission ratio. For that reason the engine-management module has multiple parameter sets for idle-speed control. When defining those parameters, changes in the drivetrain response over its service life must also be taken into account.

Power take-off (PTO) drives

Many commercial vehicles have PTO drives that are used to drive cranes, lifting platforms, pumps, etc. These often require the diesel engine to run at a virtually constant, higher operating speed that is unaffected by load. This can be governed by the EDC system using the "intermediate-speed control" function. Once again, the control function parameters can be adapted to the requirements of the driven machine.

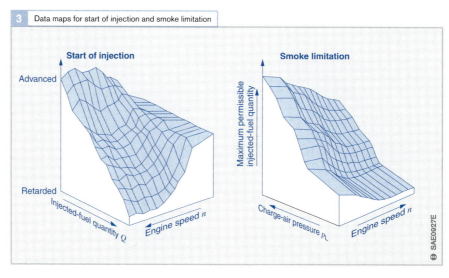

3   Data maps for start of injection and smoke limitation

Engine response characteristics

In the process of adaptation, engine response characteristics, i.e. the way in which accelerator-pedal position is translated into injected-fuel quantity and engine torque output, are to a large extent infinitely variable through control-unit configuration. It ultimately depends on the application as to whether an "RQ characteristic [6])" or "RQV characteristic [7])" engine response is programmed, or a mixture of the two.

Communication

The EDC control unit on a commercial vehicle is normally part of a network of multiple electronic control units. The exchange of data between vehicle, transmission, brake and engine control units takes place via an electronic data bus (usually a CAN). Correct interaction between the various control units involved cannot be fully tested and optimized until they are installed in the vehicle, as the process of basic configuration on the engine test bench usually involves only the engine-management module on its own.

A typical example of the interaction between two vehicle control units is the process of changing gear with an automatic transmission. The transmission control unit sends a request via the data bus for a reduction in injection quantity at the optimum point in the gear-shifting operation. The engine control unit then makes the requested reduction – without input from the driver – thus enabling the transmission control unit to disengage the current gear. If necessary, the transmission control unit may request an increase in engine speed at the appropriate point to facilitate engagement of the new gear. Once the operation is complete, control over the injected fuel quantity is passed back to the driver.

Electromagnetic compatibility

The large number of electronic vehicle systems and the wide use of other electronic communications equipment (e.g. radio telephones, two-way radios, GPS navigation systems) in commercial vehicles make it necessary to optimize the electromagnetic compatibility (EMC) of the engine-management module and all its connecting leads in terms both of immunity to external interference and of emission of interference signals. Of course, a large proportion of this optimization work is carried out during the development of the control units and sensors concerned. Since, however, the dimensioning (e.g. length of cable runs, type of shielding) and routing of the wiring looms in the actual vehicle has a major influence on immunity to and creation of interference, testing and, if necessary, optimization of the complete vehicle inside an EMC room is absolutely essential.

Fault diagnosis

The diagnostic capabilities demanded of commercial-vehicle systems are also very extensive. Reliable diagnosis of faults ensures maximum possible vehicle availability.

The engine control unit constantly checks that the signals from all connected sensors and actuators are within the specified limits and also tests for loose contacts, short circuits to ground or to battery voltage, and for plausibility with other signals. The signal range limits and plausibility criteria must be defined by the application developer. As with car engines, those limits must on the one hand be sufficiently broad to ensure that extreme conditions (e.g. hot or cold weather, high altitudes) do not produce false diagnoses, and on the other, sufficiently narrow to provide adequate sensitivity to real faults. In addition, fault response procedures must be defined which specify whether and in what way the engine may continue to be operated if a specific fault is detected. Finally, detected faults have to be stored in a fault memory in order that service technicians can quickly locate and remedy the problem.

[6]) Control function for minimum and maximum speed or maximum speed only

[7]) Variable-speed or incremental control function

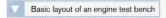

Engine test bench

A fuel-injection system is tested on an engine test bench as part of its development process. Engine test benches are designed to allow easy access to the various parts of the engine.

By conditioning the supply fluids such as intake air, fuel and engine coolant, (i.e. controlling their temperature and/or pressure) reproducible results can be obtained.

In addition to measurements under static operating conditions, dynamic tests with rapid load and engine-speed changes are increasingly demanded. For such purposes there are test benches with electric dynamometers (18). They can not only retard but also drive the test vehicle (e.g. in order to simulate overrun when traveling downhill). Using appropriate simula-

tion software, the statutory emission control tests can then be run on the test bench rather than on a vehicle tester with the engine in situ.

The test-bench computer (20) is responsible for controlling and monitoring the engine and the testing equipment. It also takes care of data recording and storage. With the aid of automation software, calibration operations (e.g. data-map measurements) can be carried out very efficiently.

Using a suitable quick-change system (8), the pallets with the engines to be tested can be changed over within about twenty minutes. This increases test-bench capacity utilization.

Basic layout of an engine test bench

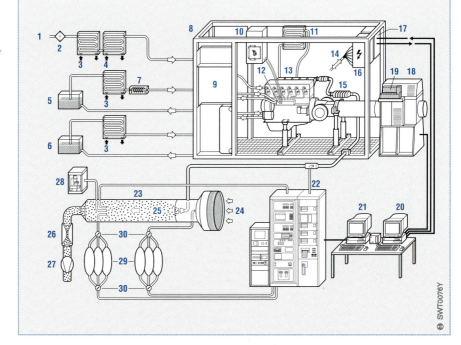

SWT0076Y

Calibration tools

The traditional calibration tools (for car and commercial-vehicle applications) include
- The "transparent" engine (usually a single-cylinder engine which has small windows and mirrors that allow the combustion process to be observed)
- The engine test bench
- The EMC room and
- A wide variety of special devices such as microphones for measuring sound levels or strain gauges for measuring mechanical stress

Computer simulation of hardware and software components is also becoming increasingly important. A large part of the adaptation work, however, is carried out using PC-based calibration tools. Such programs allow developers to modify the engine-management software. One such calibration tool is the INCA (**In**tegrated **C**alibration and **A**cquisition System) program, compromising a number of different tools. It is made up of the following components:
- The *Core System* incorporates all measurement and adjustment functions.
- The *Offline Tools (standard specification)* comprise the software for analysis of measured data and management of adjustment data, and the programming tool for the Flash EPROM.

The use and function of the calibration tools can be illustrated by the description below of a typical calibration process.

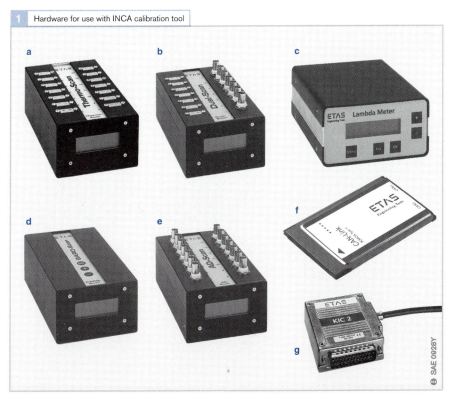

1 Hardware for use with INCA calibration tool

a
b
c
f
d
e
g

Fig. 1
a *Thermo-Scan*
   Interface module for temperature sensors
b *Dual-Scan*
   Interface module for analog signals and temperature sensors
c *Lambda Meter*
   Interface module for broadband oxygen sensor
d *Baro-Scan*
   Testing module for pressures
e *AD-Scan*
   Interface module for analog signals
f *CAN-link card*
g *KIC 2*
   Calibration module for diagnostic interface

SAE 0928Y

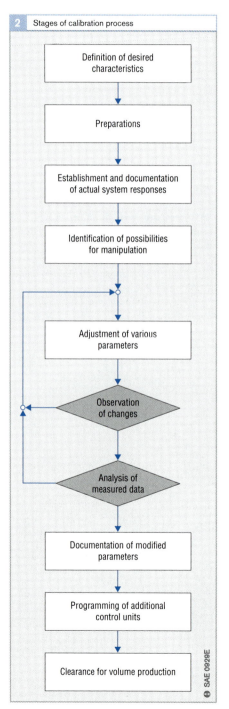

2   Stages of calibration process

Definition of desired characteristics

Preparations

Establishment and documentation of actual system responses

Identification of possibilities for manipulation

Adjustment of various parameters

Observation of changes

Analysis of measured data

Documentation of modified parameters

Programming of additional control units

Clearance for volume production

SAE 0929E

Software calibration process

Defining the desired characteristics

The desired characteristics (e.g. dynamic response, noise output, exhaust composition) are defined by the engine manufacturer and the (exhaust emissions) legislation. The aim of calibration is to alter the characteristics of the engine so that those requirements are met. This necessitates testing on the engine test bench and in the vehicle.

Preparations

Special electronic engine control units are used for calibration. Compared with the control units used on the production models, they allow the alteration of parameters that are fixed for normal operation. An important aspect of the preparations is choosing and setting up the appropriate hardware and/or software interface.

Additional measuring equipment (e.g. temperature sensors, flow meters) enables the recording of other physical variables for special tests.

Establishing and documenting the actual system responses

The recording of specific measured data is carried out using the INCA core system. The information concerned can be displayed on the screen and analyzed in the form of numerical values or graphs.

The measured data can not only be viewed after the measurements have been taken but while measurement is still in progress. In that way, the response of the engine to changes (e.g. in the exhaust-gas recirculation rate) can be investigated. The data can also be recorded for subsequent analysis of transient processes (e.g. engine starting).

Identifying possibilities for manipulation

With the help of the control-unit software documentation (data framework) it is possible to identify which parameters are best suited to altering system behavior in the manner desired.

Alteration of selected parameters

The parameters stored in the control-unit software can be displayed as numerical values (in tables) or as graphs (curves) on the PC and altered. Each time an alteration is made, the system response is observed.

All parameters can be altered while the engine is running so that the effects are immediately observable and measurable.

In the case of short-lived or transient processes (e.g. engine starting) it is effectively impossible to alter the parameters while the process is in progress. In such cases, therefore, the process has to be recorded during the course of a test, the measured data saved in a file and then the parameters that are to be altered identified by analyzing the recorded data.

Further tests are performed in order to evaluate the success of the adjustments made or to learn more about the process.

Analyzing measured data

Analysis and documentation of the measured data is performed with the aid of the offline tool MDA (Measured Data Analyzer). This stage of the calibration process involves comparing and documenting the system behavior before and after alteration of parameters. Such documentation encompasses improvements as well as problems and malfunctions.

Documentation is important because several people will be involved in the process of engine optimization at different times.

Documenting the modified parameters

The changes to the parameters are also compared and documented. This is done with the offline tool ADM (Application Data Manager), sometimes also called CDM (Calibration Data Manager).

The calibration data obtained by various technicians is compared and merged into a single data record.

Programming additional control units

The new parameter settings arrived at can also be used on other engine control units for further calibration. This necessitates reprogramming of the Flash EPROMs of those control units. This is carried out using the INCA core system tool PROF (**Programming of** Flash EPROM).

Depending on the extent of the calibration and the design innovations, multiple looping of the steps described above may take place.

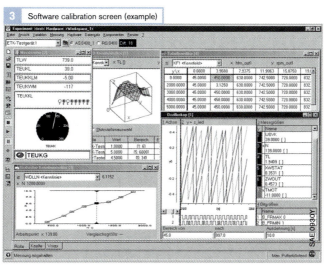

3   Software calibration screen (example)

Sensors

Sensors register operating states (e.g. engine speed) and setpoint/desired values (e.g. accelerator-pedal position). They convert physical quantities (e.g. pressure) or chemical quantities (e.g. exhaust-gas concentration) into electric signals.

Automotive applications

Sensors and actuators represent the interfaces between the ECUs, as the processing units, and the vehicle with its complex drive, braking, chassis, and bodywork functions (for instance, the Engine Management, the Electronic Stability Program ESP, and the air conditioner). As a rule, a matching circuit in the sensor converts the signals so that they can be processed by the ECU.

The field of mechatronics, in which mechanical, electronic, and data-processing components are interlinked and cooperate closely with each other, is rapidly gaining in importance in the field of sensor engineering. These components are integrated in modules (e.g. in the crankshaft CSWS (Composite Seal with Sensor) module complete with rpm sensor).

Since their output signals directly affect not only the engine's power output, torque, and emissions, but also vehicle handling and safety, sensors, although they are becoming smaller and smaller, must also fulfill demands that they be faster and more precise. These stipulations can be complied with thanks to mechatronics.

Depending upon the level of integration, signal conditioning, analog/digital conversion, and self-calibration functions can all be integrated in the sensor (Fig. 1), and in future a small microcomputer for further signal processing will be added. The advantages are as follows:

- Lower levels of computing power are needed in the ECU
- A uniform, flexible, and bus-compatible interface becomes possible for all sensors
- Direct multiple use of a given sensor through the data bus
- Registration of even smaller measured quantities
- Simple sensor calibration

Figure 1
SE Sensor(s)
SA Analog signal
     conditioning
A/D Analog-digital
     converter
SG Digital ECU
MC Microcomputer
     (evaluation
     electronics)

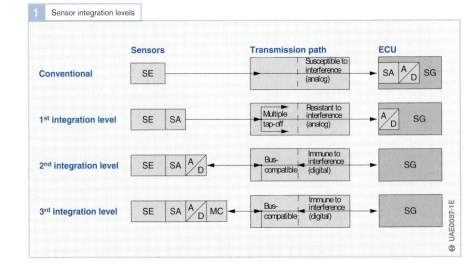

1 Sensor integration levels

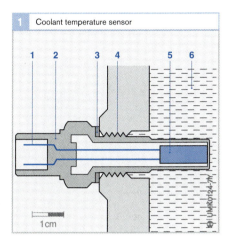

1 Coolant temperature sensor

1 cm

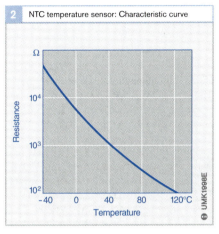

2 NTC temperature sensor: Characteristic curve

Resistance (Ω)

Temperature (°C)

Fig. 1
1 Electrical connections
2 Housing
3 Gasket
4 Thread
5 Measuring resistor
6 Coolant

Temperature sensors

Applications

Engine-temperature sensor
This is installed in the coolant circuit (Fig. 1). The engine management uses its signal when calculating the engine temperature (measuring range – 40...+130 °C).

Air-temperature sensor
This sensor is installed in the air-intake tract. Together with the signal from the boost-pressure sensor, its signal is applied in calculating the intake-air mass. Apart from this, desired values for the various control loops (e.g. EGR, boost-pressure control) can be adapted to the air temperature (measuring range –40...+120°C).

Engine-oil temperature sensor
The signal from this sensor is used in calculating the service interval (measuring range –40...+170 °C).

Fuel-temperature sensor
Is incorporated in the low-pressure stage of the diesel fuel circuit. The fuel temperature is used in calculating the precise injected fuel quantity (measuring range –40...+120°C).

Exhaust-gas temperature sensor
This sensor is mounted on the exhaust system at points which are particularly critical regarding temperature. It is applied in the closed-loop control of the systems used for exhaust-gas treatment. A platinum measuring resistor is usually used (measuring range –40...+1,000°C).

Design and operating concept
Depending upon the particular application, a wide variety of temperature sensor designs are available. A temperature-dependent semiconductor measuring resistor is fitted inside a housing. This resistor is usually of the NTC (Negative Temperature Coefficient, Fig. 2) type. Less often a PTC (Positive Temperature Coefficient) type is used. With NTC, there is a sharp drop in resistance when the temperature rises, and with PTC there is a sharp increase.

The measuring resistor is part of a voltage-divider circuit to which 5 V is applied. The voltage measured across the measuring resistor is therefore temperature-dependent. It is inputted through an analog to digital (A/D) converter and is a measure of the temperature at the sensor. A characteristic curve is stored in the engine-management ECU which allocates a specific temperature to every resistance or output-voltage.

Micromechanical pressure sensors

Application

Manifold-pressure or boost-pressure sensor

This sensor measures the absolute pressure in the intake manifold between the supercharger and the engine (typically 250 kPa or 2.5 bar) and compares it with a reference vacuum, not with the ambient pressure. This enables the air mass to be precisely defined, and the boost pressure exactly controlled in accordance with engine requirements.

Atmospheric-pressure sensor

This sensor is also known as an ambient-pressure sensor and is incorporated in the ECU or fitted in the engine compartment. Its signal is used for the altitude-dependent correction of the setpoint values for the control loops. For instance, for the exhaust-gas recirculation (EGR) and for the boost-pressure control. This enables the differing densities of the surrounding air to be taken into account. The atmospheric-pressure sensor measures absolute pressure (60...115 kPa or 0.6...1.15 bar).

Oil and fuel-pressure sensor

Oil-pressure sensors are installed in the oil filter and measure the oil's absolute pressure. This information is needed so that engine loading can be determined as needed for the Service Display. The pressure range here is 50...1,000 kPa or 0.5...10.0 bar. Due to its high resistance to media, the measuring element can also be used for pressure measurement in the fuel supply's low-pressure stage. It is installed on or in the fuel filter. Its signal serves for the monitoring of the fuel-filter contamination (measuring range: 20... 400 kPa or 0.2...4 bar).

Version with the reference vacuum on the component side

Design and construction

The measuring element is at the heart of the micromechanical pressure sensor. It is com-

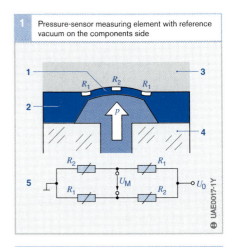

1   Pressure-sensor measuring element with reference vacuum on the components side

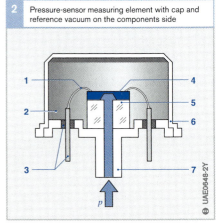

2   Pressure-sensor measuring element with cap and reference vacuum on the components side

3   Pressure-sensor measuring element with cap and reference vacuum on the components side

prised of a silicon chip (Fig. 1, Pos. 2) in which a thin diaphragm has been etched micromechanically (1). Four deformation resistors (R_1, R_2) are diffused on the diaphram. Their electrical resistance changes when mechanical force is applied. The measuring element is surrounded on the component side by a cap which at the same time encloses the reference vacuum (Figs. 2 and 3). The pressure-sensor case can also incorporate an integral *temperature sensor* (Fig. 4, Pos. 1) whose signals can be evaluated independently. This means that at any point a single sensor case suffices to measure temperature and pressure.

Method of operation

The sensor's diaphragm deforms more or less (10 ... 1,000 μm) according to the pressure being measured. The four deformation resistors on the diaphragm change their electrical resistances as a function of the mechanical stress resulting from the applied pressure (piezoresistive effect).

The four measuring resistors are arranged on the silicon chip so that when diaphragm deformation takes place, the resistance of two of them increases and that of the other two decreases. These deformation resistors form a Wheatstone bridge (Fig. 1, Pos. 5), and a change in their resistances leads to a change in the ratio of the voltages across them. This leads to a change in the measurement voltage U_M. This unamplified voltage is therefore a measure of the pressure applied to the diaphragm.

The measurement voltage is higher with a bridge circuit than would be the case when using an individual resistor. The Wheatstone bridge circuit thus permits a higher sensor sensitivity.

The component side of the sensor to which pressure is not supplied is subjected to a reference vacuum (Fig. 2, Pos. 2) so that it measures the absolute pressure.

The signal-conditioning electronics circuitry is integrated on the chip. Its assignment is to

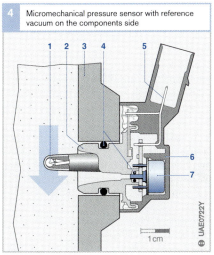

4 Micromechanical pressure sensor with reference vacuum on the components side

Fig. 4
1 Temperature sensor (NTC)
2 Lower section of case
3 Manifold wall
4 Seal rings
5 Electrical terminal (plug)
6 Case cover
7 Measuring element

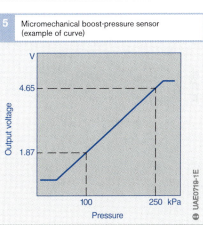

5 Micromechanical boost-pressure sensor (example of curve)

Output voltage
V
4.65
1.87
100   250 kPa
Pressure

amplify the bridge voltage, compensate for temperature influences, and linearise the pressure curve. The output voltage is between 0...5 V and is connected through electrical terminals (Fig. 4, Pos. 5) to the engine-management ECU which uses this output voltage in calculating the pressure (Fig. 5).

Version with reference vacuum in special chamber
Design and construction

The *manifold or boost-pressure sensor* version with the reference vacuum in a special chamber (Figs. 6 and 7) is easier to install than the version with the reference vacuum on the components side of the sensor element. Similar to the pressure sensor with cap and reference vacuum on the components side of the sensor element, the sensor element here is formed from a silicon chip with four etched deformation resistors in a bridge circuit. It is attached to a glass base. In contrast to the sensor with the reference vacuum on the components side, there is no passage in the glass base through which the measured pressure can be applied to the sensor element. Instead, pressure is applied to the silicon chip from the side on which the evaluation electronics is situated. This means that a special gel must be used at this side of the sensor to protect it against environmental influences (Fig. 8, Pos. 1). The reference vacuum is enclosed in the chamber between the silicon chip (6) and the glass base (3). The complete measuring element is mounted on a ceramic hybrid (4) which incorporates the soldering surfaces for electrical contacting inside the sensor.

A *temperature sensor* can also be incorporated in the pressure-sensor case. It protrudes into the air flow, and can therefore respond to temperature changes with a minimum of delay (Fig. 6, Pos. 4).

Operating concept

The operating concept, and with it the signal conditioning and signal amplification together with the characteristic curve, corresponds to that used in the pressure sensor with cap and reference vacuum on the sensor's structure side. The only difference is that the measuring element's diaphragm is deformed in the opposite direction and therefore the deformation resistors are "bent" in the other direction.

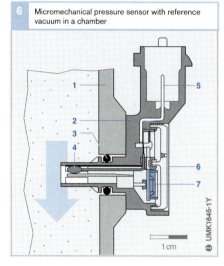

6   Micromechanical pressure sensor with reference vacuum in a chamber

1 cm

UMK1845-1Y

Fig. 6
1 Manifold wall
2 Case
3 Seal ring
4 Temperature sensor (NTC)
5 Electrical connection (socket)
6 Case cover
7 Measuring element

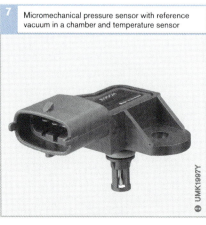

7   Micromechanical pressure sensor with reference vacuum in a chamber and temperature sensor

UMK1997Y

Fig. 8
1 Protective gel
2 Gel frame
3 Glass base
4 Ceramic hybrid
5 Chamber with reference volume
6 Measuring element (chip) with evaluation electronics
7 Bonded connection
p Measured pressure

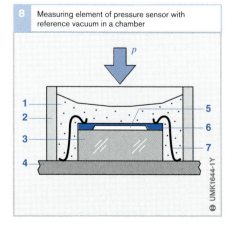

8   Measuring element of pressure sensor with reference vacuum in a chamber

p

UMK1644-1Y

Rail-pressure sensors

Application
In the Common-Rail injection system and the gasoline direct-injection system MED-Motronic, these sensors are used to measure the fuel pressure in the high-pressure fuel accumulator (or rail, from which the system took its name). Strict compliance with the stipulated fuel pressure is of extreme importance with regard to emissions, noise, and engine power. The fuel pressure is regulated in a control loop, and deviations from the desired pressure level are compensated for by a pressure-control valve.

The rail-pressure sensors feature very tight tolerances, and in the main measuring range the measuring accuracy is better than 2 % of the measuring range.

These rail-pressure sensors are used in the following engine systems:

- *Common-Rail diesel injection system (CRS)*
  Maximum operating pressure p_{max} (rated pressure) is 160 MPa (1,600 bar).

- *Gasoline direct injection MED-Motronic*
  For this gasoline direct-injection system, operating pressure is a function of load and rotational speed, and is 5...12 MPa (50...120 bar).

Design and operating concept
The heart of this sensor is a steel diaphragm on which measuring resistors in the form of a bridge circuit have been vapor-deposited (Fig. 1, Pos. 3). The sensor's measuring range is a function of diaphragm thickness (thicker diaphragms for higher pressures, thinner diaphragms for lower pressures). As soon as the pressure to be measured is applied to the diaphragm through the pressure connection (Fig. 1, Pos. 4), this bends and causes a change in the resistance of the measuring resistors (approx. 20 µm at 1,500 bar). The output voltage generated by the bridge is in the range 0...80 mV and is inputted to the evaluation circuit (2) in the sensor. This amplifies the bridge signal to 0...5 V and transmits it to the ECU which uses it together with a stored characteristic curve to calculate the pressure (Fig. 2).

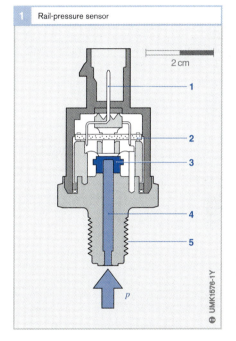

1   Rail-pressure sensor

2 cm

P

Fig. 1
1 Electrical plug-in connection
2 Evaluation circuit
3 Steel diaphragm with measuring resistors
4 Pressure connection
5 Mounting thread

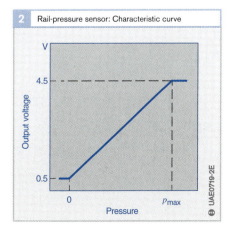

2   Rail-pressure sensor: Characteristic curve

Output voltage

V

4.5

0.5

0

p_{max}

Pressure

Inductive engine-speed sensors

Applications

Such engine-speed sensors are used for measuring:

- Engine rpm
- Crankshaft position (for information on the position of the engine pistons)

The rotational speed is calculated from the sensor's signal frequency. The output signal from the rotational-speed sensor is one of the most important quantities in electronic engine management.

Design and operating concept

The sensor is mounted directly opposite a ferromagnetic trigger wheel (Fig. 1, Pos. 7) from which it is separated by a narrow air gap. It has a soft-iron core (pole pin) (4), which is enclosed by the solenoid winding (5). The pole pin is also connected to a permanent magnet (1), and a magnetic field extends through the pole pin and into the trigger wheel. The level of the magnetic flux through the winding depends upon whether the sensor is opposite a trigger-wheel tooth or gap. Whereas the magnet's stray flux is concentrated by a tooth and leads to an increase in the working flux through the winding, it is weakened by a gap. When the trigger wheel rotates therefore, this causes a fluctuation of the flux which in turn generates a sinusoidal voltage in the solenoid

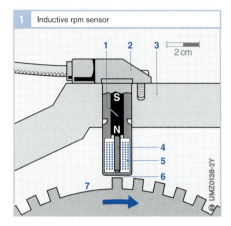

1   Inductive rpm sensor

Fig. 1
1 Permanent magnet
2 Sensor housing
3 Engine block
4 Pole pin
5 Solenoid winding
6 Air gap
7 Trigger wheel with reference-mark gap

winding which is proportional to the rate of change of the flux (Fig. 2). The amplitude of the AC voltage increases strongly along with increasing trigger-wheel speed (several mV...>100 V). At least about 30 rpm are needed to generate an adequate signal level.

The number of teeth on the trigger wheel depends upon the particular application. On solenoid-valve-controlled engine-management systems for instance, a 60-pitch trigger wheel is normally used, although 2 teeth are omitted (7) so that the trigger wheel has $60 - 2 = 58$ teeth. The very large tooth gap is allocated to a defined crankshaft position and serves as a reference mark for synchronizing the ECU.

There is another version of the trigger wheel which has one tooth per engine cylinder. In the case of a 4-cylinder engine, therefore, the trigger wheel has 4 teeth, and 4 pulses are generated per revolution.

The geometries of the trigger-wheel teeth and the pole pin must be matched to each other. The evaluation-electronics circuitry in the ECU converts the sinusoidal voltage, which is characterized by strongly varying amplitudes, into a constant-amplitude square-wave voltage for evaluation in the ECU microcontroller.

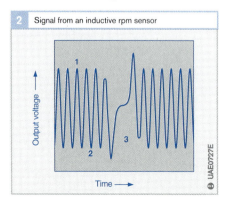

2   Signal from an inductive rpm sensor

Output voltage ——→

Time ——→

Fig. 2
1 Tooth
2 Tooth gap
3 Reference mark

Rotational-speed (rpm) sensors and incremental angle-of-rotation sensors

Application

The above sensors are installed in distributor-type diesel injection pumps with solenoid-valve control. Their signals are used for:

- The measurement of the injection pump's speed
- Determining the instantaneous angular position of pump and camshaft
- Measurement of the instantaneous setting of the timing device

The pump speed at a given instant is one of the input variables to the distributor pump's ECU which uses it to calculate the triggering time for the high-pressure solenoid valve, and, if necessary, for the timing-device solenoid valve.

The triggering time for the high-pressure solenoid valve must be calculated in order to inject the appropriate fuel quantity for the particular operating conditions. The cam plate's instantaneous angular setting defines the triggering point for the high-pressure solenoid valve. Only when triggering takes place at exactly the right cam-plate angle, can it be guaranteed that the opening and closing points for the high-pressure solenoid valve are correct for the particular cam lift. Precise triggering defines the correct start-of-injection point and the correct injected fuel quantity.

The correct timing-device setting as needed for timing-device control is ascertained by comparing the signals from the camshaft rpm sensor with those of the angle-of-rotation sensor.

Design and operating concept

The rpm sensor, or the angle-of-rotation sensor, scans a toothed pulse disc with 120 teeth which is attached to the distributor pump's driveshaft. There are tooth gaps, the number of which correspond to the number of engine cylinders, evenly spaced around the disc's circumference. A double differential magnetoresistive sensor is used.

Magnetoresistors are magnetically controllable semiconductor resistors, and similar in design to Hall-effect sensors. The double differential sensor has four resistors connected to form a full bridge circuit.

The sensor has a permanent magnet, and the magnet's pole face opposite the toothed pulse disc is homogenized by a thin ferromagnetic wafer on which are mounted the four magnetoresistors, separated from each other by half a tooth gap. This means that alternately there are two magnetoresistors opposite tooth gaps and two opposite teeth (Fig. 1). The magnetoresistors for automotive applications are designed for operation in temperatures of $\leq 170\,°C$ ($\leq 200\,°C$ briefly).

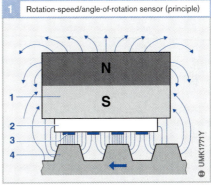

1   Rotation-speed/angle-of-rotation sensor (principle)

Fig. 1
1 Magnet
2 Homogenization wafer (Fe)
3 Magnetoresistor
4 Toothed pulse disc

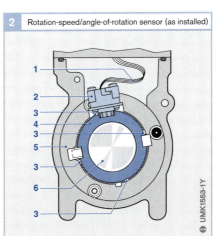

2   Rotation-speed/angle-of-rotation sensor (as installed)

Fig. 2
1 Flexible conductive foil
2 Rotation-speed (rpm)/angle-of-rotation sensor
3 Tooth gap
4 Toothed pulse wheel (trigger wheel)
5 Rotatable mounting
6 Driveshaft

Hall-effect phase sensors

Application

The engine's camshaft rotates at half the crankshaft speed. Taking a given piston on its way to TDC, the camshaft's rotational position is an indication as to whether the piston is in the compression or exhaust stroke. The phase sensor on the camshaft provides the ECU with this information.

Design and operating concept

Hall-effect rod sensors

As the name implies, such sensors (Fig. 2a) make use of the Hall effect. A ferromagnetic trigger wheel (with teeth, segments, or perforated rotor, Pos. 7) rotates with the camshaft. The Hall-effect IC is located between the trigger wheel and a permanent magnet (Pos. 5) which generates a magnetic field strength perpendicular to the Hall element.

If one of the trigger-wheel teeth (Z) now passes the current-carrying rod-sensor element (semiconductor wafer), it changes the magnetic field strength perpendicular to the Hall element. This causes the electrons, which are driven by a longitudinal voltage across the element to be deflected perpendicularly to the direction of current (Fig. 1, angle α).

This results in a voltage signal (Hall voltage) which is in the millivolt range, and which is independent of the relative speed between sensor and trigger wheel. The evaluation electronics integrated in the sensor's Hall IC conditions the signal and outputs it in the form of a rectangular-pulse signal (Fig. 2b "High"/"Low").

Differential Hall-effect rod sensors

Rod sensors operating as per the differential principle are provided with two Hall elements. These elements are offset from each other either radially or axially (Fig. 3, S1 and S2), and generate an output signal which is proportional to the difference in magnetic flux at the element measuring points. A two-

track perforated plate (Fig. 3a) or a two-track trigger wheel (Fig. 3b) are needed in order to generate the opposing signals in the Hall elements (Fig. 4) as needed for this measurement.

Such sensors are used when particularly severe demands are made on accuracy. Further advantages are their relatively wide air-gap range and good temperature-compensation characteristics.

Fig. 1

I	Wafer current
I_H	Hall current
I_V	Supply current
U_H	Hall voltage
U_R	Longitudinal voltage
B	Magnetic induction
α	Deflection of the electrons by the magnetic field

Fig. 2

a	Positioning of sensor and single-track trigger wheel
b	Output signal characteristic U_A
1	Electrical connection (plug)
2	Sensor housing
3	Engine block
4	Seal ring
5	Permanent magnet
6	Hall-IC
7	Trigger wheel with tooth/segment (Z) and gap (L)
a	Air gap
φ	Angle of rotation

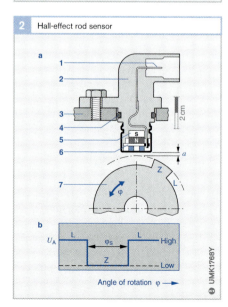

1   Hall element (Hall-effect vane switch)

2   Hall-effect rod sensor

3 Differential Hall-effect rod sensors

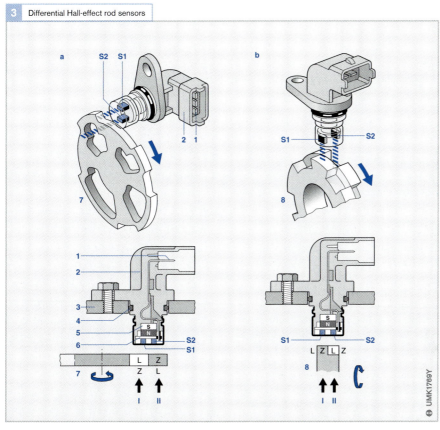

Fig. 3
a Axial tap-off
   (perforated plate)
b Radial tap-off
   (two-track trigger
   wheel)

1 Electrical connection
   (plug)
2 Sensor housing
3 Engine block
4 Seal ring
5 Permanent magnet
6 Differential Hall-IC
   with Hall elements S1
   and S2
7 Perforated plate
8 Two-track trigger
   wheel

I   Track 1
II Track 2

4 Characteristic curve of the output signal U_A from a differential Hall-effect rod sensor

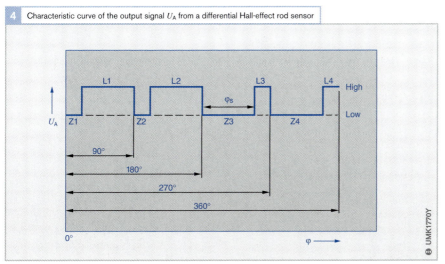

Fig. 4
Output signal "Low":
Material (Z) in front of
S1, gap (L) in front of S2

Output signal "High":
Gap (L) in front of S1,
material (Z) in front of S2

φ_S signal width

Half-differential short-circuiting-ring sensors

Application

These sensors are also known as HDK (taken from the German) sensors, and are applied as position sensors for travel or angle, They are wear-free, as well as being very precise, and very robust, and are used as:

- Rack-travel sensors (RWG) for measuring the control-rack setting on in-line diesel injection pumps, and as
- Angle-of-rotation sensors in the injected-fuel-quantity actuators of diesel distributor pumps

Design and operating concept

These sensors (Figs. 1 and 2) are comprised of a laminated soft-iron core on each limb of which are wound a measuring coil and a reference coil.

Alternating magnetic fields are generated when the alternating current from the ECU flows through these coils. The copper rings surrounding the limbs of the soft-iron cores screen the cores, though, against the effects of the magnetic fields. Whereas the reference short-circuiting rings are fixed in position, the measuring short-circuiting rings are attached to the control rack or control-collar shaft (in-line pumps and distributor pumps respectively), with which they are free to move (control-rack travel s, or adjustment angle φ).

When the measuring short-circuiting ring moves along with the control rack or control-collar shaft, the magnetic flux changes and, since the ECU maintains the current constant (load-independent current), the voltage across the coil also changes.

The ratio of the output voltage U_A to the reference voltage U_{Ref} (Fig. 3) is calculated by an evaluation circuit. This ratio is proportional to the deflection of the measuring short-circuiting ring, and is processed by the ECU. Bending the reference short-circuiting ring adjusts the gradient of the characteristic curve, and the basic position of the measuring short-circuiting ring defines the zero position.

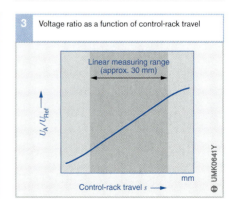

1   Design of the half-differential short-circuiting-ring sensor for diesel distributor pumps

2   Design of the rack-travel sensor (RWG) for diesel in-line injection pumps

3   Voltage ratio as a function of control-rack travel

Linear measuring range (approx. 30 mm)

Measured variables on diesel engines

Continuing efforts to improve the performance of diesel engines while simultaneously reducing harmful exhaust-gas emissions and fuel consumption mean that the number of sensors on and around the engine is constantly growing. The illustration below provides an overview of the parameters and variables that can be measured on the engine while it is running.

Some of this information is only collected and analyzed in the course of engine development or when it is being serviced (*). Of the remaining data, only a certain proportion is recorded when the engine is operational. The specific data that is required depends on the engine design, the fuel-injection system and the equipment fitted on the vehicle.

The parameters and variables are detected by sensors. The degree of accuracy and rate of detection required are determined by the type of application for which the engine is intended.

Measured variables on diesel engines

p **Pressures**
- Intake air,
- Air upstream/downstream of turbocharger,
- Recirculated exhaust upstream/downstream of cooler*,
- Exhaust upstream/downstream of turbocharger*,
- Exhaust upstream of catalytic converter,
- Exhaust downstream of catalytic converter*,
- Combustion chamber*,
- High-pressure fuel pipe*,
- Fuel supply,
- Fuel return*,
- Engine coolant*,
- Engine oil.

s **Travel**
- Needle stroke (for injection point),
- Governor settings,
- Injection timing adjuster setting,
- Valve positions.

t **Times**
- Injection period *,
- Delivery point,
- Delivery period.

U, I **Control signals**
- Injectors,
- Actuators,
- Valves (e.g. exhaust recirculation, wastegate),
- Flaps,
- Auxiliary systems.

Noise emission*

n **Speeds**
- Crankshaft
- Camshaft
- Turbocharger*,
- Auxiliary units*.

M **Torque***

Exhaust constituents
- Carbon dioxide (CO_2)*,
- Carbon monoxide (CO)*,
- Methane (CH_4)*,
- Nitrogen oxides (NO_x)*,
- Oxygen (O_2),
- Aldehydes*,
- Hydrocarbons (HC)*,
- Particulates (smoke index, soot concentration, exhaust opacity)*,
- Sulfur dioxide (SO_2)*.

T **Temperatures**
- Intake air,
- Air upstream/downstream of turbocharger,
- Recirculated exhaust upstream/downstream of cooler*,
- Exhaust upstream/downstream of turbocharger,
- Exhaust upstream/downstream of catalytic converter,
- Fuel supply,
- Fuel return,
- Engine coolant,
- Engine oil.

a **Acceleration (vibration) of components***

m **Masses**
- Intake air,
- Fuel,
- Recirculated exhaust*,
- Blow-By ([piston-ring] blow-by)*

UAE0754-1E

Accelerator-pedal sensors

Application

In conventional engine-management systems, the driver transmits his/her wishes for acceleration, constant speed, or lower speed, to the engine by using the accelerator pedal to intervene mechanically at the throttle plate (gasoline engine) or at the injection pump (diesel engine). Intervention is transmitted from the accelerator pedal to the throttle plate or injection pump by means of a Bowden cable or linkage.

On today's electronic engine-management systems, the Bowden cable and/or linkage have been superseded, and the driver's accelerator-pedal inputs are transmitted to the ECU by an accelerator-pedal sensor which registers the accelerator-pedal travel, or the pedal's angular setting, and sends this to the engine ECU in the form of an electric signal. This system is also known as "drive-by-wire". The accelerator-pedal module (Figs. 2b, 2c) is available as an alternative to the individual accelerator-pedal sensor (Fig. 2a). These modules are ready-to-install units comprising accelerator pedal and sensor, and make adjustments on the vehicle a thing of the past.

Design and operating concept

Potentiometer-type accelerator-pedal sensor

The heart of this sensor is the potentiometer across which a voltage is developed which is a function of the accelerator-pedal setting. In the ECU, a programmed characteristic curve is applied in order to calculate the accelerator-pedal travel, or its angular setting, from this voltage.

A second (redundant) sensor is incorporated for diagnosis purposes and for use in case of malfunctions. It is a component part of the monitoring system. One version of the accelerator-pedal sensor operates with a second potentiometer. The voltage across this potentiometer is always half of that across the first potentiometer. This provides two independent signals which are used for trouble-shooting (Fig. 1). Instead of the second potentiometer, another version uses a low-idle switch which provides a signal for the ECU when the accelerator pedal is in the

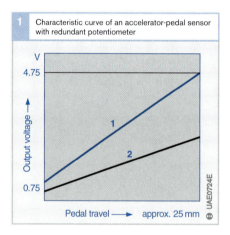

Fig. 1 Characteristic curve of an accelerator-pedal sensor with redundant potentiometer

Output voltage — V

4.75

1

2

0.75

Pedal travel ⟶ approx. 25 mm

UAE0724E

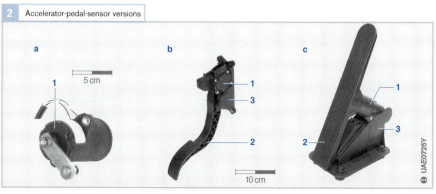

Fig. 2 Accelerator-pedal-sensor versions

a

5 cm

1

b

1

3

2

c

1

3

2

10 cm

UAE0725Y

idle position. For automatic transmission vehicles, a further switch can be incorporated for a kick-down signal.

Hall-effect angle-of-rotation sensors

The ARS1 (**A**ngle of **R**otation **S**ensor) is based on the movable-magnet principle. It has a measuring range of approx. 90° (Figs. 3 and 4).

A semicircular permanent-magnet disc rotor (Fig. 4, Pos. 1) generates a magnetic flux which is returned back to the rotor via a pole shoe (2), magnetically soft conductive elements (3) and shaft (6). In the process, the amount of flux which is returned through the conductive elements is a function of the rotor's angle of rotation φ. There is a Hall-effect sensor (5) located in the magnetic path of each conductive element, so that it is possible to generate a practically linear characteristic curve throughout the measuring range.

The ARS2 is a simpler design without magnetically soft conductive elements. Here, a magnet rotates around the Hall-effect sensor. The path it takes describes a circular arc. Since only a small section of the resulting sinusoidal characteristic curve features good linearity, the Hall-effect sensor is located slightly outside the center of the arc. This causes the curve to deviate from its sinusoidal form so that the curve's linear section is increased to more than 180°.

Mechanically, this sensor is highly suitable for installation in an accelerator-pedal module (Fig. 5).

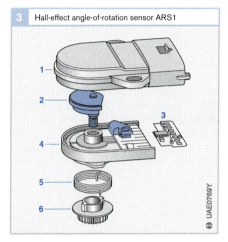

3 Hall-effect angle-of-rotation sensor ARS1

4 Hall-effect angle-of-rotation sensor ARS1 (shown with angular settings a...d)

Fig. 3
1   Housing cover
2   Rotor (permanent magnet)
3   Evaluation electronics with Hall-effect sensor
4   Housing base
5   Return spring
6   Coupling element (e.g. gear)

Fig. 4
1   Rotor (permanent magnet)
2   Pole shoe
3   Conductive element
4   Air gap
5   Hall-effect sensor
6   Shaft (magnetically soft)

φ Angle of rotation

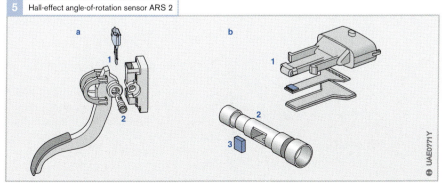

5 Hall-effect angle-of-rotation sensor ARS 2

Fig. 5
a   Installation in the accelerator-pedal module
b   Components

1   Hall-effect sensor
2   Pedal shaft
3   Magnet

Hot-film air-mass meter HFM5

Application

For optimal combustion as needed to comply with the emission regulations imposed by legislation, it is imperative that precisely the necessary air mass is inducted, irrespective of the engine's operating state.

To this end, part of the total air flow which is actually inducted through the air filter or the measuring tube is measured by a hot-film air-mass meter. Measurement is very precise and takes into account the pulsations and reverse flows caused by the opening and closing of the engine's intake and exhaust valves. Intake-air temperature changes have no effect upon measuring accuracy.

Design and construction

The housing of the HFM5 **h**ot-**fi**lm air-mass **m**eter (Fig. 1, Pos. 5) projects into a measuring tube (2) which, depending upon the engine's air-mass requirements, can have a variety of diameters (for 370...970 kg/h). This tube is installed in the intake tract downstream from the air filter. Plug-in versions are also available which are installed inside the air filter.

The most important components in the sensor are the sensor element (4), in the air intake (8), and the integrated evaluation electronics (3). The partial air flow as required for measurement flows across this sensor element.

Vapor-deposition is used to apply the sensor-element components to a semiconductor substrate, and the evaluation-electronics (hybrid circuit) components to a ceramic substrate. This principle permits very compact design. The evaluation electronics are connected to the ECU through the plug-in connection (1). The partial-flow measuring tube (6) is shaped so that the air flows past the sensor element smoothly (without whirl effects) and back into the measuring tube via the air outlet (7). This method ensures efficient sensor operation even in case of extreme pulsation, and in addition to forward flow, reverse flows are also detected (Fig. 2).

Operating concept

The hot-film air-mass meter is a "thermal sensor" and operates according to the following principle:

A micromechanical sensor diaphragm (Fig. 3, Pos. 5) on the sensor element (3) is heated by a centrally mounted heater resistor and held at a constant temperature. The temperature drops sharply on each side of this controlled heating zone (4).

The temperature distribution on the diaphragm is registered by two temperature-dependent resistors which are mounted upstream and downstream of the heater resistor so as to be symmetrical to it (measuring points M_1, M_2). Without the flow of incoming air, the temperature characteristic (1) is the same on each side of the heating zone ($T_1 = T_2$).

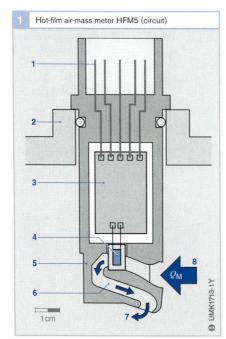

1   Hot-film air-mass meter HFM5 (circuit)

Fig. 1
1   Electrical plug-in connection
2   Measuring tube or air-filter housing wall
3   Evaluation electronics (hybrid circuit)
4   Sensor element
5   Sensor housing
6   Partial-flow measuring tube
7   Air outlet for the partial air flow Q_M
8   Intake for partial air flow Q_M

Q_M

1 cm

As soon as air flows over the sensor element, the uniform temperature distribution at the diaphragm changes (2). On the intake side, the temperature characteristic is steeper since the incoming air flowing past this area cools it off. Initially, on the opposite side (the side nearest to the engine), the sensor element cools off. The air heated by the heater element then heats up the sensor element. The change in temperature distribution leads to a temperature differential (ΔT) between the measuring points M_1 und M_2.

The heat dissipated to the air, and therefore the temperature characteristic at the sensor element is a function of the air mass flow. Independent of the absolute temperature of the air flowing past, the temperature differential is a measure of the air mass flow. Apart from this, the temperature differential is directional, which means that the air-mass meter not only registers the mass of the incoming air but also its direction.

Due to its very thin micromechanical diaphragm, the sensor has a highly dynamic response (<15 ms), a point which is of particular importance when the incoming air is pulsating heavily.

The evaluation electronics (hybrid circuit) integrated in the sensor convert the resistance differential at the measuring points M_1 and M_2 into an analog signal of 0...5 V which is suitable for processing by the ECU. Using the sensor characteristic (Fig. 2) programmed into the ECU, the measured voltage is converted into a value representing the air mass flow [kg/h].

The shape of the characteristic curve is such that the diagnosis facility incorporated in the ECU can detect such malfunctions as an open-circuit line. A temperature sensor for auxiliary functions can also be integrated in the HFM5. It is located on the sensor element upstream of the heated zone.

It is not required for measuring the air mass. For applications on specific vehicles, supplementary functions such as improved separation of water and contamination are provided for (inner measuring tube and protective grid).

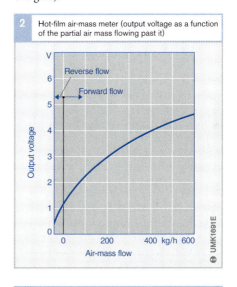

2 Hot-film air-mass meter (output voltage as a function of the partial air mass flowing past it)

UMK1691E

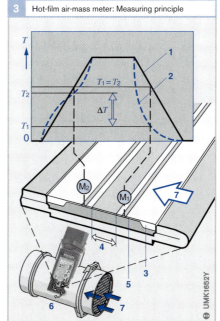

3 Hot-film air-mass meter: Measuring principle

UMK1652Y

Fig. 3
1 Temperature profile without air flow across sensor element
2 Temperature profile with air flow across sensor element
3 Sensor element
4 Heated zone
5 Sensor diaphragm
6 Measuring tube with air-mass meter
7 Intake-air flow

M_1, M_2 Measuring points

T_1, T_2 Temperature values at the measuring points M_1 and M_2

ΔT Temperature differential

LSU4 planar broad-band Lambda oxygen sensor

Application

As its name implies, the broad-band Lambda oxygen sensor is used across a very extensive range to determine the oxygen concentration in the exhaust gas. The figures provided by the sensor are an indication of the air-fuel (A/F) ratio in the engine's combustion chamber. The excess-air factor λ is used when defining the A/F ratio.

The sensor protrudes into the exhaust pipe and registers the exhaust-gas mass flow from all cylinders. Broad-band Lambda sensors make precise measurements not only at the stoichiometric point $\lambda = 1$, but also in the lean range ($\lambda > 1$) and in the rich range ($\lambda < 1$). In combination with electronic closed-loop control circuitry, these sensors generate an unmistakable, continuous electrical signal (Fig. 3) in the range from $0.7 < \lambda < \infty$ (= air with 21% O_2). This means that the broad-band Lambda sensor can be used not only in engine-management systems with two-step control ($\lambda = 1$), but also in control concepts with rich and lean air-fuel (A/F) mixtures. This type of Lambda sensor is therefore also suitable for the Lambda closed-loop control used with lean-burn concepts on gasoline engines, as well as for

diesel engines, gaseous-fuel engines, and gas-powered central heaters and water heaters (this wide range of applications led to the designation LSU: Lambda Sensor Universal (taken from the German), in other words Universal Lambda Sensor).

In a number of systems, several Lambda sensors are installed for even greater accuracy. Here, for instance, they are fitted upstream and downstream of the catalytic converter as well as in the individual exhaust tracts (cylinder banks).

Design and construction

The LSU4 broad-band Lambda sensor (Fig. 2) is a planar dual-cell limit-current sensor. It features a zirconium-dioxide/ceramic (ZrO_2) measuring cell which is the combination of a Nernst concentration cell (sensor cell which functions the same as a two-step Lambda sensor) and an oxygen-pump cell for transporting the oxygen ions. The oxygen-pump cell (Fig. 1, Pos. 8) is so arranged with respect to the Nernst concentration cell (7) that there is a 10...50 μm diffusion gap (6) between them in which are two porous platinum electrodes: A pump electrode and a Nernst measuring electrode. The gap is connected to the exhaust gas through a gas-access passage (10). The porous diffusion barrier (11) serves to limit the inflow of oxygen molecules from the exhaust gas.

Fig. 1
1 Exhaust gas
2 Exhaust pipe
3 Heater
4 Control electronics
5 Reference cell with reference-air passage
6 Diffusion gap
7 Nernst concentration cell with Nernst measuring electrode (on the diffusion-gap side), and reference electrode (on the reference-cell side)
8 Oxygen-pump cell with pump electrode
9 Porous protective layer
10 Gas-access passage
11 Porous diffusion barrier

I_P Pump current
U_P Pump voltage
U_H Heater voltage
U_{Ref} Reference voltage (450 mV, corresponds to $\lambda = 1$)
U_S Sensor voltage

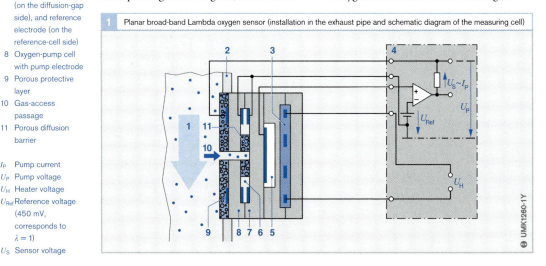

1   Planar broad-band Lambda oxygen sensor (installation in the exhaust pipe and schematic diagram of the measuring cell)

UMK1280-1Y

On the one side, the Nernst concentration cell is connected to the atmosphere by a reference-air passage (5), on the other it is connected to the exhaust gas in the diffusion gap.

An electronic closed-loop control circuit is needed in order to generate the sensor signal and for the sensor temperature control.

An integral heater (3) heats up the sensor quickly so that it soon reaches its operating temperature of 650...900 °C needed for generating a usable signal. This functions decisively reduces the effects that the exhaust-gas temperature has on the sensor signal.

Operating concept

The exhaust gas enters the actual measuring chamber (diffusion gap) of the Nernst concentration cell through the pump cell's gas-access passage. In order that the excess-air factor λ can be adjusted in the diffusion gap, the Nernst concentration cell compares the gas in the diffusion gap with that in the reference-air passage.

The complete process proceeds as follows: By applying the pump voltage U_P across the pump cell's platinum electrodes, oxygen from the exhaust gas is pumped through the diffusion barrier and into or out of the diffusion gap. With the help of the Nernst concentration cell, an electronic circuit in the ECU con-

trols the voltage (U_P) across the pump cell in order that the composition of the gas in the diffusion gap remains constant at $\lambda = 1$. If the exhaust gas is lean, the pump cell pumps the oxygen to the outside (positive pump current). On the other hand, if it is rich, due to the decomposition of CO_2 and H_2O at the exhaust-gas electrode the oxygen is pumped from the surrounding exhaust gas and into the diffusion gap (negative pump current). Oxygen transport is unnecessary at $\lambda = 1$ and pump current is zero. The pump current is proportional to the exhaust-gas oxygen concentration and is this a non-linear measure for the excess-air factor λ (Fig. 3).

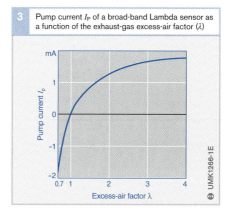

3 Pump current I_P of a broad-band Lambda sensor as a function of the exhaust-gas excess-air factor (λ)

UMK1266-1E

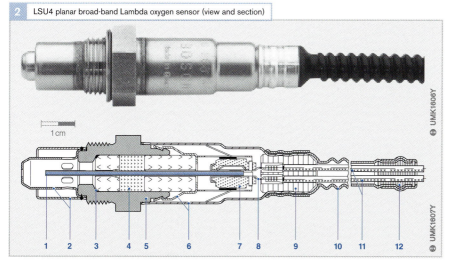

2 LSU4 planar broad-band Lambda oxygen sensor (view and section)

1 cm

UMK1606Y
UMK1607Y

Fig. 3
1 Measuring cell (combination of Nernst concentration cell and oxygen-pump cell)
2 Double protective tube
3 Seal ring
4 Seal packing
5 Sensor housing
6 Protective sleeve
7 Contact holder
8 Contact clip
9 PTFE sleeve (Teflon)
10 PTFE shaped sleeve
11 Five connecting leads
12 Seal ring

Electronic Control Unit (ECU)

Digital technology permits the implementation of a wide range of open and closed-loop control functions in the vehicle. An extensive array of influencing variables can be taken into account simultaneously so that the various systems can be operated at maximum efficiency. The ECU (Electronic Control Unit) receives the electrical signals from the sensors, evaluates them, and then calculates the triggering signals for the actuators. The control program, the "software", is stored in a special memory and implemented by a microcontroller.

Operating conditions

The ECU is subjected to very high demands with respect to
- Surrounding temperatures (during normal operation from − 40 °C to between +60 °C and +125 °C)
- Resistance to the effects of such materials as oil and fuel etc.
- Surrounding dampness
- Mechanical loading due for instance to engine vibration

Even when cranking the engine with a weak battery (cold start), the ECU must operate just as reliably as when operating voltage is at a maximum (fluctuations in on-board voltage supply).

To the same degree, very high demands apply regarding EMC (ElectroMagnetic Compatibility) and the limitation of HF interference-signal radiation.

More details on the severe standards applying to the ECU are given in the box at the end of this Chapter.

Design and construction

The pcb (printed-circuit board) with the electronic components (Fig. 1) is installed in a metal case, and connected to the sensors, actuators, and power supply through a multi-pole plug-in connector (4). The high-power driver stages (6) for the direct triggering of the actuators are integrated in the ECU case in such a manner that excellent heat dissipation to the case is ensured.

When the ECU is mounted directly on the engine, an integrated heat sink is used to dissipate the heat from the ECU case to the fuel which permanently flushes the ECU. This ECU cooler is only used on commercial vehicles. Compact, engine-mounted hybrid-technology ECUs are available for even higher levels of temperature loading.

The majority of the electronic components use SMD technology (SMD, Surface-Mounted Device). Conventional wiring is only applied at some of the power-electronics components and at the plug-in connections, so that a particularly space-saving and weight-saving design can be used.

Data processing

Input signals
In their role as peripheral components, the actuators and the sensors represent the interface between the vehicle and the ECU in its role as the processing unit. The ECU receives the electrical signals from the sensors through the vehicle's wiring harness and the plug-in connection. These signals can be of the following type:

Analog input signals
Within a given range, analog input signals can assume practically any voltage value. Examples of physical quantities which are available as analog measured values are intake-air mass, battery voltage, intake-manifold and boost pressure, coolant and intake-air temperature. An analog/digital (A/D) converter in the ECU microcontroller con-

verts these values to the digital values used by the microprocessor to perform its calculations. The maximum resolution of these signals is in steps of 5 mV per bit (approx. 1,000 steps).

Digital input signals

Digital input signals only have two states. They are either "high" or "low" (logical 1 and logical 0 respectively). Examples of digital input signals are on/off switching signals, or digital sensor signals such as the rotational-speed pulses from a Hall generator or a magnetoresistive sensor. Such signals are processed directly by the microcontroller.

Pulse-shaped input signals

The pulse-shaped signals from inductive sensors containing information on rotational speed and reference mark are conditioned in their own ECU stage. Here, spurious pulses are suppressed and the pulse-shaped signals converted into digital rectangular signals.

Signal conditioning

Protective circuitry is used to limit the input signals to a permissible maximum voltage. By applying filtering techniques, the superimposed interference signals are to a great extent removed from the useful signal which, if necessary, is then amplified to the permissible input-signal level for the microcontroller (0...5 V).

Signal conditioning can take place completely or partially in the sensor depending upon the sensor's level of integration.

Signal processing

The ECU is the system control center, and is responsible for the functional sequences of the engine management (Fig. 2, next page). The closed and open-loop control functions are executed in the microcontroller. The input signals from the sensors and the interfaces to other systems serve as the input variables, and are subjected to a further plausibility check in the computer.

---

1   ECU: Design and construction

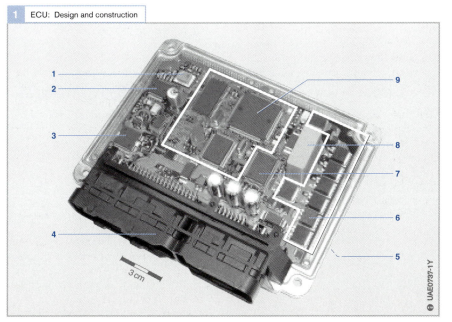

Fig. 1
1   Atmospheric-pressure sensor
2   Switched-mode power supply (SMPS) with voltage stabilization
3   Low-power driver stage
4   Plug-in connection
5   CAN interface and general input and output circuitry (underneath the pcb, therefore not visible here)
6   High-power driver stages
7   ASIC for driver-stage triggering
8   Booster-voltage store (Common Rail)
9   Microcontroller core

3 cm

UAE0737-1Y

The output signals are calculated using the program.

Microcontroller

The microcontroller is the ECU's central component and controls its operative sequence. Apart from the CPU (Central Processing Unit), the microcontroller contains not only the input and output channels, but also timer units, RAMs, ROMs, serial interfaces, and further peripheral assemblies, all of which are integrated on a single microchip. Quartz-controlled timing is used for the microcontroller.

Program and data memory

In order to carry out the computations, the microcontroller needs a program – the "software". This is in the form of binary numerical values arranged in data records and stored in a program memory.

These binary values are accessed by the CPU which interprets them as commands which it implements one after the other (refer also to the Chapter "Electronic open and closed-loop control").

This program is stored in a Read-Only Memory (ROM, EPROM, or Flash-EPROM) which also contains variant-specific data (individual data, characteristic curves, and maps). This is non-variable data which cannot be changed during vehicle operation. It is used to regulate the program's open and closed-loop control processes. The program memory can be integrated in the microcontroller and, depending upon the particular application, expanded by the addition of a separate component (e.g. by an external EPROM or a Flash-EPROM).

ROM

Program memories can be in the form of a ROM (**R**ead **O**nly **M**emory). This is a memory whose contents have been defined permanently during manufacture and thereafter remain unalterable. The ROM installed in the microcontroller only has a restricted memory capacity, which means that an additional ROM is required in case of complicated applications.

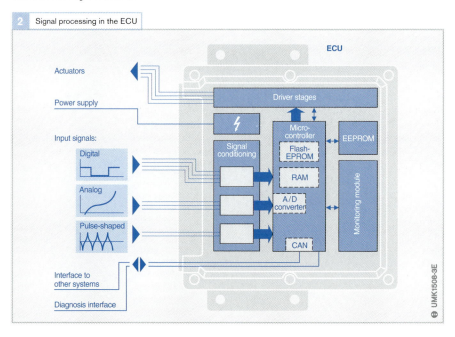

2   Signal processing in the ECU

EPROM

The data on an EPROM (**E**rasable **P**rogram-mable **ROM**) can be erased by subjecting the device to UV light. Fresh data can then be entered using a programming unit.

The EPROM is usually in the form of a separate component, and is accessed by the CPU through the Address/Data-Bus.

Flash-EPROM (FEPROM)

The Flash-EPROM is often referred to merely as a "Flash". It can be erased electrically so that it becomes possible to reprogram the ECU in the service workshops without having to open it. In the process, the ECU is connected to the reprogramming unit through a serial interface.

If the microcontroller is also equipped with a ROM, this contains the programming routines for the Flash programming. Flash-EPROMs are available which, together with the microcontroller, are integrated on a single microchip (as from EDC16).

Its decisive advantages have helped the Flash-EPROM to largely supersede the conventional EPROM.

Variable-data or main memory

Such a read/write memory is needed in order to store such variable data (variables) as the computational and signal values.

RAM

Instantaneous values are stored in the RAM (**R**andom **A**ccess **M**emory) read/write memory. If complex applications are involved, the memory capacity of the RAM incorporated in the microcontroller is insufficient so that an additional RAM module becomes necessary. It is connected to the ECU through the Address/Data-Bus.

When the ECU is switched off by turning the "ignition" key, the RAM loses its complete stock of data (volatile memory).

EEPROM (also known as the E^2PROM)

As stated above, the RAM loses its information immediately its power supply is removed (e.g. when the "ignition switch" is turned to OFF). Data which must be retained, for instance the codes for the vehicle immobilizer and the fault-store data, must therefore be stored in a non-erasable (non-volatile) memory. The EEPROM is an electrically erasable EPROM in which (in contrast to the Flash-EPROM) every single memory location can be erased individually. lt has been designed for a large number of writing cycles, which means that the EEPROM can be used as a non-volatile read/write memory.

ASIC

The ever-increasing complexity of ECU functions means that the computing powers of the standard microcontrollers available on the market no longer suffice. The solution here is to use so-called ASIC modules (**A**pplication **S**pecific **I**ntegrated **C**ircuit). These IC's are designed and produced in accordance with data from the ECU development departments and, as well as being equipped with an extra RAM for instance, and inputs and outputs, they can also generate and transmit pwm signals (see "PWM signals" below).

Monitoring module

The ECU is provided with a monitoring module. Using a "Question and Answer" cycle, the microcontroller and the monitoring module supervise each other, and as soon as a fault is detected one of them triggers appropriate back-up functions independent of the other.

Output signals

With its output signals, the microcontroller triggers driver stages which are usually powerful enough to operate the actuators directly. The driver stages can also trigger specific relays. The driver stages are proof against shorts to ground or battery voltage, as well as against destruction due to electrical or thermal overload. Such malfunctions, together with open-circuit lines or sensor faults are identified by the driver-stage IC as an error and reported to the microcontroller.

Switching signals

These are used to switch the actuators on and off (for instance, for the engine fan).

PWM signals

Digital output signals can be in the form of pwm (pulse-width modulated) signals. These are constant-frequency rectangular signals with variable on-times (Fig. 3), and are used to shift the actuators to the desired setting (e.g. EGR valve, fan, heating element, boost-pressure actuator).

Communication within the ECU

In order to be able to support the microcontroller in its work, the peripheral components must communicate with it. This takes place using an address/data bus which, for instance, the microcomputer uses to issue the RAM address whose contents are to be accessed. The data bus is then used to transmit the relevant data. For former automotive applications, an 8-bit structure sufficed whereby the data bus comprised 8 lines which together can transmit 256 values simultaneously.

The 16-bit address bus commonly used with such systems can access 65,536 addresses. Presently, more complex systems demand 16 bits, or even 32 bits, for the data bus. In order to save on pins at the components, the data and address buses can be combined in a multiplex system. That is, data and addresses are dispatched through the same lines but offset from each other with respect to time.

Serial interfaces with only a single data line are used for data which need not be transmitted so quickly (e.g. data from the fault storage).

EoL programming

The extensive variety of vehicle variants with differing control programs and data records, makes it imperative to have a system which reduces the number of ECU types needed by a given manufacturer. To this end, the Flash-EPROM's complete memory area can be programmed at the end of production with the program and the variant-specific data record (this is the so-called End-of-Line, or EoL, programming). A further possibility is to have a number of data variants available (e.g. gearbox variants), which can then be selected by special coding at the end of the line (EoL). This coding is stored in an EEPROM.

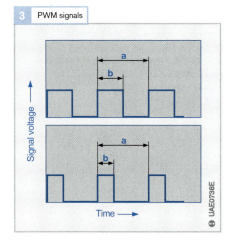

3   PWM signals

Signal voltage →

a

b

a

b

Time →

UAE0738E

Figure 3
a   Fixed frequency
b   Variable on-time

Very severe demands are made on the ECU

Basically, the ECU in the vehicle functions the same as a conventional PC. Data is entered from which output signals are calculated. The heart of the ECU is the printed-circuit board (pcb) with microcontroller using high-precision microelectronic techniques. The automotive ECU though must fulfill a number of other requirements.

Real-time compatibility

Systems for the engine and for road/traffic safety demand very rapid response of the control, and the ECU must therefore be "real-time compatible". This means that the control's reaction must keep pace with the actual physical process being controlled. It must be certain that a real-time system responds within a fixed period of time to the demands made upon it. This necessitates appropriate computer architecture and very high computer power.

Integrated design and construction

The equipment's weight and the installation space it requires inside the vehicle are becoming increasingly decisive. The following technologies, and others, are used to make the ECU as small and light as possible:

- **Multilayer:** The printed-circuit conductors are between 0.035 and 0.07 mm thick and are "stacked" on top of each other in layers.
- **SMD components** are very small and flat and have no wire connections through holes in the pcb. They are soldered or glued to the pcb or hybrid substrate, hence SMD (**S**urface **M**ounted **D**evices).
- **ASIC:** Specifically designed integrated component (**A**pplication-**S**pecific **I**ntegrated **Cir**cuit) which can combine a large number of different functions.

Operational reliability

Very high levels of resistance to failure are provided by integrated diagnosis and redundant mathematical processes (additional processes, usually running in parallel on other program paths).

Environmental influences

Notwithstanding the wide range of environmental influences to which it is subjected, the ECU must always operate reliably.

- **Temperature:** Depending upon the area of application, the ECUs installed in vehicles must perform faultlessly during continual operation at temperatures between −40°C and +60...125°C. In fact, due to the heat radiated from the components, the temperature at some areas of the substrate is considerably higher. The temperature change involved in starting at cold temperatures and then running up to hot operating temperatures is particularly severe.
- **EMC:** The vehicle's electronics have to go through severe electromagnetic compatibility testing. That is, the ECU must remain completely unaffected by electromagnetic disturbances emanating from such sources as the ignition, or radiated by radio transmitters and mobile telephones. Conversely, the ECU itself must not negatively affect other electronic equipment.
- **Resistance to vibration:** ECUs which are mounted on the engine must be able to withstand vibrations of up to 30 g (that is, 30 times the acceleration due to gravity).
- **Sealing and resistance to operating mediums:** Depending upon installation position, the ECU must withstand damp, chemicals (e.g. oils), and salt fog.

The above factors and other requirements mean that the Bosch development engineers are continually faced by new challenges.

▼ Hybrid substrate of an ECU

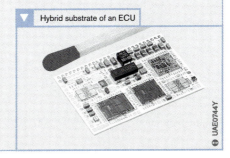

UAE0744Y

Open and closed-loop electronic control

The most important assignment of the Electronic Diesel Control (EDC) is the control of the injected fuel quantity and the instant of injection. The "Common Rail" accumulator injection system also controls the injection pressure. Furthermore, on all systems, the engine ECU also controls a number of actuators. For all components to operate efficiently, it is imperative that the EDC functions be precisely matched to every vehicle and every engine (Fig. 1).

Open and closed-loop electronic control

In both forms of control, one or more input quantities influence one or more output quantities

Open-loop control

With open-loop control, the actuators are triggered by the output signals which the ECU has calculated using the input variables, stipulated data, characteristic maps, and algorithms. The final results are not checked (open control loop). This principle is used for instance for the glow-plug sequence control.

Closed-loop control

On the other hand, as its name implies, closed-loop control is characterized by a closed control loop. Here, the actual value at the output is continually checked against the desired value, and as soon as a deviation is detected this is corrected by a change in the actuator control. The advantage of closed-loop control lies in the fact that disturbances from outside are detected and taken into account. Closed-loop control is used for instance to control the engine's idle speed.

In fact, therefore, the EDC Electronic Control Unit (ECU) is really an "open and closed-loop control unit". The term ECU "Electronic Control Unit" has become so widespread though, that it is still used even though the word "control" alone is not explicit enough.

Data processing (DP)

The ECU processes the incoming signals from the external sensors and limits them to the permissible voltage level. A number of the incoming signals are also checked for plausibility.

Using these input data, together with stored characteristic curves, the microprocessor calculates the injection's timing and its duration. This information is then converted to a signal characteristic which is aligned to the engine's piston movements. This calculation program is termed the "ECU software".

The required degree of accuracy together with the diesel engine's outstanding dynamic response necessitate high-level computing power. The output signals are applied to output stages which provide adequate power for the actuators (for instance the high-pressure solenoid valves for fuel injection, EGR positioner, or boost-pressure actuator). Apart from this, a number of other auxiliary-function components (e.g. glow relay and air conditioner) are triggered.

Faulty signal characteristics are detected by the output-stage diagnosis functions. Furthermore, signals are exchanged with other systems in the vehicle via the interfaces. The engine ECU monitors the complete injection system within the framework of a safety concept.

1   Electronic Diesel Control (EDC): Basic sequence

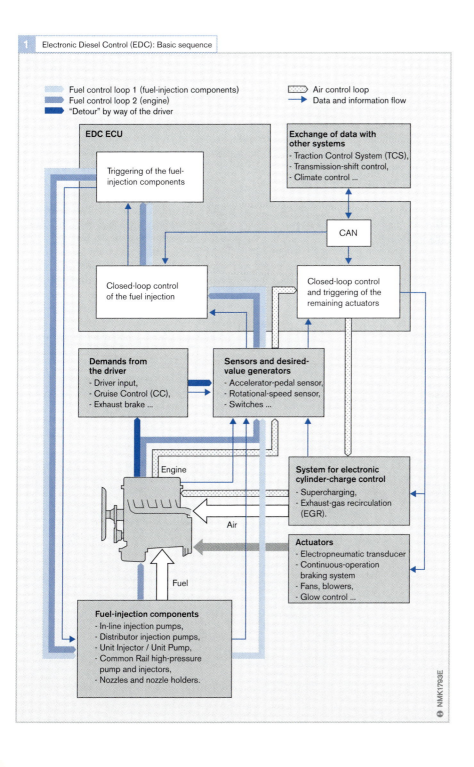

Legend:
- Fuel control loop 1 (fuel-injection components)
- Fuel control loop 2 (engine)
- "Detour" by way of the driver
- Air control loop
- Data and information flow

EDC ECU

Triggering of the fuel-injection components

Exchange of data with other systems
- Traction Control System (TCS),
- Transmission-shift control,
- Climate control ...

CAN

Closed-loop control of the fuel injection

Closed-loop control and triggering of the remaining actuators

Demands from the driver
- Driver input,
- Cruise Control (CC),
- Exhaust brake ...

Sensors and desired-value generators
- Accelerator-pedal sensor,
- Rotational-speed sensor,
- Switches ...

Engine

Air

Fuel

System for electronic cylinder-charge control
- Supercharging,
- Exhaust-gas recirculation (EGR).

Actuators
- Electropneumatic transducer
- Continuous-operation braking system
- Fans, blowers,
- Glow control ...

Fuel-injection components
- In-line injection pumps,
- Distributor injection pumps,
- Unit Injector / Unit Pump,
- Common Rail high-pressure pump and injectors,
- Nozzles and nozzle holders.

NMK1793E

Data exchange with other systems

Fuel-consumption signal

The engine ECU (Fig. 1, Pos. 3) detects the engine fuel consumption and transmits the signal via CAN to the instrument cluster, or to an independent on-board computer (6), where the driver is informed of the current fuel consumption and/or the remaining range with the fuel still in the tank. Older systems used pulse-width modulation (pwm) for the fuel-consumption signal.

Starter control

The starter (8) can be triggered from the engine ECU. This ensures that the driver cannot operate the starter with the engine already running. The starter only turns long enough for the engine to have reliably reached self-sustaining speed. This function leads to a lighter and thus lower-priced starter.

Glow control unit

The glow control unit (GZS, 5) receives information from the engine ECU on when glow is to start and for how long. It then triggers the glow plugs accordingly and monitors the glow process, as well as reporting back to the ECU on any faults (diagnosis function). The pre-glow indicator lamp is usually triggered from the ECU.

Electronic immobilizer

To prevent unauthorized starting and drive-off, the engine cannot be started before a special immobilizer (7) ECU removes the block from the engine ECU.

Either by remote control or by means of the glow-plug and starter switch ("Ignition" key), the driver can signal the immobilizer ECU that he/she is authorised to use the vehicle. The immobilizer ECU then removes the block on the engine ECU so that engine start and normal operation become possible.

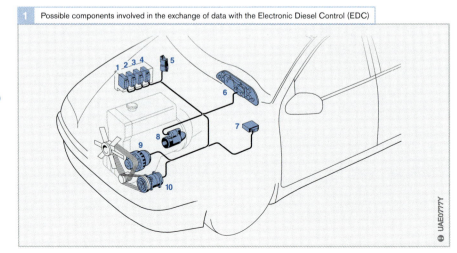

1   Possible components involved in the exchange of data with the Electronic Diesel Control (EDC)

Fig. 1
1 ESP ECU
   (with ABS and TCS)
2 ECU for transmis-
   sion-shift control
3 Engine ECU (EDC)
4 A/C ECU
5 Glow control unit
6 Instrument cluster
   with on-board
   computer
7 Immobilizer ECU
8 Starter
9 Alternator
10 A/C compressor

External torque intervention

In the case of external torque intervention, the injected fuel quantity is influenced by another (external) ECU (for instance the ECU for transmission shift, or for TCS). This informs the engine ECU whether the engine torque is to be changed, and if so, by how much (this defines the injected fuel quantity).

Alternator control

By means of a standard serial interface, the EDC can remotely control and remotely monitor the alternator (9). The regulator voltage can be controlled, just the same as the complete alternator assembly can be switched off. In case of a weak battery for instance, the alternator's charging characteristic can be improved by increasing the idle speed. It is also possible to perform simple alternator diagnosis through this interface.

Air conditioner

In order to maintain comfortable temperatures inside the vehicle when it is very hot outside, the air conditioner (A/C) cools down the air with the help of a refrigerating compressor (10). Depending upon the engine and the operating conditions, the A/C compressor can need as much as 30 % of the engine's output power.

Immediately the driver hits the accelerator pedal (in other words he/she wishes maximum torque), the compressor can be switched off briefly by the engine ECU, so that all the engine's power is available at the wheels. Since the compressor is only switched off very briefly, this has no noticeable effect upon the passenger-compartment temperature.

▶ Where does the word "Electronics" come from?

This term really originates from the ancient Greeks. They used the word electron for "amber" whose forces of attraction for wool and similar materials had already been described by Thales von Milet 2,500 years ago.

The term "electronics" originates directly from the word "electrons". The electrons, and therefore electronics as such, are extremely fast due to their very small mass and their electrical charge.

The mass of an electron has as little effect on a gram of any given substance as a 5 gram weight has on the total mass of our earth.

Incidentally, the word "electronics" is a product of the 20th century. There is no evidence available as to when the word was used for the first time. Sir John Ambrose Fleming, one of the inventors of the electron tube could have used it around 1902.

The first "Electronic Engineer" though goes back to the 19th century. He was listed in the 1888 Edition of a form of "Who's Who", published during the reign of Queen Victoria. The official title was "Kelly's Handbook of Titled, Landed and Official Classes". The Electronic Engineer is to be found under the heading "Royal Warrant Holders", that is the list of persons who had been awarded a Royal Warrant.

And what was this Electronic Engineer's job? He was responsible for the correct functiong and cleanliness of the gas lamps at court. And why did he have such a splendid title? Because he knew that "Electrons" in ancient Greece stood for glitter, shine, and sparkle.

Source:
"Basic Electronic Terms" ("Grundbegriffe der Elektronik") – Bosch publication (reprint from the "Bosch Zünder" (Bosch Company Newspaper)).

Fuel-injection control

An overview of the various control functions which are possible with the EDC control units is given in Table 1. Fig. 1 opposite shows the sequence of fuel-injection calculations with all functions, a number of which are special options. These can be activated in the ECU by the workshop when retrofit equipment is installed.

In order that the engine can run with optimal combustion under all operating conditions, the ECU calculates exactly the right injected fuel quantity for all conditions. Here, a number of parameters must be taken into account. On a number of solenoid-valve-controlled distributor pumps, the solenoid valves for injected fuel quantity and start of injection are triggered by a separate pump ECU (PSG).

1   EDC variants for road vehicles: Overview of functions

Fuel-injection system	In-line injection pumps	Helix-controlled distributor injection pumps	Solenoid-valve-controlled distributor injection pumps	Unit Injector System and Unit Pump System	Common Rail System
	PE	VE-EDC	VE-M, VR-M	UIS, UPS	CR
Function					
Injected-fuel-quantity limitation	•	•	•	•	•
External torque intervention	•[3]	•	•	•	•
Vehicle-speed limitation	•[3]	•	•	•	•
Vehicle-speed control (Cruise Control)	•	•	•	•	•
Altitude compensation	•	•	•	•	•
Boost-pressure control	•	•	•	•	•
Idle-speed control	•	•	•	•	•
Intermediate-speed control	•[3]	•	•	•	•
Active surge damping	•[2]	•	•	•	•
BIP control	–	–	•	•	–
Intake-tract switch-off	–	–	•	•[2]	•
Electronic immobilizer	•[2]	•	•	•	•
Controlled pilot injection	–	–	•	•[2]	•
Glow control	•[2]	•	•	•[2]	•
A/C switch-off	•[2]	•	•	•	•
Auxiliary coolant heating	•[2]	•	•	–	•
Cylinder-balance control	•[2]	•	•	•	•
Control of injected fuel quantity compensation	•[2]	–	•	•	•
Fan (blower) triggering	–	•	•	•	•
EGR control	•[2]	•	•	•[2]	•
Start-of-injection control with sensor	•[1,3]	•	•	–	–
Cylinder shutoff	–	–	•[3]	•[3]	•[3]

Table 1

1   Only control-sleeve in-line injection pumps
2   Passenger cars only
3   Commercial vehicles only

1   Calculation of fuel-injection process in the ECU

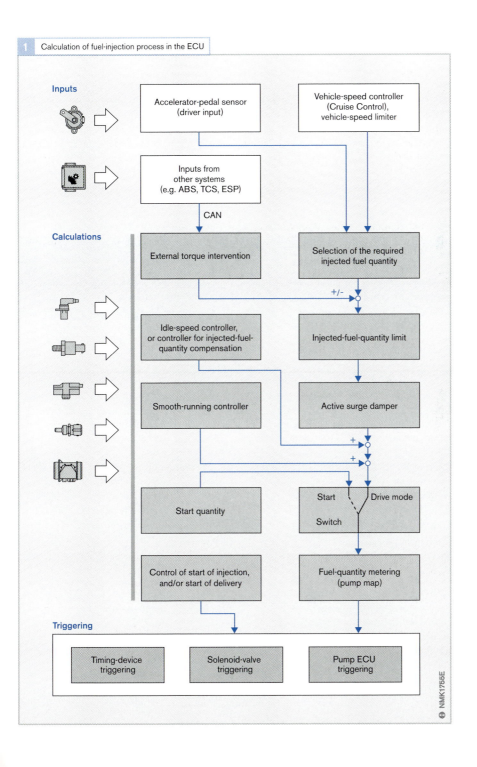

Inputs

Accelerator-pedal sensor
(driver input)

Vehicle-speed controller
(Cruise Control),
vehicle-speed limiter

Inputs from
other systems
(e.g. ABS, TCS, ESP)

CAN

Calculations

External torque intervention

Selection of the required
injected fuel quantity

+/−

Idle-speed controller,
or controller for injected-fuel-
quantity compensation

Injected-fuel-quantity limit

Smooth-running controller

Active surge damper

+

+

Start quantity

Start   Drive mode

Switch

Control of start of injection,
and/or start of delivery

Fuel-quantity metering
(pump map)

Triggering

Timing-device
triggering

Solenoid-valve
triggering

Pump ECU
triggering

NMK1755E

Start quantity

For starting, the injected fuel quantity is calculated as a function of coolant temperature and cranking speed. Start-quantity signals are generated from the moment the starting switch is turned (Fig. 1, switch in "Start" position) until a given minimum engine speed is reached.

The driver cannot influence the start quantity.

Drive mode

When the vehicle is being driven normally, the injected fuel quantity is a function of the accelerator-pedal setting (accelerator-pedal sensor) and of the engine speed (Fig. 1, switch in "Drive" position). Calculation depends upon maps which also take other influences into account (e.g. fuel and intake-air temperature). This permits best-possible alignment of the engine's output to the driver's wishes.

Idle-speed control

When the accelerator is not depressed, it is the job of the idle-speed control to ensure that a given idle speed is maintained. This can vary depending upon the engine's particular operating mode. For instance, with the engine cold the idle speed is usually set higher than when it is warm. There are further instances when the idle speed is held somewhat higher. For instance when the vehicle's electrical-system voltage is too low, when the air-conditioner is switched on, or when the vehicle is rolling freely. When the vehicle is driven in stop-and-go traffic, together with stops at traffic lights, the engine runs a lot of the time at idle. Considerations concerning emissions and fuel consumption dictate therefore that idle speed should be kept as low as possible. This of course is a disadvantage with respect to smooth-running and pull-away.

When adjusting the stipulated idle speed, the idle-speed control must cope with heavily fluctuating requirements. The input power needed by the engine-driven auxiliary equipment varies extensively.

At low electrical-system voltages for instance, the alternator consumes far more power than it does when the voltages are higher. In addition, the power demands from the A/C compressor, the steering pump, and the high-pressure generation for the diesel injection system must all be taken into account. Added to these external load moments is the engine's internal friction torque which is highly dependent upon engine temperature, and which must also be compensated for by the idle-speed control.

In order to regulate the desired idle speed, the controller continues to adapt the injected fuel quantity until the actual engine speed corresponds to the desired idle speed.

Maximum-rpm control

The maximum-rpm control ensures that the engine does not run at excessive speeds. To avoid damage to the engine, the engine manufacturer stipulates a permissible maximum speed which may only be exceeded for a very brief period.

Above the rated-power operating point, the maximum-rpm controller reduces the injected fuel quantity continually, until just above the maximum-rpm point fuel-injection stops completely. In order to prevent engine surge, a ramp function is used to ensure that the drop off in fuel injection is not too abrupt. This becomes all the more difficult the nearer the rated-power point is to the maximum-rpm point.

Intermediate-speed control

The intermediate-speed control is used only for trucks and small commercial vehicles with auxiliary power take-offs (e.g. for crane operation) or for special vehicles (e.g. ambulances with electrical power generator). With the control in operation, the engine is regulated to a load-independent intermediate speed. With the vehicle stationary, the intermediate-speed control is activated via the Cruise Control operator panel.

A fixed rotational speed can be called up from the data store at the push of a button. In addition, this operator panel can be used for preselecting specific engine speeds. The intermediate-speed control is also applied on passenger cars with automated gearboxes (e.g. Tiptronic) to control the engine speed during gearshifts.

Vehicle-speed controller (Cruise Control)

The Cruise Control is taken into operation when the vehicle is to be driven at a constant speed. It controls the vehicle speed to that selected by the driver without him/her needing to press the accelerator pedal. The driver can input the required speed either through an operating lever or through the steering-wheel keypad. The injected fuel quantity is either increased or decreased until the desired (set) speed is reached.

On some Cruise Control applications, the vehicle can be accelerated beyond the current set speed by pressing the accelerator pedal. As soon as the accelerator pedal is released again, the Cruise Control regulates the speed back down to the previously set speed.

If the driver depresses the clutch or brake pedal while the Cruise Control is activated, the control is terminated. On some applications, the control can be switched off by the accelerator pedal.

If the Cruise Control has been switched off, the driver only needs to shift the lever to the reactivate setting in order to again select the last speed which had been set.

The operator's controls can also be used for a step-by-step change of the selected speed.

Vehicle-speed limiter
Variable limitation

The vehicle-speed limiter limits the vehicle's maximum speed to a set value even if the accelerator is depressed further. On very quiet vehicles, in which the engine can hardly be heard, this is a particular help for the driver who can then no longer exceed speed limits inadvertently.

The vehicle-speed limiter keeps the injected-fuel quantity down to a limit which is in line with the selected maximum speed. It can be switched off by the lever or by the kick-down switch. In order to again select the last speed which had been set, the driver only needs to shift the lever to the reactivate setting. The operator's controls can also be used for a step-by-step change of the selected speed.

Fixed limitation

In a number of countries, fixed maximum speeds are mandatory for certain classes of vehicles (for instance, for heavy trucks). The vehicle manufacturers also limit the maximum speeds of their heavy vehicles by installing a fixed speed limit which cannot be switched off.

In the case of special vehicles, the driver can also select from a range of fixed, programmed speed limits (for instance, when there are workers on the garbage truck's rear platform).

Active surge damping

Sudden engine-torque changes excite the vehicle's drivetrain, which as a result goes into surge oscillation. These oscillations are registered by the vehicle's occupants as unpleasant periodic changes in acceleration (Fig. 2, a). It it the job of the active surge damper to reduce them (b). Two separate methods are used:

- In case of sudden changes in the torque required by the driver (through the accelerator pedal), a precisely matched filter function reduces the drivetrain excitation (1).
- The speed signals are used to detect drivetrain oscillations which are then damped by an active control. In order to counteract the drivetrain oscillations (2), the active control reduces the injected fuel quantity when rotational speed increases, and increases it when speed drops.

Smooth-running control (SRC)/Control of injected-fuel-quantity compensation (MAR)

Presuming the same duration of injection, not all of the engine's cylinders generate the same torque. This can be due to differences in cylinder-head sealing, as well as differences in cylinder friction and in the hydraulic injection components. These differences in torque output lead to rough engine running and an increase in toxic emissions.

The smooth-running control (SRC), or the control of injected-fuel-quantity compensation (MAR), use the resulting rotational-speed fluctuations when detecting such torque fluctuations. By selected variation of the injected fuel quantities at the cylinders concerned, they compensate for the torque variation. Here, the rotational speed at a given cylinder after injection is compared to a mean speed. If the particular cylinder's speed is too low the injected fuel quantity is increased, and if it is too high the fuel quantity is reduced (Fig. 3).

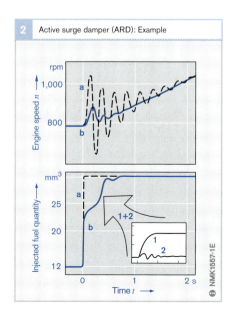

a   Without active surge
    damper
b   With active surge
    damper

1   Filter function
2   Active correction

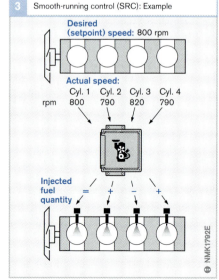

The smooth-running control is a comfort function, the primary object of which is to ensure that the engine runs smoothly in the vicinity of idle. The injected-fuel-quantity compensation function is aimed at not only improving comfort at idle but also at reducing the emissions in the medium speed ranges by ensuring identical injected fuel quantities for all cylinders. On commercial vehicles, the smooth-running control is also known as the AZG (adaptive cylinder equalization).

Injected-fuel-quantity limit

There are a number of reasons why the fuel quantity actually wished for by the driver, or that which is physically possible, should not always be injected. The injection of such fuel quantities could have the following effects:
- Excessive emissions
- Excessive soot
- Mechanical overloading due to high torque or excessive engine speed
- Thermal overloading due to excessive temperatures of the exhaust gas, coolant, oil, or turbocharger
- Thermal overloading of the solenoid valves as a result of them being triggered too long

To avoid these negative effects, a number of input variables (for instance intake-air quantity, engine speed, and coolant temperature) are used in generating this limitation figure. The result is that the maximum injected fuel quantity is limited and with it the maximum torque.

Engine-brake function

When a truck's engine brake is applied, the injected fuel quantity is either reduced to zero or the idle fuel quantity is injected. For this purpose, the ECU registers the setting of the engine-brake switch.

Altitude compensation

Atmospheric pressure drops along with increasing altitude so that the cylinder is charged with less combustion air. This means that the injected fuel quantity must be reduced accordingly, otherwise excessive soot will be emitted.

In order that the injected fuel quantity can be reduced at high altitudes, the atmospheric pressure is measured by the ambient-pressure sensor in the ECU. Atmospheric pressure also has an effect upon boost-pressure control and torque limitation.

Cylinder shutoff

If less torque is required at high engine speeds, very little fuel must be injected. As an alternative, cylinder shutoff can be applied for torque reduction. Here, half of the injectors are switched off (commercial-vehicle UIS, UPS and CRS). The remaining injectors then inject correspondingly more fuel which can be metered with even higher precision.

When the injectors are switched on and off, special software algorithms ensure smooth transitions without noticeable torque changes.

Start-of-injection control

The start of injection has a critical effect on power output, fuel consumption, noise, and emissions. The desired value for start of injection depends on engine speed and injected fuel quantity, and it is stored in the ECU in special maps. Adaptation is possible as a function of coolant temperature and ambient pressure.

Tolerances in manufacture and in the pump mounting on the engine, together with changes in the solenoid valve during its lifetime, can lead to slight differences in the solenoid-valve switching times which in turn lead to different starts of injection. The response behaviour of the nozzle-and-holder assembly also changes over the course of time. Fuel density and temperature also have an effect upon start of injection. This must be compensated for by some form of control strategy in order to stay within the prescribed emissions limits. The following closed-loop controls are employed (Table 2):

2 Start-of-injection control			
Closed-loop control	Control using needle-motion sensor	Start-of-delivery control	BIP control
Injection system			
In-line injection pumps	●	–	–
Helix-controlled distributor pumps	●	–	–
Solenoid-valve-controlled distributor pumps	●	●	–
Common Rail	–	–	–
Unit Injector/Unit Pump	–	–	●

Table 2

The start-of-injection control is not needed with the Common Rail System, since the high-voltage triggering used in the CRS permits highly reproducible starts of injection.

Closed-loop control using the needle-motion sensor

The inductive needle-motion sensor is fitted in an injection nozzle (reference nozzle, usually cylinder 1). When the needle opens (and closes) the sensor transmits a pulse (Fig. 4). The needle-opening signal is used by the ECU as confirmation of the start of injection. This means that inside a closed control loop the start of injection can be precisely aligned to the desired value for the particular operating point.

The needle-motion sensor's untreated signal is amplified and interference-suppressed before being converted to precision square-wave pulses which can be used to mark the start of injection for a reference cylinder.

The ECU controls the actuator mechanism for the start of injection (for in-line pumps the solenoid actuator, and for distributor pumps the timing-device solenoid valve) so that the actual start of injection always corresponds to the desired/setpoint start of injection.

The start-of-injection signal can only be evaluated when fuel is being injected and when the engine speed is stable. During starting and overrun (no fuel injection), the needle-motion sensor cannot provide a signal which is good enough for evaluation. This means that the start-of-injection control loop cannot be closed because there is no signal available confirming the start-of-injection.

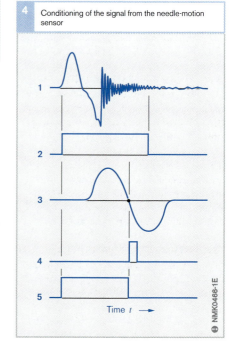

4 Conditioning of the signal from the needle-motion sensor

Time *t* →

NMK0466-1E

Fig. 4
1 Untreated signal from the needle-motion sensor (NBF),
2 Signal derived from the NBF signal,
3 Untreated signal from the inductive engine-speed sensor
4 Signal derived from untreated engine-speed signal,
5 Evaluated start-of-injection signal

In-line injection pumps

On in-line pumps, a special digital current controller improves the control's accuracy and dynamic response by aligning the current to the start-of-injection controller's setpoint value practically without any delay at all.

In order to ensure start-of-injection accuracy in open-loop-controlled operation too, the start-of-delivery solenoid in the control-sleeve actuator mechanism is calibrated to compensate for the effects of tolerances. The current controller compensates for the effects of the temperature-dependent solenoid-winding resistance. All these measures ensure that the setpoint value for current as derived from the start map leads to the correct stroke of the start-of-delivery solenoid and to the correct start of injection.

Start-of-delivery control using the incremental angle/time signal (IWZ)

On the solenoid-valve-controlled distributor pumps (VP30, VP44), the start of injection is also very accurate even without the help of a needle-motion sensor. This high level of accuracy was achieved by applying positioning control to the timing device inside the distributor pump. This form of closed-loop control serves to control the start of delivery and is referred to as start-of-delivery control. Start of delivery and start of injection have a certain relationship to each other and this is stored in the so-called wave-propagation-time map in the engine ECU.

The signal from the crankshaft-speed sensor and the signal from the incremental angle/time system (IWZ signal) inside the pump, are used as the input variables for the timing-device positioning control.

The IWZ signal is generated inside the pump by the rotational-speed or angle-of-rotation sensor (1) on the trigger wheel (2) attached to the driveshaft. The sensor shifts along with the timing device (4) which, when it changes position, also changes the position of the tooth gap (3) relative to the

TDC pulse of the crankshaft-speed sensor. The angle between the tooth gap, or the synchronization pulse generated by the tooth gap, and the TDC pulse is continually registered by the pump ECU and compared with the stored reference value. The difference between the two angles represents the timing device's actual position, and this is continually compared with its setpoint/desired position. If the timing-device position deviates, the triggering signal for the timing-device solenoid valve is changed until actual and setpoint position coincide with each other.

Since all cylinders are taken into account, the advantage of this form of start-of-delivery control lies in the system's rapid response. It has a further advantage in that it also functions during overrun when no fuel injection takes place which means that the timing device can be preset for when the next injection event occurs.

In case even more severe demands are made on the accuracy of the start of injection, the start-of-delivery control can have an optional start-of-injection control with needle-motion sensor superimposed upon it.

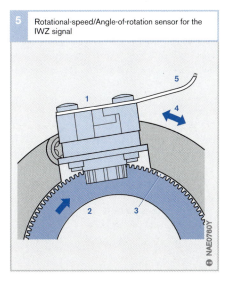

5   Rotational-speed/Angle-of-rotation sensor for the IWZ signal

Fig. 5
1   Rotational-speed/ angle-of-rotation sensor inside the injection pump
2   Trigger wheel
3   Trigger-wheel tooth gap
4   Shift due to timing device
5   Electrical plug-in connection

BIP control

BIP control is used with the solenoid-valve-controlled Unit Injector System (UIS) and Unit Pump System (UPS). The start of delivery – or BIP (Begin of Injection Period) – is defined as the instant in time in which the solenoid closes. As from this point, pressure buildup starts in the pump high-pressure chamber. The nozzle opens as soon as the nozzle-opening pressure is exceeded, and injection can commence (start of injection). Fuel metering takes place between start of delivery and end of solenoid-valve triggering. This period is termed the delivery period.

Since there is a direct connection between the start of delivery and the start of injection, all that is needed for the precise control of the start of injection is information on the instant of the start of delivery.

So as to avoid having to apply additional sensor technology (for instance, a needle-motion sensor), electronic evaluation of the solenoid-valve current is used in detecting the start of delivery. Around the expected instant of closing of the solenoid valve, constant-voltage triggering is used (BIP window, Fig. 6, Pos. 1). The inductive effects when the solenoid valve closes result in the curve having a specific characteristic which is registered and evaluated by the ECU. For each injection event, the deviation of the solenoid-valve closing point from the theoretical setpoint is registered and stored, and applied for the following injection sequence as a compensation value.

If the BIP signal should fail, the ECU changes over to open-loop control.

Shutoff

The "auto-ignition" principle of operation means that in order to stop the diesel engine it is only necessary to cut off its supply of fuel. With EDC (Electronic Diesel Control), the engine is switched off due to the ECU outputting the signal "Fuel quantity zero" (that is, the solenoid valves are no longer triggered, or the control rack is moved back to the zero-delivery setting).

There are also a number of redundant (supplementary) shutoff paths (for instance, the electrical shutoff valve (ELAB) on the port-and-helix controlled distributor pumps).

The UIS and UPS are intrinsically safe, and the worst thing that can happen is that one single unwanted injection takes place. Here, therefore, supplementary shutoff paths are not needed.

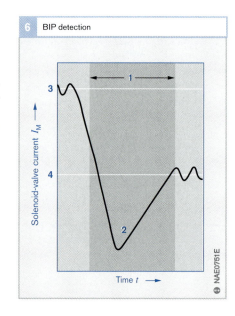

6   BIP detection

Solenoid-valve current I_M →

3

4

1

2

Time t →

NAE0751E

Fig. 6
1   BIP window
2   BIP signal
3   Level of pickup
    current
4   Holding-current level

Lambda closed-loop-control for passenger-car diesel engines

Applications

The lawmakers are continually increasing the severity of the legislation governing the exhaust-emission limit values for diesel engines. Apart from the measures taken to optimize the engine's internal combustion, the open and closed-loop control of those functions which are relevant with regard to the exhaust emissions are continuing to gain in importance. The introduction of the Lambda closed-loop control opens up immense potential for reducing the diesel engine's exhaust emissions.

The broad-band Lambda oxygen sensor in the exhaust pipe (Fig. 1, Pos. 7) measures the residual oxygen in the exhaust gas. This is an indicator for the A/F ratio (excess-air-factor Lambda l). A high level of signal accuracy is ensured throughout the sensor's service life by adapting the Lambda-sensor signal during actual operations. The Lambda-sensor signal is used as the basis for a number of Lambda functions which will be described in more detail in the following.

A Lambda closed-loop control circuit is imperative for the regeneration of NO_X accumulator-type catalytic converters.

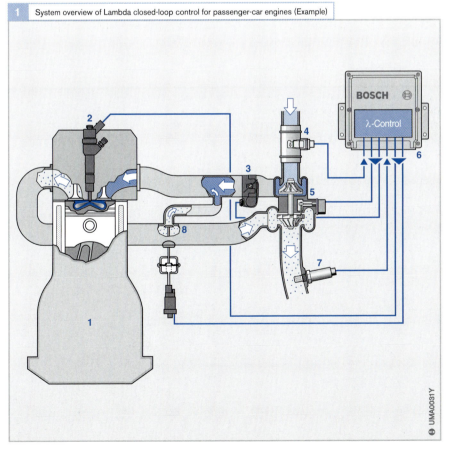

1   System overview of Lambda closed-loop control for passenger-car engines (Example)

Fig. 1
1   Diesel engine
2   Diesel injection
    component
    (here Common
    Rail injector)
3   Throttle valve
4   Hot-film air-mass
    meter
5   Exhaust-gas
    turbocharger
    (here, VTG version)
6   Engine ECU for
    EDC
7   Broad-band Lambda
    oxygen sensor
8   EGR valve

The Lambda closed-loop-control is suitable for all passenger-car diesel injection systems controlled by the EDC16 generation.

Basic functions

Pressure compensation

The untreated Lambda-sensor signal is a function of the oxygen concentration in the exhaust gas and of the exhaust-gas pressure at the sensor's installation point. The influence of the pressure on the sensor signal must therefore be compensated for.

The pressure-compensation function incorporates two maps, one for the exhaust-gas pressure and one for the pressure-dependence of the lambda sensor's output signal. These two maps are used for the correction of the sensors output signal with reference to the particular operating point.

Adaptation

In the overrun (trailing throttle) mode, the Lambda sensor's adaptation takes into account the deviation of the measured oxygen concentration from the fresh-air oxygen concentration (approx. 21 %). As a result, the system "learns" a correction value which at every engine operating point is used to correct the measured oxygen concentration. This leads to a precise, drift-compensated Lambda output signal throughout the sensor's service life.

Lambda-based EGR control

Regarding emissions, compared with the conventional EGR method based on air mass, using the Lambda oxygen sensor to measure the residual oxygen in the exhaust gas permits a tighter tolerance range for the complete vehicle fleet. In the MNEFZ (modified new European driving cycle) exhaust-gas test, therefore, from the exhaust-emissions viewpoint this equates to an improvement of 10...20 % with regard to the fleet as a whole.

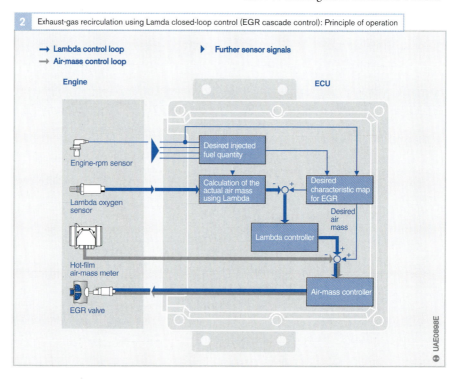

2   Exhaust-gas recirculation using Lamda closed-loop control (EGR cascade control): Principle of operation

→ Lambda control loop
→ Air-mass control loop
▶ Further sensor signals

Engine

ECU

Engine-rpm sensor

Lambda oxygen sensor

Hot-film air-mass meter

EGR valve

Desired injected fuel quantity

Calculation of the actual air mass using Lambda

Desired characteristic map for EGR

Desired air mass

Lambda controller

Air-mass controller

UAE0898E

Cascade control

With the cascade control (Fig. 2), the conventional air-mass control loop as used in present-day series production has a Lambda control loop superimposed upon it (refer to the section "Control and triggering of the remaining actuators").

The dynamic response is excellent when air-mass control is used (that is, the air-mass meter responds far more quickly). The external Lambda closed-loop control circuit improves the EGR-system accuracy.

The actual air-mass figure is calculated from the Lambda oxygen sensor signal and from the desired value for the injected fuel quantity. The system deviation between the calculated air mass and the desired air mass taken from the EGR desired-value characteristic map, is compensated for by the Lambda controller.

Notwithstanding the change to the calculated air mass as the reference (command) variable for the EGR, in its physical effects, this control architecture acts as a Lambda closed-loop control (in contrast to the Lambda closed-loop control used on gasoline engines in which the fuel mass is taken as the reference (command) variable).

Fuel-quantity mean-value adaptation

The fuel-quantity mean-value adaptation (MMA) provides a precise injected-fuel-quantity signal for the setpoint generation as needed for those control loops which are relevant for exhaust-gas emission (e.g. EGR control, boost-pressure control, and start-of-injection control). The MMA operates in the lower part-load range and determines the average deviation in the injected fuel quantity of all cylinders together.

Fig. 3 shows the basic structure of the MMA and its intervention in the control loops which are relevant for the exhaust-gas emissions.

3   Fuel-quantity mean-value adaptation in the "indirect control" mode: Principle of operation

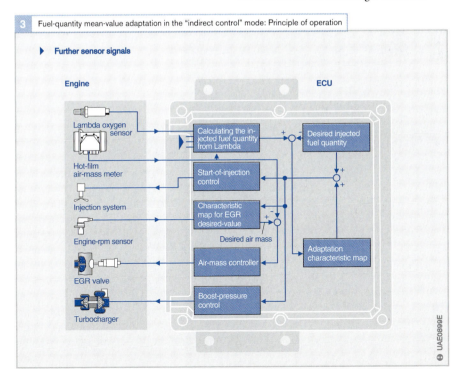

UAE0899E

The Lambda-sensor signal and the air-mass signal are used in calculating the actually injected fuel mass which is then compared with the desired injected fuel mass. Differences are then stored in an adaptation map in defined "learning points". This procedure ensures that when the operating point necessitates an injected fuel quantity correction, this can be implemented without delay even during dynamic changes of state.

These correction quantities are stored in the EEPROM of the ECU and are available immediately the engine is started.

Basically speaking, there are two MMA operating modes. These differ in the way they apply the detected deviations in injected fuel quantity:

Operating mode: Indirect control
In the "indirect control" mode (Fig. 3), a precise desired injected fuel quantity is used as an input variable in the desired-value characteristic maps which are relevant for exhaust emissions. The injected fuel quantity is not corrected during the fuel-metering process.

Operating mode: Direct control
In the "direct control" mode, in order that the amount of fuel actually injected corresponds as closely as possible to the desired quantity, the fuel-quantity deviation is used to correct the injected fuel quantity during the fuel-metering process. In this case, this is (indirectly) a closed fuel-quantity loop.

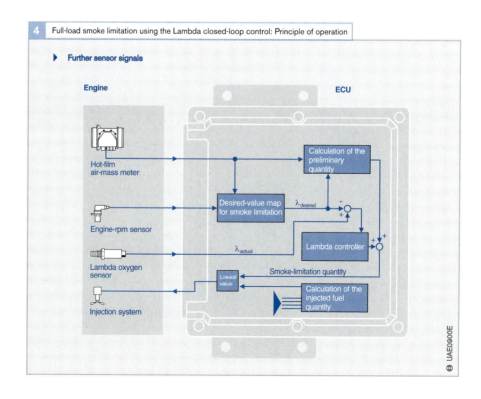

4   Full-load smoke limitation using the Lambda closed-loop control: Principle of operation

▶ **Further sensor signals**

Engine

ECU

Hot-film air-mass meter

Engine-rpm sensor

Lambda oxygen sensor

Injection system

Calculation of the preliminary quantity

Desired-value map for smoke limitation

$\lambda_{desired}$

λ_{actual}

Lambda controller

Lowest value

Smoke-limitation quantity

Calculation of the injected fuel quantity

UAE0900E

Full-load smoke limitation

Fig. 4 shows the block diagram of the control structure for full-load smoke limitation using a Lambda oxygen sensor. The objective here is to determine the maximum amount of fuel which may be injected without exceeding a given smoke value.

The signals from the air-mass meter and the engine rpm sensor are applied together with a smoke-limitation map in determining the desired Lambda value $\lambda_{DESIRED}$. This, in turn, is applied together with the air mass in order to calculate the preliminary value for the maximum permissible injected fuel quantity.

This form of control is already in series production, and has a Lambda closed-loop control imposed upon it. Using the difference between $\lambda_{DESIRED}$ and the actually measured residual oxygen in λ_{ACTUAL} the Lambda controller calculates the correction fuel quantity. The maximum full-load injected fuel quantity is the total of the preliminary quantity and the correction quantity.

This control architecture permits a high level of dynamic response due to the pilot control, and improved precision due to the superimposed Lambda control loop.

Detection of undesirable combustion

During overrun (trailing throttle), if the Lambda-sensor signal drops below a given, specially calculated threshold this indicates that undesirable combustion is taking place. In this case, the engine can be switched off by closing a control flap and the EGR valve. The detection of undesirable combustion represents an additional engine safeguard function.

Summary

Using Lambda-based EGR it is possible to considerably reduce a vehicle fleet's exhaust-gas emissions scatter. Here, either fuel-quantity mean-value adaptation (MMA) can be used or cascade control.

MMA provides a precise injected-fuel-quantity signal for generating the desired (setpoint) values for the control loops which are relevant for emissions (for instance, for EGR control, boost-pressure control, start-of-injection control). The precision of these control loops is increased as a result.

In addition, the application of Lambda closed-loop control permits the precise definition of the full-load smoke quantity as well as the detection of undesirable combustion in the overrun (trailing throttle) mode.

Furthermore, the Lambda sensor's high-precision signal can be used in a Lambda control loop for the regeneration of NO_X catalytic converters.

▶ Racing trucks

The diesel engines and fuel-injection systems for trucks tuned for racing – also known as racing trucks – are adapted to the specials requirements of racing sport. For instance, the engine of a production truck with a power output of about 300 kW (410 bhp) is increased by a factor of 3.7 to about 110 kW (1,500 bhp)!

This means higher engine revs, greater cylinder charges (air mass), and therefore larger injected-fuel quantities within shorter periods of time.

Just as in a production vehicle, the electronic systems have the task of providing highly precise control. The competition rules stipulate that the top speed of 100 mph (160 km/h) may only be exceeded by a maximum of 1.25 mph (2 km/h), that is, 1.25%! This requires special features when it comes to speed-regulation breakaway.

In all other aspects, the Electronic Diesel Control (EDC) is identical to production models.

NMM0596Y

During the races, the engines are driven within the range of $\lambda = 1$. It means even greater injected-fuel quantities, and this requires larger plunger-and-barrel assemblies and special nozzles. Even the injection cams – if fitted – must have a more pointed shape.

Of course, these modifications have a negative impact on fuel consumption, exhaust-gas emissions, and the service life of all components.

▶ MAN racing-truck in-line 6-cylinder engine with Bosch common-rail system

Two turbochargers supply the air mass required. The turbines glow red-hot at exhaust-gas temperatures of about 1,000°C (1,832°F).

(Source: MAN)

NMM0597Y

Further special adaptations

In addition to those described here, EDC permits a wide range of other functions. For instance, these include:

Drive recorder
On commercial vehicles, the Drive Recorder is used to record the engine's operating conditions (for instance, how long was the vehicle driven, under what temperatures and loads, and at what engine speeds). This data is used in drawing up an overview of operational conditions from which, for instance, individual service intervals can be calculated.

Special application engineering for competition trucks
On race trucks, the 160 km/h maximum speed may be exceeded by no more than 2 km/h. On the other hand, this speed must be reached as soon as possible. This necessitates special adaptation of the ramp function for the vehicle-speed limiter.

Adaptations for off-highway vehicles
Such vehicles include diesel locomotives, rail cars, construction machinery, agricultural machinery, boats and ships. In such applications, the diesel engine(s) is/are far more often run in the full-load range than is the case with road vehicles (90 % full-load operation compared with 30 %). The power output of such engines must therefore be reduced in order to ensure an adequate service life.

The mileage figures which are often used as the basis for the service interval on road vehicles are not available for such equipment as agricultural or construction machinery, and in any case if they were available they would have no useful significance. Instead, the Drive Recorder data is used here.

Port-and-helix-controlled fuel-injection systems: Triggering

Electronically controlled in-line injection pumps, PE-EDC
As with the mechanically (flyweight) governed in-line fuel-injection pumps, the injected fuel quantity here is also a function of the control-rack position and the engine speed. The control rack is shifted to the desired position by the linear magnet of the actuator mechanism directly attached to the pump (Fig. 1 on the next page, Pos. 3).

On the control-sleeve in-line pump, an additional electrical actuator in connection with start-of-injection control can be used to arbitrarily adjust the injected fuel quantity and the start of delivery. This necessitates an extra actuator mechanism (4).

Triggering the control-rack actuator mechanism
With the solenoid de-energized, a spring forces the control rack to the stop position and thus interrupts the fuel supply. When the current through the solenoid increases, the solenoid gradually overcomes the force of the spring and control-rack travel increases so that more fuel is injected.

This means that the level of current permits the continuous adjustment of control-rack travel between zero and maximum delivery quantity (pwm signal = pulse-width-modulated signal).

The corresponding pump characteristic map is programmed into the ECU. Using this map, and depending on engine speed, the control-rack travel appropriate to the desired fuel quantity is calculated. In order to improve driveability, a control characteristic can be provided which is familiar from the mechanical (flyweight) RQ and RQV governors.

Using a sensor (rack-travel sensor (RWG)), the position control in the ECU registers the actual control-rack setting so that the system deviation can be calculated.

This enables the position control to quickly and accurately correct the rack setting.

Triggering of the control-sleeve actuator mechanism

A power stage which is triggered directly from the processor with a pwm signal provides the power needed to energize the start-of-delivery solenoids. A low-level current causes minor travel of the start-of-delivery solenoid, and retards the start of delivery or start of injection. On the other hand, a high level of current advances the start-of-delivery point.

The injected-fuel-quantity controls as applied to the control-sleeve in-line pumps and the conventional in-line pumps (both with EDC) are identical. The more extensive functional scope in the ECU for the control-sleeve pump results to a great extent from expanded programs.

Key stop

The key stop function using the "ignition" key supersedes the conventional mechanical shutoff device. It stops the supply of fuel by interrupting the power supply to the electrical shutoff valve (ELAB, redundant shutoff path) and to the control-rack linear solenoids.

Port-and-helix-controlled axial-piston distributor pumps, VE-EDC

Solenoid actuator for injected-fuel-quantity control

The solenoid actuator (moving-magnet actuator) engages the delivery-piston control collar via a shaft (Fig. 2, Pos. 3). Similar to the mechanically governed distributor pumps, the cutoff bores are exposed sooner or later depending upon the position of the control collar.

The injected fuel quantity can be continually varied between zero and maximum. By means of an angle sensor (Hall short-circuiting-ring sensor, HDK), information on the angle of rotation of the actuator, and therefore on the setting of the control collar with respect to the cutoff ports, is reported back to the ECU and used to calculate the correct fuel quantity as a function of engine speed.

In the de-energized state, the fuel delivery is set to zero by return springs on the actuator mechanism.

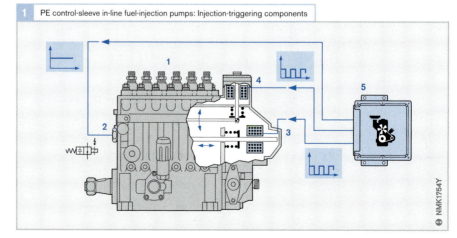

1 PE control-sleeve in-line fuel-injection pumps: Injection-triggering components

Fig. 1
1   In-line injection pump
2   Electrical shutoff valve (ELAB)
3   Control-rack actuator mechanism (injected fuel quantity)
4   Control-sleeve actuator mechanism (start of delivery)
5   Engine ECU

NMK1754Y

2 VE-EDC axial-piston distributor pumps: Injection-triggering components

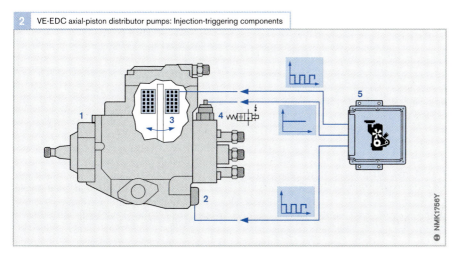

Fig. 2
1 Distributor pump
2 Timing-device
   solenoid valve
   (start of injection)
3 Moving-magnet
   actuator (injected
   fuel quantity)
4 Electric shutoff
   valve (ELAB)
5 Engine ECU

Solenoid valve for start-of-injection control

The pump's internal pressure is proportional to the pump speed and, similar to the mechanical timing device, is effective at the timing-device piston. This pressure is applied to the timing-device pressure side and is modulated by the clocked timing-device solenoid valve (2). The on/off ratio (ratio of the solenoid's open time to its closed time) for the triggering of the solenoid valve is taken from a programmed control map.

A permanently opened solenoid valve (pressure reduction) results in later start-of-injection points, and with the solenoid valve permanently closed (pressure increase) the start-of-injection points take place earlier. In between these two extremes, infinite variation of the on/off ratio can be implemented by the ECU.

Deviations between actual and desired start-of-injection points, as detected with the help of the needle-motion sensor, result in a change of the on/off ratio for triggering the timing-device solenoid valve. This ratio is changed continually until the system deviation is "zero", and ensures a dynamic response which is comparable to that of the mechanical start-of-injection adjustment.

Shutoff

As a rule, shutoff is by means of the injected-fuel-quantity actuator ("zero" fuel delivery). The redundant electrical shutoff valve (ELAB) provides a further degree of safety (4).

On the distributor pump, the redundant electrical shutoff valve is mounted on the upper side of the pump's distributor head. When switched on (that is, with the engine running), the solenoid keeps the inlet port to the high-pressure chamber open (the armature with sealing cone is pulled in). When switch off takes place using the "ignition switch", the solenoid winding is de-energized and the sealing cone is forced back onto its seat by a spring so that the inlet port to the high-pressure chamber is interrupted.

On marine engines, the ELAB is open when de-energized. This means that the engine can still run even though the on-board power supply has failed. Apart from this, the number of electrical consumers is kept to a minimum since continuous current aggravates the effects of saltwater corrosion.

Solenoid-valve-controlled injection systems: Triggering

The solenoid-valve-controlled injection-system family includes the following:
- Axial-piston distributor pumps, VE-M (VP30)
- Radial-piston distributor pumps, VR (VP44)
- Common Rail System CRS
- Unit Injector System UIS
- Unit Pump System UPS

These solenoid-valve-controlled injection systems are all triggered by similar signals. The characteristic features of the individual systems are described in the following sections, together with the differences between them and conventional injection systems.

Due to better EMC (ElectroMagnetic Compatibility) characteristics, the high-pressure solenoid valves are triggered by analog signals. Since the triggering signal must feature steep current edges in order to ensure narrow tolerances and a high degree of reproducibility for the injected fuel quantity, this form of triggering makes severe demands on the output stages.

In addition, the triggering process must ensure a minimum of power loss in the ECU and in the high-pressure solenoid valve. In other words, the triggering currents must be as low as possible.

Irrespective of operating range, injection control must be extremely accurate in order that the injection pump and solenoid injector inject precisely and with a high level of reproducibility.

The injection system must respond extremely quickly to changes, which means that the calculations in the microcontroller and the triggering-signal implementation in the output stages must take place at very high speed. Data processing is therefore referred to as being "real-time compatible" (resolution time 1 μs).

Solenoid-valve-controlled distributor pumps

On the distributor pumps equipped with a high-pressure fuel-quantity solenoid valve (VP44 and VP30), a pump ECU (PSG) attached to the pump housing is responsible for triggering the solenoid valves at the calculated start-of-delivery point (Fig. 1a, Pos. 1a). This takes place in accordance with the requirements of the engine ECU (6) which is responsible for the engine and vehicle functions. The two ECUs communicate with each other through the CAN bus.

The latest generation of the VP44 system features only a single ECU (PSG 16) which has united all the EDC functions in a single unit, and which is situated directly on the pump (Fig. 1b, Pos. 1b).

The timing-device solenoid valve (4) used on the distributor pumps is triggered using a pulse-width-modulated (pwm) signal. In order to avoid malfunctions due to resonance effects, the clock frequency is not held constant throughout the complete rotational-speed range but instead, at given speed ranges, it is switched to a different frequency (window technique).

The high-pressure fuel-quantity solenoid valve is triggered via current control (Fig. 2)

Fig. 1
a With separate pump and engne ECU
b With integral pump and engine ECU

1a Pump ECU (PSG5)
1b Pump ECU (PSG16)
2 Distributor pump
3 Timing device
4 Timing-device solenoid valve
5 High-pressure fuel-quantity solenoid valve
6 Engine ECU (MSG)

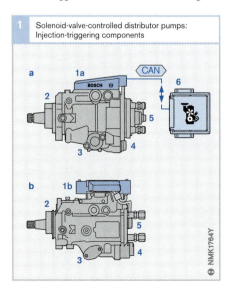

1 Solenoid-valve-controlled distributor pumps: Injection-triggering components

NMK1764Y

which subdivides the triggering process into a pickup-current phase (a) of approx. 18 A, and a holding-current phase (c) of approx. 10 A. At the beginning of the controlled holding-current phase (after $200...250\ \mu s$), the BIP evaluation circuit detects the solenoid-valve needle closing against the valve seat (BIP = Beginning of the Injection Period).

The latest generation uses the PSG16 pump ECU, and is also provided with a BIP current (b) between the pickup and holding-current phases which is at the optimum level for the BIP detection function.

In order that the pump's injection characteristics are always reproducible irrespective of operating conditions, the triggering circuitry as a whole, and the current control, must be extremely accurate. Furthermore, they must keep the power loss in ECU and solenoid valve down to a minimum.

Defined, rapid opening of the solenoid valve is required at the end of the injection process. To this end, high-speed quenching (d) using a high quenching voltage (1) is applied at the valve to dissipate the energy stored in its solenoid.

The solenoid valve can also be used to control pilot injection (PI) for reduction of combustion noise. Here, the solenoid valve is operated ballistically between the PI point and the MI (main injection) point. In other words it is only partially opened, which means that it can be closed again very quickly. The resulting injection spacing is very short so that even at high rotational speeds, adequate cam pitch remains for the main injection process.

The subdivision into the individual triggering phases is calculated by the microcontroller in the pump ECU.

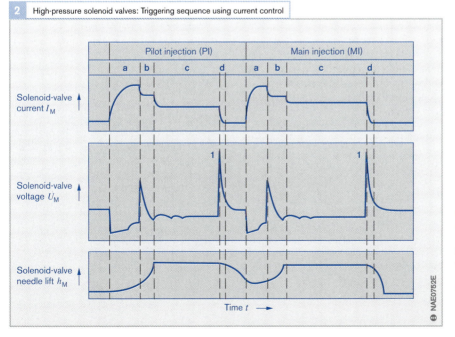

2 High-pressure solenoid valves: Triggering sequence using current control

Fig. 2
a Pickup-current phase
b BIP detection
c Holding-current phase
d High-speed quenching

1 Quenching voltage

Common Rail System CRS

In the Common Rail System (CRS), the fuel pressure in the rail (Fig. 3, Pos. 3) and with it the injection pressure, are determined by the rail-pressure valve (Pos. 8), and if required by a throttle upstream of the high-pressure pump (1). The injector's (6) high-pressure solenoid valve (7) defines the injection point and the duration of injection in accordance with the various operating parameters. This means, therefore, that injection pressure is decoupled from injection point and duration of injection. Decoupling the injection pressure from injection point and duration of injection means that in addition to the main injection (MI), which is responsible for the generation of torque, other injection processes independent of the injection pressure can also be triggered. On the one hand these are for the most part pilot injections (PI) with the principle objective of reducing the combustion noise, and on the other secondary injection processes (post injection (POI))which serve to reduce exhaust emissions. The injected fuel quantity is the product of injection pressure and duration of injection.

Post injection involves minute quantities of fuel. Injection is not carried out individually at stipulated cylinders, but rather the minute quantities concerned are added together in the ECU until the smallest quantity is reached which can be handled by the injector. When this is reached, it is injected at the next possible opportunity.

Rail-pressure control

The permanent pressure in the rail is generated by a continuously operating high-pressure pump. The closed control loop for the rail pressure comprises the rail-pressure sensor (4) the engine ECU (5) and the rail-pressure control valve (8).

The microcontroller in the engine ECU receives the sensor signal and uses it to calculate the desired pressure. This is outputted to a driver stage which triggers the rail-pressure control valve by means of a pwm signal. The level of the applied current corresponds to the desired pressure. The higher the triggering current the higher the rail pressure. The microcontroller compares the actual pressure at the sensor with the desired pressure and takes appropriate closed-loop control action in case of deviation.

In a number of versions, at high speeds and low levels of required fuel, one of the high-pressure pump's three pumping elements can be switched off (element shutoff). In this case, a triggering current flows through the element-shutoff solenoid valve. Switching off one of the elements reduces the strain on the pump as well as increasing the engine's efficiency.

3 Common Rail System (CRS): Injection-triggering components

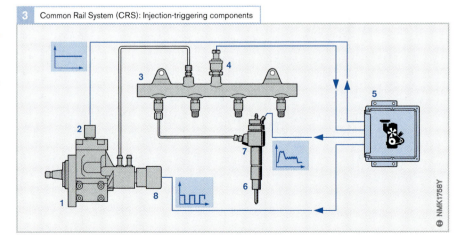

4   Triggering sequence of a high-pressure solenoid valve for a single injection event

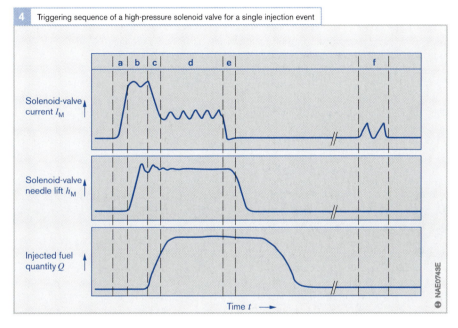

Fig. 4
a   Opening phase
b   Pickup-current
     phase
c   Transition to holding-
     current phase
d   Holding-current
     phase
e   Switch off,
f   Recharge

Injector triggering

In the inoperative mode, the injector's high-pressure solenoid valve is not triggered and is therefore closed.

The injector injects when the solenoid valve opens. Solenoid-valve triggering is subdivided into five phases (Figs. 4 and 5).

Opening phase
Initially, in order to ensure tight tolerances and high levels of reproducibility for the injected fuel quantity, the current for opening the valve features a steep, precisely defined flank and increases rapidly up to approx. 20 A. This is achieved with a so-called "booster voltage" of up to as much as 100 V which is generated in the ECU and stored in a capacitor (boost-voltage store). When this voltage is applied across the solenoid valve, the current increases several times faster than it does when only battery voltage is used.

Pickup-current phase
During the pickup-current phase, battery voltage is applied to the solenoid valve, and

assists in opening it quickly. Current control limits pickup current to approx. 20 A.

Holding-current phase
In order to reduce the power loss in ECU and injector, the current is dropped to approx. 12 A in the holding-current phase. The energy which becomes available when pickup current and holding current are reduced is routed to the booster-voltage store.

Switch off
When the current is switched off in order to close the valve, the surplus energy is also routed to the booster-voltage store.

Recharge
Between the actual injection events, a saw-tooth waveform is applied to the injectors which are not injecting (Fig. 5, f_1). Maximum current level is so low that there is no danger of the injector opening. Energy stored in the solenoid valve as a result is then routed to the booster-voltage store (Fig. 5, f_2), which it recharges until the original voltage is reached as required for opening the solenoid valve.

5 | Common Rail System: Block diagram of the triggering phase in the control for a single cylinder group

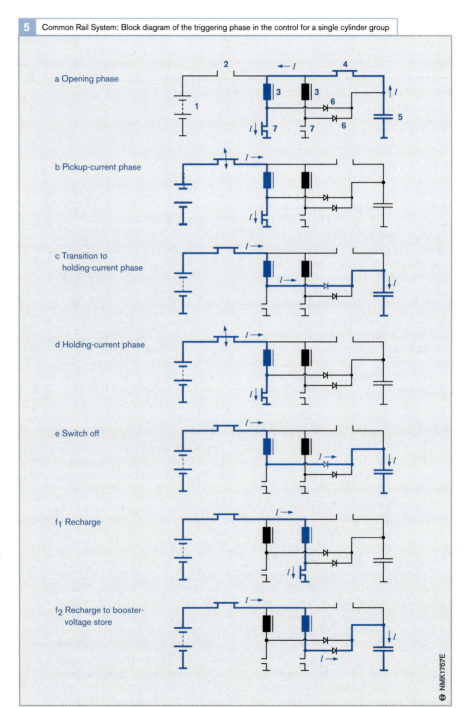

a Opening phase

b Pickup-current phase

c Transition to
  holding-current phase

d Holding-current phase

e Switch off

f_1 Recharge

f_2 Recharge to booster-
  voltage store

Fig. 5

1   Battery
2   Current control
3   Solenoid windings
    of the high-pressure
    solenoid valves
4   Booster switch
5   Booster-voltage
    store (capacitor)
6   Free-wheeling
    diodes for energy
    recovery and high-
    speed quenching
7   Cylinder selector
    switch

I   Current flow

NMK1757E

Unit Injector Systems and Unit Pump Systems (UIS/UPS)

The triggering of the high-pressure solenoid valve places severe demands on the output stages. Close tolerances and high reproducibility of the injected fuel quantity demand that the triggering signal features a particularly steep edge.

The Unit Injector and Unit Pump high-pressure solenoid valves are triggered in a similar manner using current control (Figs. 6 through 8) which divides the triggering process into the pickup-current phase (a) and the holding-current phase (c). This form of triggering permits very fast switching times and reduces the power loss. For a brief period between these two phases, constant triggering current is applied to permit the detection of the solenoid-valve closing point (refer to section "BIP control", b).

In order to ensure high-speed, defined opening of the solenoid valve at the end of the injection event, a high voltage is applied across the terminals (d) for rapid quenching of the energy stored in the solenoid valve.

The microcontroller is responsible for the calculation of the individual triggering phases. An ASIC module (gate array) with high computing power assists the microcontroller by generating two digital triggering signals (MODE signal and ON signal).

6   Unit Injector and Unit Pump Systems (UIS/UPS): Triggering components

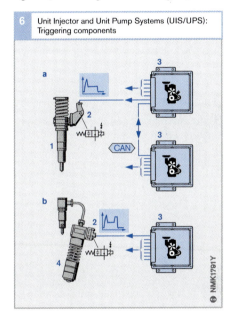

Fig. 6
a   Unit Injector System (UIS) with 2 ECUs
b   Unit Pump System (UPS)

1   Unit Injector (UI)
2   High-pressure solenoid valve
3   Engine ECU
4   Unit Pump (UP)

7   High-pressure solenoid valve: Triggering sequences

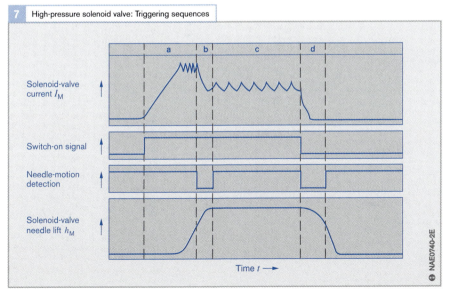

Solenoid-valve current I_M

Switch-on signal

Needle-motion detection

Solenoid-valve needle lift h_M

Time $t \longrightarrow$

Fig. 7
a   Pickup current (commercial-vehicle UIS/UPS: 12...20 A; passenger-car UIS: 20 A)
b   BIP detection,
c   Holding current (commercial-vehicle UIS/UPS: 8...14 A; passenger-car UIS: 12 A)
d   High-speed quenching

These triggering signals then instruct the output-stage drivers to generate the required triggering-current sequence.

Boot-shaped injection characteristic

Future systems will provide the possibility of a special "boot-shaped" injection characteristic (with the main injection (MI) immediately following the pilot injection (PI Fig. 8). Here, the solenoid-valve current is held at a precise intermediate level (approx. 4...6 A, c_1) between the pickup-current and holding-current phases. This leads to the solenoid valve being held in an intermediate position so that a "boot-shaped" injection characteristic is the result.

ECU coupling

The passenger-car Unit Injector System can also be used on engines with more than 6 cylinders. In order to be able to cope with the requirement for more driver stages for injector triggering, and for increased microcontroller computing power, such engines can be equipped with two ECUs coupled together to form a "master-slave alliance". Similar to the Common Rail System (CRS), the ECUs are coupled together by means of an internal CAN-Bus operating at a baud rate of up to 1 Mbaud (1,000,000 bit/s).

Some of the functions are allocated to a specific ECU which is then solely responsible for their processing (for instance, the injected-fuel-quantity compensation). Other assignments though can be flexibly processed in this configuration by either of the ECUs (for instance, the registration of sensor signals). This configuration cannot be changed during operation.

Using this master-slave concept, it is possible to implement both Biturbo control and active exhaust-gas treatment.

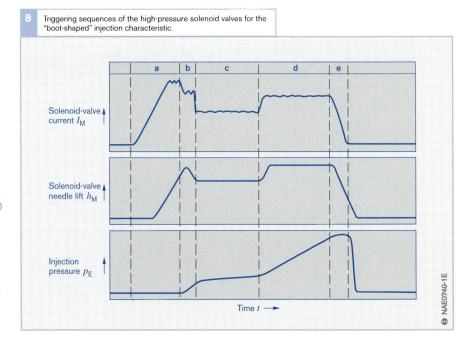

8 Triggering sequences of the high-pressure solenoid valves for the "boot-shaped" injection characteristic

Solenoid-valve current I_M

Solenoid-valve needle lift h_M

Injection pressure p_E

Time t ⟶

NAE0740-1E

Fig. 8
a   Pickup current (commercial-vehicle UIS/UPS: 12...20 A)
b   BIP detection
c1 Holding current for "boot-shaped" injection characteristic
c2 Holding current (commercial-vehicle UIS/UPS: 8...14 A)
d   High-speed quenching

Control and triggering of the remaining actuators

In addition to the fuel-injection components themselves, EDC is responsible for the control and triggering of a large number of other actuators. These are used for cylinder-charge control, or for the control of engine cooling, or are used in diesel-engine start-assist systems. Here too, as is the case with the closed-loop control of injection, the inputs from other systems (such as TCS) are taken into account.

A variety of different actuators are used, depending upon the vehicle type, its area of application and the type of fuel injection. This chapter deals with a number of examples, and further actuators are covered in the Chapter "Actuators".

A variety of different methods are used for triggering:
- The actuators are triggered directly from an output (driver) stage in the engine ECU using appropriate signals (e.g. the EGR valve).
- If high currents are involved (for instance for fan control), the ECU triggers a relay.
- The engine ECU transfers signals to an independent ECU, which is then used to trigger or control the remaining actuators (for instance, for glow control).

The advantage of incorporating all engine-control functions in the EDC ECU lies in the fact that not only the injected fuel quantity and instant of injection can be taken into account in the engine control concept, but also other engine functions such as EGR and boost-pressure control. This leads to a considerable improvement in engine management. Apart from this, the engine ECU has a vast amount of information at its disposal as needed for other functions (for instance, engine and intake-air temperature as used for glow control on the diesel engine).

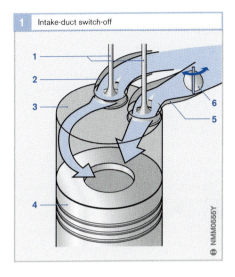

1   Intake-duct switch-off

Fig. 1
1 Intake valve
2 Turbulence duct
3 Cylinder
4 Piston
5 Intake duct
6 Flap

Auxiliary coolant heating
High-performance diesel engines are very efficient, and under certain circumstances do not generate enough waste heat to adequately heat the vehicle's interior. One solution for overcoming this problem is to install auxiliary coolant heating using glow plugs. Depending upon the power available from the alternator, this system is triggered in a number of steps. It is controlled by the engine ECU as used for EDC.

Intake-duct switch-off
In the lower engine-rpm ranges and at idle, a flap (Fig. 1, Pos. 6) operated by an electropneumatic transducer closes one of the intake ducts (5). Fresh air in then only inducted through the turbulence duct (2). This leads to improved air turbulence in the lower rpm ranges which in turn results in more efficient combustion. In the higher rpm ranges, the engine's volumetric efficiency is improved thanks to the open intake duct (5) and the power output increases as a result.

Boost-pressure control

Boost-pressure control applied to the exhaust-gas turbocharger improves the engine's torque curve in full-load operation, and its exhaust and refill cycle in the part-load range.The optimum (desired) boost pressure is a function of engine speed, injected fuel quantity, coolant and fuel temperature, and the surrounding air pressure. This optimum (desired) boost pressure is compared with the actual value registered by the boost-pressure sensor and, in the case of deviation, the ECU either operates the bypass valve's electropneumatic transducer or the guide blades of the VTG (**V**ariable **T**urbine **G**eometry) exhaust-gas turbocharger (refer also to the Chapter "Actuators").

Fan triggering

When a given engine temperature is exceeded, the engine ECU triggers the engine cooling fan, which continues to rotate for a brief period after the engine is switched off. This run-on period is a function of the coolant temperature and the load imposed on the engine during the preceding driving cycle.

Exhaust-gas recirculation (EGR)

In order to decrease the NO_X emissions, exhaust gas is directed into the engine's intake duct through a channel, the cross section of which can be varied by an EGR valve. The EGR valve is triggered by an electropneumatic transducer or by an electric actuator.

Due to the high temperature of the exhaust gas and its high proportion of contamination, it is difficult to precisely measure the exhaust-gas flow which is recirculated back into the engine. Control, therefore, takes place indirectly through an air-mass meter located in the flow of fresh intake air. The meter's output signal is then compared in the ECU with the engine's theoretical air requirement which has been calculated from a variety of data (e.g. engine rpm).

The lower the measured mass of the incoming fresh air compared to the theoretical air requirement, the higher is the proportion of recirculated exhaust gas.

Currently, EGR is only used on passenger cars, although work is proceeding on the development of a commercial-vehicle version.

Substitute functions

If individual input signals should fail, the ECU is without the important information it needs for calculations. In such cases, substitute functions are used. Two examples are given below:

Example 1: The fuel temperature is needed for calculation of the injected fuel quantity. If the fuel-temperature sensor fails, the ECU uses a substitute value for its calculations. This must be selected so that excessive soot formation is avoided, although this can lead to a reduction of engine power in certain operating ranges.

Example 2: Should the camshaft sensor fail, the ECU applies the crankshaft-sensor signal as a subsitute. Depending on the vehicle manufacturer, there are a variety of different concepts for using the crankshaft signal to determine when cylinder 1 is in the compression cycle. The use of substitute functions leads to engine restart taking slightly longer.

Substitute functions differ according to vehicle manufacturer, so that many vehicle-specific functions are possible.

The diagnosis function stores data on all malfunctions that occur. This data can then be accessed in the workshop (refer also to the Chapter "Electronic Diagnosis (OBD)").

Torque-controlled EDC systems

The engine-management system is continually being integrated more closely into the overall vehicle system. Through the CAN-Bus, vehicle dynamics systems such as TCS, and comfort and convenience systems such as Cruise Control, have a direct influence on the Electronic Diesel Control (EDC). Apart from this, much of the information registered and/or calculated in or by the engine management system must be passed on to other ECUs through the CAN-Bus.

In order to be able to incorporate the EDC even more efficiently in a functional alliance with other ECUs, and implement other changes rapidly and effectively, it was necessary to make far-reaching changes to the newest-generation controls. These changes resulted in the torque-controlled EDC which was introduced with the EDC16. The main feature is the change over of the module interfaces to the parameters as commonly encountered in practice in the vehicle.

Engine parameters

Essentially, an IC engine's output can be defined using the three parameters: Power P, rpm n, and torque M.

For 2 diesel engines. Fig. 1 compares typical curves of torque and power as a function of engine rpm. Basically speaking, the following formula applies:

$$P = 2 \cdot \pi \cdot n \cdot M$$

In other words, it suffices to use the torque as the reference (command) variable. Engine power then results from the above formula. Since power output cannot be measured directly, torque has turned out to be a suitable reference (command) variable for engine management.

Torque control

When accelerating, the driver uses the accelerator pedal (sensor) to directly demand a given torque from the engine. At the same time, but independent of the driver's requirements, via the interfaces other vehicle systems submit torque demands resulting from the power requirements of the particular component (e.g. air conditioner, alternator). Using these torque-requirement inputs, the engine management calculates the output torque to be generated by the engine and controls the fuel-injection and air-system actuators accordingly. This method has the following advantages:

- No single system (for instance, boost pressure, fuel injection, pre-glow) has a direct effect upon the engine management. This enables the engine management to also take into account higher-level optimization criteria (such as exhaust emissions and fuel consumption) when processing external requirements, and thus control the engine in the most efficient manner,
- Many of the functions which do not directly concern the engine management can be designed to function identically for diesel and gasoline engines.
- Extensions to the system can be implemented quickly.

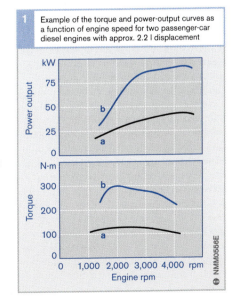

1 Example of the torque and power-output curves as a function of engine speed for two passenger-car diesel engines with approx. 2.2 l displacement

Fig. 1
a   Year of manufacture 1968
b   Year of manufacture 1998

Engine-management sequence

Fig. 2 shows (schematically) the processing of the setpoint inputs in the engine ECU. In order to be able to fulfill their assignments efficiently, the engine management's control functions all require a wide range of sensor signals and information from other ECUs in the vehicle.

Propulsion torque

The driver's input (that is, the signal from the accelerator-pedal sensor), is interpreted by the engine management as the request for a propulsive torque. The inputs from the Cruise Control and the Vehicle-Speed Limiter are processed in exactly the same manner.

Following this selection of the desired propulsive torque, should the situation arise, the vehicle-dynamics system (TCS, ESP) increases the desired torque value when there is the danger of wheel lockup and decreases it when the wheels show a tendency to spin.

Further external torque demands

The drivetrain's torque adaptation must be taken into account (drivetrain transmission ratio). This is defined for the most part by the ratio of the particular gear, or by the torque-converter efficiency in the case of automatic gearboxes. On vehicles with an automatic-gearbox, the gearbox control stipulates the torque requirement during the actual gear shift. Apart from reducing the load on the gearbox, reduced torque at this point results in a comfortable, smooth gear shift. In addition, the torque required by other engine-powered units (for instance, air-conditioner compressor, alternator, servo pump) is determined. This torque requirement is calculated either by the units themselves or by the engine management.

Calculation is based on unit power and rotational speed, and the engine management adds up the various torque requirements. The vehicle's driveability remains unchanged notwithstanding varying requirements from the auxiliary units and changes in the engine's operating status.

Internal torque demands

At this stage, the idle-speed control and the active surge damper intervene.

For instance, if demanded by the situation, in order to prevent mechanical damage, or excessive smoke due to the injection of too much fuel, the torque limitation reduces the internal torque requirement. In contrast to the previous engine-management systems, limitations are no longer only applied to the injected fuel-quantity, but instead, depending upon the required effects, also to the particular physical quantity involved.

The engine's losses are also taken into account (e.g. friction, drive for the high-pressure pump). The torque represents the engine's measurable effects to the outside. The engine management though can only generate these effects in conjunction with the correct fuel injection together with the correct injection point, and the necessary marginal conditions as apply to the air-intake system (e.g. boost pressure and EGR rate). The required injected fuel quantity is determined using the current combustion efficiency. The calculated fuel quantity is limited by a protective function (for instance, protection against overheating), and if necessary can be varied by the smooth-running control (SRC). During engine start, the injected fuel quantity is not determined by external inputs such as those from the driver, but rather by the separate "start-quantity control" function.

Actuator triggering

Finally, the desired values for the injected fuel quantity are used to generate the triggering data for the injection pump and/or the injectors, and for defining the optimum operating point for the intake-air system.

2   Engine-management sequence for torque-controlled diesel injection

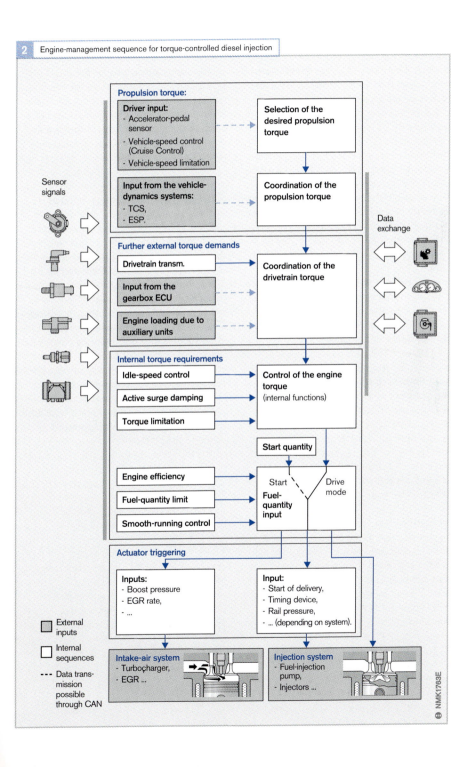

Sensor signals

Data exchange

Propulsion torque:

Driver input:
- Accelerator-pedal sensor
- Vehicle-speed control (Cruise Control)
- Vehicle-speed limitation

Selection of the desired propulsion torque

Input from the vehicle-dynamics systems:
- TCS,
- ESP.

Coordination of the propulsion torque

Further external torque demands

Drivetrain transm.

Input from the gearbox ECU

Engine loading due to auxiliary units

Coordination of the drivetrain torque

Internal torque requirements

Idle-speed control

Active surge damping

Torque limitation

Control of the engine torque (internal functions)

Start quantity

Engine efficiency

Fuel-quantity limit

Smooth-running control

Start   Drive mode
Fuel-quantity input

Actuator triggering

Inputs:
- Boost pressure
- EGR rate,
- ...

Input:
- Start of delivery,
- Timing device,
- Rail pressure,
- ... (depending on system).

Intake-air system
- Turbocharger,
- EGR ...

Injection system
- Fuel-injection pump,
- Injectors ...

☐ External inputs

☐ Internal sequences

--- Data transmission possible through CAN

NMK1763E

Data transfer between automotive electronic systems

Today's vehicles are being equipped with a constantly increasing number of electronic systems. Along with their need for extensive exchange of data and information in order to operate efficiently, the data quantities and speeds concerned are also increasing continuously.

For instance, in order to guarantee perfect driving stability, the Electronic Stability Program (ESP) must exchange data with the engine management system and the transmission-shift control.

System overview

Increasingly widespread application of electronic communications systems, and electronic open and closed-loop control systems, for automotive functions such as
- Electronic engine-management (EDC and Motronic)
- Electronic transmission-shift control (GS)
- Antilock braking system (ABS)
- Traction control system (TCS)
- Electronic Stability Program (ESP)
- Adaptive Cruise Control (ACC) and
- Mobile multimedia systems together with their display instrumentation

has made it vital to interconnect the individual ECUs by means of networks.

The conventional point-to-point exchange of data through individual data lines has reached its practical limits (Fig. 1), and the complexity of current wiring harnesses and the sizes of the associated plugs are already very difficult to manage. The limited number of pins in the plug-in connectors has also slowed down ECU development work.

To underline this point:
Apart from being about 1 mile long, the wiring harness of an average middle-class vehicle already includes about 300 plugs and sockets with a total of 2,000 plug pins. The only solution to this predicament lies in the application of specific vehicle-compatible Bus systems. Here, CAN has established itself as the standard.

Serial data transfer (CAN)

Although CAN (Controller Area Network) is a linear bus system (Fig. 2) specifically designed for automotive applications, it has already been introduced in other sectors (for instance, in building installation engineering).

Data is relayed in serial form, that is, one after another on a common bus line. All CAN stations have access to this bus, and via a CAN interface in the ECUs they can receive and transmit data through the CAN bus line. Since a considerable amount of data can be exchanged and repeatedly accessed on a single bus line, this networking results in far fewer lines being needed.

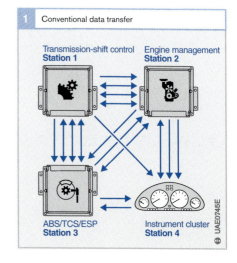

1   Conventional data transfer

Transmission-shift control
Station 1

Engine management
Station 2

ABS/TCS/ESP
Station 3

Instrument cluster
Station 4

UAE0745E

Applications in the vehicle

For CAN in the vehicle there are four areas of application each of which has different requirements. These are as follows:

Multiplex applications

Multiplex is suitable for use with applications controlling the open and closed-loop control of components in the sectors of body electronics, and comfort and convenience. These include climate control, central locking, and seat adjustment. Transfer rates are typically between 10 kbaud and 125 kbaud (1 kbaud = 1 kbit/s) (low-speed CAN).

Mobile communications applications

In the area of mobile communications, CAN networks such components as navigation system, telephone, and audio installations with the vehicle's central display and operating units. Networking here is aimed at standardizing operational sequences as far as possible, and at concentrating status information at one point so that driver distraction is reduced to a minimum. With this application, large quantities of data are transmitted, and data transfer rates are in the 125 kbaud range. It is impossible to directly transmit audio or video data here.

Diagnosis applications

The diagnosis applications using CAN are aimed at applying the already existing network for the diagnosis of the connected ECUs. The presently common form of diagnosis using the special K line (ISO 9141) then becomes invalid. Large quantities of data are also transferred in diagnostic applications, and data transfer rates of 250 kbaud and 500 kbaud are planned.

Real-time applications

Real-time applications serve for the open and closed-loop control of the vehicle's movements. Here, such electronic systems as engine management, transmission-shift control, and electronic stability program (ESP) are networked with each other. Commonly, data transfer rates of between 125 kbaud and 1 Mbaud (high-speed CAN) are needed to guarantee the required real-time response.

Bus configuration

Configuration is understood to be the layout and interaction between the components in a given system. The CAN bus has a linear bus topology (Fig. 2) which in comparison with other logical structures (ring bus and/or star bus) features a lower failure probability. If one of the stations fails, the bus still remains fully accessible to all the other stations. The stations connected to the bus can be either ECUs, display devices, sensors, or actuators. They operate using the Multi-Master principle, whereby the stations concerned all have equal priority regarding their access to the bus. It is not necessary to have a higher-order administration.

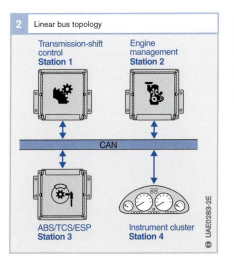

2   Linear bus topology

Transmission-shift control
Station 1

Engine management
Station 2

CAN

ABS/TCS/ESP
Station 3

Instrument cluster
Station 4

UAE0283-2E

Content-based addressing

The CAN bus system does not address each station individually according to its features, but rather according to its message contents. It allocates each "message" a fixed "*identifier*" (message name) which identifies the contents of the message in question (for instance, engine speed). This identifier has a length of 11 bits (standard format) or 29 bits (extended format).

With content-based addressing each station must itself decide whether it is interested in the message or not ("message filtering" Fig. 3). This function can be performed by a special CAN module (Full-CAN), so that less load is placed on the ECU's central microcontroller. Basic CAN modules "read" all messages. Using content-based addressing, instead of allocating station addresses, makes the complete system highly flexible so that equipment variants are easier to install and operate. If one of the ECUs requires new information which is already on the bus, all it needs to do is call it up from the bus. Similarly, provided they are receivers, new stations can be connected (implemented) without it being necessary to modify the already existing stations.

Bus arbitration

The identifier not only indicates the data content, but also defines the message's priority rating. An identifier corresponding to a low binary number has high priority and vice versa. Message priorities are a function for instance of the speed at which their contents change, or their significance with respect to safety. There are never two (or more) messages of identical priority in the bus.

Each station can begin message transmission as soon as the bus is unoccupied. Conflict regarding bus access is avoided by applying bit-by-bit identifier arbitration (Fig. 4), whereby the message with the highest priority is granted first access without delay and without loss of data bits (nondestructive protocol).

The CAN protocol is based on the logical states "dominant" (logical 0) and "recessive" (logical 1). The "Wired And" arbitration principle permits the dominant bits transmitted by a given station to overwrite the recessive bits of the other stations. The station with the lowest identifier (that is, with the highest priority) is granted first access to the bus.

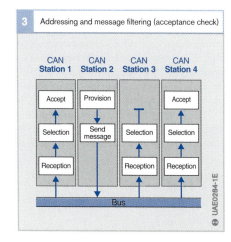

3 Addressing and message filtering (acceptance check)

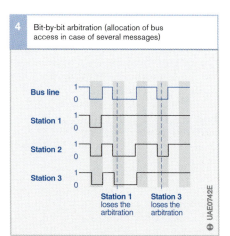

4 Bit-by-bit arbitration (allocation of bus access in case of several messages)

The transmitters with low-priority messages automatically become receivers, and repeat their transmission attempt as soon as the bus is vacant again.

In order that all messages have a chance of entering the bus, the bus speed must be appropriate to the number of stations participating in the bus. A cycle time is defined for those signals which fluctuate permanently (e.g. engine speed).

Message format

CAN permits two different formats which only differ with respect to the length of their identifiers. The standard-format identifier is 11 bits long, and the extended-format identifier 29 bits. Both formats are compatible with each other and can be used together in a network. The data frame comprises seven consecutive fields (Fig. 5) and is a maximum of 130 bits long (standard format) or 150 bits (extended format).

The bus is recessive at idle. With its dominant bit, the *"Start of frame"* indicates the beginning of a message and synchronises all stations.

The *"Arbitration field"* consists of the message's identifier (as described above) and an additional control bit. While this field is being transmitted, the transmitter accompanies the transmission of each bit with a check to ensure that it is still authorized to transmit or whether another station with a higher-priority message has accessed the Bus. The control bit following the identifier is designated the RTR-bit (**R**emote **T**ransmission **R**equest). It defines whether the message is a "Data frame" (message with data) for a receiver station, or a "Remote frame" (request for data) from a transmitter station.

The *"Control field"* contains the IDE bit (**I**dentifier **E**xtension **B**it) used to differentiate between standard format (IDE = 0) and extended format (IDE = 1), followed by a bit

reserved for future extensions. The remaining 4 bits in this field define the number of data bytes in the next data field. This enables the receiver to determine whether all data has been received.

The *"Data field"* contains the actual message information comprised of between 0 and 8 bytes. A message with data length = 0 is used to synchronise distributed processes. A number of signals can be transmitted in a single message (e.g. engine rpm and engine temperature).

The *"CRC Field"* (**C**yclic **R**edundancy **C**heck) contains the frame check word for detecting possible transmission interference.

The *"ACK Field"* contains the **ack**nowledgement signals used by the receiver stations to confirm receipt of the message in non-corrupted form. This field comprises the ACK slot and the recessive ACK delimiter. The ACK slot is also transmitted recessively and overwritten "dominantly" by the receivers upon the message being correctly received. Here, it is irrelevant whether the message is of significance or not for the particular receiver in the sense of the message filtering or acceptance check. Only correct reception is confirmed.

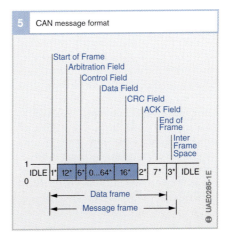

5   CAN message format

Start of Frame
Arbitration Field
Control Field
Data Field
CRC Field
ACK Field
End of Frame
Inter Frame Space

IDLE | 1* | 12* | 6* | 0...64* | 16* | 2* | 7* | 3* | IDLE

Data frame
Message frame

Fig. 5
0   Dominant level
1   Recessive level
*   Number of bits

© UAE0285-1E

The *"End of frame"* marks the end of the message and comprises 7 recessive bits.

The *"Inter-frame space"* comprises three bits which serve to separate successive messages. This means that the bus remains in the recessive IDLE mode until a station starts a bus access.

As a rule, a sending station initiates data transmission by sending a "data frame". It is also possible for a receiving station to call in data from a sending station by transmitting a "remote frame".

Detecting errors
A number of control mechanisms for detecting errors are integrated in the CAN protocol.

In the *"CRC field"*, the receiving station compares the received CRC sequence with the sequence calculated from the message.

With the *"Frame check"*, frame errors are recognized by checking the frame structure. The CAN protocol contains a number of fixed-format bit fields which are checked by all stations.

The *"ACK check"* is the receiving stations' confirmation that a message frame has been received. Its absence signifies for instance that a transmission error has been detected.

"Monitoring" indicates that the sender observes (monitors) the bus level and compares the differences between the bit that has been sent and the bit that has been checked.

Compliance with *"Bitstuffing"* is checked by means of the "Code check". The stuffing rule stipulates that in every *"data frame"* or *"remote frame"*, a maximum of 5 successive equal-priority bits may be sent between the *"Start of frame"* and the end of the *"CRC field"*. As soon as five identical bits have been transmitted in succession, the sender inserts an opposite-priority bit. The receiving sta-

tion erases these opposite-polarity bits after receiving the message. Line errors can be detected using the "bitstuffing" principle.

If one of the stations detects an error, it interrupts the actual transmission by sending an "Error frame" comprising six successive dominant bits. Its effect is based on the intended violation of the stuffing rule, and the object is to prevent other stations accepting the faulty message.

Defective stations could have a derogatory effect upon the bus system by sending an "error frame" and interrupting faultless messages. To prevent this, CAN is provided with a function which differentiates between sporadic errors and those which are permanent, and which is capable of identifying the faulty station. This takes place using statistical evaluation of the error situations.

Standardization
The International Organization for Standardization (ISO) and SAE (Society of Automotive Engineers) have issued CAN standards for data exchange in automotive applications:
- For low-speed applications up to 125 kbit/s: ISO 11519-2, and
- For high-speed applications above 125 kbit/s: ISO 11898 and SAE J 22584 (passenger cars) and SAE J 1939 (trucks and buses).
- Furthermore, an ISO Standard on CAN Diagnosis (ISO 15765 – Draft) is being prepared.

Prospects

Along with the increasing levels of system-component performance and the rise in function integration, the demands made on the vehicle's communication system are also on the increase. And new systems are continually being introduced, for instance in the consumer-electronics sector. All in all, it is to be expected that a number of bus systems will establish themselves in the vehicle, each of which will be characterized by its own particular area of application.

In addition to electronic data transmission, optical transmission systems will also come into use in the multimedia area. These are very-high-speed bus systems and can transmit large quantities of data as needed for audio and video components.

Individual functions will be combined by networking to form a system alliance covering the complete vehicle, in which information can be exchanged via data buses. The implementation of such overlapping functions necessitates binding agreements covering interfaces and functional contents. The CARTRONIC® from Bosch is the answer to these stipulations, and has been developed as a priority-override and definition concept for all the vehicle's closed and open-loop control systems. The possible sub-division of the functions which are each controlled by a central coordinator can be seen in Fig. 1. The functions can be incorporated in various ECUs.

The combination of components and systems can result in completely novel functions. For instance, the exchange of data between the transmission-shift control and the navigation equipment can ensure that a change down is made in good time before a gradient is reached. With the help of the navigation facility, the headlamps will be able to adapt their beam of light to make it optimal for varying driving situations and for the route taken by the road (for instance at road intersections). Car radios, sound-carrier drives, TV, telephone, E-mail, Internet, as well as the navigation and terminal equipment for traffic telematics will be networked to form a multimedia system.

1 CARTRONIC®: Design schematic

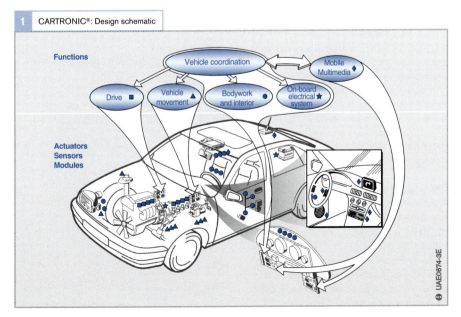

Actuators

Actuators are the devices which convert the electric output signals from the control unit into physical quantities (e.g. position of the exhaust-gas recirculation valve or the control flap).

Electropneumatic converters

Exhaust-gas recirculation valve
On a vehicle with exhaust-gas recirculation, a proportion of the exhaust-gas flow is returned to the intake manifold in order to reduce the level of pollutant emissions.

An electropneumatic valve that provides a connection between the exhaust manifold and the intake manifold controls the amount of exhaust gas that is recirculated. In future, electric valves will also be used.

Wastegate actuator
The turbocharger is designed to deliver a high charge-air pressure even at low engine speeds in order to enable high engine-torque output right from the lower end of the speed range. Thus, in order to prevent the charge-air pressure rising to excessive levels at high speeds, the charge-air pressure control function operates a wastegate actuator which uses an electropneumatic bypass valve (the wastegate) to divert a certain amount of the exhaust-gas flow away from the turbocharger turbine (Figure 1).

Systems with variable turbine geometry (VTG) turbochargers also adjust the turbocharger output. In such cases, an electric or electropneumatic valve alters the angle of the deflector blades in the turbocharger inlet channel.

Swirl valve actuator
The swirl control function on car engines controls the swirling motion of the intake air inside the cylinder. Swirl is generally induced by means of spiral-shaped intake ports. Swirl is a determining factor in the efficiency with which the fuel and air are mixed in the combustion chamber and therefore has a major effect on the quality of combustion. As a rule, a large degree of swirl is induced at low engine speeds and a lesser degree at high speeds.

The degree of swirl is controlled by means of a swirl valve actuator which operates a flap or valve near to the inlet valve.

Intake shut-off valve
UIS systems for cars incorporate an electropneumatically controlled intake shut-off valve which cuts off the air supply when the engine is switched off. This reduces the amount of air being compressed and the engine cuts out more smoothly.

Control flap (throttle valve)
The control flap operated by an electropneumatic valve on a diesel engine has an entirely different function from the throttle valve of a gasoline engine – it is used to increase the exhaust-gas recirculation rate by lowering pressure in the intake manifold. The control flap control function is only active at low engine loads and speeds.

1 Charge-air pressure control using wastegate actuator

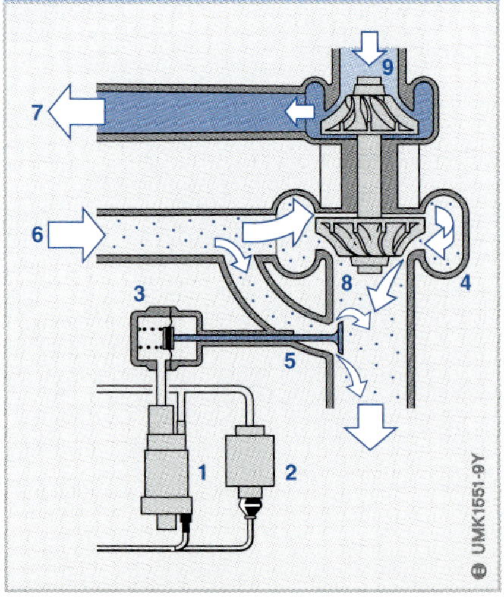

Fig. 1
1 Wastegate actuator
2 Vacuum pump
3 Pressure actuator
4 Turbocharger
5 Bypass valve
6 Exhaust-gas flow
7 Intake air flow
8 Turbine
9 Compressor

Continuous-operation braking systems

Braking systems of this type for heavy commercial vehicles are non-wearing systems that can reduce the speed of the vehicle – but not bring it to a standstill. In contrast with service-brake systems which use friction brakes on the road wheels, continuous braking systems are primarily suited to retardation on long descents as the heat generated can still be effectively dissipated in the course of long periods of braking. As a consequence, the friction brakes are used significantly less and so remain cool and fully effective if needed in an emergency. The continuous braking systems are controlled by the engine management module.

Engine brake

When the engine brake (also called an exhaust brake) is switched on, the fuel-injection system cuts off the fuel supply to the engine and an electropneumatic valve moves a slide valve or a flap in the exhaust pipe. This makes it more difficult for the "exhaust" (because the fuel has been shut off it is actually only air) to be expelled by the engine through the exhaust system. The resulting air cushion inside the cylinder slows down the piston during the compression and exhaust strokes. The engine brake is not capable of graduated application – it is either on or off.

Supplementary engine brake

When the engine needs to be braked, an electrohydraulically operated valve-actuating device opens the exhaust valve at the end of the compression stroke. The cylinder pressure is thus released and energy is lost from the system. The actuating fluid is engine oil.

Retarder

A retarder is a continuous braking device that is independent of the engine. It is fitted to the drivetrain between the gearbox and the wheels and is therefore effective even when the drive between engine and gearbox is disengaged. There are two types as described below.

Hydrodynamic retarder

This type of retarder consists of a rotating turbine (the rotor) and a similar, but static, bladed component (the stator) facing opposite it. The rotor is mechanically linked to the vehicle's drivetrain. When the retarder is operated, the rotor and stator chambers are filled with oil. The rotor accelerates the oil flow while the stator slows it down. The kinetic energy is converted into heat and is dissipated by the engine coolant. The braking effect is infinitely variable by controlling the quantity of oil in the retarder.

Electrodynamic retarder

This type of retarder consists of an air-cooled soft-iron disk that rotates in a controllable electromagnetic field created by the power supply from the battery. The resulting eddy currents retard the disk and therefore the road wheels. The braking effect is infinitely variable.

Fan control function

The engine control unit switches the radiator fan on and off as required according to the temperature of the engine coolant. This is done by means of an electromagnetic clutch.

Electric fans are also increasingly employed. As they do not have to be driven by a belt running off an engine pulley, they offer greater scope with regard to choice of location.

Start-assist systems

Compared with petrol, diesel fuel is very easily combustible. That is why a warm diesel engine will start spontaneously and a direct-injection (DI) model will do so even when started from cold at temperatures ≥ 0 °C. The spontaneous ignition temperature of 250 °C is achieved with the engine turning at starting speed. Indirect-injection (IDI) engines always require assistance when starting from cold, while direct-injection engines only need help at temperatures below 0 °C. Engines with precombustion or swirl chambers have a glow plug in the precombustion/swirl chamber to initiate combustion. On smaller direct-injection engines (up to 1 l/cylinder), it is placed at the edge of the combustion chamber. Large direct-injection engines for commercial-vehicles sometimes also have an intake-air preheating system or use a special more easily combustible fuel for starting which is injected into the intake system.

Intake-air preheating system
Flame glow plug
A flame glow plug heats up the intake air by burning fuel in the intake port. The fuel is usually fed to the flame glow plug by the fuel-injection system's injection pump via a solenoid valve. In the union of the flame glow plug, there is a filter and a metering device which is set to allow a specific amount of fuel to pass through as required by the engine in which it is fitted. The fuel vaporizing inside a vapourizing tube around the glow plug element and mixes with the intake air. The fuel-and-air mixture then ignites at the front end of the flame glow plug as it passes over the glow-plug element that is heated to over 1,000 °C. The amount of heat produced is limited by the fact that the heating flame is only allowed to burn a certain proportion of the oxygen required for combustion inside the cylinder.

Electric heater (grid heater)
A relay switches a series of heater elements in the air intake system on and off.

Glow plug
The element of the glow plug is permanently sealed inside the gas-tight glow plug body (Figure 1, Item 3). It consists of a hot-gas and corrosion-resistant element sheath (4) which encloses a filament surrounded by compressed magnesium oxide powder (6). That filament is made up of two resistors connected in series – the heating filament (7) located in the tip of the sheath, and the control filament (5). Whereas the heating filament has an electrical impedance that is virtually independent of temperature, the control filament has a positive temperature coefficient (PTC).

1　Type GSK2 sheathed-element glow plug

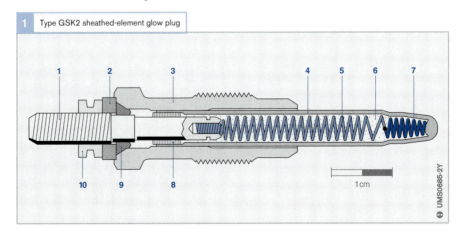

Fig. 1
1　Connector
2　Insulating washer
3　Glow-plug body
4　Element sheath
5　Control filament
6　Packing powder
7　Heating filament
8　Element seal
9　Double seal
10　Threaded collar

1 cm

UMS0685-2Y

In the latest generation of glow plugs (Type GSK2), its impedance increases even more steeply as the temperature rises than with the older designs (Type S-RSK). The new Type GSK2 glow plugs are faster at reaching the temperature required for ignition (850 °C in 4 s) and also have a lower steady-state temperature. This means that the temperature is kept below the critical level for the glow plug. Consequently, it can remain in operation for up to three minutes after the engine has started. This post-glow function results in a more effective engine warm-up phase with substantially lower noise and emission output.

Glow-plug control unit

A Type GZS glow plug control unit controls the glow plugs via a power relay. It receives its starting signal from the engine management module or a temperature sensor.

The glow-plug control unit controls how long the glow plugs remain switched on and also performs safety and monitoring functions. Advanced glow-plug control units can use the diagnosis functions to detect failure of individual glow plugs. The fault is then indicated to the driver. The control signal inputs are in the form of multi-connectors.

Operating sequence

The glow plug and starting sequence is (in similar fashion to a gasoline engine) governed by the glow-plug/starter switch ("ignition switch"). The glow-plug preheating phase begins when the key is turned to the "Ignition On" position (Figure 3). When the glow-plug indicator lamp on the instrument cluster goes out, the glow plugs are hot enough for the engine to be started. In the subsequent starting phase, droplets of injected fuel vaporize and ignite on contact with the hot, compressed air. The heat released further assists in the propagation of combustion.

A post-glow phase after the engine has started helps to prevent misfiring during the warm-up phase, thereby reducing smoke emission and combustion noise while the engine is below normal operating temperature. If, for any reason, the engine is not started after the ignition is switched on, a safety cut-out for the glow-plugs prevents the battery from discharging.

If the glow-plug control unit is linked with the control unit for the EDC (Electronic Diesel Control), the latter can use the information at its disposal to effect optimum control of the glow plugs under different engine operating conditions. This provides a further means of minimizing blue smoke and noise emission.

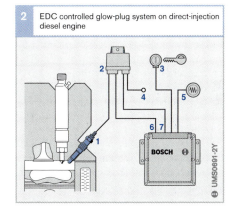

2   EDC controlled glow-plug system on direct-injection diesel engine

UMS0691-2Y

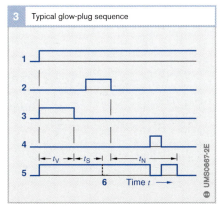

3   Typical glow-plug sequence

6   Time t →

UMS0667-2E

Fig. 2
1   Glow-plug
2   Glow-plug control unit
3   Glow-plug/ starter switch ("ignition switch")
4   To battery
5   Indicator lamp
6   Control line to engine management module
7   Diagnosis lead

Fig. 3
1   Glow-plug/ starter switch ("ignition switch")
2   Starter
3   Indicator lamp
4   Load switch
5   Glow-plug ON period
6   Point from which engine runs independently
τ_V Glow-plug preheating phase
τ_S Engine ready for starting
τ_N Post-glow phase

Electronic diagnosis

ECU-integrated diagnostics belong to the basic scope of electronic engine-management systems. During normal vehicle operation, input and output signals are checked by monitoring algorithms, and the overall system is checked for malfunctions and faults. If faults are discovered in the process, these are stored in the ECU. When the vehicle is checked in the workshop, this stored information is retrieved through a serial interface and provides the basis for rapid and efficient trouble-shooting and repair.

Operating concept

Originally, it was intended that the self-diagnosis of the engine-management system (on-board diagnostics/OBD) should merely be a help in rapid and efficient fault-finding in the workshop. Increasingly severe legal stipulations, and the more wide-ranging functional scope of the vehicle's electronic systems though, led to the emergence of a more extensive diagnostics system within the engine-management system.

Input-signal monitoring
Here, the analysis of the input signals is applied for monitoring the sensors and their connection lines to the ECU (Table 1). These checks serve to uncover not only sensor faults, but also short-circuits to the battery voltage U_{Batt} and to ground, as well as open circuits in lines. The following processes are applied:
- Monitoring the sensors' power supply.
- Checking that the measured values are inside the correct range (e.g. engine temperature −40 °C...+150 °C).
- If auxiliary information is available, the registered value is subjected to a plausibility check (e.g. the camshaft and/or crankshaft speed).
- Important sensors (such as the accelerator-pedal sensor) are designed to be redundant which means that their signals can be directly compared with each other.

Output-signal monitoring
Here, in addition to the connections to the ECU, the actuators are also monitored. Using the results of these checks, open-circuits and short-circuits in the lines and connections can be detected in addition to actuator faults. The following processes are applied here:
- Hardware monitoring of the output-signal circuit using the driver stage. The circuit is checked for open circuit, and for short circuits to battery voltage U_{Batt} and to ground.
- The actuator's influence on the system is checked for plausibility. In the case of exhaust-gas recirculation (EGR) for instance, a check is made whether intake-manifold pressure is within given limits and that it reacts accordingly when the actuator is triggered.

Monitoring ECU communication
As a rule, communication with the other ECUs takes place via CAN-Bus (Controller Area Network). The diagnostics integrated in the CAN module are described in the Chapter "Data transfer between electronic systems". A number of other checks are also performed in the ECU. Since most CAN messages are sent at regular intervals by the particular ECUs, monitoring the time intervals concerned leads to the detection of ECU failure.

In addition, when redundant information is available in the ECU, the received signals are checked the same as all input signals.

Monitoring the internal ECU functions

In order that the functional integrity of the ECU is ensured at all times, monitoring functions are incorporated in the hardware (e.g. "intelligent driver-stage modules") and in the software.

These check the individual ECU components (e.g. microcontroller, Flash-EPROM, RAM). Many of these checks are performed immediately the engine is switched on. During normal operations, further checks are performed regularly so that the failure of a component is detected immediately. Sequences which require extensive computing capacity (for instance for checking the EPROM), are run through immediately the engine is switched off (only possible at present on gasoline engines).

This method ensures that the other functions are not interfered with. On the diesel engine, the switch-off paths are checked in the same period.

1 Monitoring the most important input signals	
Signal path	**Monitoring**
Accelerator-pedal sensor	Check of the power supply and the signal range
	Plausibility with regard to a redundant signal
	Plausibility with regard to the brakes
Crankshaft-rpm sensor	Checking the signal range
	Plausibility with the camshaft-rpm sensor
	Checking the changes as a function of time (dynamic plausibility)
Engine-temperature sensor	Checking the signal range
	Logical plausibility as a function of rpm
	and injected fuel quantity or engine load
Brake-pedal switch	Plausibility with regard to redundant brake contact
Speed signal	Checking the signal range
	Plausibility with regard to rpm and injected fuel quantity or engine load
EGR positioner	Checks for short circuits and open circuits in lines
	EGR control
	Checking the system's reaction to valve triggering
Battery voltage	Checking the signal range
	Plausibility with regard to engine rpm (at present, only possible with gasoline (SI) engines)
Fuel-temperature sensor	Checking the signal range (at present, only possible with diesel engines)
Boost-pressure sensor	Checking the power supply and the signal range
	Plausibility with regard to ambient-pressure sensor and/or further signals
Boost-pressure actuator	Checking for short circuits and open-circuit lines
	Control deviation of boost-pressure control
Air-mass meter	Checking the power supply and the signal range
	Logical plausibility
Air-temperature sensor	Checking the signal range
	Logical plausibility with regard to the engine-temperature sensor for instance
Clutch-signal sensor	Plausibility with regard to vehicle speed
Ambient-pressure sensor	Checking the signal range
	Logical plausibility of the intake-manifold-pressure sensor

Table 1

Dealing with a fault

Fault recognition

When a fault remains in a signal path longer than a defined period, the signal path is classified as being defective. Until it is finally classified as defective, the last valid value is utilised in the system. Normally, as soon as it is classified as defective, a substitute function is triggered (refer to the Chapter "Closed-loop and open-loop electronic control").

Most faults can be revoked and the signal path classified as serviceable again provided the signal path remains without fault for a defined period of time.

Fault storage

Each fault is stored as a malfunction code in the non-volatile area of the data memory. In the fault entry, each malfunction code is accompanied by auxiliary information in the so-called "freeze-frame" containing the operating and environmental conditions at the moment the fault occurred (e.g. engine rpm, engine temperature). Information is also stored on the fault type (for instance short circuit, conductor open circuit) and the fault status (in other words, permanent fault or sporadic fault).

The lawmaker has prescribed specific malfunction codes for many of the faults which have an effect on the vehicle's toxic emissions. Further fault information not covered by legislation can also be stored for retrieval by the vehicle workshop.

Following storage of the fault entry, the diagnosis for the system or component concerned continues. If the fault does not occur again in the further course of the diagnosis (in other words it was a sporadic fault), it is then erased from the fault memory provided that certain conditions are complied with.

Fault retrieval

The faults can be retrieved from the fault memory with a specified workshop tester provided by the vehicle manufacturer, or by using a System Tester (e.g. Bosch KTS500), or a scan tool. Once the fault information has been retrieved in the workshop and the fault repaired, the fault memory can be cleared again using the tester.

Diagnostic interface

A communication interface is needed for "Off-board testers" in order for them to be able evaluate the "On-Board Diagnosis". This serial interface is mandatory and is defined in ISO 9141 (diagnostic interface through the K-line). The interface operates with a transfer rate (baud rate) of between 10 baud and 10 kbaud, and is either a single-wire interface with common send and receive lines, or a two-wire interface with separate data line (K-line) and initiate line (L-line). A number of ECUs can be connected together at a single diagnostic connector.

The tester sends an initiate address to all the ECUs, one of which recognises this address and replies with a baud-rate identification word. Using the interval between the pulse edges, the tester determines the baud rate, adjusts itself accordingly and sets up the communication with the ECUs.

Actuator diagnosis

An actuator diagnosis facility is incorporated in the ECU so that the workshops can selectively actuate individual actuators and check their correct functioning. This test mode is triggered by the tester and only functions with the vehicle at standstill, and with the engine running at below a given speed or with it stopped. Actuator functioning is checked either acoustically (e.g. clicking of the valve), visually (e.g. movement of a flap), or by other uncomplicated methods.

On-Board-Diagnosis (OBD)

In the past years there has been a continual reduction in toxic emissions per vehicle. In order for the emission limits defined by the vehicle manufacturers to be maintained during continuous in-field operations, it is necessary that the engine and its components are monitored continually. This was the reason for the lawmaker defining specifications to regulate the diagnosis scope for the exhaust-gas-relevant components and systems in the vehicle.

1988 marked the coming into force of OBD I in California, that is, the first stage of CARB legislation (California Air Resources Board). All newly registered vehicles in California were forced to comply with these statutory regulations. OBD II, that is the second stage, came into force in 1994.

Since 1994, in the remaining US States the laws of the Federal Authority EPA (Environmental Protection Agency) have applied. The scope of these diagnostics comply for the most part with the CARB legislation (OBD II), although the requirements for compliance with the emissions limits are less severe.

The OBD adapted to European requirements is known as the EOBD and is based on the EPA-OBD. At present, the EOBD stipulations are even less severe than those of the EPA-OBD.

OBD I

The electrical components which are relevant for exhaust-gas emissions are checked for short circuits and breaks by the first stage of the CARB-OBD. The resulting electrical signals must remain within the stipulated plausibility limits.

When a fault/error is discovered, the driver is warned by a lamp in the instrument cluster. Using "on-board devices" (for instance, a diagnosis lamp which displays a blink code), it must be possible to read out which component has failed.

OBD II

The diagnostic procedure for Stage 2 of the CARB-OBD is far more extensive than OBD I. Monitoring no longer stops at the check of the electrical signals from the components, but has been extended to include the check of correct system functioning. For instance, it is no longer sufficient to check that the signal from the engine-temperature sensor does not exceed certain fixed limits. OBD II registers a fault if the engine temperature remains too low (for instance below 10 °C) for a longer period of time (plausibility check).

OBD II demands that all those systems and components be checked which in case of malfunction can lead to a noticeable increase in toxic emissions. In addition, all the components which are actually used for OBD must also to be checked, and every detected fault must be stored. The driver must be warned of malfunctions by a lamp in the instrument cluster. The stored faults are then retrieved by testers which are connected for trouble-shooting.

OBD II legislation stipulates the standardization of the information stored in the fault memory in accordance with the ruling of the SAE (Society of Automotive Engineers). The means that provided they comply with the standards, commercially available testers can retrieve fault information from the fault memory (so-called "scan tools").

Diagnosis-sequence control

As a rule, the diagnostic functions for all the systems and components to be tested must be run through at least once during each exhaust-gas test cycle (e.g. ECE/EU test cycle).

Depending on the driving status, the diagnostics-system management can dynamically change the order of the diagnosis functions. The target here is that all diagnosis functions are run through often enough during everyday operations.

Service technology

When car drivers need help, they can count on more than 10,000 Bosch Service centers located in 132 countries. As these centers are not associated with any specific automotive manufacturer, they can provide neutral, impartial assistance. Fast assistance is always available, even in the sparsely populated regions of South America and Africa. A single set of quality standards applies everywhere. It is no wonder, therefore, that the Bosch service warranty is valid throughout the world.

Overview

The specifications and performance data of Bosch components and systems are precisely matched to the requirements of each individual vehicle. Bosch also develops and designs the test equipment, special tools and diagnosis technology needed for tests and inspections.

Bosch universal testers – ranging from the basic battery tester to the complete vehicle test stand – are being used in automotive repair shops and by inspection agencies all over the world.

Service personnel receive training in the efficient use of this test technology as well as information on a range of automotive systems. Meanwhile, feedback from our customers constantly flows back to the development of new products.

AWN service network
Test technology
It is still possible to test mechanical systems in motor vehicles using relatively basic equipment. But mastering the increasingly complex electronic systems found in modern

---

[1] Bosch service technology stems from development activities carried out by the Bosch AWN service network. The "asanetwork GmbH" is responsible for advanced development and marketing under the "AWN" name.

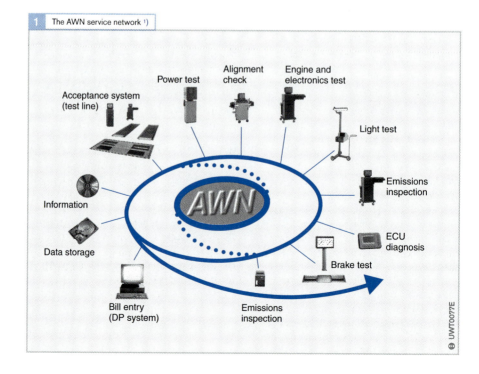

1   The AWN service network [1]

Acceptance system (test line)

Power test

Alignment check

Engine and electronics test

Light test

Emissions inspection

ECU diagnosis

Brake test

Emissions inspection

Bill entry (DP system)

Information

Data storage

AWN

UWT0077E

vehicles means using new test methods that rely on electronic data processing. The future belongs to a technology that links every IT system in every service center in a single, unified network, the AWN Asanet WorkshopNetwork (Fig.1). In 1998 Bosch received the Automechanika Innovation Prize in the Shop and Service category for this innovation.

Test process

When a vehicle arrives for a service inspection, the job-order processing system database provides immediate access to all the available information on the vehicle. The moment the vehicle enters the shop, the system offers access to the vehicle's entire service history, including all service and repairs that it has received in the past.

Individual diagnostic testers provide the data needed for direct comparisons of set-point values and actual measured values, with no need for supplementary entries. All service procedures and replacement components are recorded to support the billing process. After the final road test, the bill is produced simply by striking a few keys. The system also provides a clear and concise printout with the results of the vehicle diagnosis. This offers the customer a full report detailing all of the service operations and materials that went into the vehicle's repair.

Electronic Service Information (ESI[tronic])

Even in the past the wide variety of vehicle makes and models made the use of IT systems essential (for part numbers, test specifications, etc.) Large data records, such as those containing information on spare parts, are contained on microfiche cards. Microfiche readers provide access to these microfiche libraries and are still standard equipment in every automotive service facility.

In 1991 ESI[tronic] (Electronic Service Information), intended for use with a standard PC, was introduced to furnish data on CDs. As ESI[tronic] can store much more data than a conventional microfiche system,

it accommodates a larger range of potential applications. It can also be incorporated in electronic data processing networks.

Application

The ESI[tronic] software package supports service personnel throughout the entire vehicle-repair process by providing the following information:

- Spare component identification (correlating spare part numbers with specific vehicles, etc.)
- Flat rates
- Repair instructions
- Circuit diagrams
- Test specifications and
- Test data from vehicle diagnosis

Service technicians can select from various options for diagnosis problems and malfunctions: The KTS500 is a high-performance portable system tester, or the KTS500C, which is designed to run on the PCs used in service shops (diagnostic stations). The KTS500C consists of a PC adapter card, a plugin card (KTS) and a test module for measuring voltage, current and resistance. An interface allows ESI[tronic] to communicate with the electronic systems in the vehicle, such as the engine control unit. Working at the PC, the user starts by selecting the SIS (Service Information System) utility to initiate diagnosis of on-board control units and access the engine control unit's fault storage. ESI[tronic] uses the results of the diagnosis as the basis for generating specific repair instructions. The system also provides displays with other information, such as component locations, exploded views of assemblies, diagrams showing the layouts of electrical, pneumatic and hydraulic systems, etc. Working at the PC, users can then proceed directly from the exploded view to the parts list with part numbers to order the required replacement components.

Testing EDC systems

Every EDC system has a self-diagnosis capability which allows comprehensive testing of the entire fuel-injection system.

Testing equipment

Effective testing of the system requires the use of special testing equipment. Whereas in the past an electronic fuel-injection system could be tested using simple electronic testers (e.g. a multimeter), the continuous advancement of EDC systems means that complex testing devices are essential today.

The system testers in the KTS series are widely used by vehicle repairers. The KTS500 (Fig. 1) offers a wide range of capabilities for use in the workshop, enhanced in particular by its graphical display of data, such as test results. In the descriptions that follow, these system testers are sometimes also referred to as "engine testers".

Functions of KTS500

The KTS500 offers a wide variety of functions which are selected by means of buttons and menus on the large display screen. The list below details the most important functions offered by the KTS500.

Reading the fault memory: The KTS500 can read out faults stored in the fault memory during system operation by the self-diagnosis function and display them on screen in plain text.

Reading actual data: Actual data that is used in calculations by the engine ECU can be read and displayed as a physical quantity (e.g. engine speed in rpm).

Actuator diagnosis: The function of the electrical actuators can be tested.

Engine test: The system tester initiates programmed test sequences for checking the engine ECU or the engine (e.g. compression test).

Multimeter function: Electrical current, voltage and resistance can be tested in the same way as with a conventional multimeter.

Time graph display: The measured data recorded over time is displayed as a signal graph as on an oscilloscope.

Additional information: Specific additional information relevant to the faults/components displayed can also be shown (e.g. location, component test data, electrical circuit diagrams).

Printout: All data (e. g. list of actual data) can be printed out on standard PC printers.

Programming: The software on the engine ECU can be reprogrammed with the KTS500 (software update).

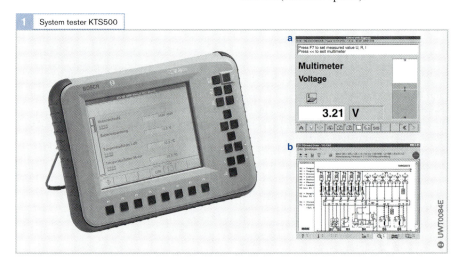

1   System tester KTS500

Fig. 1
a Graphical display of multimeter function
b Graphical display of an electrical terminal diagram

The extent to which the capabilities of the KTS500 can be utilized depends on the system under test. Not all engine ECUs support its full range of functions.

Workshop procedures (standard procedures)

The diagnosis procedure is the same for all systems with an electronic diesel-engine control (EDC) system. The most important tool is the engine tester which is connected to the engine ECU via the diagnosis interface.

Vehicle identification

First of all, the vehicle must be identified. The engine tester has to know which model of vehicle is under test so that it can refer to the correct data.

Reading the fault memory

The self-diagnosis function of the EDC system checks the electrical components for malfunctions. If a fault is detected, a record is permanently stored in the fault memory detailing
- The fault source (e.g. engine temperature sensor)
- The failure mode (e.g. short circuit to ground, implausible signal)
- The fault status (e.g. constantly present, sporadically present)
- The ambient conditions (readings at the time the fault was recorded, e.g. engine speed, temperature, etc.)

Using the menu option "Fault Memory", the fault data stored on the engine ECU can be transferred to the engine tester. The fault data is displayed on the tester in plain text giving details of the fault source, location, status, etc.

Fault localization

For some types of engine fault, the self-diagnosis function is unable to establish the cause. Maintenance technicians must be able to identify and rectify these types of fault quickly and reliably.

For both types of fault – those recorded in the fault memory and those that are not – the Electronic Service Information ESI[tronic] provides assistance with fault localization. It offers instructions for fault localization for every conceivable problem (e.g. engine stutter) or malfunction (e.g. engine-temperature sensor short-circuit).

Fault rectification

Once the cause of the fault has been localized with the aid of the ESI[tronic] information, the fault can be rectified.

Clearing the fault memory

When the fault has been rectified, the record of the fault must be deleted from the fault memory. This function can be performed using the menu option "Clear Fault Memory" on the engine tester.

Road test

In order to ensure that the fault has really been eliminated, a road test is carried out. During the road test, the self-diagnosis function checks the system and records any faults that are still present.

Rechecking the fault memory

After the road test, the contents of the fault memory are checked again. If the problem has been solved, no faults should have been recorded. Assuming that is the case, the repair procedure is then successfully completed.

Other testing methods

The KTS 500 system tester offers other EDC diagnosis facilities in addition to the standard diagnosis functions. The additional functions are initiated by the engine tester and then carried out by the engine ECU.

Actuator diagnosis

Many ECU functions (e.g. exhaust-gas recirculation) only operate under certain conditions when the vehicle is being driven. For this reason, it is not possible to check the function of the relevant actuators (e.g. the exhaust-gas recirculation valve) in the workshop without a means of operating them selectively.

The engine tester can be used to initiate actuator diagnosis in order to check the function of actuators in the workshop. The correct functioning of the components concerned is then indicated by audible or visual feedback.

The actuator diagnosis function tests the entire electrical path from the engine ECU via the wiring harness to the actuator. However, even that is insufficient to make a definitive statement about the complete functional capability of the component.

Actuator diagnosis is generally carried out with the vehicle stationary. It extends only to those actuators that do not affect the function of important components. This means that actuators that could be damaged or could cause engine damage if the actuator diagnosis were to be incorrectly carried out (e.g. nozzle solenoid valves) are excluded from the test.

Signal testing

Where there is a component malfunction, an oscilloscope function allows the progression of an actuator control signal to be checked over time. This applies in particular to actuators that cannot be included in the actuator diagnosis function (e.g. nozzles or injectors).

Engine test function

Faults that the self-diagnosis function cannot identify can be localized by means of supplementary functions (engine-testing functions). There are testing routines stored on the engine ECU that can be initiated by the engine tester (e.g. compression test). These tests last for a specific period of time. The start and finish of the test are indicated by the engine tester. The results are transmitted to the engine tester by the engine ECU in the form of a list.

In the case of the compression test, the fuel-injection system is deactivated while the engine is turned over by the starter motor. The engine ECU records the crankshaft speed pattern. From the fluctuations in crankshaft speed, i.e. the difference between the highest and lowest speeds, conclusions can be drawn as to the compression in individual cylinders and, therefore, about the condition of the engine.

Engine-speed and injected-fuel quantity comparison tests

Different injected-fuel quantities in different cylinders produce varying, cylinder-specific torque levels and, therefore, uneven engine running. The engine ECU measures the momentary speeds and transmits the readings to the engine tester. The engine tester then shows the speeds for each cylinder. Large divergences in speed and injected-fuel quantity between individual cylinders indicate fuel-metering problems.

The engine-smoothness function on the engine ECU evens out engine-speed fluctuations by adjustments to the injected-fuel quantity for each cylinder. For this reason, the engine-smoothness function is deactivated while the speed-comparison test is in progress.

The quantity-comparison test is carried out with the engine-smoothness function active. The engine tester displays the injected-fuel quantities for each cylinder required to produce smooth engine running.

The assessment is based on a comparison of the cylinder-specific speeds and injected-fuel quantities.

▶ Global service

"Once you have driven an automobile, you will soon realize that there is something unbelievably tiresome about horses (...). But you do require a conscientious mechanic for the automobile (...)".

Robert Bosch wrote these words to his friend Paul Reusch in 1906. In those days, it was indeed the case that breakdowns could be repaired on the road or at home by an employed chauffeur or mechanic. However, with the growing number of motorists driving their own cars after the First World War, the need for workshops offering repair services in-

creased rapidly. In the 1920s Robert Bosch started to systematically create a nationwide customer-service organization. In 1926 all the repair centers were uniformly named "Bosch Service" and the name was registered as a trademark.

Today's Bosch Service agencies retain the same name. They are equipped with the latest electronic equipment in order to meet the demands of 21st-century automotive technology and the quality expectations of the customers.

1   A repair shop in 1925 (photo: Bosch)

2   A Bosch service in 2001, carried out with the very latest electronic testing equipment

Fuel-injection pump test benches

Accurately tested and precisely adjusted fuel-injection pumps and governor mechanisms are key components for obtaining optimized performance and fuel economy from diesel engines. They are also crucial in ensuring compliance with increasingly strict exhaust-gas emission regulations. The fuel-injection pump test bench (Fig. 1) is a vital tool for meeting these requirements.

The main specifications governing both test bench and test procedures are defined by ISO standards; particularly demanding are the specifications for rigidity and geometrical consistency in the drive unit (5).

As time progresses, so do the levels of peak pressure that fuel-injection pumps are expected to generate. This development is reflected in higher performance demands and power requirements for pump test benches. Powerful electric drive units, a large flyweight and precise control of rotational speed guarantee stability at all engine speeds. This stability is an essential requirement for repeatable, mutually comparable measurements and test results.

Flow measurement methods

An important test procedure is to measure the fuel pumped each time the plunger moves through its stroke. For this test, the fuel-injection pump is clamped on the test bench support (1), with its drive side connected to the test bench drive coupling. Testing proceeds with a standardized calibrating oil at a precisely monitored and controlled temperature. A special, precision-calibrated nozzle-and-holder assembly (3) is connected to each pump barrel. This strategy ensures mutually comparable measurements for each test. Two test methods are available.

Glass gauge method (MGT)

The test bench features an assembly with two glass gauges (Fig. 2, Pos. 5). A range of gages with various capacities are available for each cylinder. This layout can be used to test fuel-injection pumps for engines of up to 12 cylinders.

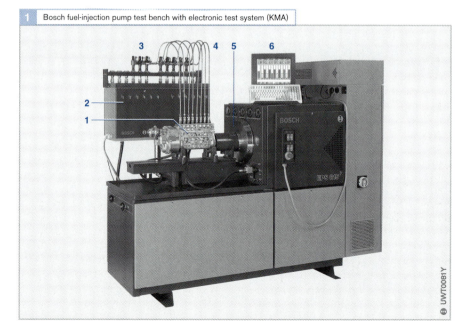

1   Bosch fuel-injection pump test bench with electronic test system (KMA)

Fig. 1
1 Fuel-injection pump on test bench
2 Quantity test system (KMW)
3 Test nozzle-and-holder assembly
4 High-pressure test line
5 Electric drive unit
6 Control, display and processing unit

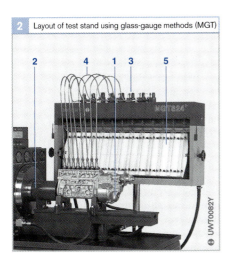

2 Layout of test stand using glass-gauge methods (MGT)

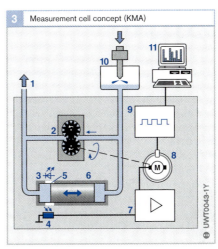

3 Measurement cell concept (KMA)

Fig. 2
1 Fuel-injection pump
2 Electric drive unit
3 Test nozzle-and-holder assembly
4 High-pressure test line
5 Glass gages

Fig. 3
1 Return line to calibrating oil tank
2 Gear pump
3 LED
4 Photocell
5 Window
6 Plunger
7 Amplifier with electronic control circuitry
8 Electric motor
9 Pulse counter
10 Test nozzle-and-holder assembly
11 Monitor (PC)

In the first stage, the discharged calibrating flows past the glass gages to return directly to the oil tank. As soon as the fuel-injection pump reaches the rotational speed indicated in the test specifications, a slide valve opens, allowing the calibrating oil from the fuel-injection pump to flow to the glass gages. Supply to the glass containers is then interrupted when the pump has executed the preset number of strokes.

The fuel quantity delivered to each cylinder in cm³ can now be read from each of the glass gages. The standard test period is 1,000 strokes, making it easy to interpret the numerical result in mm³ per stroke of delivered fuel. The test results are compared with the setpoint values and entered in the test record.

Electronic flow measurement system (KMA)
This system replaces the glass gauges with a control, display and processor unit (Fig. 1, Pos. 6). While this unit is usually mounted on the test bench, it can also be installed on a cart next to the test bench.

This test relies on continuous measuring the delivery capacity (Fig. 3). A control plunger (6) is installed in parallel with the input and output sides of a gear pump (2). When the pump's delivery quantity equals the quantity of calibrating oil emerging from the

test nozzle (10), the plunger remains in its center position. If the flow of calibrating oil is greater, the plunger moves to the left – if the flow of calibrating oil is lower, the plunger moves to the right. This plunger motion controls the amount of light traveling from an LED (3) to a photocell (4). The electronic control circuitry (7) records this deviation and responds by varying the pump's rotational speed until its delivery rate again corresponds to the quantity of fluid emerging from the test nozzle. The control plunger then returns to its center position. The pump speed can be varied to measure delivery quantity with extreme precision.

Two of these measurement cells are present on the test bench. The computer connects all of the test cylinders to the two measurement cells in groups of two, proceeding sequentially from one group to the next (multiplex operation). The main features of this test method are:
● Highly precise and reproducible test results
● Clear test results with digital display and graphic presentation in the form of bar graphs
● Test record for documentation
● Supports adjustments to compensate for variations in cooling and/or temperature

Testing in-line fuel-injection pumps

The test program for fuel-injection pumps involves operations that are carried out with the pump fitted to the engine in the vehicle (system fault diagnosis) as well as those performed on the pump in isolation on a test bench or in the workshop. This latter category involves
- Testing the fuel-injection pump on the pump test bench and making any necessary adjustments
- Repairing the fuel-injection pump/governor and subsequently resetting them on the pump test bench

In the case of in-line fuel-injection pumps, a distinction has to be made between those with mechanical governors and those which are electronically controlled. In either case, the pump and its governor/control system are tested in combination, as both components must be matched to each other.

The large number and variety of in-line fuel-injection pump designs necessitates variations in the procedures for testing and adjustment. The examples given below can, therefore, only provide an idea of the full extent of workshop technology.

Adjustments made on the test bench

The adjustments made on the test bench comprise
- Start of delivery and cam offset for each individual pump unit
- Delivery quantity setting and equalization between pump units
- Adjustment of the governor mounted on the pump
- Harmonization of pump and governor/control system (overall system adjustment)

For every different pump type and size, separate testing and repair instructions and specifications are provided which are specifically prepared for use with Bosch pump test benches.

The pump and governor are connected to the engine lube-oil circuit. The oil inlet connection is on the fuel-injection pump's camshaft housing or the pump housing. For each testing sequence on the test bench, the fuel-injection pump and governor must be topped up with lube oil.

Testing delivery quantity

The fuel-injection pump test bench can measure the delivery quantity for each individual cylinder (using a calibrated tube apparatus or computer operating and display terminal, see "Fuel-injection pump test benches"). The individual delivery quantity figures obtained over a range of different settings must be within defined tolerance limits. Excessive divergence of individual delivery quantity figures would result in uneven running of the engine. If any of the delivery quantity figures are outside the specified tolerances, the pump barrel(s) concerned must be readjusted. There are different procedures for this depending on the pump model.

Governor/control system adjustment

Governor

Testing of mechanical governors involves an extensive range of adjustments. A dial gauge is used to check the control-rack travel at defined speeds and control-lever positions on the fuel-injection pump test bench. The test results must match the specified figures. If there are excessive discrepancies, the governor characteristics must be reset. There are a number of ways of doing this, such as changing the spring characteristics by altering spring tension, or by fitting new springs.

Electronic control system

If the fuel-injection pump is electronically controlled, it has an electromechanical actuator that is operated by an electronic control unit instead of a directly mounted governor. That actuator moves the control rack and thus controls the injected fuel quantity. Otherwise, there is no difference in the mechanical operation of the fuel-injection pump. During the tests, the control rack is held at a specific position. The control-rack travel must be calibrated to match the voltage signal of the rack-travel sensor. This done by adjusting the rack-travel sensor until its signal voltage matches the specified signal level for the set control-rack travel.

In the case of control-sleeve in-line fuel-injection pumps, the start-of-delivery solenoid is not connected for this test in order to be able to obtain a defined start of delivery.

Adjustments with the pump in situ

The pump's start of delivery setting has a major influence on the engine's performance and exhaust-gas emission characteristics. The start of delivery is set, firstly, by correct adjustment of the pump itself, and secondly, by correct synchronization of the pump's camshaft with the engine's timing system. For this reason, correct mounting of the injection pump on the engine is extremely important. The start of delivery must therefore be tested with the pump mounted on the engine in order to ensure that it is correctly fitted.

There are a number of different ways in which this can be done depending on the pump model. The description that follows is for a Type RSF governor.

On the governor's flyweight mount, there is a tooth-shaped timing mark (Figure 1). In the governor housing, there is a threaded socket which is normally closed off by a screw cap. When the piston that is used for calibration (usually no. 1 cylinder) is in the start-of-delivery position, the timing mark is exactly in line with the center of the threaded socket. This "spy hole" in the governor housing is part of a sliding flange.

Fitting the fuel-injection pump
Locking the camshaft
The fuel-injection pump leaves the factory with its camshaft locked (Figure 1a) and is mounted on the engine when the engine's crankshaft is set at a defined position. The pump lock is then removed. This tried and tested method is economical and is adopted increasingly widely.

Start-of-delivery timing mark
Synchronizing the fuel-injection pump with the engine is performed with the aid of the start-of-delivery timing marks, which have to be brought into alignment. Those marks are to be found on the engine as well as on the fuel-injection pump (Figure 2 overleaf). There are several methods of determining the start of delivery depending on the pump type.

Normally, the adjustments are based on the engine's compression stroke for cylinder no. 1 but other methods may be adopted for reasons related to specific engine designs. The engine manufacturer's instructions must therefore always be observed. On most diesel engines, the start-of-delivery timing mark is on the flywheel, the crankshaft pulley or the vibration damper. The vibration damper is generally mounted on the crankshaft in the position normally occupied by the V-belt pulley, and the pulley then bolted to the vibration damper. The complete assembly then looks rather like a thick V-belt pulley with a small flywheel.

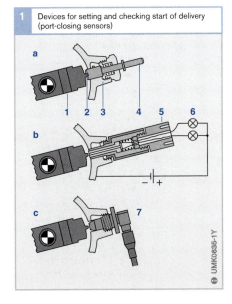

| 1 | Devices for setting and checking start of delivery (port-closing sensors) |

Fig. 1
Illustration shows Type RSF governor; other types have a sliding flange

a   Locked in position by locking pin
b   Testing with an optical sensor (indicator-lamp sensor)
c   Testing with an inductive sensor (governor signal method)

1   Governor flyweight mount
2   Timing mark
3   Governor housing
4   Locking pin
5   Optical sensor
6   Indicator lamp
7   Inductive speed sensor

UMK0635-1Y

Checking static start of delivery
Checking with indicator-lamp sensor
The tooth-shaped timing mark can be located with the aid of an optical sensor, the indicator-lamp sensor (Figure 1b), which is screwed into the socket in governor housing. When it is opposite the sensor, the two indicator lamps on the sensor light up. The start of delivery in degrees of crankshaft rotation can then be read off from the flywheel timing marks, for example.

High-pressure overflow method
The start-of-delivery tester is connected to the pressure outlet of the relevant pump barrel (Figure 3). The other pressure outlets are closed off. The pressurized fuel flows through the open inlet passage of the pump barrel and exits, initially as a jet, into the observation vessel (3). As the engine crankshaft rotates, the pump plunger moves towards its top dead center position. When it reaches the start-of-delivery position, the pump plunger closes off the barrel's inlet passage. The injection jet entering the observation vessel thus dwindles and the fuel flow is reduced to a drip. The start of delivery in degrees of crank shaft rotation is read off from the timing marks.

Checking dynamic start of delivery
Checking with inductive sensor
An inductive sensor that is screwed into the socket in the governor housing (Figure 1c) supplies an electrical signal every time the governor timing mark passes when the engine is running. A second inductive sensor supplies a signal when the engine is at top dead center (Figure 4). The engine analyzer, to which the two inductive sensors are connected, uses those signals to calculate the start of delivery and the engine speed.

Checking with a piezoelectric sensor and a stroboscopic timing light
A piezoelectric sensor is fixed to the high-pressure delivery line for the cylinder on which adjustment is to be based. As soon as the fuel-injection pump delivers fuel to that cylinder, the high-pressure delivery line expands slightly and the piezoelectric sensor transmits an electrical signal. This signal is received by an engine analyzer which uses it to control the flashing of a stroboscopic timing light. The timing light is pointed at the timing marks on the engine. When illuminated by the flashing timing light, the flywheel timing marks appear to be stationary. The angular value in degrees of crankshaft rotation can then be read off for start of delivery.

2 Timing marks on the engine used for setting the fuel-injection pump

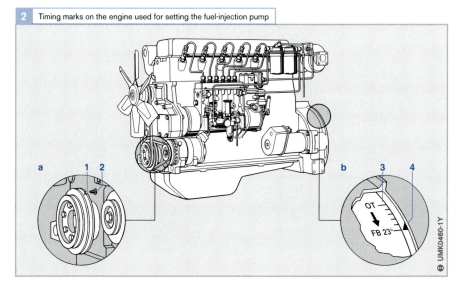

Fig. 2
a V-belt pulley timing marks
b Flywheel timing marks

1 Notch in V-belt pulley
2 Marker point on cylinder block
3 Graduated scale on flywheel
4 Timing mark on crankcase

Venting

Air bubbles in the fuel impair the proper operation of the fuel-injection pump or disable it entirely. Therefore, if the system has been temporarily out of use it should be carefully vented before being operated again. There is generally a vent screw on the fuel-injection pump overflow or the fuel filter for this purpose.

Lubrication

Fuel-injection pumps and governors are normally connected to the engine lube-oil circuit as the fuel-injection pump then requires no maintenance.

Before being used for the first time, the fuel-injection pump and the governor must be filled with the same type of oil that is used in the engine. In the case of fuel-injection pumps that are not directly connected to the engine lube-oil circuit, the pump is filled through the filler cap after removing the vent flap or filter. The oil level check takes place at the same time as the regular engine oil changes and is performed by removing the oil check plug on the governor. Excess oil (from leak fuel) is then drained off or the level topped up if required. Whenever the fuel-injection pump is removed or the engine overhauled,

the oil must be changed. Fuel-injection pumps and governors with separate oil systems have their own dipsticks for checking the oil level.

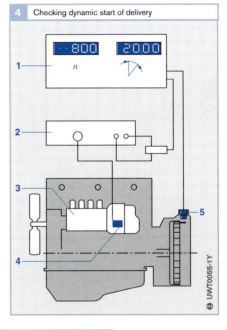

4 Checking dynamic start of delivery

Fig. 4
Schematic diagram of in-line fuel-injection pump and governor using port-closing sensor system

1 Engine analyzer
2 Adaptor
3 In-line fuel-injection pump and governor
4 Inductive speed sensor (port-closing sensor)
5 Inductive speed sensor (TDC sensor)

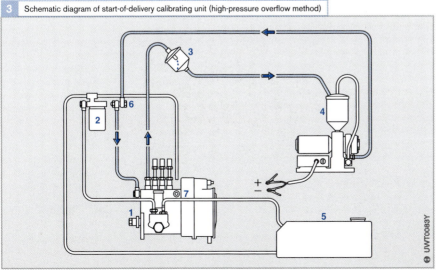

3 Schematic diagram of start-of-delivery calibrating unit (high-pressure overflow method)

Fig. 3
1 Fuel-injection pump
2 Fuel filter
3 Observation vessel
4 Start-of-delivery calibrating unit
5 Fuel tank
6 Oversize banjo bolt and nut
7 Screw cap

Testing helix and port-controlled distributor injection pumps

Good engine performance, high fuel economy and low emissions depend on correct adjustment of the helix and port-controlled distributor injection pump. This is why compliance with official specifications is absolutely essential during testing and adjustment operations on fuel-injection pumps.

One important parameter is the start of delivery (in service bay), which is checked with the pump installed. Other tests are conducted on the test bench (in test area). In this case, the pump must be removed from the vehicle and mounted on the test bench. Before the pump is removed, the engine crankshaft should be rotated until the reference cylinder is at TDC. The reference cylinder is usually cylinder No. 1. This step eases subsequent assembly procedures.

Test bench measurements

The test procedures described here are suitable for use on helix and port-controlled axial-piston distributor pumps with electronic and mechanical control, but not with solenoid-controlled distributor injection pumps.

Test bench operations fall into two categories:
● Basic adjustment and
● Testing

The results obtained from the pump test are entered in the test record, which also lists all the individual test procedures. This document also lists all specified minimum and maximum results. The test readings must lie within the range defined by these two extremes.

A number of supplementary, special-purpose test steps are needed to assess all the different helix and port-controlled axial-piston distributor pumps; detailed descriptions of every contingency, however, extend beyond the bounds of this chapter.

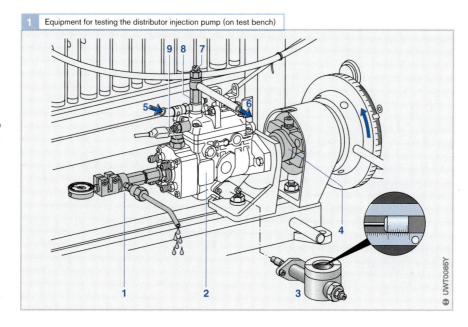

1   Equipment for testing the distributor injection pump (on test bench)

Fig. 1
1 Test layout with drain hose and dial gauge
2 Distributor injection pump
3 Timing device travel tester with vernier scale
4 Pump drive
5 Calibrating oil inlet
6 Return line
7 Overflow restrictor
8 Adapter with connection for pressure gauge
9 Electric shutoff valve (ELAB) (energized)

UWT0086Y

Basic adjustments

The first step is to adjust the distributor injection pump to the correct basic settings. This entails measuring the following parameters under defined operating conditions.

LPC adjustment

This procedure assesses the distributor plunger lift between bottom dead center (BDC) and the start of delivery. The pump must be connected to the test-bench fuel supply line for this test. The technician unscrews the 6-point bolt from the central plug fitting and then installs a test assembly with drain tube and gauge in its place (Fig. 1, Pos. 1).

The gauge probe rests against the distributor plunger, allowing it to measure lift. Now the technician turns the pump's input shaft (4) by hand until the needle on the gauge stops moving. The control plunger is now at top dead center (TDC).

A supply pressure of roughly 0.5 bar propels the calibrating oil into the plunger chamber behind the distributor plunger (5). For this test, the solenoid-operated shutoff valve (ELAB) (9) is kept energized to maintain it in its open position. The calibrating oil thus flows from the plunger chamber to the test assembly before emerging from the drain hose.

Now the technician manually rotates the input shaft in its normal direction of rotation. The calibrating oil ceases to flow into the plunger chamber once its inlet passage closes. The oil remaining in the chamber continues to emerge from the drain hose. This point in the distributor plunger's travel marks the start of delivery.

The lift travel between bottom dead center (BDC) and the start of delivery indicated by the gauge can now be compared with the setpoint value. If the reading is outside the tolerance range, it will be necessary to dismantle the pump and replace the cam mechanism between cam disk and plunger.

Supply-pump pressure

As it affects the timing device, the pressure of the supply pump (internal pressure) must also be tested. For this procedure, the overflow restrictor (7) is unscrewed and an adapter with a connection to the pressure gauge (8) is installed. Now the overflow restrictor is installed in an adapter provided in the test assembly. This makes it possible to test the pump's internal chamber pressure upstream of the restrictor.

A plug pressed into the pressure-control valve controls the tension on its spring to determine the pump's internal pressure. Now the technician continues pressing the plug into the valve until the pressure reading corresponds to the setpoint value.

Timing device travel

The technician removes the cover from the timing device to gain access for installing a travel tester with a vernier scale (3). This scale makes it possible to record travel in the timing device as a function of rotational speed; the results can then be compared with the setpoint values. If the measured timing device travel does not correspond with the setpoint values, shims must be installed under the timing spring to correct its initial spring tension.

Adjusting the basic delivery quantity

During this procedure the fuel-injection pump's delivery quantity is adjusted at a constant rotational speed for each of the following four conditions:

- Idle (no-load)
- Full-load
- Full-load governor regulation and
- Starting

Delivery quantities are monitored using the MGT or KMA attachment on the fuel-injection pump test bench (refer to section on "Fuel-injection pump test benches").

First, with the control lever's full-load stop adjusted to the correct position, the full-load governor screw in the pump cover is adjusted to obtain the correct full-load delivery quantity at a defined engine speed. Here, the governor adjusting screw must be turned back to prevent the full-load stop from reducing delivery quantity.

The next step is to measure the delivery quantity with the control lever against the idle-speed stop screw. The idle-speed stop screw must be adjusted to ensure that the monitored delivery quantity is as specified.

The governor screw is adjusted at high rotational speed. The measured delivery quantity must correspond to the specified full-load delivery quantity.

The governor test also allows verification of the governor's intervention speed. The governor should respond to the specified rpm threshold by first reducing and then finally interrupting the fuel flow. The breakaway speed is set using the governor speed screw.

There are no simple ways to adjust the delivery quantity for starting. The test conditions are a rotational speed of 100 rpm and the control lever against its full-load shutoff stop. If the measured delivery quantity is below a specified level, reliable starting cannot be guaranteed.

Testing

Once the basic adjustment settings have been completed, the technician can proceed to assess the pump's operation under various conditions. As during the basic adjustment procedure, testing focuses on

- Supply-pump pressure
- Timing device travel
- Delivery quantity curve

The pump operates under various specific conditions for this test series, which also includes a supplementary procedure.

Overflow quantity

The vane-type supply pump delivers more fuel than the nozzles can inject. The excess calibrating oil must flow through the overflow restriction valve and back to the oil tank. It is the volume of this return flow that is measured in this procedure. A hose is connected to the overflow restriction valve; de-

pending on the selected test procedure. The other end is then placed in a glass gauge in the MGT assembly, or installed on a special connection on the KMA unit. The overflow quantity from a 10-second test period is then converted to a delivery quantity in liters per hour.

If the test results fail to reach the setpoint values, this indicates wear in the vane-type supply pump, an incorrect overflow valve or internal leakage.

Dynamic testing of start of delivery

A diesel engine tester (such as the Bosch ETD 019.00) allows precise adjustment of the distributor injection pump's delivery timing on the engine. This unit registers the start of delivery along with the timing adjustments that occur at various engine speeds with no need to disconnect any high-pressure delivery lines.

Testing with piezo-electric sensor and stroboscopic timing light

The piezo-electric sensor (Fig. 2, Pos. 4) is clamped onto the high-pressure delivery line leading to the reference cylinder. Here, it is important to ensure that the sensor is mounted on a straight and clean section of tubing with no bends; the sensor should also be positioned as close as possible to the fuel-injection pump.

The start of delivery triggers pulses in the fuel-injection line. These generate an electric signal in the piezo-electric sensor. The signal controls the light pulses generated by the timing light (5). The timing light is now aimed at the engine's flywheel. Each time the pump starts delivery to the reference cylinder, the timing light flashes, lighting up the TDC mark on the flywheel. This allows correlation of timing to flywheel position. The flashes occur only when delivery to the reference cylinder starts, producing a static image. The degree markings (6) on the crankshaft or flywheel show the crankshaft position relative to the start of delivery.

Engine speed is also indicated on the diesel engine tester.

Setting start of delivery

If the results of this start-of-delivery test deviate from the test specifications, it will be necessary to change the fuel-injection pump's angle relative to the engine.

The first step is to switch off the engine. Then the technician rotates the crankshaft until the reference cylinder's piston is at the point at which delivery should start. The crankshaft features a reference mark for this operation; the mark should be aligned with the corresponding mark on the bellhousing. The technician now unscrews the 6-point screw from the central plug screw. As for basic adjustment process on the test bench, the technician now installs a dial-gauge assembly in the opening. This is used to observe distributor plunger travel while the crankshaft is being turned. As the crankshaft is turned counter to its normal direction of rotation (or in the normal direction on some engines), the plunger retracts in the pump. The technician should stop turning the crankshaft once the needle on the gauge stops moving. The plunger is now at bottom dead center. Now the dial gauge is reset to zero. The crankshaft is then rotated in its normal direction of rotation as far as the TDC mark. The dial gauge now indicates the travel executed by the distributor plunger on

its way from its bottom dead center position to the TDC mark on the reference cylinder. It is vital to comply with the precise specification figure for this travel contained in the fuel-injection pump's datasheet. If the dial gauge reading is not within the specification, it will be necessary to loosen the attachment bolt on the pump flange, turn the pump housing and repeat the test. It is important to ensure that the cold-start accelerator is not active during this procedure.

Measuring the idle speed

The idle speed is monitored with the engine heated to its normal operating temperature, and in a no-load state, using the engine tester. The idle speed can be adjusted using the idle-speed stop screw.

2 Checking start of delivery with piezo-electric sensor and timing light

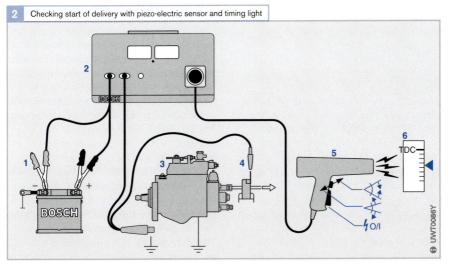

Fig. 2
1 Battery
2 Diesel tester
3 Distributor injection pump
4 Piezo-electric sensor
5 Stroboscopic timing light
6 Angle and TDC marks

Nozzle tests

The nozzle-and-holder assembly consists of the nozzle and the holder. The holder includes all of the required filters, springs and connections.

The nozzle affects the diesel engine's output, fuel economy, exhaust-gas composition and operating refinement. This is why the nozzle test is so important.

An important tool for assessing nozzle performance is the nozzle tester.

Nozzle tester

Keep your hands away from the nozzle jet. Spray from the nozzle stings and penetrates the skin. There is a risk of blood poisoning.

Wear safety goggles.

The nozzle tester is basically a manually operated fuel-injection pump (Fig. 1). For testing, a high-pressure delivery line (4) is used to connect the nozzle-and-holder assembly (3) to the tester. The calibrating oil is contained in a tank (5). The required pressure is generated using the hand lever (8). The pressure gage (6) indicates the pressure of the calibrating oil; a valve (7) can be used to disconnect it from the high-pressure circuit for specific test procedures.

1 Nozzle tester with nozzle-and-holder assembly

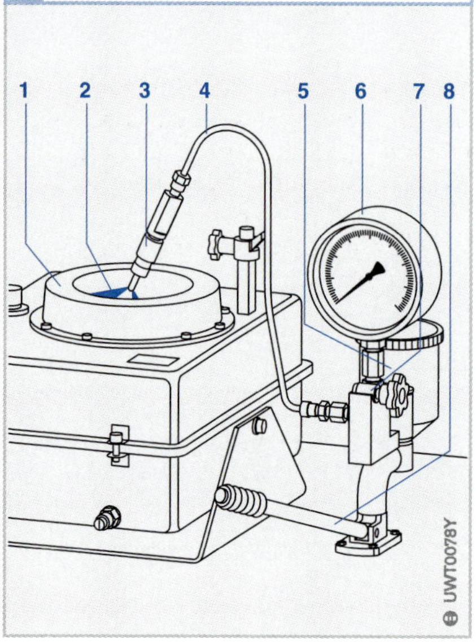

Fig. 1
1 Suction equipment
2 Injection jet
3 Nozzle-and-holder assembly
4 High-pressure test line
5 Calibrating oil tank with filter
6 Pressure gage
7 Valve
8 Hand lever

The EPS100 (0684200704) nozzle tester is specified for testing nozzles of Sizes P, R, S and T. It conforms to the standards defined in ISO 8984. The prescribed calibrating oil is defined in ISO standard 4113. A calibration case containing all the components is required to calibrate inspect the nozzle tester.

This equipment provides the basic conditions for reproducible, mutually compatible test results.

Test methods

Ultrasonic cleaning is recommended for the complete nozzle-and-holder assemblies once they have been removed from the engine. Cleaning is mandatory on nozzles when they are submitted for warranty claims.

Important: Nozzles are high-precision components. Careful attention to cleanliness is vital for ensuring correct operation.

The next step is to inspect the assembly to determine whether any parts of the nozzle or holder show signs of mechanical or thermal wear. If signs or wear are present, it will be necessary to replace the nozzle or nozzle-and-holder assembly.

The assessment of the nozzle's condition proceeds in four test steps, with some variation depending on whether the nozzles are pintle or hole-type units.

Chatter test

The chatter test provides information on the smoothness of action of the needle. During injection, the needle oscillates back and forth to generate a typical chatter. This motion ensures efficient dispersion of the fuel particles.

The pressure gage should be disconnected for this test (close valve).

Pintle nozzle

The lever on the nozzle tester is operated at a rate of one to two strokes per second. The pressure of the calibrating oil rises, ultimately climbing beyond the nozzle's opening pressure. During the subsequent discharge, the nozzle should produce an audible chatter; if it fails to do so, it should be replaced.

When installing a new nozzle in its holder, always observe the official torque specifications, even on hole-type nozzles.

Hole-type nozzle
The hand lever is pumped at high speed. This produces a hum or whistling sound, depending on the nozzle type. No chatter will be present in some ranges. Evaluation of chatter is difficult with hole-type nozzles. This is why the chatter test is no longer assigned any particular significance as an assessment tool for hole-type nozzles.

Spray pattern test
High pressures are generated during this test. Always wear safety goggles.

The hand lever is subjected to slow and even pressure to produce a consistent discharge plume. The spray pattern can now be evaluated. It provides information on the condition of the injection orifices. The prescribed response to an unsatisfactory spray pattern is to replace the nozzle or nozzle-and-holder assembly.

The pressure gage should also be switched off for this test.

Pintle nozzle
The spray should emerge from the entire periphery of the injection orifice as even tapered plume. There should be no concentration on one side (except with flatted pintle nozzles).

Hole-type nozzle
An even tapered plume should emerge from each injection orifice. The number of individual plumes should correspond to the number of orifices in the nozzle.

Checking the opening pressure
Once the line pressure rises above the opening pressure, the valve needle lifts from its seat to expose the injection orifice(s). The specified opening pressure is vital for correct operation of the overall fuel-injection system.

The pressure gage must be switched back on for this test (valve open).

Pintle nozzle and hole-type nozzle
with single-spring nozzle holder
The operator slowly presses the lever downward, continuing until the gage needle indicates the highest available pressure. At this point, the valve opens and the nozzle starts to discharge fuel. Pressure specifications can be found in the "nozzles and nozzle-holder components" catalog.

Opening pressures can be corrected by replacing the adjustment shim installed against the compression spring in the nozzle holder. This entails extracting the nozzle from the nozzle holder. If the opening pressure is too low, a thicker shim should be installed; the response to excessive opening pressures is to install a thinner shim.

Hole-type nozzle with two-spring nozzle holder
This test method can only be used to determine the initial opening pressure on two-spring nozzle-and-holder assemblies.

The is no provision for shim replacement on some nozzle-and-holder assemblies. The only available response with these units is to replace the entire assembly.

Leak test
The pressure is set to 20 bar above the opening pressure. After 10 seconds, formation of a droplet at the injection orifice is acceptable, provided that the droplet does not fall.

The prescribed response to an unsuccessful leak test is to replace the nozzle or nozzle-and-holder assembly.

Emissions measurement concept

Emissions testers are employed to examine exhaust-gas composition in the field (AU emissions inspection) and ensure compliance with legally mandated emissions limits. They also represent essential service tools. The emissions tester provides the data needed for optimal adjustment of induction and injection systems, and is an essential diagnosis tool.

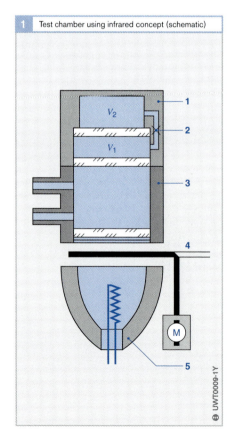

1   Test chamber using infrared concept (schematic)

V_2

V_1

1

2

3

4

M

5

UWT0009-1Y

Fig. 1

1 Receiver chamber with compensation volumes V_1 and V_2
2 Flow sensor
3 Measuring cell
4 Rotating chopper with motor
5 Infrared emitter

Test procedure

Required is the ability to measure specific exhaust components with a high degree of accuracy. Laboratories employ complex and extensive procedures for these measurements. Over the course of time, the infrared process has advanced to assume the status of the method of choice in vehicle service operations. The concept exploits the fact that individual substances absorb infrared light at different rates determined by each gas' characteristic wavelength.

Both single (for instance, for CO) and multiple-component (for CO/HC, CO/CO$_2$, CO/HC/CO$_2$, etc.) testers are available.

Emissions-tester measurement chamber

Infrared radiation is discharged from an emitter (Figure 1, Pos. 5) which has been heated to approx. 700 °C. The radiation then passes through a measuring cell (3) and into the receiver chamber (1).

Example of CO measurement
For CO measurements the sealed receiver chamber is charged with a gas of a defined CO content. This gas absorbs a portion of the CO-specific radiation. The absorption process heats the gas, and generates a current of gas which flows from volume V_1 through the flow sensor (2), and into the compensation volume V_2. A rotating chopper disk (4) induces a rhythmic interruption in the beam to produce an alternating flow pattern between the two volumes V_1 and V_2. The flow sensor converts this motion into an alternating electrical signal.

When a test gas with a variable CO content flows through the measuring cell, it absorbs radiant energy in quantities proportional to its CO content; this energy is then no longer available in the receiver chamber. The result is a reduction of the base flow in the receiver chamber. The deviation from the alternating base signal serves as an index of the CO content in the test gas.

Gas flow in the tester

A probe (Figure 2, Pos. 1) is used to extract exhaust gases from the test vehicle. The tester's integral diaphragm pump (7) draws the gas through a coarse-mesh filter (2) and into the condensation separator (3). The separator filters coarse particulates and ingested condensation from the test gas before it proceeds to yet another filter for further cleaning. A second diaphragm pump (8) conducts the condensation to the condensation discharge (16).

The next stop for the test gas is gas-analysis chamber GA1 (10), where CO_2 and CO levels are measured. The gas then proceeds to gas-analysis chamber GA2 (11) for an assessment of HC content. Before leaving the tester through the gas discharge (15), the test gas passes over electrochemical sensors (13 and 14) for measurements of oxygen (O_2) and nitrous oxide (NO) content.

A solenoid valve (6) upstream from the gas pump (7) switches the test chamber's supply from exhaust gas to fresh air for automatic zero-balance calibrations.

The activated-charcoal filter (5) in the air-inlet passage (4) prevents hydrocarbons in the surrounding air from entering the tester.

A pressure sensor (9) detects leakage at any point in the gas passage. The remaining pressure sensor (12) monitors atmospheric pressure, which is also included as a parameter in the unit's calculations.

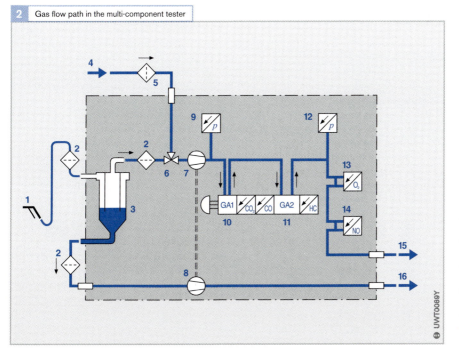

2   Gas flow path in the multi-component tester

Fig. 2
1 Extraction probe
2 Coarse-mesh filter (wet filter)
3 Condensation separator
4 Air
5 Activated-charcoal filter
6 Solenoid valve
7 Diaphragm pump (gas pump)
8 Diaphragm pump (condensation pump)
9 Pressure sensor (internal pressure/ leaks)
10 GA1 gas analyser (CO_2, CO test chamber)
11 GA2 gas analyser (HC test chamber)
12 Pressure sensor (barometric pressure)
13 Electrochemical sensor (O_2 sensor)
14 Electrochemical sensor (NO sensor)
15 Gas discharge
16 Condensation discharge

Turbidity testing

Long before the introduction of legislation to control gaseous emissions, separate regulations were in force to control smoke emission from diesel-engined vehicles. They still remain in force today. In countries that have statutory diesel smoke-emission controls, the testing methods applied are not uniform. All existing smoke tests are closely linked to the testing equipment used. One measure of smoke (soot/particulate) emission is the smoke number. There are essentially two methods for measuring it:

- The *absorption method* (turbidity test) uses the attenuation of a light beam that passes through the exhaust gas to measure exhaust-gas turbidity (used for design type approval and exhaust-gas emissions testing).
- In the *filter method* (reflected-light test), a defined quantity of exhaust gas is passed through a filter. The blackening of the filter then serves as a measure of the soot content in the exhaust gas (only used for development purposes).

Measurement of diesel-smoke emission is only of any practical use if performed with the engine under load as only under such conditions are significant amounts of particulates emitted. Once again, there are two commonly used methods:

- Measurement under WOT, e.g. on a chassis dynamometer, or on a defined test circuit under the load of the vehicle's brakes.
- Measurement under unrestricted acceleration using a defined throttle burst under the load of the engine flywheel.

The results of the diesel smoke-emission tests are dependent not only on the testing method but also on the type of engine loading and are generally not directly comparable.

Turbidity tester (absorption method)

During unrestricted acceleration, a certain amount of exhaust gas is taken from the vehicle's exhaust pipe (without vacuum assistance) using an exhaust-gas sampling probe and a hose leading to the measuring chamber. This method avoids impacts arising from exhaust-gas back-pressure and its fluc-

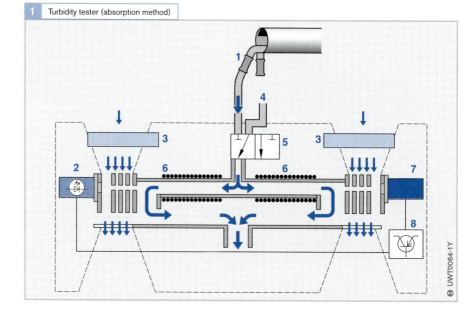

1   Turbidity tester (absorption method)

Fig. 1
1   Exhaust-sample probe
2   Green LED
3   Fan
4   Purge air for calibration
5   Calibrating valve
6   Heater
7   Receiver
8   Electronic analyzer and display
→   Path followed by exhaust gas

UWT0064-1Y

tuations on the test results since pressure and temperature are controlled (Hartridge tester).

In the measuring chamber, a light beam passes through the diesel exhaust gas. The attenuation of the light is measured photo-electrically and displayed as a percentage turbidity on an instrument T or absorption coefficient k. A precisely defined measuring-chamber length and keeping the optical window clear of soot (by air curtains, i.e. tangential air flows) are basic requirements for high levels of accuracy and reproducibility of the measurements.

When testing under load, the levels are measured and displayed continuously. In the unrestricted acceleration test, the test graph can be digitally recorded. The tester automatically analyzes the peak level and calculates the average level of several accelerator bursts.

Smoke-emission test equipment (filter method)

The tester draws in a predefined quantity (e.g. 0.1 or 1 l) of diesel exhaust gas through a strip of filter paper. As a requirement for the high-precision reproducibility of results, the volume drawn is recorded for every test sequence and converted to the specified quantity. Pressure and temperature impacts and the dead volume between the exhaust-sample probe and the filter paper are taken into account.

The blackened filter paper is analyzed optoelectronically using a reflex photometer. The results are generally indicated as the Bosch smoke number or mass concentration (mg/m³).

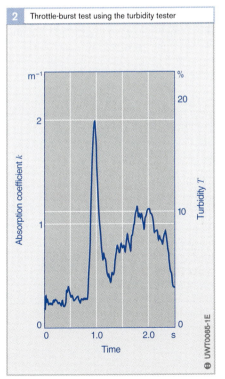

2 Throttle-burst test using the turbidity tester

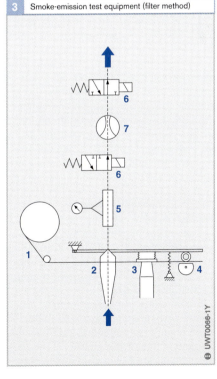

3 Smoke-emission test equipment (filter method)

Fig. 3
1 Filter paper
2 Gas penetration
3 Reflex photometer
4 Paper transport
5 Volume measuring
   device
6 Purge-air switch-over
   valves
7 Pump

Exhaust-gas emissions

Increasing energy consumption, which is mainly covered by the energy contained in fossil fuels, have made air quality a critical issue. The quality of the air we breathe depends on a wide range of factors. In addition to emissions from industry, homes and power plants, the exhaust gas generated by motor vehicles also plays a significant role. In developed countries, this accounts for about 20 % of total emissions.

Overview

The statutory limits restricting pollutant emissions from motor vehicles have been progressively tightened in recent years. In order to achieve compliance with these limits, vehicles have been equipped with supplementary emissions-control systems.

Combustion of the air-fuel mixture
A basic rule that applies to all internal-combustion engines is that absolutely complete combustion does not occur inside the engine's cylinders. This rule remains valid even when the combustion mixture contains excess air. Less efficient combustion leads to an increase in levels of toxic components containing carbon in the exhaust gas. In addition to a high percentage of non-toxic elements, the internal-combustion engine's gas also contains by-products which – at least when present in high concentrations – represent potential sources of environmental damage. These are classified as pollutants.

Positive crankcase ventilation
Additional emissions stem from the engine's crankcase ventilation system. Combustion gases travel along the cylinder walls and into the crankcase. From there, they are returned to the intake manifold for renewed combustion within the engine.

Since nothing more than pure air is compressed in the diesel's compression stroke, diesel engines generate only negligible amounts of these bypass emissions. The gases that make their way into the crankcase con-

tain only about 10 % as much pollution as the bypass gases in a gasoline engine. Despite this fact, closed crankcase-ventilation systems are now mandatory on diesel engines.

Evaporative emissions
Additional emissions can escape from vehicles powered by gasoline engines when volatile components in the fuel evaporate and emerge from the fuel tank, regardless of whether the vehicle is moving or parked. These emissions consist primarily of hydrocarbons. To prevent these gases from evaporating directly to the atmosphere, vehicles must be equipped with an evaporative-emissions control system designed to store them for subsequent combustion in the engine.

Evaporative emissions from diesels are not a major concern, as diesel fuel possesses virtually no high-volatility components.

Major components

Assuming the presence of sufficient oxygen, ideal, complete combustion of pure fuel would be possible in the following chemical reaction:

$$n_1 \, C_xH_y + m_1 \, O_2 \rightarrow n_2 \, H_2O + m_2 \, CO_2$$

The absence of ideal conditions for combustion combines with the composition of the fuel itself to produce a certain number of toxic components in addition to the primary combustion products water (H_2O) and carbon dioxide (CO_2) (Figure 1).

Water (H_2O)
During combustion, the water chemically bound within the fuel is transformed into water vapor, most of which condenses when its cools. This is the source of the exhaust plume visible on cold days. The proportion of water in the exhaust gas is dependent on the operating point in diesel engines.

Carbon dioxide (CO_2)

In complete combustion, the hydrocarbons in the fuel's chemical bonds are transformed to carbon dioxide (CO_2). Its proportion is also dependent on the operating point. Here again, the proportion depends on the engine operating conditions. The amount of converted carbon dioxide in the exhaust gas is directly proportional to fuel consumption. Thus the only way to reduce carbon-dioxide emissions when using standard fuels is to reduce fuel consumption.

Carbon dioxide is a natural component of atmospheric air, and the CO_2 contained in automotive exhaust gases is not classified as a pollutant. However, it is one of the substances responsible for the greenhouse effect and the global climate change that this causes. In the period since 1920, atmospheric CO_2 has risen continuously, from roughly 300 ppm to approx. 450 ppm in the year 2001. This renders efforts to reduce carbon-dioxide emissions and fuel consumption more important than ever.

Nitrogen (N_2)

Nitrogen is the primary constituent (78%) of the air drawn in by the engine. Although it is not directly involved in the combustion process, it is the largest single component within the exhaust gas, at approximately 69...75%.

| 1 | Exhaust-gas composition of internal-combustion engines without exhaust-gas treatment systems (untreated emissions) |

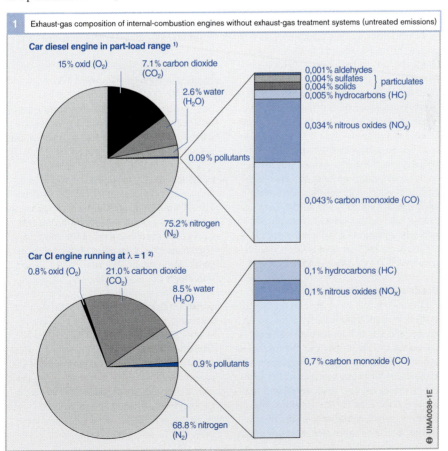

Car diesel engine in part-load range [1]

15% oxid (O_2)
7.1% carbon dioxide (CO_2)
2.6% water (H_2O)
0.09% pollutants
75.2% nitrogen (N_2)

0,001% aldehydes
0,004% sulfates
0,004% solids } particulates
0,005% hydrocarbons (HC)
0,034% nitrous oxides (NO_x)
0,043% carbon monoxide (CO)

Car CI engine running at $\lambda = 1$ [2]

0.8% oxid (O_2)
21.0% carbon dioxide (CO_2)
8.5% water (H_2O)
0.9% pollutants
68.8% nitrogen (N_2)

0,1% hydrocarbons (HC)
0,1% nitrous oxides (NO_x)
0,7% carbon monoxide (CO)

Fig. 1
Figures in percent by weight

The concentrations of exhaust-gas components, especially pollutants, may vary. Among other factors, they depend on the engine operating conditions and the ambient conditions (e.g. air humidity).

[1] NO_x and particulate emissions can be reduced by more than 90% by the use of NO_x storage catalytic converters and particulate filters.
[2] Catalytic converters, which are standard equipment today, can reduce pollutant emissions by up to 99%.

UMA0036-1E

Combustion by-products

During combustion, the air-fuel mixture generates a number of by-products. The most significant of these combustion by-products are:
- Carbon monoxide (CO)
- Hydrocarbons (HC) and
- Oxides of nitrogen (NO_x)

Engine modifications and exhaust-gas treatment can reduce the amount of pollutants produced.

Since they always operate with excess air, diesel engines inherently produce much smaller amounts of CO and HC than gasoline engines. The main emphasis is therefore on NO_X and particulate emissions. Both of these types of emissions can be reduced by more than 90 % by using modern NO_X storage catalytic converters and particulate filters.

Carbon monoxide (CO)

Carbon monoxide results from incomplete combustion in rich air-fuel mixtures due to an air deficiency.

Although carbon monoxide is also produced during operation with excess air, the concentrations are minimal, and stem from brief periods of rich operation or inconsistencies within the air-fuel mixture. Fuel droplets that fail to vaporize form pockets of rich mixture that do not combust completely.

Carbon monoxide is an odorless and tasteless gas. In humans, it inhibits the ability of the blood to absorb oxygen, thus leading to asphyxiation.

Hydrocarbons (HC)

Hydrocarbons, or HC, is a generic designation for the entire range of chemical compounds uniting hydrogen H with carbons C. HC emissions are the result of inadequate oxygen being present to support complete combustion of the air-fuel mixture. The combustion process also produces new hydrocarbon compounds not initially present in the original fuel (by separating extended molecular chains, etc.).

Aliphatic hydrocarbons (alkanes, alkenes, alkines and their cyclical derivatives) are virtually odorless. Cyclic aromatic hydrocarbons (such as benzol, toluol and polycyclic hydrocarbons) emit a discernible odor.

Some hydrocarbons are considered to be carcinogenic in long-term exposure. Partially oxidized hydrocarbons (aldehydes, ketones, etc.) emit an unpleasant odor. The chemical products that result when these substances are exposed to sunlight are also considered to act as carcinogens under extended exposure to specified concentrations.

Nitrous oxides (NO_x)

Nitrous oxides, or oxides of nitrogen, is the generic term embracing chemical compounds consisting of nitrogen and oxygen. They result from secondary reactions that occur in all combustion processes where air containing nitrogen is burned. The primary forms that occur in the exhaust gases of internal-combustion engines are nitrogen oxide (NO) and nitrogen dioxide (NO_2), with dinitrogen monoxide (N_2O) also present in minute concentrations.

Nitrogen oxide (NO) is colorless and odorless. In atmospheric air, it is gradually converted to nitrogen dioxide (NO_2). Pure NO_2 is a poisonous, reddish-brown gas with a penetrating odor. NO_2 can induce irritation of the mucous membranes when present in the concentrations found in highly-polluted air.

Nitrous oxides contribute to forest damage (acid rain) and also act in combination with hydrocarbons to generate photochemical smog.

Sulfur dioxide (SO_2)

Sulfurous compounds in exhaust gases – primarily sulfur dioxide – are produced by the sulfates contained in fuels. A relatively small proportion of these pollutant emissions stem from motor vehicles. These emissions are not restricted by official emission limits.

It is not possible to use a catalytic converter to convert sulfur dioxide. Sulfur forms deposits within catalytic converters, reacting with the active chemical layer and inhibiting the catalytic converter's ability to remove other pollutants from the exhaust gases. While sulfur contamination can be reversed in the NO_X storage catalytic converters used for emissions control with direct-injection gasoline engines, this process requires a considerable amount of energy, and consequently negate the fuel-economy benefits achieved by direct injection.

The earlier limits on sulfur concentrations within fuel of 500 ppm (parts per million, 1,000 ppm = 0.1 %), valid until the end of 1999, have now been tightened by EU legislation. The new limits, valid from 2000 onward, are 150 ppm for gasoline and 350 ppm for diesel fuels. A further reduction to 50 ppm for both types of fuel is slated for 2005. In practice, however, sulfur-free fuel will be introduced sooner. Gasoline and diesel fuel with a sulfur content of ≤ 10 ppm will already be available throughout Germany in 2003 (throughout the EU by 2005).

Particulates

The problem of particulate emissions is primarily associated with diesel engines. Levels of particulate emissions from gasoline engines with multipoint injection systems are negligible.

Particulates result from incomplete combustion. While exhaust-gas composition varies as a function of the combustion process and engine operating condition, these particulates basically consist of hydrocarbon chains (soot) with an extremely extended specific surface ratio. Uncombusted and partly combusted hydrocarbons form deposits on the soot, where they are joined by aldehydes, with their penetrating odor. Aerosol components (minutely dispersed solids or fluids in gases) and sulfates bond to the soot. The sulfates result from the sulfur content in the fuel. Consequently, these pollutants do not occur if sulfur-free fuel is used.

▶ **Ozone and smog**

Exposure to the sun's radiation splits nitrogen-dioxide molecules (NO_2). The products are nitrogen oxide (NO) and atomic oxygen (O), which combine with the ambient air's atomic oxygen (O_2) to form ozone (O_3). Ozone formation is also promoted by volatile organic compounds. This is why higher ozone levels must be anticipated on hot, windless summer days when high levels of air pollution are present.

In normal concentrations, ozone is essential for human life. However, in higher concentrations it leads to coughing, irritation of the throat and sinuses, and burning eyes. It adversely affects lung function, reducing performance potential.

There is no direct contact or mutual movement between the ozone formed in this way at ground level, and the stratospheric ozone that reduces the amount of ultraviolet radiation penetrating the earth's atmosphere.

Smog is not limited to the summer. It can also occur in winter in response to atmospheric layer inversions and low wind speeds. The temperature inversion in the air layers prevents the heavier, colder air containing the higher pollutant concentrations from rising and dispersing.

Smog leads to irritation of the mucous membranes, eyes and respiratory system. It can also impair visibility. This last factor explains the origin of the term smog, which combines "smoke" and "fog".

Reduced emissions

Mixture formation and combustion

Compared with gasoline engines, diesel engines run on fuel with a higher boiling point (approx. 160...360 °C) which is more easily ignitable (cetane number approx. 50).

The mixture formation process in a diesel engine can be divided into two phases:
- The "premixed" phase which takes place during ignition lag and
- The diffusion phase which takes place during combustion

In a diesel engine, the air-fuel mixture is heterogeneous. In order to prevent high levels of soot, CO and HC emissions, and high fuel consumption, there must always be an excess amount of air ($\lambda > 1.0$). The factors that impact on mixture formation include the following:
- The right number of nozzle jets to produce the air swirl required
- The right jet direction to match the shape of the combustion chamber, start of injection and injection time
- The right injection pressure (droplet size and distribution) and
- A tight rate-of-discharge curve (injection of fuel at the right time)

The variable engine operating parameters are engine speed, injection mass, pressures and temperatures, and the excess-air factor λ. The resulting engine variables are:
- Specific fuel consumption
- NO_X emission (NO and NO_2)
- Particulate emission
- HC emission
- CO emission, and
- Combustion noise

The power/torque versus engine-speed curve is modified according to the specifications of the engine manufacturer. All mixture-formation parameters have an impact on the resulting variables. NO_X formation is promoted by high combustion temperatures and excess air combined with the movement of air at the beginning of the combustion phase. Soot is produced by (localized) air shortage and an inadequate A/F mixture.

Engine modifications

The design of the combustion chamber, the choice of compression ratio and the motion of the air – as well as the mixture-formation parameters described above – influence the emission and fuel-consumption results achieved by the engine. The valve layout on engines with four valves per cylinder allows nozzles to be placed in a central position in the cylinder head. This creates optimized conditions for an even dissipation of the fuel jet in the combustion chamber. A turbocharger with variable turbine geometry allows a greater air charge across a wide band of the engine load/speed map. It shortens the time required to build up dynamic turbocharger pressure, and reduces exhaust-gas back pressure. An intercooler reduces NO_X emissions and fuel consumption.

Exhaust-gas recirculation

Exhaust-gas recirculation is an effective method of reducing NO_X emissions. It has been a standard feature of car diesel engines for many years. It is also used on commercial vehicles to achieve low NO_X emissions. In accordance with legal requirements, exhaust-gas recirculation is only used at low engine loads and speeds on cars. On the other hand, it must operate across virtually the entire load/speed range of commercial (heavy-duty) vehicles.

The NO_X-reducing effect of exhaust-gas recirculation is based on three mechanisms:
- Reduces oxygen concentration in the combustion chamber
- Reduces the amount of exhaust gas expelled and
- Reduces exhaust-gas temperature by the higher specific heat of the inert gases H_2O and CO_2 (gases that take no part in the reaction)

A particularly effective result is the recirculation of cooled exhaust gas. Recirculation rates can be up to 50 % on cars and 5...25 % on commercial vehicles. On cars, the pressure difference required for exhaust-gas recirculation prevails at low engine loads between points upstream of the turbocharger turbine and downstream of the turbocharger compressor (the turbocharger has either a wastegate or variable turbine geometry, VTG) . On commercial vehicles, the exhaust-gas pressure upstream of the turbocharger turbine is normally lower than the turbocharger pressure downstream of the turbocharger compressor and intercooler at high engine loads. For this reason, the turbocharger must be suitably modified to operate in exhaust-gas recirculation conditions, or a VTG turbocharger that can generate the required pressure differential must be used. Another alternative is the use of a venturi tube (lower pressure at the constriction point) in the bypass passage to the air intake.

The exhaust-gas recirculation system is controlled by regulating the differential air mass by means of an air-mass meter.

Influence of fuel injection

The most important factors that determine engine operating results are start of injection, injection pressure, the rate-of-discharge curve as well as pre-injection and secondary injection.

Cars

Low NO_X emissions are achieved by retarding the start of injection under the exhaust-emission cycles specified by statutory requirements. At high engine loads and speeds, the start of injection is advanced to achieve low fuel consumption and high power output. When the engine is cold, start of injection is advanced (HC).

Commercial vehicles

Low NO_X emissions are achieved by retarding the start of injection at medium and high engine loads. However, this increases fuel consumption. Advanced start of injection is required at low engine loads and when the engine is cold (HC).

On modern fuel-injection systems with EDC (Electronic Diesel Control), start of injec-

tion is an infinitely variable parameter. A start-of-injection function controls start of injection/delivery to within $< \pm 1°$ of crankshaft rotation. The common-rail system controls start of injection with a similar degree of accuracy by means of an electronically controlled solenoid valve.

Injection pressure

Higher injection pressures normally reduce soot emission. Advanced start of injection produces lower fuel consumption than retarded start of injection. If injection pressure is increased with a retarded start of injection, fuel consumption can be reduced a little further but NO_X emissions are higher. The injection pressure (peak pressure or fuel-rail pressure) is, therefore, a variable that has to be carefully optimized along with start of injection and the exhaust-gas recirculation rate. The analysis variables within the test range are fuel consumption and soot emission on the one hand, and NO_X emissions on the other. The task is to find a practical compromise between these parameters. This becomes clear when fuel consumption and soot levels are plotted against NO_X emissions. Cars with diesel engines require very high injection pressures at high engine loads/speeds in order to achieve high performance and low specific fuel consumption.

Rate-of-discharge curve

The rate-of-discharge curve refers to the rate of discharge of fuel (mass flow rate) over time during the period of injection.

A low rate of fuel discharge during ignition lag is beneficial to low NO_X emissions and low combustion noise.

A precisely metered pre-injection quantity substantially reduces combustion noise. Pre-injection also reduces NO_X emissions and, at high engine loads, fuel consumption as well, while black smoke generally increases slightly.

A well atomized secondary injection immediately following the main injection can help to lower soot emissions without substantially altering other engine characteristics. Systems with secondary injection are already in volume production.

Exhaust-gas treatment systems

In order to be able to comply with ever stricter exhaust-gas emission limits, emission control will become increasingly important for diesel engines in the future despite the advances in internal engine design. This is particularly true for larger cars and all types of commercial vehicle. There are a number of different systems under development. Which of them will eventually become established is as yet an unanswered question.

Exhaust-gas treatment systems for diesel engines aim primarily to reduce two types of harmful emission:
- Particulates, which are caused by heterogeneous mixture distribution in the combustion chamber and
- Nitrogen oxides (NO_X), which result from the high temperatures at which diesel combustion takes place

The untreated emission of such substances has already been drastically reduced in recent years by developments such as high-pressure fuel-injection systems.

Diesel oxidation-type catalytic converter

The diesel oxidation-type catalytic converter (DOC) is fitted in the exhaust-gas system close to the engine (Figure 1, Item 9) so that it reaches its optimum operating temperature as quickly as possible. It reduces hydrocarbon (HC) and carbon monoxide (CO) emissions, together with some of the volatile components of the particulate emissions. It converts these exhaust-gas products into water (H_2O) and carbon dioxide (CO_2).

Oxidation-type catalytic converters are already in use on volume-production vehicles. Special catalytic-converter designs can also simultaneously reduce nitrogen oxides (NO_X) together with the HC and CO components, though the NO_X conversion is limited to 5...10%.

Particulate filter

A particulate filter (Figure 1, Item 10) filters out the particulates from the exhaust gas. The pressure drop across the particulate filter is a possible indicator of the amount of soot retained. Above a specific retention volume, the filter has to be regenerated. The temperatures of over 600°C required to burn off the soot are not achieved by the diesel engine under normal operation. Engine modifications such as retarded injection and intake air-flow constriction can increase the temperature of the exhaust gas.

At present, filters made of porous ceramic materials are preferred. They are already in use on volume-production cars.

Additive system
The use of an additive that is mixed with the fuel in the tank can reduce the temperature required to burn off the particulates in the particulate filter by about 100°C. Nevertheless, the exhaust-gas back pressure gradually increases over time as the non-combustible deposits (the additive ash) remain in the filter. This increases fuel consumption and limits the life of the filter.

CRT system
With the CRT (Continuous Regeneration Trap) system, an oxidation catalytic converter is fitted upstream of the particulate filter and oxidizes the NO in the exhaust to NO_2. The soot that collects in the filter is then continuously burned with NO_2 as soon as the temperature exceeds 250°C – in other words, at a temperature substantially lower than required by conventional particulate filters where combustion with O_2 takes place.

Temperature sensors, a differential-pressure sensor and a soot sensor downstream of the particulate filter monitor the operation of the system. The CRT system is currently being tested on selected fleets of busses with a view to use in commercial vehicles.

Because of the sensitivity to sulfur of the required oxidation catalytic converters, low-sulfur fuel is stipulated.

By using a catalytic coating on the filter, the oxidation-type catalytic converter and the particulate filter can be incorporated in a single unit. This type of filter is called a Catalyzed Soot Filter (CSF). It is also sometimes referred to as a CDPF (Catalyzed Diesel Particulate Filter) system.

NO$_X$ accumulator-type catalytic converter

A diesel engine always operates with an air excess (lean mixture, $\lambda > 1$). Because of that, a three-way catalytic converter as used for gasoline engines with manifold fuel injection cannot be used to reduce the nitrogen oxides (NO$_X$) in the exhaust. When there is an air excess, CO and HC react with the excess oxygen in the exhaust gas to form CO$_2$ and H$_2$O, and are thus not available for reducing the NO$_X$ to nitrogen (N$_2$).

The NO$_X$ accumulator-type catalytic converter is being developed as a means of reducing the nitrogen-oxide emissions on car diesel engines. It breaks down the nitrogen oxides in a different way – this is because it is able to store nitrogen oxides and then convert them. This process involves the following two stages:
- NO$_X$ storage in high-oxygen exhaust ($\lambda > 1$; from 30 seconds to several minutes)
- NO$_X$ release and conversion in low-oxygen exhaust ($\lambda < 1$; 2...10 seconds)

NO$_X$ storage
When the oxygen content of the exhaust is high, nitrogen oxides combine with metal oxides on the surface of the NO$_X$ accumulator-type catalytic converter to form nitrates (Figure 2 overleaf). This process is assisted by an oxidation-type catalytic converter (3) which is either upstream of or integrated in the NO$_X$ accumulator-type catalytic converter and which oxidizes the NO exhaust component to NO$_2$.

As the amount of stored nitrogen oxide (the charge) increases, the ability to continue to bind nitrogen oxides decreases. There are two ways of detecting when the catalytic converter is charged to such a degree that the storage phase needs to be terminated.
- A model-based procedure calculates the quantity of stored nitrogen oxides on the basis of the temperature of the catalytic converter.

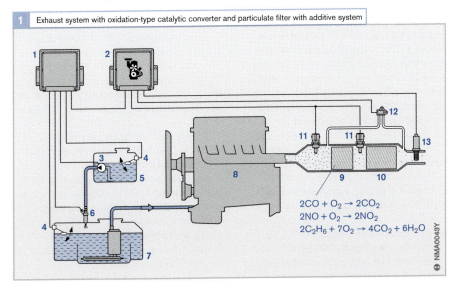

1   Exhaust system with oxidation-type catalytic converter and particulate filter with additive system

$$2CO + O_2 \rightarrow 2CO_2$$
$$2NO + O_2 \rightarrow 2NO_2$$
$$2C_2H_6 + 7O_2 \rightarrow 4CO_2 + 6H_2O$$

Fig. 1
1 Additive control unit
2 Engine control unit
3 Additive pump
4 Fluid-level sensor
5 Additive tank
6 Additive metering unit
7 Fuel tank
8 Diesel engine
9 Oxidation-type catalytic converter (DOC)
10 Particulate filter
11 Temperature sensor
12 Differential-pressure sensor
13 Soot sensor

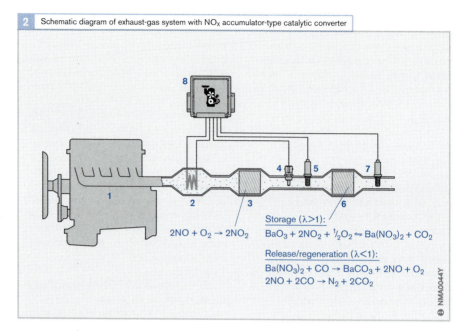

Fig. 2
1   Diesel engine
2   Exhaust heater
     (optional)
3   Oxidation-type
     catalytic converter
     (optional)
4   Temperature sensor
5   Broadband oxygen
     sensor Type LSU
6   NO$_X$ accumulator-
     type catalytic
     converter
7   NO$_X$ sensor or
     oxygen sensor
8   Engine control unit

2   Schematic diagram of exhaust-gas system with NO$_X$ accumulator-type catalytic converter

$2NO + O_2 \rightarrow 2NO_2$

Storage ($\lambda > 1$):
$BaO_3 + 2NO_2 + {}^1\!/_2O_2 \leftrightarrow Ba(NO_3)_2 + CO_2$

Release/regeneration ($\lambda < 1$):
$Ba(NO_3)_2 + CO \rightarrow BaCO_3 + 2NO + O_2$
$2NO + 2CO \rightarrow N_2 + 2CO_2$

NMA0044Y

- An NO$_X$ sensor downstream of the NO$_X$ accumulator-type catalytic converter measures the nitrogen-oxide concentration in the exhaust.

NO$_X$ release and conversion

Once a certain charge is reached, the NO$_X$ accumulator-type catalytic converter has to be regenerated, i.e. the stored nitrogen oxides have to be released and converted into N$_2$. In order for this to take place, the engine is briefly run with a rich mixture ($\lambda \approx 0.95$). Regeneration takes place in two stages involving the production of carbon dioxide (CO$_2$) and nitrogen (N$_2$) (Figure 2).

There are two different methods of detecting when the regeneration phase is complete.
- The model-based procedure calculates the amount of nitrogen oxides remaining in the NO$_X$ accumulator-type catalytic converter.
- An oxygen sensor (Figure 2, Item 7) downstream of the catalytic converter measures the oxygen concentration in the exhaust gas and indicates by means of the

change of its signal from "high oxygen" to "low oxygen" that the regeneration phase is complete (CO breakthrough).

In order that good NO$_X$ reduction rates are also achieved when the engine is started from cold, an electric exhaust-gas heater (2) can be used.

Sulfur from the fuel and from lubricants "contaminates" the catalytic converter. This is because the sulfur uses up the storage capacity for NO$_X$. For that reason, fuel with as low a sulfur content as possible (< 10 ppm) is required.

By heating the exhaust gas to around 650 °C at $\lambda \approx 1$, sulfur contamination can be reversed to a large degree (desulfurization). However, because of the frequency of desulphation, a high sulfur content in the fuel has an adverse effect on fuel consumption.

The NO$_X$ accumulator-type catalytic converter is sometimes also called an NO$_X$ Storage Catalyst (NSC).

SCR principle

In exhaust-gas denitrification processes based on the SCR (**S**elective **C**atalytic **Re**duction) principle, a reducing agent, such as a dilute urea solution with a concentration of 32.5 % by weight, is added to the exhaust gas in very precisely metered quantities. A hydrolyzing catalytic converter then extracts ammonia from the urea solution (Figure 3).

The ammonia reacts with NO_X in the SCR catalytic converter to form nitrogen and water. Modern SCR catalytic converters can also perform the function of the hydrolyzing converter so that a separate unit is not required.

An oxidation-type catalytic converter upstream of the reducing-agent injection point increases the efficiency of the system. An oxidation-type catalytic converter (NH_3 blocking catalytic converter) downstream of the SCR converter prevents NH_3 emission.

Because of the high NO_X reduction rates (up to 90 % in the European transient test cycle for commercial vehicles), consumption-optimized engine calibration is possible. As a result, such systems can save as much as 10 % on fuel consumption. SCR systems for commercial-vehicle applications are close to being ready for volume production.

Combination systems

In order to be able to comply with future exhaust-gas emission limits, many diesel vehicles will require exhaust-gas treatment systems that enable both particulate filtration and highly efficient reduction of NO_X content. Such systems will also be known as *four-way systems* because they will limit not only NO_X and particulate output but also HC and CO emissions.

Combination systems demand very powerful engine control units. Systems combining NO_X accumulator-type catalytic converters with particulate filters, or SCR catalytic converters with particulate filters are currently in the course of development.

Example system

Soot is continuously oxidized by a catalyzed diesel particulate filter (CDPF) while the downstream SCR system reduces the NO_X emissions. A reducing agent is injected in quantities based on a stored data map and exhaust-gas temperature or on the basis of the detected NO_X concentration upstream of the catalytic converter. The function of the overall system is monitored by gas sensors (for NO_X and/or NH_3) and temperature testing points.

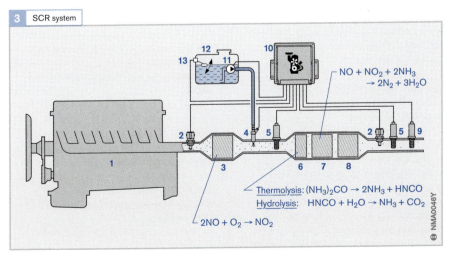

3   SCR system

NO + NO$_2$ + 2NH$_3$ → 2N$_2$ + 3H$_2$O

Thermolysis: (NH$_3$)$_2$CO → 2NH$_3$ + HNCO
Hydrolysis: HNCO + H$_2$O → NH$_3$ + CO$_2$

2NO + O$_2$ → NO$_2$

Fig. 3
1 Diesel engine
2 Temperature sensor
3 Oxidation-type
   catalytic converter
4 Injector for reducing
   agent
5 NO_X sensor
6 Hydrolyzing catalytic
   converter
7 SCR catalytic
   converter
8 NH_3 blocking
   catalytic converter
9 NH_3 sensor
10 Engine control unit
11 Reducing-agent
   pump
12 Reducing-agent tank
13 Fluid-level sensor

Emissions-control legislation

Over the years traffic density has displayed a marked increase, creating a corresponding negative array of consequences for the environment. The repercussions have been especially conspicuous in urban areas. As a result, it has become imperative to place legal limits on exhaust-gas emissions from motor vehicles. Both component limits and the procedures for verifying compliance are defined in legislation. Each new vehicle model must comply with the prevailing regulations.

Overview

The state of California assumed a pioneering role in efforts to restrict toxic emissions emanating from motor vehicles. This development arises from the fact that the geography of cities like Los Angeles prevents wind from dispersing exhaust gases, fostering formation of smog layers that encompass the city. The resulting smog not only has substantial negative effects on the health of the residents, but also impairs visibility.

California introduced the first regulations restricting emissions levels from motor vehicles in the 1960s. These directives became progressively more stringent in the ensuing years. In the intervening period, regulations governing exhaust-gas emissions have been adopted in all industrialized nations. These laws impose mandatory limits on emissions from gasoline and diesel engines, while also defining the test procedures employed to confirm compliance.

The most important legal restrictions on exhaust-gas emissions are:
● CARB regulations
  (California Air Resources Board)
● EPA regulations
  (Environmental Protection Agency)
● EU regulations (European Union)
● Japanese regulations

Test procedures
Japan and the European Union have followed the lead of the United States by defining test procedures for certifying compliance with emissions limits. These procedures have been adopted in modified or unrevised form by other countries.

Legal requirements prescribe any of three different test procedures according to vehicle class and the object of the test:
● Type test for homologation approval
● Random testing of vehicles from serial production conducted by the approval authorities and
● Field monitoring of specified exhaust-gas components from vehicles in highway operation

The most extensive test procedures are those used for type approval. The procedures employed for field monitoring are simpler.

Classifications
Countries with legal limits on emissions from motor vehicles divide these vehicles into various classes:
● Passenger cars: Testing is conducted on a chassis dynamometer.
● Light-duty trucks: The upper limit lies at an approved gross vehicle weight of between 3.5 and 3.8 metric tons, varying according to country. Testing is carried out on a chassis dynamometer (as with passenger cars).
● Heavy-duty trucks: Approved gross vehicle weights in excess of 3.5...3.8 metric tons. Testing is performed on an engine dynamometer, with no provision for in-vehicle testing.
● Off-highway (e.g. construction-industry, agricultural and forestry vehicles): Tested on engine test bench as for heavy-duty trucks.

Type test
Vehicles must successfully absolve emissions testing as a condition for receiving homologation approval for each specific engine and vehicle type. This process entails proving compliance with stipulated emissions limits in defined test cycles. Different countries have defined individual test cycles and emissions limits (Fig. 1).

Test cycles

Dynamic test cycles are specified for passenger cars and light-duty trucks. The country-specific differences between the two procedures are rooted in their respective origins:

- Test cycles designed to mirror conditions recorded in actual highway operation (FTP test cycle in the USA, etc.) and
- Synthetically generated test cycles consisting of phases with constant cruising speeds and acceleration rates (MNEDC in Europe)

The mass of the toxic emissions from each vehicle is determined by operating it in conformity with speed curves precisely defined for the test cycle. During this test cycle, the exhaust gases are collected for subsequent analysis to determine the mass of the pollutants emitted during testing. For heavy-duty trucks, steady-state exhaust-gas tests (e.g. 13-stage test in the EU) or dynamic tests (e.g. Transient Cycle in the USA) are carried out on the engine test bench.

Testing serial-production vehicles

This testing is usually conducted by the vehicle manufacturer in the quality control checks that accompany the production process. The authorities responsible for granting homologation approval can demand confirmation testing as often as deemed necessary. EU and ECE directives (Economic Commission of Europe) take account of production tolerances by carrying out random testing on between 3 and a maximum of 32 vehicles. The most stringent requirements are encountered in the USA, and particularly in California, where the authorities require what is essentially comprehensive and total quality monitoring.

On-Board Diagnosis

Emissions legislation also defines the processes to be employed in confirming conformity with the specified limits. The engine ECU incorporates diagnostic functions (software algorithms) designed to detect emissions-relevant malfunctions within the system. OBD functions (On-Board Diagnosis) monitor performance of all components in which malfunctions could lead to higher levels of exhaust-gas emissions. Different countries have defined their own specific emissions limits. When the vehicle exceeds these limits the malfunction indicator lamp lights up to alert the driver.

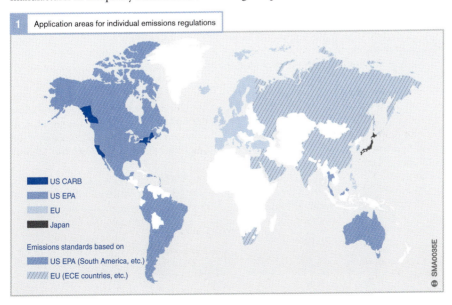

1   Application areas for individual emissions regulations

US CARB
US EPA
EU
Japan

Emissions standards based on
US EPA (South America, etc.)
EU (ECE countries, etc.)

SMA0035E

Fig. 1
Test cycles prescribed by various emissions regulations:
- FTP 75 cycle (CARB and EPA)
- Highway cycle (CARB and EPA, for determining fleet averages only)
- MNEDC (EU/ECE test cycle)
- 10·15-mode cycle (both in Japan)

Other test cycles are in the introductory phase in the US:
- SC03 cycle and
- US06 cycle

CARB legislation (Cars/LDT)

The CARB, or California Air Resources Board emissions limits for passenger cars and light-duty trucks and vans (LDT, Light-Duty Trucks) are defined in statutes governing exhaust emissions:
- LEV I and
- LEV II

The LEV I standard applies to passenger cars and light-duty trucks with an approved gross vehicle weight of up to 6,000 lbs manufactured in the model years 1994...2003. From the 2004 model year onwards, the LEV II standard will be introduced for all new vehicles with an approved gross vehicle weight of up to 8,500 lbs.

Emissions limits
The CARB regulations define limits on:
- Carbon monoxide (CO)
- Nitrous oxides (NO_x)
- NMOG (non-methane organic gases)
- Formaldehyde (LEV II only) and
- PM particulate emissions (diesel: LEV I and LEV II; gasoline-engines planned for LEV II)

Actual emission levels are determined using the FTP 75 driving cycle (Federal Test Procedure). Limits are defined in relation to distance and specified in grams per mile.

Emissions categories
Automotive manufacturers can apply various vehicle concepts for classification in the following categories according to their respective emissions of NMOG, CO, NO_x and particulates:
- Tier 1
- TLEV (Transitional Low-Emission Vehicle)
- LEV (Low-Emission Vehicle), applying to both exhaust and evaporative emissions
- ULEV (Ultra-Low-Emission Vehicle)
- SULEV (Super Ultra-Low-Emission Vehicle)
- ZEV (Zero-Emission Vehicle), vehicles without exhaust gas or evaporative emissions and
- PZEV (Partial ZEV), which is basically SULEV, but with more stringent limits on evaporative emissions and stricter long-term performance criteria

Fig. 1

[1] For Tier 1, NMHC limit value applies instead of NMOG limit value
[2] Limit value in each case for "full useful life" (10 years/ 100,000 miles with LEV I or 120,000 miles with LEV II)
[3] Limit value in each case for "intermediate useful life" (5 years/ 50,000 miles)
[4] Only limit values for "full useful life"

LEV I:
for passenger cars and light trucks/vans up to 3,750 lbs
Model years 1994 to 2003

LEV II:
for all vehicles up to 8,500 lbs
From 2004 model year only

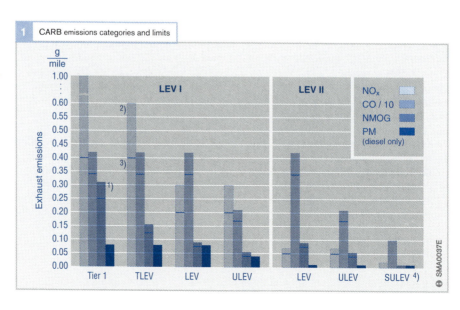

1 CARB emissions categories and limits

The essential categories for LEV I are Tier 1, TLEV, LEV and ULEV. Figure 1 illustrates the limits on NO_x, CO, NMOG and PM in the individual categories. Figure 2 is a graphic illustration of the limits on NO_x and NMOG in the various emissions categories.

 The LEV II emissions standards come into effect in 2004. At the same time Tier 1 and TLEV will be replaced by SULEV with its substantially lower limits. The LEV and ULEV classifications remain in place. The CO and NMOG limits from LEV I remain unchanged, but the NO_x limit is substantially lower for LEV II.

The LEV II standard also includes new, supplementary limits governing formaldehyde.

Long-term compliance

To obtain approval for each vehicle model the manufacturer must prove compliance with the official emissions limits over a period of:
- 50,000 miles or 5 years ("intermediate useful life") or
- 100,000 miles (LEV I)/120,000 miles (LEV II) or 10 years ("full useful life")

The applicable figures for the PZEV emissions category are 150,000 miles and 15 years.

Manufacturers also have the option of certifying vehicles for 150,000 miles using the same limits that apply to 120,000 miles. The manufacturer then receives a bonus when the NMOG fleet average is defined (refer to section on "Fleet averages" on next page).

For this type of approval test the manufacturer must furnish two vehicle fleets from series production:
- One fleet in which each vehicle must cover 4,000 miles prior to testing.
- One fleet for long-term testing, in which the deterioration factors for individual components are defined.

Long-term testing entails subjecting the vehicles to specific driving programs over periods of 50,000 and 100,000 miles. Exhaust-gas emissions are tested at intervals of 5,000 miles. Service inspections and maintenance are restricted to the standard prescribed intervals.

Countries that base their regulations on the US test cycles (such as Switzerland) allow application of defined deterioration factors to simplify the certification process.

2   Graphics showing CARB limits for emissions categories

Phase-in

Following introduction of the LEV II standards in 2004 compliance will be mandatory for at least 25% of new vehicles being registered for the first time in that year. The phase-in rule stipulates that an additional 25% of vehicles will then be required to conform to the LEV II standards in each consecutive year. All new vehicles will be required to meet the LEV II standards starting in 2007.

Fleet averages

Each vehicle manufacturer must ensure that exhaust-gas emissions for its total vehicle fleet do not exceed a specified average. NMOG emissions serve as the reference category for assessing compliance with these averages. The fleet average is determined based on average emission levels produced by all of the manufacturer's vehicles in complying with the NMOG limits. Different fleet averages apply to passenger cars and light-duty trucks and vans.

The compliance limits for the NMOG fleet average are lowered in each subsequent year (Figure 3). To meet the lower limits, manufacturers must produce progressively more "clean" vehicles in the more stringent emissions categories in each consecutive year. This phase-in rule does not affect the fleet averages.

Average fleet fuel consumption

Each manufacturer's fleet must comply with US legislation specifying maximum average fuel consumption in miles per gallon. The current CAFE value (Corporate Average Fuel Economy) for passenger cars is 27.5 miles per gallon. This corresponds to 8.55 liters per 100 kilometers in metric terms. The value for LDT is 20.3 miles per gallon or 11.6 liters per 100 kilometers. There are no values for heavy-duty trucks.

At the end of each year the average fuel economy for each manufacturer is calculated based on the numbers of individual vehicle models that have been sold. The manufacturer must remit a penalty fee of $ 5.50 per vehicle for each 0.1 miles per gallon by which its fleet exceeds the target. Buyers also pay a "gas-guzzler" tax on vehicles with especially high fuel consumption. Here, the limit is 22.5 miles per gallon (corresponding to 10.45 liters per 100 kilometers in metric terms).

These penalties are intended to spur development of vehicles offering high levels of fuel economy.

In order to measure fuel consumption, the Highway Cycle is completed in addition to the FTP 75 Test Cycle.

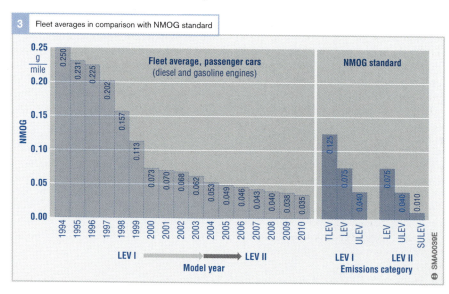

3 Fleet averages in comparison with NMOG standard

On-Board Diagnosis

With the introduction of OBD II (1994), all new passenger cars and light-duty trucks and vans with an approved gross vehicle weight of up to 8,500 lbs (3.85 metric tons) are now required to incorporate diagnostic capabilities that can detect all problems with the potential to affect vehicle emissions

The system's response threshold is defined as 1.5 percent of the official emission limit in each exhaust category. Once this threshold is crossed, a warning lamp must be triggered after completion of two driving cycles at the absolute latest. This malfunction lamp can extinguish again if three subsequent driving cycles elapse with no detected errors.

Field monitoring

Unscheduled testing

Random emissions testing is conducted on in-use vehicles using the FTP 75 test procedure. Only vehicles with milages of less than 50,000 or 75,000 miles (varies according to the certification status of the individual vehicle model) are selected for testing.

Vehicle monitoring by the manufacturer

Official reporting of problems and damage related to defined emissions-relevant components and systems has been mandatory for vehicle manufacturers since the 1990 model year. The reporting obligation remains in force for a period of 5 or 10 years, or 50,000 or 100,000 miles, depending on the length of the warranty applying to the component or assembly.

The reporting procedure consists of three stages:
- Emissions Warranty Information Report (EWIR)
- Field Information Report (FIR) and
- Emission Information Report (EIR)

The difference between them is that progressively more detailed data is obtained at each consecutive level. Information concerning:
- Problem reports
- Malfunction statistics

- Defect analysis and
- The effects on emissions

is then forwarded to the emissions authorities. The authorities use the FIR as the basis for issuing mandatory recall orders to the manufacturer.

Zero-emission vehicles

Starting in 2003 10 % of the new vehicles being registered for the first time will have to meet requirements for classification as ZEV, or Zero-Emission Vehicles. These vehicles must emit no exhaust gas or evaporative emissions during operation. This category essentially applies to electric cars.

While vehicles in the PZEV (Partial Zero-Emission Vehicles) category are not absolutely free of emissions, they do emit extremely low levels of pollutants. PZEV vehicles can also be used to comply with the standard mandating that 10 % of each vehicle fleet meet the ZEV standards. These vehicles are counted using a rating factor of 0.2...1 depending on the extent to which emissions have been reduced. The minimum weighting factor of 0.2 is granted when the following demands are met:
- SULEV certification indicating long-term compliance extending over 150,000 miles or 15 years
- Warranty coverage extending over 150,000 miles or 15 years on all emissions-relevant components
- No evaporative emissions from the fuel system (0 EVAP, Zero **Evap**oration), achieved through extensive encapsulation of tank and fuel system

Special regulations apply to hybrid vehicles with diesel engines and electric motors. These vehicles can also contribute to achieving compliance with the 10 % limit.

EPA regulations (Cars/LDT)

The EPA (Environment Protection Agency) regulations apply to the 49 states outside California, where the CARB stipulations are in force. The EPA regulations for passenger cars and LDT (Light-Duty Trucks) are not as strict as the CARB requirements. The individual states also have the option of adopting the CARB emissions regulations. This step has already been taken in some states, such as Maine, Massachusetts and New York.

The legislative foundation is provided by the "Clean Air Act," which contains an action catalog with measures to protect the environment but does not specify actual limits.

The EPA regulations currently in force conform to the "Tier 1" standard. The next stage, "Tier 2", is slated to enter effect in 2004.

The NLEV (National Low Emission Vehicle) program is a voluntary undertaking aimed at reducing emissions in the 49 states (except California). Vehicles are classified in four emissions categories: Tier 1, TLEV, LEV and ULEV. As in California, average fleet fuel economy is then calculated based on NMOG emissions.

The NLEV program lapses upon introduction of the "Tier 2" emissions standards.

Limits

EPA regulations define limits on emissions of the following pollutants:
- Carbon monoxide (CO)
- Nitrous oxides (NO_x)
- Non-methane organic gases (NMOG)
- Formaldehyde (HCHO) and
- Particulates

Pollutant emissions are determined using the FTP 75 driving cycle. Limits are defined in relation to distance and specified in grams per mile.

With the introduction of the Tier 2 standards, vehicles with diesel and spark-ignition engines will be subject to a single set of emissions standards.

Emissions categories

The Tier 1 standard defines limits on each regulated emissions component. Tier 2 (Figure 1) classifies limits according to the emission standards (bins) 10 (passenger cars) or 11 (HLDT). Bins 9...11 are interim bins that will be discontinued from 2007.

The transition to Tier 2 will produce the following changes:
- Introduction of fleet averages for NO_x
- Formaldehydes (HCHO) will be subject to individual pollutant limits
- Passenger cars and light-duty trucks with AGVW up to 6,000 lbs (2.72 metric tons) will be combined in a single vehicle class
- Introduction of an extra vehicle category, the MDPV (Medium Duty Passenger Vehicle, previously included under HDV)
- The "full useful life" will be extended to 120,000 miles (192,000 kilometers)

Phase-in

At least 25 % of all new passenger cars and LLDT registered for the first time will be required to conform to the Tier 2 standards once they take effect in 2004. The phase-in rule stipulates that an additional 25 % of vehicles will then be required to conform to the Tier 2 standards in each consecutive year. All vehicles will be required to conform to the Tier 2 standard starting in 2007. For the category HLDT/MDPV, the phase-in period ends in 2009.

Fleet averages

NO_x emissions will be used in determining fleet averages for individual manufacturers under EPA regulations. This procedure is at variance with the CARB procedure, in which fleet averages are based on NMOG emissions.

Average fleet fuel consumption

The regulations defining average fleet fuel consumption in the 49 states are the same as those applied in California. Again, the limit applicable to passenger cars is 27.5 miles per gallon (8.55 liters per 100 kilometers). Beyond this figure manufacturers are required to pay a penalty. The purchaser also pays a penalty tax on vehicles providing less than 22.5 miles per gallon.

On-Board Diagnosis

The on-board diagnosis utility employed to detect emissions-relevant malfunctions under EPA regulations is basically the same as that prescribed by the CARB mandates.

Field monitoring

Unscheduled testing

The EPA regulations mirror the CARB laws by requiring random FTP 75 emissions inspections on vehicles in highway operation (in-use vehicles). Testing is conducted on low-milage vehicles (10,000 miles, roughly one year old) and higher milages (50,000 miles, and at least one vehicle per test group with 75,000/90,000 miles). The number of vehicles tested varies according to the number sold.

Vehicle monitoring by the manufacturer

Since the 1972 model year, manufacturers have been obligated to submit mandatory reports detailing all known defects in defined emissions-relevant components and systems. Mandatory reports are required when defects occur in at least 25 similar emissions-relevant components in any model year. Reporting periods expire five years after the end of each model year. In addition to indicating the relevant components, the reports also contain descriptions of the defects and their effects on emissions as well as information concerning the remedial action undertaken by the manufacturer. The environmental authorities use this information as the basis for determining whether to issue recall orders to the manufacturer.

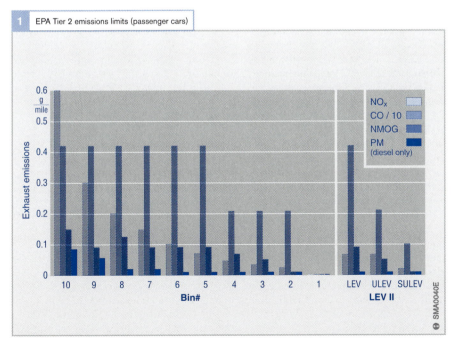

1 EPA Tier 2 emissions limits (passenger cars)

Fig. 1
The indicated figures apply to the "full useful life" (entire service life of 120,000 miles, 10 years). The figures for the "intermediate useful life" of 50,000 miles or 5 years are lower for Bin 10...5, no data are specified for Bin 4...1 Bin 1 applies only for zero-emissions vehicles (electric vehicles, etc.)

EU regulations (Cars/LDT)

The regulations contained in European Union directives are defined by the EU Commission. The emissions regulations defining emissions limits for passenger cars and light-duty trucks (LDT) are:

- EU 1 (as from 1 July, 1992)
- EU 2 (as from 1 January, 1996)
- EU 3 (as from 1 January, 2000) and
- EU 4 (slated to come into effect on 1 January, 2005)

New emissions regulations are generally introduced in two stages. In the first stage, compliance with the newly defined emissions limits is required in vehicle models submitted for initial homologation approval certification (TA, Type Approval). In the second stage – usually one year later – every new vehicle must comply with the new limits at initial registration (FR, First Registration). The authorities can also inspect vehicles from series production to verify compliance with emissions limits (COP, Conformity of Production).

The individual nations within the European Union can adopt the regulations defined in the EU 1 and EU 2 directives as national law. In Germany, this proviso led to the creation of the D 3 and D 4 emissions levels. The earlier D 3 standards were stricter than the EU 2 regulations. Within the EU, Germany assumes the role of leader in implementation of new standards.

The EU 3 standards superseded the several national regulatory instruments then effective within the individual countries when it entered effect on January 1, 2000. The national regulations lapsed on this date. EU 4 assumes legal force in January, 2005.

Aside from the emissions standards, Germany also has vehicle tax rates based on emissions. EU directives allow "tax incentives" for vehicles that comply with upcoming standards before these actually become law.

Limits

The EU standards define limits for the following pollutants:

- Carbon monoxide (CO)
- Hydrocarbons (HC)
- Nitrous oxides NO_x and
- Particulates, although these limits are initially restricted to diesel vehicles

The limits are defined based on milage and indicated in grams per kilometer (g/km) (Figure 1). Since EU 3, emissions are measured on a chassis dynamometer using the MNEDC (Modified New European Driving Cycle). This exhaust-gas test also measures the emissions during the starting sequence. Earlier test procedures excluded the starting process, postponing actual monitoring to 40 seconds after the engine start.

Although the limits in effect for diesel and spark-ignition engines currently differ, they are slated for harmonization at a future date.

The limits for the LDT category are not uniform. There are three classes (1 to 3) into which LDTs are subdivided depending on the vehicle reference weight (unladen weight + 100 kg). The limits for Class 1 are the same as for cars.

Type approval testing

While type approval testing basically corresponds to the US procedures, deviations are encountered in the following areas: Measurements of the pollutants HC, CO, NO_x are supplemented by particulate and exhaust-gas opacity measurements on diesel vehicles. Test vehicles absolve an initial break-in period of 3,000 kilometers before being subjected to testing. Deterioration factors for use in assessing test results are defined in the legislation; manufacturers are also allowed to present documentation confirming lower factors following specified long-term durability testing programs extending over 80,000 km (100,000 km starting with EU 4).

Compliance with the defined limits must be maintained over a distance of 80,000 km (EU 3) or 100,000 km (EU 4) or after 5 years. Verification of compliance is part of the certification test.

Directives

These emissions standards for passenger cars/ LDT are based on EU Directive 70/220/EC dating from the year 1970. This directive placed the first official limits on exhaust-gas emissions. The data have since been subjected to repeated updates.

Type tests

This directive defines six different test procedures. Of those, the Type I test and the Type V test are applied to diesel-engined vehicles.

The type I test evaluates exhaust-gas emissions immediately following cold starts. Exhaust-gas opacity is also assessed on vehicles with diesel engines. While compliance with the EU 3 regulations is currently mandatory for new vehicles, some are already able to meet the limits defined in EU 4 (which comes into effect in 2005).

The Type V test assesses the long-term durability of the emissions-reducing equipment. It may involve a specific testing sequence or alternatively subjection to deterioration factors specified by the legislation.

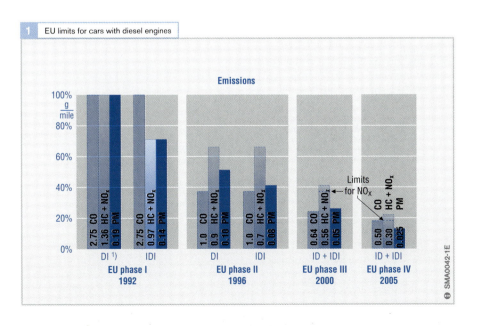

1   EU limits for cars with diesel engines

CO_2 emissions

Data indicating CO_2 emissions in grams per kilometer are specified for new vehicles being registered for the first time in the EU countries. Although no legal restrictions limit emissions of CO_2, which would also equate with a fuel-economy regulation, the vehicle manufacturers (ACEA, Association des Constructeurs Européen d'Automobiles) have united in promoting a voluntary program. The objective is to achieve CO_2 emissions of 140 g/km (5.8 l/100 km fuel consumption) by 2008 from Class M1 vehicles. By the year 2003, CO_2 emissions are not to exceed an intermediate goal of 165...170 grams per kilometer, corresponding to a fuel consumption rate of 6.8...7.0 l/100 km.

Vehicles with extremely low CO_2 emissions currently enjoy tax advantages in Germany.

On-Board Diagnosis

Introduction of the EU 3 emissions standards has been accompanied by the advent of EOBD (European On-Board Diagnosis) for spark-ignition engines. These regulations call for a diagnosis system capable of detecting all malfunctions with the potential to affect emissions in all new passenger cars and light-duty trucks and vans with up to 9 seats and an approved gross vehicle weight of up to 3.5 metric tons. The EOBD requirement has been extended to embrace vehicles with diesel power plants since 1.1.2003.

The following absolute emissions limits are defined as error thresholds for pollutant concentrations:
● Carbon monoxide (CO): 3.2 g/km
● Hydrocarbons (HC): 0.4 g/km
● Nitrogen oxides (NO_x): 0.6 g/km (gasoline) or 2 g/km (diesel)
● Particulates: 0.18g/km (diesel)

The EOBD system must respond to detection of malfunctions causing the vehicle to exceed the specified limits by triggering the error lamp after no more than three driving cycles. The system also starts to record milage when the alert is issued.

This malfunction lamp can extinguish again should three subsequent driving cycles elapse with no detected errors.

Field monitoring

EU legislation also calls for conformity-verification testing on in-use vehicles as part of the Type I test regimen. The minimum number of vehicles to be tested is three, while the maximum number varies according to the test procedure.

Vehicles selected for testing must meet specific criteria:
● The vehicle model must have been granted previous type approval in accordance with applicable regulations, and certification of conformity must be present.
● The milage and vehicle age must lie between 15,000 km/6 months and 80,000 km/5 years (starting with Euro 3 in 2000) or 100,000 km (Euro 4, 2005).
● Proof of regular periodic service inspections as specified by the manufacturer must be available.
● The vehicle is to display no indications of non-standard use (manipulation, major repairs, etc.).

If emissions from an individual vehicle fail substantially to comply with the standards, the source of the high emissions must be determined. If more than one vehicle from a series displays excessive emissions in random testing, then the results of the test are classified as negative. As long as the maximum number of vehicles specified for the random testing series has not been exceeded, an additional vehicle may be subjected to testing in response to various scenarios.

If the authorities arrive at the conclusion that a particular vehicle model does not conform with the legal requirements, they can respond by demanding a remedial-action plan from the manufacturer. The action catalog must be applicable to all vehicles displaying the same defect. Implementation of the action plan can also entail a vehicle recall.

Periodic emissions inspections (AU)

Within the Federal Republic of Germany, all passenger cars and light-duty trucks and vans are required to undergo emissions inspections (AU) three years after their initial registration, and then at subsequent intervals of two years. For gasoline-engined vehicles the main focus is on CO levels, while for diesel vehicles, the turbidity test is the main criterion.

Japanese legislation (cars/LDTs)

In Japan, the present emission limits will be replaced by stricter requirements at the end of 2002. Further tightening of the regulations is planned for 2005.

In addition to the cars category (up to 10 occupants), Japanese legislation also provides for the categories LDV (Light-Duty Vehicles, up to 1.7 t gross weight) and MDV Medium-Duty Vehicles, up to 2.5 t). Slightly higher limits for NO_X and particulates apply to the MDV category compared with the other two vehicle classes.

Emission limits

Japanese legislation specifies limits for the following emissions:
- Carbon monoxide (CO)
- Nitrogen oxides (NO_X)
- Hydrocarbons (HC)
- Particulates (diesel vehicles only)
- Smoke (diesel vehicles only)

Emission levels are measured during the $10 \cdot 15$ Mode Test. A modified $10 \cdot 15$ Mode Test including a cold start to be introduced in 2005 is under discussion.

OBD

As of 10/2000 all new cars must be fitted with an OBD system as well as all new cars from 9/2002. Functions specified by the Japanese OBD requirements include monitoring of the fuel system, the exhaust-gas recirculation system and the fuel-injection system.

Fleet consumption

In Japan, measures for reducing the CO_2 emissions of cars are planned. One proposal envisages setting the average fuel consumption for the entire fleet of cars at 33.5 miles per gallon for the year 2010. Another proposal bases the figure on the vehicle weight.

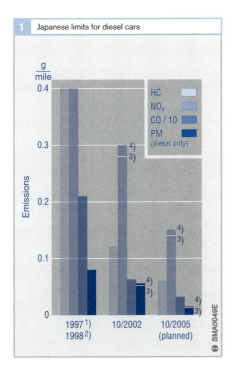

1 Japanese limits for diesel cars

Legend:
HC
NO_X
CO / 10
PM (diesel only)

y-axis: Emissions — $\frac{g}{mile}$ — 0, 0.1, 0.2, 0.3, 0.4

x-axis: 1997[1] 1998[2] | 10/2002 | 10/2005 (planned)

SMA0049E

Fig. 1
[1] For vehicles with an unladen weight of up to 1,265 kg
[2] For vehicles with an unladen weight of over 1,265 kg
[3] Limit for vehicles up to 1,265 kg
[4] Limit for vehicles over 1,265 kg

US legislation (commercial vehicles)

Heavy-duty trucks are defined by the EPA legislation as vehicles with a gross weight rating of over 8,500 lbs (equivalent to 3,850 kg). With the introduction of Tier 2 (from 2004), vehicles with a gross weight of between 8,500 and 10,000 lbs intended for passenger transport (MDPVs, Medium Duty Passenger Vehicles) will be classified as light-duty trucks and accordingly certified on the dynamometer. In California, all vehicles over 14,000 lbs (equivalent to 6,350 kg) are heavy-duty trucks. Californian legislation is to a large extent identical to EPA legislation but has separate requirements for urban busses.

Emission limits

The US standards for diesel engines define limits for:

- Hydrocarbons (HC)
- NMHCs in some cases
- Carbon monoxide (CO)
- Nitrogen oxides (NO_X)
- Particulates and
- Exhaust-gas turbidity

The permissible limits are related to engine power output and specified in $g/kW \cdot h$. The emissions are measured on the engine test bench during the dynamic test cycle with cold starting sequence (HDTC, Heavy-Duty Transient Cycle) and the exhaust-gas turbidity measured by the Federal Smoke Test.

The limits for the model year 1998 are set until 2003. Over the same period, a voluntary program ("Clean Fuel Fleet Program") is in operation, offering the incentive of tax benefits for achieving lower emissions standards. As of model year 2004, the next phase of emission limits will come into force with significantly reduced NO_X limits. Non-methane hydrocarbons and nitrogen oxides are grouped together as a combined total ($NMHC + NO_X$). CO and particulate emission limits remain at the same level as for model year 1998.

Another very drastic tightening of emission restrictions comes into force as of model year 2007. The new NO_X and particulate emission limits are 10 times lower than the previous levels. This is not achievable without the use of emission-control systems (e.g. NO_X catalytic converters and particulate filters).

Fig. 1

[1] PM limits of 0.04...0.05 $g/kW \cdot h$ under discussion

[2] Self-imposed undertaking by engine manufacturers: one engine type per manufacturer as of 2003

[3] Averaging, banking, trading (basic principle of US legislation)

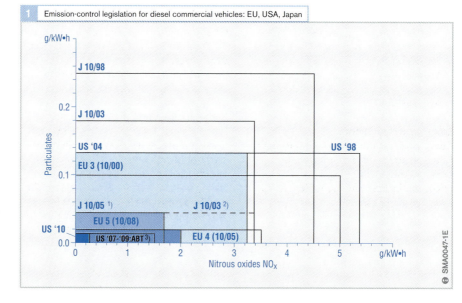

1 Emission-control legislation for diesel commercial vehicles: EU, USA, Japan

The NO_X and NMHC limits will be introduced in stages over a phase-in period between model years 2007 and 2010. In order to make emission restrictions achievable, the maximum permissible sulfur content in diesel fuel will be reduced from the present 500 ppm to 15 ppm as of mid-2006.

For heavy-duty trucks – in contrast with cars and LDTs – there are no limits specified for average fleet emissions and fleet consumption.

Consent Decree

In 1998 a legal agreement was reached between EPA, CARB and a number of engine manufacturers that provided for sanctions against the manufacturers for illegal optimized-consumption adaptations on engines in highway operation resulting in higher NO_X emissions. The main components of that "Consent Decree" were:

- The engine manufacturers concerned were required to meet the limits for model year 2004 from October 2002 onwards.
- The emission limits have to be satisfied not only in the dynamic test but also in the European 13-stage steady-state test. Furthermore, emissions are not allowed to exceed the limits for model year 2004 by more than 25 % regardless of driving mode within a specified engine-speed/torque range ("not-to-exceed" zone). These additional tests are applicable to all diesel commercial vehicles as of model year 2007.

Long-term compliance

Compliance with the emission limits must be demonstrated over a defined milage or period of time. Three weight classes are defined with increasingly stricter long-term compliance requirements:

- Light-duty trucks and vans from 8,500 (EPA) or 14,000 (CARB) to 19,500 lbs
- Medium-duty trucks from 19,500 to 33,000 lbs and
- Heavy-duty trucks over 33,000 lbs

In the case of heavy-duty trucks, for example, long-term emission-limit compliance over a period of 8 years or 290,000 miles is presently under test. As of model year 2004, the requirement will increase to 13 years or 435,000 miles.

EU legislation (commercial vehicles)

In Europe, all vehicles with a permissible gross weight of over 3,500 kg or which are designed to carry more than 9 persons are classed as heavy-duty trucks. The emissions regulations are set down in Directive 88/77/EEC, which is continually updated.

As for cars and light-duty trucks, new emission limits for heavy-duty trucks are introduced in two stages. In the first stage, new engine designs must meet the new emission limits as part of the Type Approval (TA) process. One year later, compliance with the new limits will be a condition for issuing a general vehicle approval. The authorities can check conformity of production (COP) by selecting engines from a new production batch and testing them for compliance with the new exhaust-gas emission limits.

Emission limits

For commercial-vehicle diesel engines, the EU standards define emission limits for hydrocarbons (HC), NMHCs in certain cases, carbon monoxide (CO), nitrogen oxides (NO_X), particulates and exhaust-gas turbidity. The permissible limits are related to engine power output and specified in $g/kW \cdot h$.

The EU 3 level of emission limits has been in force since October 2000 for all new approved engine designs and since October 2001 for all new production vehicles. The emissions are measured during the 13-stage European Steady-State Cycle (ESC) and the exhaust-gas turbidity in the supplementary European Load Response (ELR) test. Diesel engines that are fitted with "advanced systems" for emissions control (e.g. NO_X catalytic converter or particulate filter) also have to be tested over the dynamic European Transient Cycle (ETC). The European test cycles are performed with the engine at normal operating temperature.

In the case of small engines, i.e. engines with a capacity of less than 0.75*l* per cylinder and a rated speed of over 3,000 rpm, slightly higher particulate emission levels are permitted than for large engines. There are separate emission limits for the ETC – for example, particulate limits are approximately 50% higher than specified for the ESC because of the soot-emission peaks expected under dynamic operating conditions.

In October 2005 the EU 4 levels will come into force, initially for new approvals and one year later for production models. All emission limits are significantly lower than specified by EU 3, but the biggest increase in severity applies to particulates for which the limits have been reduced by approximately 80%. The introduction of EU 4 will also bring about the following changes:
- The dynamic exhaust-gas emission test (ETC) will be obligatory – in addition to the ESC and ELR – for all diesel engines.
- All new vehicle models will have to be fitted with an on-board diagnosis (OBD) system.
- The continued functioning of emissions-related components must be demonstrable over the entire service life of the vehicle.

The EU 5 level of emission limits will be introduced in October 2008 for all new approved engine designs and one year later for all new production vehicles. Compared with EU 4, only the NO_X limits will be tightened. As a result, the reduction in permissible NO_X limits as compared with the EU 3 levels will – as with the particulate limits – be of the order of around 80%.

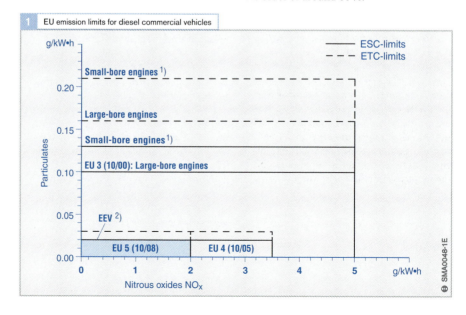

Fig. 1
[1]) $V_{cyl} \leq 0.75$ l,
$n_{rated} \geq 3,000$ rpm
[2]) Enhanced Environmentally-Friendly Vehicle (voluntary limits)

1 EU emission limits for diesel commercial vehicles

ESC-limits
ETC-limits

Small-bore engines [1])

0.20

Large-bore engines

0.15

Small-bore engines [1])

EU 3 (10/00): Large-bore engines

0.10

0.05

EEV [2])

0.00

EU 5 (10/08)   EU 4 (10/05)

Particulates   g/kW•h

0   1   2   3   4   5   g/kW•h

Nitrous oxides NO_X

SMA0048-1E

Very low-emission vehicles
The EU Directives allow for tax incentives for early compliance with the limits specified by a particular phase of the EU standards and for EEVs (Enhanced Environmentally Friendly Vehicles). Voluntary emission limits are defined for the EEV category for the ESC, ETC, and ELR tests. The NO_X and particulate limits are equivalent to those specified by EU 5 for the ESC. The standards for HC, NMHC, CO and exhaust-gas turbidity are stricter than required by EU 5.

Japanese legislation (commercial vehicles)

In Japan, all vehicles with a permissible gross weight of over 2,500 kg or which are designed to carry more than 10 persons are classed as heavy-duty trucks.

Emission limits
The standards currently in force in Japan were introduced between 1997 and 1999. They specify limits for HC, NO_X, CO, particulates and exhaust-gas turbidity. The emission levels are measured by the Japanese 13-stage test (warm engine) and the exhaust-gas turbidity by the Japanese smoke test. Long-term compliance with the limits must be demonstrated over a distance of 45,000 km.

As of October 2003, the "New Short-Term Regulations" apply which specify lower emission limits and tougher standards for long-term compliance (80,000...650,000 km depending on gross vehicle weight). According to a self-imposed undertaking on the part of Japanese engine manufacturers, one engine design per manufacturer must satisfy the next stage of emission limits.

The "New Long-Term Regulations" are due to come into force in October 2005. The precise details have not yet been agreed upon. A general reduction of emissions by 50 % as compared with 2003 is envisaged, while particulates are to be lowered by as much as 75 %. The introduction of a Japanese dynamic test cycle is also under discussion for the new phase.

Regional programs
In addition to the nationwide regulations for new vehicles, there are also regional requirements for the overall vehicle population which are aimed at reducing the existing emission levels by replacing or upgrading old diesel vehicles.

The "Vehicle NO_X Law" applies as of 2003 within, among other places, the greater urban area of Tokyo to vehicles with a permissible gross weight of over 3,500 kg. It states that 8...12 years after a vehicle is first registered, the NO_X and particulate emission levels of the relevant preceding phase of emission limits must be met (e.g. the 1998 limits after 2003). The particulates legislation of the municipal government in Tokyo operates according to the same principle. However, in this case the regulations for particulate emissions apply as soon as a vehicle has been registered for 7 years.

US test cycles

FTP 75 test cycle

The FTP 75 test cycle (**Federal Test Proce-dure**) consists of three phases, and repre-sents the speeds and conditions actually recorded in morning commuter traffic in the US city of Los Angeles (Figure 1a):

Preconditioning

The vehicle to be tested is first conditioned (allowed to stand with engine off for 12 hours at a room temperature of 20...30 °C), then started and run through the prescribed test cycle:

Collecting pollutants

The emitted pollutants are collected sepa-rately during various phases.

Phase ct:
During the cold transition phase, diluted exhaust gases are collected in bag 1 for the CVS test.

Phase s:
Exhaust gases are diverted to sample bag 2 at the beginning of the stabilized phase (after 505 s) without any interruption in the dri-ving cycle. Upon termination of phase s, after a total of 1,365 seconds, the engine is switched off for a period of 600 seconds.

Phase ht:
The engine is restarted for hot testing, which employs the speed curve used for the cold transition phase (Phase ct) in unmodified form. Exhaust gases are collected in a third sample bag.

Analysis

The bag samples from the previous phases are analyzed during the pause before the hot test, as samples should not remain in the bags for longer than 20 minutes.

The sample exhaust gases contained in the third bag are also subjected to analysis fol-lowing completion of the driving cycle. The results of the three individual phases are added using the weighting factors 0.43 (ct phase), 1 (s phase) and 0.57 (ht phase). The test distance is then incorporated in the calculations, and the weighted sums of the emissions (HC, CO and NO_x) from all three bags are converted into emissions per mile.

Outside of the USA and California this test is also employed in various other coun-tries (in South America, etc.).

SFTP schedules

The tests defined by the SFTP standard are being introduced in stages between 2001 and 2004. These are composites including the following driving cycles:
- FTP 75
- SC03 and
- US06

This extended test routine allows assessment of the following vehicle operating conditions (Figure 1b, c):
- Aggressive driving
- Radical changes in vehicle speed
- Engine start and acceleration from a standing start
- Operation with frequent minor variations in speed
- Periods with vehicle parked and
- operation with air conditioner on

Following preconditioning, the SC03 and US06 cycles proceed through the ct phase from FTP 75 without exhaust-gas collection. Other conditioning procedures may also be used.

The SC03 cycle is carried out at a temperature of 35 °C and 40 % relative humidity (vehicles with air conditioning only). The individual driving schedules are weighted as follows:
● Vehicles with air conditioning:
  35 % FTP 75 + 37 % SC03 + 28 % US06
● Vehicles without air conditioning:
  72 % FTP 75 + 28 % US06

The SFTP and FTP 75 test cycles must be successfully completed on an individual basis.

Test cycles for determining fleet averages

Each vehicle manufacturer is required to provide data on fleet averages. Manufacturers that fail to comply with the specified limits are required to pay penalties. A bonus is awarded for figures that lie below specified levels. Fuel consumption is determined based on exhaust emissions produced during two test cycles: the FTP 75 test cycle (55 %) and the highway test cycle (45 %).

An unmeasured highway test cycle (Figure 3d) is conducted once after preconditioning (vehicle allowed to stand with engine off for 12 hours at 20...30 °C). The exhaust-gas emissions from a second test run are then collected. The emissions can be used to calculate fuel consumption.

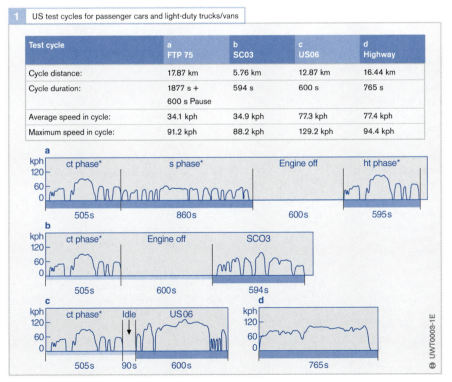

1 US test cycles for passenger cars and light-duty trucks/vans

Test cycle	a FTP 75	b SC03	c US06	d Highway
Cycle distance:	17.87 km	5.76 km	12.87 km	16.44 km
Cycle duration:	1877 s + 600 s Pause	594 s	600 s	765 s
Average speed in cycle:	34.1 kph	34.9 kph	77.3 kph	77.4 kph
Maximum speed in cycle:	91.2 kph	88.2 kph	129.2 kph	94.4 kph

Fig. 1
* ct transitional phase
s stabilized phase
ht hot test
■ Exhaust-gas collection phases
▨ Preconditioning (may consist of other driving cycles)

European test cycle

The EU/ECE test cycle (Economic Commission of Europe) – also known as the European test cycle – employs a driving curve (Figure 1) representing a close approximation of urban operation (UDC, Urban Driving Cycle). In 1993 the cycle was extended to include a rural component with speeds of up to 120 km/h (EUDC, Extra-Urban Driving Cycle). The composite test cycle uniting these two tests is referred to as the NEDC (New European Driving Cycle).

The 40-second waiting period prior to the start of actual measurements was deleted for EU 3 testing (2000) (NMEDC, Modified New European Driving Cycle). It thus includes measurements of emissions immediately following starting.

Preconditioning
Prior to emissions testing, vehicles must remain parked for a period of at least 6 hours at a defined temperature. This temperature is currently 20...30 °C.

Urban cycle
The urban cycle consists of four sections, all lasting 195 seconds, performed in immediate and uninterrupted sequence. The distance is 4.052 km, which produces an average speed of 18.7 kph. The maximum speed is 50 km/h.

Extra-urban cycle
The urban cycle is followed by operation at speeds of up to 120 kph. This section lasts 400 seconds, and extends over a distance of 6.955 kilometers.

Analysis
During measurement, exhaust gas is collected in a sample bag using the CVS method. The mass of the pollutants ascertained in analysis of the bags' contents is converted to mass per unit of distance.

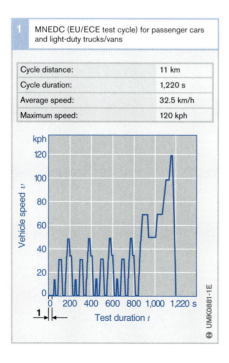

1 MNEDC (EU/ECE test cycle) for passenger cars and light-duty trucks/vans

Cycle distance:	11 km
Cycle duration:	1,220 s
Average speed:	32.5 km/h
Maximum speed:	120 kph

Japanese test cycles for cars and LDTs

The 10·15-mode test cycle (Figure 1) is conducted once with a hot start. This test cycle, simulating typical operating conditions in Tokyo, has been expanded to include a high-speed component. However, the top speed is lower than that used in the European test cycle, as the higher traffic densities in Japan result in generally lower driving speeds.

The preconditioning procedure for the hot test includes the mandatory idle emissions test. The routine is as follows: After the vehicle is allowed to warm up during approximately 15 minutes of operation at 60 kph, the concentrations of HC, CO and CO_2 are measured in the exhaust pipe. The 10·15-mode hot test commences after a second warm-up phase consisting of 5 minutes at 60 km/h. The exhaust-gas analysis is performed using CVS equipment. Exhaust gases are diluted and collected in bags during each test. The pollutants are defined relative to distance, and are indicated in grams per kilometer.

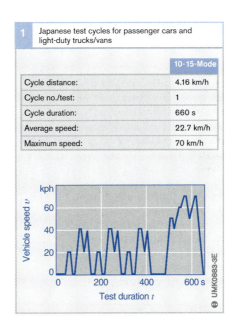

1   Japanese test cycles for passenger cars and light-duty trucks/vans

	10·15-Mode
Cycle distance:	4.16 km/h
Cycle no./test:	1
Cycle duration:	660 s
Average speed:	22.7 km/h
Maximum speed:	70 km/h

UMK0883-3E

Test cycles for commercial vehicles

For commercial vehicles, all test cycles are run on the engine test bench. For dynamic test cycles, the exhaust gases are collected and analyzed using the CVS method, while the untreated exhaust gases are measured for the steady-state test cycles. The emission levels are related to engine power output and specified in g/kW·h.

Europe

For vehicles with a permissible gross weight of more than 3.5 t or more than 9 seats, the 13-stage test as per ECE R49 was applied until the year 2000 in Europe. Since the introduction of EU 3 (October 2000), the new ESC (European Steady-State Cycle) 13-stage test with different operating levels applies (Figure 1). The operating levels are determined from the engine's full-load curve.

The test procedure specifies a sequence of 13 steady-state operating statuses. The gaseous and particulate emissions measured at each operating level are weighted according to certain factors. This also apples to power output. The test results in g/kW·h are obtained by calculating the sum total of the weighted emission levels divided by the sum total of the weighted power output.

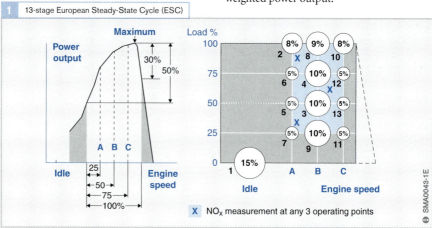

1   13-stage European Steady-State Cycle (ESC)

X   NO$_x$ measurement at any 3 operating points

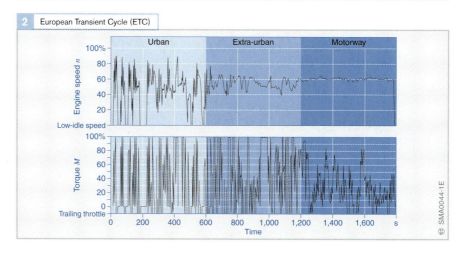

2   European Transient Cycle (ETC)

For certification, an additional three NO_X measurements can be taken over the tested range. The NO_X emissions must not vary by a significant degree from the levels measured at the adjacent operating points. The aim of the additional measurements is to prevent modification of engines specifically for testing purposes.

As well as the ESC, EU 3 also introduced the ETC (European Transient Cycle, Fig. 2) for determining gaseous emissions and particulate levels, and the ELR (European Load Response) test for establishing exhaust-gas turbidity. Under the EU 3 standards, the ETC applies only to commercial vehicles with "advanced" emission-control equipment (particulate filters, NO_X catalytic converters); as of EU 4 (10/2005), it will be obligatory for all vehicles.

The test cycle is designed to simulate real road-driving patterns and is subdivided into three sections, an urban section, an extra-urban section and a motorway section. The length of the test is 30 minutes and the periods of time for which engine speeds and torque levels must be maintained are specified in seconds.

All European test cycles are performed with the engine at normal operating temperature.

Japan
The pollutant emissions are measured under the Japanese 13-stage steady-state test (warm engine). The engine operating levels, the sequence that they follow and their weighting are different from those defined by the European 13-stage test, however. The main focus of the test is on low engine speeds and loads compared with the ESC. This is due to the high traffic density in Japan.

The introduction of a Japanese dynamic test cycle is also under discussion for the new phase due to start in 2005.

USA
Since 1987, engines for heavy-duty trucks have been tested on an engine test bench according to a steady-state test cycle (Transient Cycle) including a cold-starting sequence (Fig. 3). The test cycle attempts to simulate engine operation under real road-traffic conditions. It includes significantly more idling sections than the ETC.

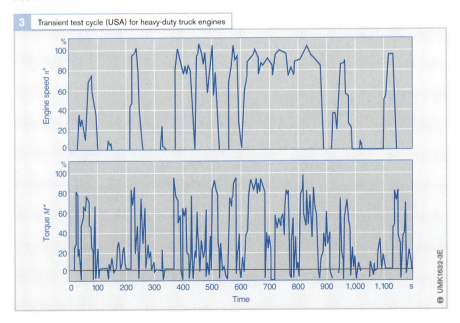

3 Transient test cycle (USA) for heavy-duty truck engines

UMK1632-3E

Fig. 3
Both standardized engine speed n^* and standardized torque M^* are figures specified by the legislation

An additional test (Federal Smoke Cycle) measures exhaust-gas turbidity under dynamic and quasi steady-state conditions.

As of model year 2007 (October 2002 for signatories to the "Consent Decree"), the US limits also must be met under the European 13-stage test (ESC).

Emission testing for type approval

Vehicle testing must reflect real-world conditions to allow precise assessment of vehicle emissions. Compared to highway driving, operation in test cells offers the advantage of allowing tests to be conducted at precisely predefined speeds, without distortions arising from variations in traffic flow patterns. This is essential in obtaining reproducible test results suitable for comparison purposes.

Test structure

The test vehicle is parked on a chassis dynamometer with its drive wheels on the rollers (Figure 1, Pos. 3). The various forces applied to the vehicle – such as inertia, rolling resistance and aerodynamic drag – must be simulated to ensure that the emissions generated on the chassis dynamometer correspond to those produced in actual highway operation. Asynchronous motors, DC generators and eddy-current brakes can

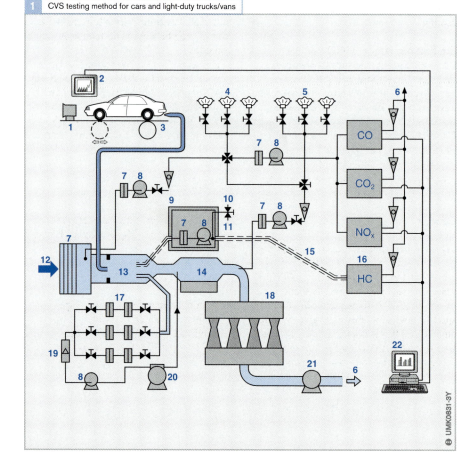

1   CVS testing method for cars and light-duty trucks/vans

Fig. 1
1 Cooling fan
2 Monitor (for test-cycle sequence)
3 "Rolling road" with dynamometer
4 Air bag
5 Exhaust-gas bag
6 Extraction
7 Filter
8 Pump
9 Kiln
10 Calibrating gas
11 Zero gas
12 Air
13 Dilution tunnel
14 Heat exchanger/heater
15 Heated pipe
16 Gas analyzers
17 Measuring filter
18 Quadruple venturi tubes
19 Flow meter
20 Gas meter
21 Fan
22 PC with monitor

be used to generate loads that reflect those encountered at various vehicle speeds. These forces act upon the rollers and must be overcome by the vehicle. On more modern systems, electronic simulation of the centrifugal or flywheel mass is used for the inertia simulation. Older test stands use actual oscillating masses which can be connected to vehicles with rapid-action couplings to obtain simulated vehicle mass. Precise adherence to the curve for load over vehicle speed, and maintenance of required inertial masses, are vital. Deviations lead to inaccurate test results. Environmental factors such as humidity, temperature and barometric pressure also affect the results.

A fan (1) placed immediately forward of the vehicle provides the required engine cooling.

Dilution procedure (CVS)

The CVS dilution method (Constant Volume Sampling) represents one procedure for collecting the exhaust gas emitted by the engine. Originally introduced for passenger cars and light-duty trucks and vans in the US in 1972, CVS technology has since evolved through several stages. Europe converted to the CVS method in 1982. This means that a single concept for collecting exhaust-gas samples is now in global use.

Concept

The concept employed in the CVS method is as follows: the exhaust gases emerging from the test vehicle are diluted with surrounding air (12) at a mean ratio of $1:5 \dots 1:10$. The gases are then extracted by a special system of pumps (7, 8) designed to maintain the respective flow rates of exhaust gases and fresh air at a specific volumetric ratio. The system thus regulates air-feed rate in line with the vehicle's instantaneous exhaust-gas volume. Throughout the test, a constant proportion of the diluted exhaust gas is extracted for storage in several sample bags (5).

The pollutant concentration in the sample bags at the end of the test cycle corresponds precisely to the mean concentration in the total quantity of fresh-air/exhaust-gas mixture which has been extracted. Because the total volume of the fresh-air/exhaust-gas mixture can be defined, pollutant concentrations can be used as the basis for calculating the pollutant masses produced during the course of the test.

Advantages of the CVS method

Due to the dilution, condensation of the water vapor contained in the exhaust gases is avoided. This provides a substantial reduction in the rate of nitrous-oxide loss during the residence time in the bag. In addition, dilution greatly inhibits the tendency of the exhaust components (especially hydrocarbons) to support mutual secondary reactions.

However, dilution does mean that pollutant concentrations decrease proportionally as a function of the mean dilution ratio, necessitating the use of more sensitive analysis equipment. Standardized equipment is available for analysis of the pollutants in the bags.

Properties of the CVS test method

The CVS method is distinguished by the following:
- Results based on actual exhaust-gas volumes generated by the engine during testing
- Accurate registration of all stationary and non-stationary vehicle operating conditions
- Avoids condensation of water vapor and unburned hydrocarbons and
- Provides technically precise measurements of particulate emissions

Dilution systems

One of two different but equally acceptable pump arrangements is generally used to maintain a constant flow volume during the test. In the first, a standard blower extracts a mixture of exhaust gas and fresh air through a venturi nozzle. In the second, a special rotary-piston fan (Roots blower) is employed. Either method is capable of measuring the flow volume with an acceptable degree of accuracy.

Testing diesel vehicles

In the USA, the CVS method has also been applied in testing diesel-powered vehicles since 1975. Modifications in both sampling modalities and analysis equipment used in measuring hydrocarbons had to be introduced for this application. The entire system used to extract the samples must be heated to 190°C to prevent condensation of heavy, low-volatility hydrocarbons in the gas samples. Heating also inhibits recondensation of the condensed hydrocarbons initially present in the diesel's exhaust.

Inclusion of particulate limits in the emissions legislation also led to modification of the CVS method. A dilution tunnel (13) with increased internal flow turbulence (Reynolds number > 40,000) and the required filter test points (17) were integrated in the system to allow collection of particulate emissions.

Measurement concepts

All of the nations that have adopted the CVS method for verifying compliance with their emissions legislation, employ standard test concepts for analysis of exhaust and other emissions:

- Concentrations of CO and CO_2 using non-dispersive infrared (NDIR) analyzers
- Concentrations of NO_x based on the CLD chemiluminescence principle
- Gravimetric quantification of particulate emissions (particulate filters conditioned and weighed before and after collection of sample charges)
- Concentrations of total hydrocarbons using the flame ionisation method (FID)

Testing commercial vehicles

The transient test method for measuring emissions from diesel engines in heavy-duty trucks over 8,500 lbs that has been required in the USA since model year 1986 and is applicable in Europe as of 2000/2005 for vehicles over 3.5 t is performed on dynamic engine test benches and also uses the CVS testing method. However, the size of the engines demands a test setup with a substantially higher throughput capacity in order to keep to same the dilution ratios as for CVS testing of cars and light-duty trucks/vans. The double dilution (by means of a secondary tunnel) allowed by the legislation helps to limit the cost of the equipment.

The volumetric flow rate of the diluted exhaust gas under critical conditions can be controlled either with a calibrated Rootes blower or a venturi tube.

Greenhouse effect

Shortwave solar radiation penetrates the earth's atmosphere and continues to the ground, where it is absorbed. This process promotes warming in the ground, which then radiates long-wave heat, or infrared energy. A portion of this radiation is reflected by the atmosphere, causing the earth to warm.

Without this natural greenhouse effect the earth would be an inhospitable planet with an average temperature of −18°C. Greenhouse gases within the atmosphere (water vapor, carbon dioxide, methane, ozone, dinitrogen oxide, aerosols and particulate mist) raise average temperatures to approximately +15°C. Water vapor, in particular, retains substantial amounts of heat.

Carbon dioxide has risen substantially since the dawn of the industrial age more than 100 years ago. The primary source of this increase has been the combustion of coal and petroleum products. In this process, the carbon bound in the fuels is released in the form of carbon dioxide.

The processes that influence the greenhouse effect within the earth's atmosphere are extremely complex. While some scientists maintain that anthropogenic (of human origin) emissions are the primary source of climate change, this theory is challenged by other experts, who believe that the warming of the earth's atmosphere is being caused by increased solar activity.

There is, however, a large degree of unanimity in calling for reductions in energy use to lower carbon-dioxide emissions and combat the greenhouse effect.

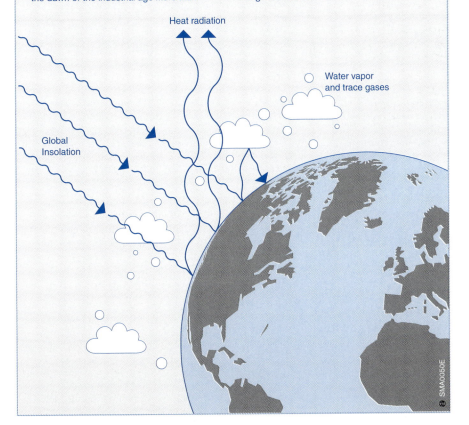

Heat radiation

Water vapor and trace gases

Global Insolation

SMA0050E

Index of technical terms

Abbreviations

A

ABS: Antilock Braking System

A/C: Air Conditioner

ACC: Adaptive Cruise Control

ACEA: Association des Constructeurs Européen d'Automobiles (Association of european automobile constructors)

ACK: Acknowledgement

A/D: Analog/Digital converter

ADA: Atmospheric-pressure sensitive full-load stop (German: Atmosphärendruckabhängiger Volllastanschlag)

ADM: Application Data Manager

ALDA: Manifold-pressure compensator (German: Ladedruck-abhängiger Volllastanschlag, absolut messend)

ALFB: Load-dependent start of delivery with deactivation feature (German: Abschaltbarer, lastabhängiger Förderbeginn)

ARD: Active surge dumping (German: Aktive Ruckeldämpfung)

ARS: Angle of Rotation Sensor

ASIC: Application Specific Integrated Circuit

ATDC: After Top Dead Center (piston/crankshaft)

ATL: Exhaust-gas turbocharger (German: Abgasturbolader)

AU: German emissions inspection (German: Abgasuntersuchung)

AWN: Bosch workshop network

AZG: Adaptive cylinder equalization (German: Adaptive-Zylindergleichstellung)

B

BDC: Bottom Dead Center (piston/crankshaft)

bhp: Brake horse power (1 bhp = 0.7355 kW)

BIP: Begin of Injection Period

BTDC: Before Top Dead Center (piston/crankshaft)

C

CA: Camshaft

CAD: Computer-Aided Design

CAFE: Corporate Average Fuel Economy

CAN: Controller Area Network

CARB: California Air Resources Board

CCRS: Current Control Rate Shaping

CDM: Calibration Data Manager

CDPF: Catalyzed Diesel Particulate Filter

CFPP: Cold Filter Plugging Point

cks: Crankshaft

CN: Cetane Number

COP: Conformity of Production

CPU: Central Processing Unit

CR: Common Rail

CRC: Cyclic Redundancy Check

CRS: Common Rail System

CRT: Continuously Regenerating Trap

CSF: Catalyzed Soot Filter

CVS: Constant Volume Sampling

D

DDS: Diesel-engine immobilizers (German: Diesel-Diebstahl-Schutz)

DHK: Nozzle-and-holder assembly (German: Düsenhalterkombination)

DI: Diesel engine

DI: Direct Injection

DOC: Diesel Oxidation Catalyst

DWS: Angle-of-rotation sensor (German: Drehwinkelsensor)

DZG: Speed sensor (German: Drehzahlgeber (Drehzahlsensor))

E

EAB: Solenoid-operated shutoff valve (German: Elektrisches Abstellventil)

EC: End of Combustion

EC: Exhaust Closes

ECE: Economic Commission for Europe

ECM: Electrochemical Machining

ECU: Electronic Control Unit

EDC: Electronic Diesel Control

EDR: Maximum-speed governor (German: Enddrehzahlregelung)

EEV: Enhanced Environmentally-Friendly Vehicle

EGR: Exhaust-Gas Recirculation

EGS: Electronic transmission-shift control (German: Elektronische Getriebesteuerung)

EHAB: Electrohydraulic shutoff device (German: Elektro-Hydraulische Abstellvorrichtung)

EI: End of Ignition

EIR: Emission Information Report

ELAB: Solenoid-operated shutoff valve (German: Elektrisches Abstellventil)

ELR: Electronic idle-speed control system (German: Elektronische Leerlaufregelung)

ELR: European Load Response

EMC: Electromagnetic Compatibility

EN: European standard (German: Europäische Norm)

EO: Exhaust Opens

EOBD: European On-Board-Diagnosis

EoL: End of Line-programming

EPROM: Erasable Programmable Read Only Memory

EPA: Environment Protection Agency

ESC: European Steady-State Cycle

ESI[tronic]: Electronic Service Information

ESP: Electronic Stability Program

ETC: European Transient Cycle

EU: European Union

EUDC: Extra Urban Driving Cycle

EWIR: Emissions Warranty Information Report

F

FAME: Fatty Acid Methyl Ester

FD: Delivery Period (German: Förderdauer)

FGB: Vehicle-speed limitation (German: Fahrgeschwindigkeitsbegrenzung)

FGR: Vehicle-speed controler (Cruise control) (German: Fahrgeschwindigkeitsregelung)

FIR: Field Information Report

Flash-EPROM: Flash-Erasable Programmable Read Only Memory

FR: First Registration

FSS: Delivery-signal sensor (German: Fördersignalsensor)

FTP: Federal Test Procedure

G

GDI: Gasoline Direct Injection

GDV: Constant-pressure valve (German: Gleichdruckventil)

GRV: Constant-volume valve (German: Gleichraumventil)

GSK: Glow plug (German: Glühstiftkerze)

GST: Graduated (or adjustable) start quantity (German: Gestufte Startmenge)

GZS: Glow plug control unit (German: Glühzeitsteuergerät)

H

HBA: Hydraulically controlled torque control device (German: Hydraulisch betätigte Angleichung)

HDK: Sensor with semidifferential short-circuiting ring (German: Halb-Differenzial-Kurzschlussring)

HFM: Hot-film air-mass meter

HFRR: High Frequency Reciprocating Rig

HGB: Maximum-speed limiter (German: Höchstgeschwindigkeitsbegrenzung)

HSV: Hydraulic start-quantity locking device (German: Hydraulische Startmengenverriegelung)

I
IC: Inlet Closes
IDI: Indirect Injection
IDE: Identifier Extension Bit
IGL: Ignition Lag
IL: Injection Lag
INCA: Integrated Calibration and Acquisition System
IO: Inlet Opens
IP: Injection Point
ISO: International Organization for Standardization
IWZ: Incremental angle-time signal (German: Inkrementales-Winkel-Zeit-Signal)

K
KMA: Electronic flow measurement system (German: Kontinuierliche Mengenanalyse)
KSB: Cold-start accelerator (German: Kaltstartbeschleuniger)
KTS: Small tester series (German: Klein-Tester-Serie)

L
LDA: Manifold-pressure compensator (German: Ladedruck-abhängiger Vollastanschlag)
LDT: Light-Duty Trucks
LED: Light-Emitting Diode
LEV: Low-Emission Vehicle
LFB: Load-sensitive start of delivery (German: Lastabhängiger Förderbeginn)
LFG: Idle-speed spring attached to governor housing (German: Leerlauffeder – gehäusefest)
LSU: Broadband oxygen sensor Type LSU (German: (Breitband-) Lambda-Sonde-Universal)

M
MAB: Fuel cutout (German: Mengenabschaltung)
MAR: Control of injected-fuel-quantity compensation (German: Mengenausgleichsregelung)
MBEG: Fuel limitation (German: Mengenbegrenzung)
MDA: Measure Data Analyzer
MDPV: Medium Duty Passenger Vehicle

MGT: Glass gauge method (German: Messglas-Technik)
MI: Main Injection
MIL: Malfuction Indicator Lamp
MMA: Fuel-quantity mean-value adaptation (German: Mengen-mittelwertadaption)
MNEDC: Modified New European Driving Cycle
MSG: Engine ECU (German: Motorsteuergerät, Motor-ECU)
MV: Solenoid valve (German: Magnetventil)

N
NBS: Needle-motion sensor (German: Nadelbewegungssensor)
NEDC: New European Driving Cycle
NLEV: National Low Emission Vehicle
NMOG: Non-Methane Organic Gases
NSC: NO$_x$ Storage Catalyst
NTC: Negative Temperature Coefficient
NYCC: New York City Cycle

O
OBD: On-Board-Diagnosis

P
PCB: Printed-Circuit Board
PF: Particulate Filter
PI: Pre Injection
PLA: Pneumatic idle-speed increase (German: Pneumatische Leerlaufanhebung)
PM: Particular Matter
PNAB: Pneumatic shutoff device (German: Pneumatische Abstellvorrichtung)
POI: Post Injection
ppm: Parts per million (1,000 ppm = 0.1%)
PROF: Programming of Flash-EPROM
PSG: Pump ECU (German: Pumpensteuergerät)
PTC: Positive Temperature Coefficient
PWM: Pulse-Width Modulation
PZEV: Partial Zero-Emission Vehicle

R
RAM: Random Access Memory
RDV: Orifice check valve (German: Rückströmdrosselventil)
RME: Rape seed oil Methyl Ester
ROM: Read Only Memory
RSD: Orifice check valve (German: Rückströmdrosselventil)

RTR: Remote Transmission Request
RWG: Control-rack travel sensor (German: Regelweggeber)

S
SAE: Society of Automotive Engineers
SC: Start of Combustion
SCR: Selective Catalytic Reduction
SD: Start of Delivery
SFTP: Supplemental Federal Test Procedure
SI: Start of Ignition
SMD: Surface Mounted Device
SOC: Start of Combustion
SRC: Smooth Running Control
SULEV: Super Ultra-Low-Emission Vehicle

T
TA: Type Approval
TAS: Temperature-compensating start-quantity stop (German: Temperaturabhängiger Startanschlag)
TCS: Traction Control System
TDC: Top Dead Center (piston/crankshaft)
TLA: Temperature-controlled idle-speed increase (German: Temperaturabhängige Leerlaufanhebung)
TLEV: Transitional Low-Emission Vehicle
TQ signal: Fuel-consumption signal

U
UDC: Urban Driving Cycle
UDDS: Urban Dynamometer Driving Schedule
UI: Unit Injector
UIS: Unit Injector System
ULEV: Ultra-Low-Emission Vehicle
UP: Unit Pump
UPS: Unit Pump System

V
V_h: Swept volume of an engine cylinder
V_H: Overall cylinder capacity of an engine
VST: Variable Sleeve Turbine
VTG: Variable Turbine Geometry

W
WSD: Wear Scar Diameter

Z
ZDR: Intermediate-speed control (German: Zwischendrehzahlregelung)
ZEV: Zero-Emission Vehicle

Bosch reference books –
First-hand technical knowledge

Gasoline-Engine Management

Starting with a brief review of the beginnings of automotive history, this book discusses the basics relating to the method of operation of gasoline-engine control systems. The descriptions of cylinder-charge control systems, fuel-injection systems (intake manifold and gasoline direct injection), and ignition systems provide a comprehensive, firsthand overview of the control mechanisms indispensable for operating a modern gasoline engine. The practical implementation of engine management and control is described by the examples of various Motronic variants, and of the control and regulation functions integrated in this particular management system. The book concludes with a chapter describing how a Motronic system is developed.

Contents

- Basics of the gasoline engine
- Gasoline-engine control system
- History of gasoline-engine management system development
- Cylinder-charge control systems
- Fuel supply and delivery
- Intake-manifold fuel injection
- Gasoline direct injection
- Inductive ignition system
- Ignition coils
- Spark plugs
- Motronic engine management
- Sensors
- Electronic control unit
- Electronic control and regulation
- Electronic diagnostics (on-board diagnostics, OBD)
- Data transmission between electronic systems
  - Pollutant reduction
  - Catalytic emission-control systems
  - Emission-control legislation
  - Workshop technology
  - Electronic control unit development

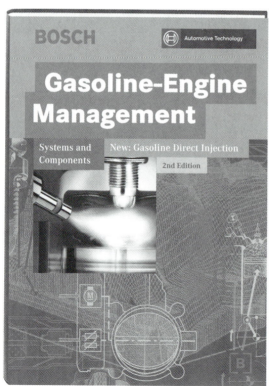

Hardcover,
17 x 24 cm format,
2nd Edition, completely revised and extended,
418 pages,
hardback,
with numerous illustrations.

ISBN
0-8376-1052-4

Automotive Electrics / Automotive Electronics

The rapid pace of development in automotive electrics and electronics has had a major impact on the equipment fitted to motor vehicles. This simple fact necessitated a complete revision and amendment of this authoritative technical reference work. The 4th Edition goes into greater detail on electronics and their application in the motor vehicle. The book was amended by adding sections on "Microelectronics" and "Sensors". As a result, the basics and the components used in electronics and microelectronics are now part of this book. It also includes a review of the measured quantities, measuring principles, a presentation of the typical sensors, and finally a description of sensor-signal processing.

Contents
- Automotive electrical systems, including calculation of wire dimensions, plug-in connections, circuit diagrams and symbols
- Electromagnetic compatibility and interference suppression
- Batteries
- Alternators
- Starters
- Lighting technology
- Windshield and rear-window cleaning
- Microelectronics
- Sensors
- Data processing and transmission in motor vehicles.

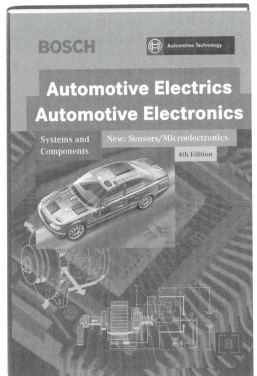

Hardcover,
17 x 24 cm format,
4th Edition, completely revised and extended,
503 pages,
hardback,
with numerous illustrations.

ISBN
0-8376-1050-8

Automotive Handbook

The Automotive Handbook is a comprehensive 960-page reference work. The 5th edition has again been completely revised and updated by experts from Bosch and from the automotive industry.

The state-of-the-art in automotive technology is presented comprehensibly, in a well laid out and concise form. A wide variety of system schematics, illustrations, and tables provide an insight into a fascinating technological world.

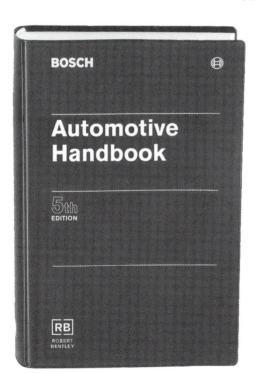

Added subjects:
- Fuel cells
- Cartronic
- EHB for passenger cars
- Natural gas as engine fuel
- Adaptive cruise control (ACC)
- Display instruments
- Fuel filters
- MED Motronic
- Plug-in connections

Completely revised subjects:
- Engine management for spark-ignition engines
- Engine management for diesel engines
- Emissions-control engineering
- Safety systems
- Lighting technology
- Electronics
  - Materials science
  - Vibration and oscillation
  - and many more subjects.

The Bosch "Blue Book" has in the meantime advanced to become a bestseller in the area of technical literature.

Hardcover,
12 x 18 cm format,
5th Edition, completely revised and extended,
962 pages,
softcover,
with numerous illustrations.

ISBN
0-8376-0614-4

621.436
STL